AVIAN BIOLOGY
Volume V

CONTRIBUTORS

Peter Berthold

Rudolf Drent

Stephen T. Emlen

Eberhard Gwinner

Fernando Nottebohm

C. J. Pennycuick

François Vuilleumier

AVIAN BIOLOGY
Volume V

EDITED BY

DONALD S. FARNER

Department of Zoology
University of Washington
Seattle, Washington

JAMES R. KING

Department of Zoology
Washington State University
Pullman, Washington

TAXONOMIC EDITOR

KENNETH C. PARKES

Curator of Birds
Carnegie Museum
Pittsburgh, Pennsylvania

 1975

ACADEMIC PRESS New York San Francisco London
A Subsidiary of Harcourt Brace Jovanovich, Publishers

ACADEMIC PRESS, INC.
111 Fifth Avenue, New York, New York 10003

United Kingdom Edition published by
ACADEMIC PRESS, INC. (LONDON) LTD.
24/28 Oval Road, London NW1

Library of Congress Cataloging in Publication Data
Main entry under title:

Avian biology.

    Includes bibliographies.
    1.   Ornithology.   2.   Zoology—Ecology.   I.   Farner,
Donald Stanley, (date)   ed.   II.   Kings, James Roger,
(date)   ed.
QL673.A9   598.2   79-178216
ISBN 0–12–249405–9

PRINTED IN THE UNITED STATES OF AMERICA

These volumes are dedicated to the memory of

A. J. "JOCK" MARSHALL

(1911–1967)

whose journey among men was too short by half

CONTENTS

Chapter 1.   Mechanics of Flight

C. J. Pennycuick

Chapter 2.   Migration: Control and Metabolic Physiology

Peter Berthold

Chapter 3.   Migration: Orientation and Navigation

Stephen T. Emlen

Chapter 4.   Circadian and Circannual Rhythms in Birds

Eberhard Gwinner

Chapter 5.   Vocal Behavior in Birds

Fernando Nottebohm

Chapter 6.   Incubation

Rudolf Drent

Chapter 7.   Zoogeography

François Vuilleumier

LIST OF CONTRIBUTORS

Numbers in parentheses indicate the pages on which the authors' contributions begin.

PETER BERTHOLD (77), Max-Planck-Institut für Verhaltensphysiologie, Vogelwarte Radolfzell, Germany

RUDOLF DRENT (333), Zoological Laboratory, University of Groningen, Haren, The Netherlands

STEPHEN T. EMLEN (129), Section of Neurobiology and Behavior, Division of Biological Sciences, Cornell University, Ithaca, New York

EBERHARD GWINNER (221), Max-Planck-Institut für Verhaltensphysiologie Seewiesen und Erling-Andechs/Obb., Germany

FERNANDO NOTTEBOHM (287), The Rockefeller University, New York, New York

C. J. PENNYCUICK (1), Department of Zoology, University of Bristol, Bristol, England

FRANÇOIS VUILLEUMIER (421), Department of Ornithology, The American Museum of Natural History, New York, New York

PREFACE

The birds are the best-known of the large and adaptively diversified classes of animals. About 8600 living species are currently recognized, and it is unlikely that more than a handful of additional species will be discovered. Knowledge of the distribution of living species, although much remains to be learned, is much more nearly complete than that for any other class of animals. Other aspects of avian biology may be less well known, but in general the knowledge in these areas surpasses that available for other animals. It is noteworthy that our relatively advanced knowledge of birds is attributable to a very substantial degree to a large group of dedicated and skilled amateur ornithologists.

Because of the abundance of empirical information on distribution, habitat, life cycles, breeding habits, etc., it has been relatively easier to use birds instead of other animals in the study of the general aspects of ethology, ecology, population biology, evolutionary biology, physiological ecology, and other fields of biology of contemporary interest. Model systems based on birds have had a prominent role in the development of these fields. The function of this multivolume treatise in relation to the place of birds in biological science is therefore envisioned as twofold: to present a reasonable assessment of selected aspects of avian biology for those having this field as their primary interest, and to contribute to the broader fields of biology in which investigations using birds are of substantial significance.

Barely a decade has passed since the publication of A. J. Marshall's "Biology and Comparative Physiology of Birds," but progress in the fields included in this treatise has made most of the older chapters obsolete. Avian biology has shared in the so-called information explosion. The number of serial publications devoted mainly to avian biology has increased by about 20% per decade since 1940, and the spiral has been amplified by the parallel increase in page production and by the spread of publication into ancillary journals. By 1964, there were about 215

exclusively ornithological journals and about 245 additional serials publishing appreciable amounts of information on avian biology (P. A. Baldwin and D. E. Oehlert, *Studies in Biological Literature and Communica-
tions, No. 4. The Status of Ornithological Literature, 1964.* Biological
Abstracts, Inc., Philadelphia, 1964).

These data reflect only the quantitative acceleration in the output
of information in recent time. The qualitative changes have been much
more impressive. Avifaunas that were scarcely known except as lists
of species a decade ago have become accessible to investigation because
of improved transportation and facilities in many parts of the world.
New instrumentation has allowed the development of new fields of study
and has extended the scope of old ones. Obvious examples include the
use of radar in visualizing migration, of telemetry in studying the physiology of flying birds, and of spectrography in analyzing bird sounds. The
development of mathematical modeling, for instance in evolutionary biology and population ecology, has supplied new perspectives for old problems and a new arena for the examination of empirical data. All of
these developments—social, practical, and theoretical—have profoundly
affected many aspects of avian biology in the last decade. It is now
time for another inventory of information, hypotheses, and new
questions.

Marshall's "Biology and Comparative Physiology of Birds" was the
first treatise in the English language that regarded ornithology as consisting of more than anatomy, taxonomy, oology, and life history. This
viewpoint was in part a product of the times; but it also reflected Marshall's own holistic philosophy and his understanding that "life history"
had come to include the entire spectrum of physiological, demographic,
and behavioral adaptation. This treatise is the direct descendent of Marshall's initiative. We have attempted to preserve the view that ornithology belongs to anyone who studies birds, whether it be on the level
of molecules, individuals, or populations. To emphasize our intentions
we have called the work "Avian Biology."

It has been proclaimed by various oracles that sciences based on
taxonomic units (such as insects, birds, or mammals) are obsolete, and
that the forefront of biology is process-oriented rather than taxon-ori-
ented. This narrow vision of biology derives from an understandable
but nevertheless myopic philosophy of reductionism and from the hyper-
specialization that characterizes so much of science today. It fails to
notice that lateral synthesis as well as vertical analysis are inseparable
partners in the search for biological principles. Avian biologists of both
stripes have together contributed a disproportionately large share of
the information and thought that have produced contemporary principles

in zoogeography, systematics, ethology, demography, comparative physiology, and other fields too numerous to mention. The record speaks for itself.

In part, this progress results from the attributes of birds themselves. They are active and visible during the daytime; they have diversified into virtually all major habitats and modes of life; they are small enough to be studied in useful numbers but not so small that observation is difficult; and, not least, they are esthetically attractive. In short, they are relatively easy to study. For this reason we find in avian biology an alliance of specialists and generalists who regard birds as the best natural vehicle for the exploration of process and pattern in the biological realm. It is an alliance that seems still to be increasing in vigor and scope.

In the early planning stages of the treatise we established certain working rules that we have been able to follow with rather uneven success.

1. "Avian Biology" is the conceptual descendent of Marshall's earlier treatise, but is more than simply a revision of it. We have deleted some topics and added or extended others. Conspicuous among the deletions are embryology and the central nervous system. Avian embryology, under the new banner of developmental biology, has expanded and specialized to the extent that a significant review of recent advances would be a treatise in itself. The avian brain has been treated very extensively in a recent publication, "The Avian Brain" by Ronald Pearson (Academic Press, 1972).

2. Since we expect the volumes to be useful for reference purposes as well as for the instruction of advanced students, we have asked authors to summarize established facts and principles as well as to review recent advances.

3. We have attempted to arrange a balanced account of avian biology as it stands at the beginning of the 1970's. We have not only retained chapters outlining modern concepts of structure and function in birds, as is traditional, but have also encouraged contributions representing multidisciplinary approaches and synthesis of new points of view.

4. We have attempted to avoid a parochial view of avian biology by seeking diversity among authors with respect to nationality, age, and ornithological heritage.

5. As a corollary of the preceding point, we have not intentionally emphasized any single school of thought, nor have we sought to dictate the treatment given to controversial subjects. Our single concession to conceptual conformity is in taxonomic usage, as explained by Kenneth Parkes in the Note on Taxonomy.

We began our work with a careful plan for a logical topical development through the five volumes. Only its dim vestiges remain. Owing to belated defections by a few authors and conflicting commitments by others we have been obliged to sacrifice logical sequence in order to retain authors whom we regarded as the best for the task. In short, we gave first priority to the maintenance of general quality, trusting that each reader would supply logical cohesion by selecting chapters that are germane to his individual interests.

Donald S. Farner
James R. King

NOTE ON TAXONOMY

Early in the planning stages of "Avian Biology" it became apparent to the editors that it would be desirable to have the manuscript read by a taxonomist, whose responsibility it would be to monitor uniformity of usage in classification and nomenclature. Other multiauthored compendia have been criticized by reviewers for use of obsolete scientific names and for lack of concordance from chapter to chapter. As neither of the editors is a taxonomist, they invited me to perform this service.

A brief discussion of the ground rules that we have tried to follow is in order. Insofar as possible, the classification of birds down to the family level follows that presented by Dr. Storer in Chap. 1, Volume I.

Within each chapter, the first mention of a species of wild bird includes both the scientific name and an English name, or the scientific name alone. If the same species is mentioned by English name later in the same chapter, the scientific name is usually omitted. Scientific names are also usually omitted for domesticated or laboratory birds. The reader may make the assumption throughout the treatise that, unless otherwise indicated, the following statements apply:

1. "The duck" or "domestic duck" refers to domesticated forms of *Anas platyrhynchos*.

2. "The goose" or "domestic goose" refers to domesticated forms of *Anser anser*.

3. "The pigeon" or "domesticated pigeon" or "homing pigeon" refers to domesticated forms of *Columba livia*.

4. "The turkey" or "domestic turkey" refers to domesticated forms of *Meleagris gallopavo*.

5. "The chicken" or "domestic fowl" refers to domesticated forms of *Gallus gallus;* these are often collectively called *"Gallus domesticus"* in biological literature.

6. "Japanese Quail" refers to laboratory strains of the genus *Coturnix,*

the exact taxonomic status of which is uncertain. See Moreau and Wayre, *Ardea* 56, 209–227, 1968.

7. "Canary" or "domesticated canary" refers to domesticated forms of *Serinus canarius*.

8. "Guinea Fowl" or "Guinea Hen" refers to domesticated forms of *Numida meleagris*.

9. "Ring Dove" refers to domesticated and laboratory strains of the genus *Streptopelia*, often and incorrectly given specific status as *S. "risoria."* Now thought to have descended from the African Collared Dove (*S. roseogrisea*), the Ring Dove of today *may* possibly be derived in part from *S. decaocto* of Eurasia; at the time of publication of Volume 3 of Peters' "Check-list of Birds of the World" (p. 92, 1937), *S. decaocto* was thought to be the direct ancestor of *"risoria."* See Goodwin, *Pigeons and Doves of the World*, 129, 1967.

As mentioned above, an effort has been made to achieve uniformity of usage, both of scientific and English names. In general, the scientific names are those used by the Peters "Check-list"; exceptions include those orders and families covered in the earliest volumes for which more recent classifications have become widely accepted (principally Anatidae, Falconiformes, and Scolopacidae). For those families not yet covered by the Peters' list, I have relied on several standard references. For the New World I have used principally Mayer de Schauensee's "The Species of Birds of South America and Their Distribution" (1966), supplemented by Eisenmann's "The Species of Middle American Birds" (*Trans. Linnaen Soc. New York* 7, 1955). For Eurasia I have used principally Vaurie's "The Birds of the Palaearctic Fauna" (1959, 1965), and Ripley's "A Synopsis of the Birds of India and Pakistan" (1961). There is so much disagreement as to classification and nomenclature in recent checklists and handbooks of African birds that I have sometimes had to use my best judgment and to make an arbitrary choice. For names of birds confined to Australia, New Zealand, and other areas not covered by references cited above, I have been guided by recent regional checklists and by general usage in recent literature. English names have been standardized in the same way, using many of the same reference works. In both the United States and Great Britain, the limited size of the avifauna has given rise to some rather provincial English names; I have added appropriate (and often previously used) adjectives to these. Thus *Sturnus vulgaris* is "European Starling," not simply "Starling"; *Cardinalis cardinalis* is "North American Cardinal," not simply "Cardinal"; and *Ardea cinerea* is "Gray Heron," not simply "Heron."

Reliance on a standard reference, in this case Peters, has meant that certain species appear under scientific names quite different from those used in most of the ornithological literature. For example, the Zebra Finch, widely used as a laboratory species, was long known as *Taeniopygia castanotis*. In Volume 14 of the Peters' "Check-list (pp. 357–358, 1968), *Taeniopygia* is considered a subgenus of *Poephila*, and *castanotis* a subspecies of *P. guttata*. Thus the species name of the Zebra Finch becomes *Poephila guttata*. In such cases, the more familiar name will usually be given parenthetically.

For the sake of consistency, scientific and English names used in Volume I will be used throughout "Avian Biology," even though these may differ from names used in standard reference works that would normally be followed, but which were published after the editing of Volume I had been completed.

Strict adherence to standard references also means that some birds will appear under scientific names that, for either taxonomic or nomenclatorial reasons, would *not* be those chosen by either the chapter author or the taxonomic editor. Similarly, the standardized English name may *not* be the one most familiar to the chapter author. As a taxonomist, I naturally hold some opinions that differ from those of the authors of the Peters' list and the other reference works used. I feel strongly, however, that a general text such as "Avian Biology" should not be used as a vehicle for taxonomic or nomenclatorial innovation, or for the furtherance of my personal opinions. I therefore apologize to those authors in whose chapters names have been altered for the sake of uniformity, and offer as solace the fact that I have had my objectivity strained several times by having to use names that do not reflect my own taxonomic judgment.

KENNETH C. PARKES

CONTENTS OF OTHER VOLUMES

AVIAN BIOLOGY
Volume V

Chapter 1

MECHANICS OF FLIGHT

C. J. Pennycuick

I. Introduction

The mechanical problems of flying impose certain limitations on the anatomy of flying birds, and conversely the anatomy of a particular bird sets definite limits to the performance that can be expected of it. It is the purpose of this chapter to outline the mechanical basis of both powered and gliding flight in order to investigate this relationship between anatomy and performance. One aspect of performance concerns the energy requirements of flying, and in this the mechanical approach

1

is complementary to direct metabolic investigations but lends itself more readily to generalization.

II. Theoretical Model of a Bird in Horizontal Flight

The most direct way to estimate the power required by a flying animal is to consider the lift and drag forces on every part of the body and wings in flapping flight. Analyses of this type, of which many have been published, have been reviewed by Brown (1961). Because of the extremely complicated motion involved in flapping flight and the complex wing structure of most flying animals, such an approach leads to excessively elaborate calculations, embedded in which are a variety of assumptions and approximations whose correctness may be difficult to verify. Perhaps only in the work of Weis-Fogh (1952, 1956), Jensen (1956), and Weis-Fogh and Jensen (1956) on the locust *Schistocerca gregaria* has this direct approach been carried into sufficient detail to give a satisfactory interpretation of an animal's capabilities in flight in terms of its anatomy and physiology.

When generalizations are required, it is usually more convenient to construct a simplified model representing any flying animal and then to get performance estimates for particular species by inserting whatever experimental data are available in the model. This approach seems to have been pioneered by Fullerton (1925), who used it to estimate cruising speeds and power requirements for several species.

The basic model used in this chapter was developed by Pennycuick (1969) on the basis of classical aerodynamic theory. Subsequently, Tucker (1971, 1973) revised it to take account of several matters that were neglected in the original model. These were (1) span efficiency, (2) the variation of certain components of drag with Reynolds number, and (3) additional power terms connected with lung ventilation, circulation, and basal metabolism. In the present treatment, the considerations raised by Tucker are taken into account, but not always in exactly the way that he adopted. Tucker's modifications will be indicated below as the theory is developed, and the reasons why some are accepted while others are not will be indicated.

A. Synthesis of the Power Curve

1. General Remarks

The total mechanical power output required by a bird to fly can be estimated by adding together a number of components that are estimated separately. These components are the following.

1. *Induced power*—the power required to support the weight in air.
2. *Parasite power*—the power required to overcome the skin friction and form drag of the body.
3. *Profile power*—the power required to overcome the profile drag of the wings.
4. *Inertial power*—the power needed to accelerate the wings at each stroke.
5. *Circulation and respiration power*—the power required to pump the blood, and to ventilate the lungs.
6. *Basal metabolism*—in the present treatment this component is converted into an equivalent mechanical power so that it can be added to the other components of the power required.

In Tucker's (1973) treatment, power inputs and outputs are distinguished by subscripts. A power *output* is the rate at which mechanical work is done, while the corresponding power *input* is the rate at which chemical energy is consumed. To reduce the proliferation of subscripts, this distinction will be indicated here by Π for a power input and P for a power output. For any particular component of power, input and output are related thus:

$$\Pi = P/\eta \tag{1}$$

where η is the mechanical efficiency of the muscles and supporting systems. The various components of the power required to fly will now be considered in turn.

2. Induced Power

In steady horizontal flight, the bird's weight must be supported by the upward reaction from downward movement imparted to the air. The upward force on the bird is equal to the rate at which downward momentum is imparted to the air.

In Fig. 1, a bird is shown proceeding horizontally at a speed V, which is mechanically equivalent to a stationary bird with air flowing past it at this speed. The bird flaps its wings, sweeping out a wing disk, which has a disk area S_d. The action of the wings causes the air to accelerate downward. Part of this acceleration occurs before the air reaches the disk, being caused by reduced pressure on the dorsal (and forward) side of the disk. This results in the air attaining a downward induced velocity V_i as it passes through the disk.

The acceleration of the air is not complete at this point, however, since there is a region of increased pressure behind and below the disk, which is a mirror image of the area of reduced pressure ahead and

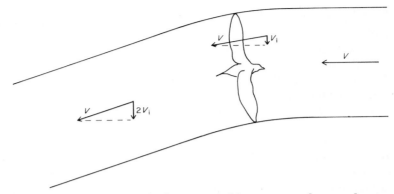

FIG. 1. The weight of a flying bird is supported by imparting downward momentum to the air (see text).

above. After leaving the disk the air therefore continues to accelerate downward until, far behind the bird, it reaches an eventual downward velocity $2V_i$.

The rate at which downward momentum is imparted to the air is the product of the mass flow (mass per unit time) passing through the disk times the eventual downward velocity $2V_i$, and this rate of change of momentum must equal the weight. Thus,

$$W = VS_d\rho \cdot 2V_i \tag{2}$$

so that

$$V_i = W/2VS_d\rho \tag{3}$$

where W is the weight and ρ is the density of the air. The power required to maintain this flow P_i is therefore

$$P_i = WV_i = W^2/2VS_d\rho \tag{4}$$

This result is actually an approximation based on the assumption that V_i is small in comparison with V. At very low speeds, this becomes a poor approximation. In hovering (zero forward speed), the induced power is not, of course, infinity, as Eq. (4) suggests, but

$$P_{ih} = (W^3/2\rho S_d)^{1/2} \tag{5}$$

The curve of induced power begins from this value at $V = 0$ and, as speed is increased, approximates ever more closely to the hyperbolic curve of Eq. (4) (Fig. 2).

To interpret Eqs. (4) and (5) in practice, the question remains, what exactly is S_d? Ideally, S_d is the area of a circle whose diameter is the wing span. It may be noted that this is true in theory for either a flapping or a gliding bird, and the practical value of S_d does not depend on the angle through which the wings are actually flapped. On the other hand, the induced power predicted if this ideal value of S_d is inserted in Eqs. (4) and (5) assumes that the air everywhere within the disk passes through it at exactly V_i, whereas that outside is unaccelerated. Any deviation from this situation leads to more power being required to support the weight than Eqs. (4) and (5) indicate. In aeronautics this deviation from the ideal result is usually represented by simply multiplying the induced power by an induced power factor k, thus

$$P_i = kW^2/2\rho V S_d \tag{4a}$$

$$P_{ih} = kW^{3/2}/(2\rho S_d)^{1/2} \tag{5a}$$

The disk area itself is arbitrarily defined as

$$S_d = \tfrac{1}{4}\pi b^2 \tag{6}$$

where b is the wing span, measured from tip to tip with the wings fully outstretched.

Airplane wings and helicopter rotors generally have values of k of about 1.1–1.2, but there are no reliable experimental determinations of this factor for bird wings. In the earlier theory of Pennycuick (1969), it was implicitly assumed that $k = 1$, and Tucker (1973) is certainly correct in regarding this as unrealistic. He chose a value of 1.43, which is perhaps rather high. In the present treatment, k is included in the analysis without assigning a numerical value to it in the hope that experimental estimates will be available in the future. However, it will be shown below that one effect of increasing k is to raise the estimates of the airspeeds for minimum power and maximum range. These estimates tend, if anything, to be on the high side even if it is assumed that $k = 1$, so that there is no immediate incentive to choose a high value of k in the absence of direct (i.e., aerodynamic) experimental evidence. If a guess is required, then $k = 1.2$ is suggested.

3. Parasite Power

The parasite power in a bird is the power required to overcome the drag of the body. It can be represented as

$$P_{pa} = D_{pa}V = \tfrac{1}{2}\rho V^3 A \tag{7}$$

where D_{pa} is the parasite drag and A is the equivalent flat-plate area of the body. A is itself equal to the actual cross-sectional area of the body at its widest point (S_{pa}) multiplied by the parasite drag coefficient $C_{D_{pa}}$

$$A = S_{pa}C_{D_{pa}} \tag{8}$$

In Pennycuick's (1969) theory, it was assumed that $C_{D_{pa}}$ remains the same for birds of any size, implying that any change of $C_{D_{pa}}$ with Reynolds number is neglected. This simplification is supported by two measurements of $C_{D_{pa}}$ made by similar methods at different Reynolds numbers (Re) in the same wind tunnel, which gave values of 0.43 in both cases. These were a domestic pigeon at $Re = 4.1 \times 10^4$ (Pennycuick, 1968a) and a Rüppell's Griffon Vulture (*Gyps rueppellii*) at $Re = 1.4 \times 10^5$ (Pennycuick, 1971b). The Reynolds numbers are based on $(S_{pa})^{1/2}$ as the reference length.

This high drag coefficient agrees with values in the literature for rounded bodies with a region of separated flow at the downstream end. The drag of such bodies is discussed at length by von Mises (1959), and also by Schmitz (1960), who show that over a range of Reynolds numbers extending from roughly 2×10^4 to 2×10^5, the drag coefficient of such a body remains more or less constant, in the region of 0.4–0.5. At still higher Reynolds numbers, the flow in the boundary layer changes from laminar to turbulent, which leads to a sharp contraction of the region of separated flow, with a corresponding reduction of the drag coefficient. It is possible that geese and swans, which fly faster than vultures and have better streamlined bodies, may operate in this region and thus get substantially lower body drag coefficients than vultures and smaller birds.

From this interpretation, the flat portion of the curve of drag coefficient against Reynolds number should extend down to a body mass of 70 gm or so. In smaller birds, the drag coefficient should be somewhat greater, but no very marked increase would be expected in the smallest birds ($Re \approx 5 \times 10^3$)—not, indeed, until mass is reduced well down into the insect range, where skin friction becomes prominent in comparison with form drag.

Tucker (1973) proposed that $C_{D_{pa}}$ should decrease with Reynolds number in the manner characteristic of a thin plate held parallel to the airflow. The drag on such a plate is entirely due to skin friction, which does not vary with the square of the speed but with some lower power. Consequently, the drag coefficient decreases with speed and with Reynolds number. This interpretation does not seem very appropriate, since, as has been seen, it seems that the drag on birds' bodies is mainly form drag. If they behave as rounded bodies with a region of separated flow behind, as is suggested here, then the skin friction would be small in comparison with the form drag over the range of Reynolds number in question.

If geometric similarity is assumed, then the cross-sectional area S_{pa} should vary with the two-thirds power of the mass, and so also would A, provided that $C_{D_{pa}}$ is independent of Reynolds number. One consequence of Tucker's proposal is that it would make A vary with some lower power of the mass. However, he presents new data on the body drag of various birds and shows that A varies almost exactly with the two-thirds power of the mass, so implying that $C_{D_{pa}}$ is independent of the Reynolds number. Tucker's drag measurements are on average 17% higher than the two earlier ones mentioned above, but this may be the result of a difference in technique. The bodies were mounted in the wind tunnel on cylindrical rather than streamlined supports, and this could produce an overestimate of this magnitude (von Mises, 1959).

Because of these objections, Tucker's proposal to make A a function of Reynolds number will not be adopted here. It will be assumed instead that A can be estimated from the formula given by Pennycuick (1972a), which can be reexpressed as

$$A = (2.85 \times 10^{-3})M^{2/3} \tag{9}$$

where A is the equivalent flat-plate area in square meters, and M is the body mass in kilograms. This formula may overestimate A for a few very large, fast birds, and would certainly underestimate it for insects, but it should be adequate for the majority of birds.

4. Power Curve for the Ideal Bird

The sum of the induced and parasite powers is the power required to fly by a bird consisting of a streamlined body supported by an ideal actuator disk, the work needed to flap the wings to and fro being neglected. This may be regarded as an ideal bird, for which the curve of power versus speed is shown in Fig. 2. It will be seen that the curve is U-shaped, reaching a minimum at the minimum power speed V_{mp}, at which the total power required is the absolute minimum power P_{am}. At some higher speed, V_{mr}, defined by the point of contact with the curve of a tangent from the origin, the ratio of power to speed is a minimum, and hence the distance traveled per unit work done is a maximum. This is the maximum range speed, and the power required to fly at this speed is P_{mr}.

Expressions for V_{mp} and P_{am} are derived by Pennycuick (1969). These are slightly modified by the introduction of k into the expression for

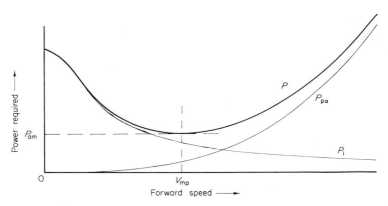

Fɪɢ. 2. Power curve for an ideal bird, that is, one with dragless wings. The total power required to fly P is the sum of the induced power P_i and the parasite power P_{pa}.

induced power and now become

$$V_{mp} = \frac{0.760 k^{1/4} W^{1/2}}{\rho^{1/2} A^{1/4} S_d^{1/4}} \tag{10}$$

$$P_{am} = \frac{0.877 k^{3/4} W^{3/2} A^{1/4}}{\rho^{1/2} S_d^{3/4}} \tag{11}$$

Similarly, the speed and power for maximum range are

$$V_{mr} = \frac{k^{1/4} W^{1/2}}{\rho^{1/2} A^{1/4} S_d^{1/4}} \tag{12}$$

$$P_{mr} = \frac{k^{3/4} W^{3/2} A^{1/4}}{\rho^{1/2} S_d^{3/4}} \tag{13}$$

V_{mr} is thus 1.32 times V_{mp}, and P_{mr} is 1.14 times P_{am}. These ratios are not altered by the introduction of k.

5. Profile Power

Of the three aerodynamic components of power output, the profile power—that component of power which is needed to overcome the profile drag of the flapping wings—is the most difficult to calculate. The calculation has been attempted on a strip-analysis basis by Pennycuick (1968b), but the results do not lend themselves to generalization. Two approximations have been proposed for incorporating profile power into the model.

1. Pennycuick (1969) suggested on the basis of the earlier strip-analysis calculation that profile power (P_{pr}) could be considered approximately constant over the speed range that is of most interest in connection with migration. It was taken to be a fixed multiple of the absolute minimum power, thus,

$$P_{pr} = X_1 P_{am} \qquad (14)$$

where X_1 is a constant, the profile power ratio. The effect of this on the power curve is shown in Fig. 3. V_{mp} is unchanged, but P'_{min}, a preliminary estimate of the power required to fly at this speed, becomes

$$P'_{min} = (1 + X_1) P_{am} \qquad (15)$$

The ratios of V_{mr} to V_{mp} and of P_{mr} to P'_{min} both increase progessively as X_1 is increased.

2. Tucker (1973) proposed that instead of being independent of speed, profile power should be proportional to the sum of the induced and parasite powers at all speeds. This assumption accentuates the U shape of the power curve, giving a very high power requirement in hovering compared to the minimum power. It also results in the ratio of V_{mr} to V_{mp} remaining at 1.32 and that of P_{mr} to P'_{min} at 1.14 irrespective of the profile power ratio. V_{mr} is thus less than would be predicted on the other assumption, but P_{mr} is not reduced in proportion. This leads to a reduction of 15–20% in the maximum lift:drag ratio, and hence also in the predicted range (see Section II,B,1).

It must be recognized that there is no reason to expect profile power to be related in any simple way to the other components of power. The most one can hope for is to achieve an approximation that is not too unrealistic and that is amenable to analysis. Tucker's assumption would considerably simplify the analysis, but it is less realistic than the assumption that profile power is independent of speed. The reason is that as the forward speed increases, so also does the relative airspeed "seen" by the flapping wing, and the lift coefficient therefore decreases. The wing profile drag coefficient, being a function of lift coefficient, also decreases. The profile power is a function of both speed and profile drag coefficient, the changes in which are in opposite directions and roughly compensate, at any rate at medium speeds. For the purposes of performance estimation, the older assumption is therefore retained here. Profile power is assumed to be independent of speed and is defined in terms of a profile power ratio X_1 as in Eq. (14).

At very low speeds and in hovering, the situation is more complicated, and no approximate theory can be relied on in this region. In cases where an accurate estimate of power consumption in hovering is of special importance, the more exact analysis developed by Weis-Fogh (1972, 1973) should be used, rather than the simplified scheme developed here.

6. Inertial Power

A further component of power associated with flapping the wings is the power required to accelerate and decelerate the wing at the beginning and end of each stroke. This problem has been intensively investigated by Weis-Fogh (1972, 1973).

In fast forward flight, work has to be done to accelerate the wing at the beginning of the downstroke, but at the end of the downstroke, the downward motion can be stopped and reversed by the aerodynamic lift. In this case, the kinetic energy removed from the wing can be transferred to the air, so contributing to lift and propulsion. This is the same work that was done on the wing during the initial acceleration, and so it does not have to be accounted for separately from the aerodynamic work already considered.

In very slow flight and hovering, Weis-Fogh shows that the relative airspeed seen by the wing is not sufficient to produce the necessary deceleration of the wing, and the kinetic energy must therefore be removed and restored by some other means. There are various ways in which this might be done:

1. Most birds, when flying very slowly, exhibit wing movements of the type analyzed by Brown (1948, 1953). The angular velocity remains almost constant until nearly the end of the downstroke, when the wing decelerates abruptly. This is accompanied by sudden straightening of the primary feathers, which have hitherto been sharply bent upward by the lift acting on them. The energy needed to bend these feathers is supplied by the flight muscles during the initial acceleration, and is released, and presumably transferred to the air, when the feathers straighten. It is possible that the reaction from this is sufficient to stop the downward swing of the wing. If this is so, then the work done in accelerating the wing eventually ends up as aerodynamic work and so once again does not need to be separately accounted for.

2. In most insects, the motion of the wings is more or less sinusoidal and quite unlike that seen in the slow flight of birds. In this case, the kinetic energy of the wings can be stored at each half-cycle in the form of elastic deformation of structures in the body

and then released to accelerate the wing again for the next stroke. Elastic energy may be stored in the cuticle, or in special elastic ligaments made of the protein resilin, or both. The wing and the elastic structures then make up a tuned system. Provided that this operates near its resonant frequency, the only power needed to maintain the oscillation is that required to offset losses in the elastic elements.

3. Weis-Fogh pointed out that hummingbirds differ from other birds in that the wing motion in both slow and fast flight is approximately sinusoidal, as is usual in insects. On the other hand he was unable to find any elastic structures in the body corresponding to those seen in insects in spite of the most careful search. Thus, it appears that hummingbirds cannot use either of the above mechanisms for recovering the work done in accelerating the wing. Weis-Fogh concluded that this work has to be done anew at each half-cycle. He calculated the total power required by a hovering hummingbird on this assumption and obtained good agreement with the experimental oxygen consumption measurements of Lasiewski (1963).

It appears that Weis-Fogh's argument applies to hummingbirds only, and then only at very low speeds. Even in this case there is a possibility that elastic energy could be stored in the contractile proteins of the flight muscles themselves. Weis-Fogh suggests this, but says that no mechanism is known on the molecular scale that could lead to this result.

For the purposes of performance prediction, it seems that the inertial power need not be separately accounted for, provided that only medium and high airspeeds are considered. It should be taken into account when considering hovering hummingbirds, but it is not clear whether this is necessary in other hovering birds. More information on the elastic properties of primary feathers is needed to clarify the latter point.

7. Basal Metabolism

This component of power is assumed to be needed at all times, irrespective of what the bird is doing. It will be assumed to be equal to the "standard metabolic rate" of Lasiewski and Dawson (1967), for which they give two equations. These can be reexpressed as

$$\Pi_m = 6.25 M^{0.724} \tag{16a}$$

for passerines, and

$$\Pi_m = 3.79 M^{0.723} \tag{16b}$$

for nonpasserines. It will be remembered that the symbol Π was earlier defined as a power input, that is, a rate of consumption of chemical energy, expressed here in watts. M is the body mass in kilograms.

Since the basal metabolism is being assumed to be independent of speed, it can be added into the total power in the same way as the profile power. To enable this to be done, it will be converted into an equivalent mechanical power P_m, where

$$P_m = \eta \Pi_m \tag{17}$$

This conversion has, of course, no physiological meaning, but is undertaken so that all components of power are expressed in the same form, so that they can be added together. P_m, like the profile power, can be expressed as a multiple of the absolute minimum power, thus,

$$P_m = X_2 P_{am} \tag{18}$$

where X_2 is the metabolic power ratio. The minimum power required to fly, P_{min}, can now be obtained by adding the metabolic power, in the form of its mechanical equivalent, to the preliminary estimate P'_{min} given above (Eq. 15), giving

$$P_{min} = P_{am}(1 + X_1 + X_2) \tag{19}$$

The metabolic power ratio would not be expected to remain constant in birds of different sizes, as some authors continue implicitly to assume. Equation (16) shows that

$$P_m \propto M^{0.72} \tag{20}$$

whereas it can be deduced from Eq. (11) that

$$P_{am} \propto M^{1.17} \tag{21}$$

Hence the metabolic power ratio

$$X_2 = P_m/P_{am} \propto M^{-0.45} \tag{22}$$

over a series of geometrically similar birds. That is, the metabolic power is a larger fraction of the total in smaller than in larger birds. Conversely, the ratio of metabolic rate in flight to basal metabolic rate is larger in large birds than in small ones.

a. Calculation of Metabolic Power Ratio. By substituting for P_m and P_{am} from Eqs. (11), (16), and (17), a relationship between the disk area and the body mass can be obtained for any particular value of the metabolic power ratio. Alternatively, we can substitute for the disk area from Eq. (6) and so get a relationship between wing span b and body mass M. If Eqs. (16) and (17) are combined to give

$$P_m = \alpha \eta M^\delta \tag{23}$$

where α and δ are constants, then the substitution yields

$$X_2 = \frac{6.03 \alpha \eta \rho^{1/2} b^{3/2} M^{(\delta - 5/3)}}{k^{3/4} g^{5/3}} \tag{24}$$

The values of α and δ, which are different for passerines and nonpasserines, are given in Eqs. (16).

The charts of Figs. 6 and 7 are provided for the graphical determination of X_2 for a bird whose body mass and wing span are known. They have been calculated with the following values for the other variables: $\eta = 0.23$, $\rho = 1.00$ kg m^{-3}, $k = 1.20$, $g = 9.81$ m sec^{-2}. The appropriate adjustment of values of X_2, drawn from the charts, for other values of these variables, is apparent from Eq. (24). The chosen value of ρ corresponds to a flying height of 2000 m (6600 feet).

b. Effect of Flying Height on X_2. It can be seen that for flying heights other than 2000 m, the value of X_2 drawn from the chart has to be corrected by multiplying by the square root of the density (in kilograms per cubic meter) at the chosen height. X_2 thus decreases with increasing height. It will be seen later that if X_1 and X_2 were independent of air density, then so also would be the effective lift:drag ratio and hence the range. The decrease of X_2 with height modifies this in the sense that the effective lift:drag ratio increases with height, so providing a positive incentive to fly high. The effect would be most marked in small birds, where the metabolic power ratio is high in any case.

c. Effect of Changing Body Mass on X_2. The concept of a metabolic power ratio is straightforward in cases where the body mass can be considered essentially constant. It is less easy to define in the case of long migratory flights, where there is a substantial reduction in the body mass in the course of the flight, due to the consumption of fuel. The way in which P_{am} varies in this situation is readily apparent from Eq. (11). It is not so clear whether or not P_m changes as the mass declines, or if it does, in what way. This point is further considered below in connection with the range attainable on long flights (Section II,B,3).

8. Ventilation and Circulation Power

According to Tucker (1973), the mechanical power required to ventilate the lungs is about 5% of the total power required for all other purposes, and about the same amount again is required for the circulation of the blood. Thus, when all the other components of power have been added together, an additional 10% must be added to allow for these two items. This can be represented by multiplying the sum of all the other components by a ventilation and circulation factor R, whose value can be set at 1.10 following Tucker's data.

9. Power Curve for the Real Bird

Assembling the various components of power from the previous sections, we can now combine Eqs. (4), (7), (14), and (18) to get the total mechanical power required to fly horizontally P as a function of the forward speed V as follows:

$$P = R[P_{am}(X_1 + X_2) + (kW^2/2\rho S_d V) + \tfrac{1}{2}\rho A V^3] \qquad (25)$$

The minimum power speed V_{mp}, which follows from this equation, is the

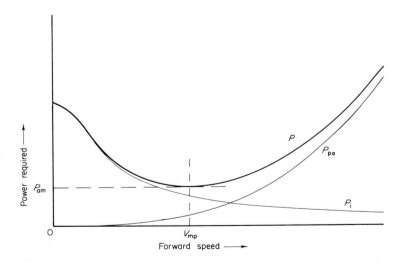

Fig. 3. Power curve for a real bird, obtained by adding a constant profile power P_{pr} and metabolic power P_m to the curve of Fig. 2. The power required for ventilation and circulation still has to be allowed for by multiplying the total power shown on the graph by a constant R.

same as that for the ideal bird (Eq. 10), and the minimum power is R times the value of P_{min} given by Eq. (19). The maximum range speed V_{mr} and the power required to fly at that speed P_{mr} are higher than the corresponding values for the ideal bird, but can be found by multiplying V_{mp} and P_{am}, respectively, by appropriate factors. Thus

$$V_{mr} = BV_{mp} \tag{26}$$

and

$$P_{mr} = RCP_{am} \tag{27}$$

B and C can be found by interpolation in Table II after values have been assigned to X_1 and X_2. R is the ventilation and circulation factor, for which a provisional value of 1.10 is assumed.

During a long flight, in which the mass declines appreciably owing to the consumption of fuel, V_{mr} also declines in the fashion indicated by Eq. (12), and the bird must therefore progressively reduce its cruising speed if maximum range is to be obtained. The average speed attained in a flight in which this is done is considered below in Section II,B,6.

10. Mechanical Efficiency

To convert the mechanical power of Eq. (25) into the chemical power input (as observed by determinations of oxygen consumption), it must be divided by the mechanical efficiency η of the whole system [Eq. (1)].

Experimental determinations of η have been made by Tucker (1972) on a Laughing Gull (*Larus atricilla*) and by Bernstein *et al.* (1973) on a Fish Crow (*Corvus ossifragus*). An incremental method, of which the principle is as follows, was used in both cases. The oxygen consumption of a bird flying horizontally at a speed V in a wind tunnel is first determined. The tunnel is then tilted through an angle θ. Suppose that the sense of the tilt is such that the air is now directed downward at the angle θ to the horizontal; then the effect on the bird is the same as though it were climbing at this angle. An increment of power output ΔP is required, such that

$$\Delta P = WV \sin \theta \tag{28}$$

where W is the body weight. A corresponding increment of power input $\Delta \Pi$ is observed, and the mechanical efficiency is estimated as

$$\eta = \Delta P / \Delta \Pi \tag{29}$$

Bernstein *et al.* (1973) measured power input (from oxygen consumption) at descent angles of 2°, 4°, and 6°, at each of five different airspeeds. From their results, a regression of power output on power input can be calculated through the three points at each speed. The five regression coefficients, which are estimates of η, ranged from 0.200 to 0.263, with a mean of 0.233. Tucker (1972) gives six estimates of η obtained at different speeds, which range from 0.19 to 0.28, with a mean of 0.235. In view of the close agreement between these two independent determinations of η, a value of 0.23 will be adopted here as the best available estimate. Earlier writers (including Tucker, 1973) have used a rough value $\eta = 0.20$ for performance calculations.

11. Variation of Profile Power Ratio with Body Mass

We are now in a position to make some empirical observations about profile power. Apart from the relationship between profile power and forward speed, which has already been considered, there remains the question as to whether the profile power would be expected to remain a fixed proportion of the whole in different birds of various sizes. In other words, should the profile power ratio X_1 vary with the body mass?

This can be answered by extending the line of reasoning developed below in Section III,A,2. It can be shown that if P_{fwd} is the component of profile power due to forward speed, and P_{flap} that due to flapping, then

$$P_{\text{fwd}} \propto \rho C_{D_{\text{p}}} S V^3 \tag{30}$$

and

$$P_{\text{flap}} \propto \rho C_{D_{\text{p}}} S \omega^3 b^3 \tag{31}$$

where ρ is the air density, $C_{D_{\text{p}}}$ is the profile drag coefficient (assumed constant for the whole wing), S is the wing area, V is the forward speed, ω is the flapping angular velocity, and b is the wing span.

It is shown in Section III,A,2 that the maintenance of the same lift coefficient in a series of geometrically similar birds implies that

$$V \propto M^{1/6}$$
$$\omega \propto M^{-1/6}$$

If these relationships are obeyed, and $C_{D_{\text{p}}}$ is also constant, then it follows that both the above components of the profile power should vary with the seven-sixths power of the mass. The same relationship applies to P_{am}. Hence, for flight at any particular lift coefficient, X_1 should be independent of the body mass, provided that the profile drag coefficient does not change.

The profile drag coefficient may, however, change as a function of Reynolds number, even though the lift coefficient is held constant, and in this case the profile power ratio would vary in a systematic way with body mass. This possibility can now be investigated empirically, using the experimental values of Π_{mr} given by Tucker (1973) for various birds and bats, for which body measurements were also given. Measured values, or reasonable guesses, are now available for all the variables in Eq. (25) except X_1, the profile power ratio. X_1 can then be chosen to make the predicted Π_{mr} agree with the measured value. It must be recognized that the calculation of metabolic power ratio in particular is a little conjectural, bats being deemed to be "nonpasserines" for this purpose. The regression coefficient of log X_1 versus log body mass for Tucker's birds and bats came out to be —0.121, but this estimate did not differ significantly from zero ($P > 0.1$).

There are therefore no strong grounds for introducing a variation of X_1 with Reynolds number (and hence with body mass)—not, at any rate, on the basis of the rather limited and heterogeneous body of data currently available. Tucker (1973) is no doubt correct in stating that the original theory underestimates the power requirements for small birds and overestimates those for large ones, but it seems that this deficiency is adequately corrected by taking account of the variation of the metabolic power ratio X_2 with body mass [proportionality (22)], a point that was omitted from the original theory. The mean of the values of X_1 calculated from Tucker's data is 1.22. It is suggested that a rounded figure of 1.2 be used for all birds until such time as this has to be modified in the light of a larger body of experimental data.

The assumption that the profile drag coefficient is independent of Reynolds number is not at variance with the aerodynamic evidence. Schmitz (1960) found that most of the airfoil sections he investigated showed a downward jump in the profile drag coefficient at some critical Reynolds number in the general neighborhood of 10^5. However the "cambered plate" (Göttingen 417a), which is the most birdlike of the sections tested by him, did not show this characteristic. His figures 53 and 54 show that the profile drag coefficient of this section remained almost exactly constant over a Reynolds number range from 20,000 to 165,000 and over a wide range of angles of attack. This Reynolds number range corresponds to birds with body masses from roughly 30 gm to 2 kg. The possibility remains that some very large, fast birds might perform significantly better than the simple theory predicts, while some very small ones might do somewhat worse, owing the increasing relative importance of skin friction at the smallest sizes. This is the same proviso which was made in the discussion of parasite drag in Section II,A,3 above.

B. CRUISING PERFORMANCE

1. Range in Terms of Lift: Drag Ratio

For ornithologists concerned with migration, the range a bird can attain on a given load of fuel is of special significance and may indeed be the only performance prediction of interest. It is not necessary to estimate a bird's speed or power consumption in order to predict its range. The key to range prediction is the effective lift:drag ratio.

Consider a bird in steady horizontal flight at a speed V. Its weight must be balanced by an average lift L', which is the time average of the actual (fluctuating) lift over an integral number of wing-beat cycles. Similarly the various components of the power required to fly can be considered as contributing to an effective drag force D', acting backward along the flight path, which has to be balanced by an effective thrust T'. The mechanical power output can now be represented as

$$P = T'V = W(D'/L')V \tag{32}$$

L'/D' is the effective lift:drag ratio, and will be written $(L/D)'$. As P is the rate at which work is being done, it is proportional to the rate at which fuel is used up. If e is the energy released on oxidizing unit mass of fuel, and η is the mechanical efficiency as before, then the rate of change of the bird's mass is

$$\frac{dM}{dt} = -\frac{P}{e\eta} = -\frac{W}{e\eta}\left(\frac{D}{L}\right)' V \tag{33}$$

and

$$\frac{1}{V}\frac{dM}{dt} = -\frac{Mg}{e\eta}\left(\frac{D}{L}\right)' \tag{33a}$$

where g is the acceleration due to gravity. If Y is the distance traveled, then $V = dY/dt$. Substituting this in Eq. (33a) and inverting,

$$\frac{dY}{dM} = -\frac{e\eta}{Mg}\left(\frac{L}{D}\right)' \tag{34}$$

This can be integrated with respect to M to give the distance traveled for a given change in the body mass:

$$
\begin{aligned}
Y &= -\frac{e\eta}{g}\left(\frac{L}{D}\right)'\int_{M_1}^{M_2}\left(\frac{1}{M}\right)dM \\
&= -\frac{e\eta}{g}\left(\frac{L}{D}\right)'[\ln M]_{M_1}^{M_2} \\
&= \frac{e\eta}{g}\left(\frac{L}{D}\right)'\ln\frac{M_1}{M_2}
\end{aligned}
\tag{35}
$$

where the body mass is M_1 at the beginning of the flight and M_2 at the end. It may sometimes be more convenient to express Eq. (35) in terms of the fuel ratio F, which is that fraction of the takeoff mass that consists of fuel that is used up in the course of the flight. In this case,

$$
Y = \frac{e\eta}{g}\left(\frac{L}{D}\right)'\ln\left(\frac{1}{1-F}\right)
\tag{35a}
$$

2. Calculation of Effective Lift: Drag Ratio

The effective lift:drag ratio can be expressed as

$$
(L/D)' = WV/P
\tag{36}
$$

It is a maximum at V_{mr}, as has been explained. Thus

$$
(L/D)'_{max} = WV_{mr}/P_{mr}
\tag{37}
$$

The ultimate effective lift:drag ratio $(L/D)'_{ult}$, is defined as the maximum value for the ideal bird of Section II,A,4. It can be found by substituting in Eq. (37) the values for V_{mr} and P_{mr} obtained from Eqs. (12) and (13). This gives

$$
\left(\frac{L}{D}\right)'_{ult} = \left(\frac{S_d}{kA}\right)^{1/2}
\tag{38}
$$

For the real bird of Section II,A,9, this becomes

$$
\left(\frac{L}{D}\right)'_{max} = \frac{D}{R}\left(\frac{S_d}{kA}\right)^{1/2}
\tag{39}
$$

where R is the ventilation and circulation factor as before and D is a factor that can be found by interpolation in Table II once values have been assigned to the profile power factor X_1 and the metabolic power factor X_2.

3. Effect of Body Mass on $(L/D)'$

It will be noticed that Eq. (39) does not explicitly involve the body weight or mass, which implies that there would be no systematic change of $(L/D)'$ with body mass if the ratios X_1 and X_2, which determine the factor D, were independent of mass. However Eq. (22) shows that for a series of geometrically similar birds, X_2 is less in the larger than in the smaller birds, and this leads in turn to a smaller $(L/D)'$ for smaller than for larger birds. With a given fraction of their takeoff mass consisting of fuel, smaller birds therefore achieve less range than larger ones because of their higher basal metabolism relative to the power required to fly. Without this effect, the (absolute) range attainable for a given fuel ratio would be independent of body mass.

A less readily calculable effect is the change in X_2 to be expected in the course of a long flight. It is not very clear whether Lasiewski and Dawson's (1967) equations should be applied literally in this situation—that is, whether the metabolic rate declines as the mass declines owing to the consumption of fuel. Even if it does, P_m would decline only with the 0.72 power of the mass, whereas P_{am} would decline with the 1.5 power, which leads to a progressive increase of X_2 (with the −0.78 power of the mass) as fuel is consumed. If, as seems more likely, P_m declines only a little or not at all as fuel is used up, then X_2 would increase even more steeply, with up to the −1.5 power of the mass. On either assumption $(L/D)'$ will decline progressively in the course of a long flight in a way whose effect on the range is not readily calculable, but which can be determined by reference to Table II from the effect on D.

On the other hand, there is another effect that changes $(L/D)'$ in the opposite direction. As the mass declines, so also does the volume of the body. As the body length does not change, the cross-sectional area has to decline in direct proportion to the mass. As the wings stay the same length, S_d does not change. Hence, it follows from Eq. (39) that the effect is to increase $(L/D)'$ in proportion to $M^{-1/2}$.

In small passerines, these two effects nearly compensate, although the second probably always predominates—that is, the effective lift:drag ratio is higher at the end of the flight than at the beginning. In larger birds, the gain may be quite large, adding as much as 20% to the range estimate that would be obtained if the both effects were neglected. In Section II,B,8, two rough-and-ready procedures are given for calculating range, including a correction for the change of effective lift:drag ratio during the flight. The first (shorter) method is for small passerines, and the second (slightly longer) for other birds.

4. Changes of Speed and Power on a Long Flight

In contrast to the effective lift:drag ratio, the speed V_{mr} at which the bird must fly is strongly affected by changes in weight, as is the corresponding power P_{mr}. It can be seen from Eqs. (12) and (13) that

$$V_{mr} \propto W^{1/2} \tag{40}$$

and

$$P_{mr} \propto W^{3/2} \tag{41}$$

The speed and power must be progressively reduced during a long flight, in the way indicated by proportionalities (40) and (41), if maximum range is to be achieved. To take an extreme case, if half the takeoff mass consists of fuel and is ultimately consumed, then the speed will have to be reduced by a factor of 1.41 and the power by a factor of 2.83 at the end of the flight as compared to the beginning.

5. The Flight Muscles as a Fuel Reserve

One consequence of the latter relationship is that if the flight muscles are equal to their task at the beginning of a long flight, then there will be an increasing amount of surplus muscle later in the flight. In addition, if it is supposed that the muscles were working at a power output near that for maximum efficiency at some point early in the flight, then they would be working far below this point toward the end of the flight. If the bird were able to metabolize the surplus muscle, it would gain two advantages—in addition to the extra fuel reserve that would be made available, the working conditions for the remaining muscle would be restored nearer to those for maximum efficiency. The amount of extra fuel energy gained would depend on the relative proportions of muscle, fat, and other body parts at takeoff, but would typically amount to 5–10% of the energy derived from oxidizing fat.

To use this stratagem, the bird would have to be able not only to break down the contractile machinery of its muscles quickly but also to replace it quickly during the next feeding period. Kendall *et al.* (1973) showed that large variations of the protein content of *Quelea quelea* pectoralis muscles occur, but considered that the protein component involved was not part of the contractile mechanism. Whether or not this interpretation is correct, the contractile proteins themselves certainly can change in amount by a large factor in mice. Goldspink (1965) found that during starvation fibers of the mouse biceps brachii changed in an all-or-nothing manner from "large phase," with a diameter of 42 μm, to "small phase" at 23 μm. The reverse change took place on restoring the food supply. The change involved a reorganization of the contractile proteins, as well as a change in their amount, and evidently occurred rapidly, as intermediate stages were seldom if ever seen. It seems likely that a similar mechanism would be found in bird muscles, and in this case consumption of the muscles in flight would be a practical and advantageous proposition.

If birds do in fact consume their muscles in flight, then one would expect to find that in a population of migrating birds the mass of the flight muscles would vary with the 1.5 power of the total body mass. The flight muscles should thus constitute a smaller proportion of the body mass at the end of a long flight than at the beginning. It should be possible to test this prediction by sampling migrants before and after a long sea crossing.

6. Flight Time and Average Speed

Because of the continually changing cruising speed on a long flight, it is not possible to calculate the flight time by assuming a particular fixed speed. The flight time can, however, be obtained by inverting Eq. (33) and integrating it with respect to the mass, assuming that the bird flies throughout at the (progressively decreasing) speed V_{mr} and that the effective lift:drag ratio does not change. This gives

$$T = \frac{2.63 \eta e (L/D)' \rho^{1/2} A^{1/4} S_d^{1/4}}{B k^{1/4} g^{3/2}} \frac{(M_1^{1/2} - M_2^{1/2})}{(M_1 M_2)^{1/2}} \tag{42}$$

where T is the flight time (in seconds) and B is the ratio V_{mr}/V_{mp}, a value for which can be found from Table II once values have been assigned to X_1 and X_2.

The average speed for the flight $\bar{V}$ is found by dividing the range from Eq. (35) by the flight time from Eq. (42), which gives

$$\bar{V} = \frac{0.380 B k^{1/4} g^{1/2}}{\rho^{1/2} A^{1/4} S_d^{1/4}} \frac{(M_1 M_2)^{1/2}}{(M_1^{1/2} - M_2^{1/2})} \ln \left(\frac{M_1}{M_2}\right) \tag{43}$$

It can be seen from this that in a series of geometrically similar birds, $\bar{V}$ would vary with the one-sixth power of the body mass, as would the characteristic speeds V_{mp} and V_{mr}.

7. Optimum Cruising Height

The air density does not appear explicitly in the equations for lift:drag ratio and range, which implies that the range attainable is not affected by the height at which the bird flies. In practice, this conclusion is modified by the variation of the metabolic power ratio X_2 with air density, which results in a progressive increase of effective lift:drag ratio with height, as explained above (Section II,A,7). This means that the optimum height to fly is the maximum at which the bird is able to maintain its maximum range speed. This speed itself increases with height, however, and so does the power needed to maintain it, as both speed and power vary inversely with the square root of the air density [Eqs. (12) and (13)]. On the other hand, the maximum rate at which oxygen can be absorbed by the lungs decreases as the partial pressure of oxygen declines. Thus, there is some definite height at which the bird can obtain oxygen just fast enough to maintain V_{mr}, and this is the optimum cruising height.

As fuel is used up in the course of a long flight, both V_{mr} and P_{mr} decrease at any particular height. The bird can then maintain P_{mr} at a progressively higher altitude as its weight declines. The appropriate tactic is the "cruise climb," in which the bird climbs very gradually, always flying at the greatest height at which it is just able to maintain V_{mr}. The principle is illustrated in Fig. 4, which shows a series of curves of power required as a function of altitude and body mass, calculated

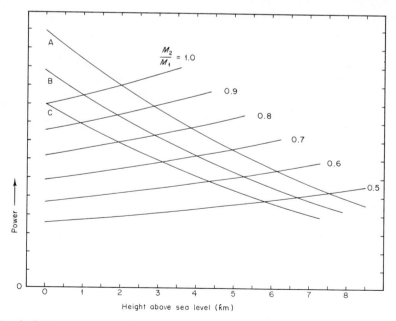

FIG. 4. Optimum cruising height (see text). The data on atmospheric density and pressure used in computing these curves were taken from von Mises (1959, Table 1).

according to Eq. (13). The power scale is arbitrary, the top curve being for takeoff mass ($M_2 = M_1$), and the lower ones for progressively decreasing M_2, as marked. The curves A, B, and C are power-available curves, calculated on the assumption (for want of a better one) that the maximum rate of oxygen consumption is proportional to the ambient partial pressure of oxygen. Curve A is for a bird that can just maintain V_{mr} at an altitude of 2000 m above sea level at takeoff mass, bird B can just do this at 1000 m, and bird C at sea level. It can be seen that even bird C should be cruising at 5800 m (19,000 feet) after consuming half of its take-off mass. As a more realistic case, bird B would start its flight at 1000 m (3300 feet). After consuming 20% of its takeoff mass, its optimum cruising height would be up to 2950 m (9700 feet). By the time 40% of the takeoff mass had been consumed, the optimum cruising height would be 5300 m (17,400 feet). The shape of the power-available curves may, of course, turn out to be different from that assumed, in which case the exact amount of the cruise climb will be modified accordingly.

8. Practical Range Calculations

Because of the great practical interest attached to estimates of range in migrating birds, a set of tables and charts is included at this point, which will allow such estimates to be made with a minimum of arithmetic. Two procedures are given. The first method is for small passerines up to about 50 gm takeoff mass, and the second is for other birds. If in doubt, use the second method. Both methods include corrections for the change in $(L/D)'$ during the flight, as discussed in Section II,B,3. Both estimate the air range for a flying height of 2000 m above sea level. The effect of wind is considered in the next section. Other flying heights require a correction to X_2 (see Section II,A,7 and Table I).

TABLE I

AIR DENSITY AS A FUNCTION OF
ALTITUDE ABOVE MEAN SEA LEVEL

Altitude (m)	Air density (kg m^{-3})
0	1.22
1000	1.11
2000	1.00
3000	0.905
4000	0.813
5000	0.732
6000	0.656

Method 1

1. Assign values to the following variables

M_1 Body mass at takeoff (including fuel) expressed in kilograms.

b   Wing span in meters, measured from tip to tip of the fully outstretched wings.

F   Fuel ratio, i.e., the mass of fuel which is to be consumed during the flight divided by M_1.

2. Assume the following values for constants, unless more reliable values are available:

$$e = 4 \times 10^7 \text{ J/kg (for fat)} \qquad R = 1.10$$
$$\eta = 0.23 \qquad\qquad\qquad\qquad g = 9.81 \text{ m sec}^{-2}$$
$$k = 1.20$$

3. Assign a value to the profile power ratio X_1. Assume $X_1 = 1.20$ unless a more reliable estimate is available.

4. Refer to Fig. 6 or 7 to find the value of X_2 that corresponds to the values assigned to M_1 and *b*.

5. Add X_1 and X_2 together, then find *D* from Table II.

6. Refer to Fig. 5 to find the value of $(S_d/A)^{1/2}$ that corresponds to the values assigned to M_1 and *b*.

7. Find the effective lift:drag ratio from the formula

$$\left(\frac{L}{D}\right)' = \frac{D}{k^{1/2}R}\left(\frac{S_d}{A}\right)^{1/2}$$

8. Increase this estimate by 10F%. (This is the correction for the change in $(L/D)'$ during the flight.)

9. Find the range from the formula

$$Y = \frac{e\eta}{g}\left(\frac{L}{D}\right)' \ln \frac{1}{(1-F)}$$

using Table III if required. The range will come out in meters and can be divided by 1000 to convert to kilometers.

Method 2

In this method $(L/D)'$ is calculated separately for the beginning and end of the flight, and the mean of the two values is taken as representative for the whole flight.

10. Assign values to M_1 and *b* as before. In addition, assign a value to M_2, the body mass at the end of the flight. This is equal to M_1 minus the mass of the fuel to be consumed (which need not necessarily be all of the fuel).

TABLE II

CONSTANTS FOR CALCULATING[a]

V_{mr}, P_{mr}, AND $(L/D)'_{max}$

$(X_1 + X_2)$	B	C	D
0.00	1.316	1.140	1.000
0.25	1.386	1.458	0.824
0.50	1.452	1.783	0.706
0.75	1.515	2.115	0.621
1.00	1.574	2.453	0.556
1.25	1.631	2.795	0.506
1.50	1.684	3.141	0.465
1.75	1.735	3.490	0.431
2.00	1.784	3.841	0.402
2.25	1.830	4.195	0.378
2.50	1.875	4.550	0.357
2.75	1.918	4.907	0.339
3.00	1.959	5.266	0.322
3.25	1.999	5.625	0.308
3.50	2.038	5.986	0.295
3.75	2.075	6.348	0.283
4.00	2.111	6.711	0.273
4.25	2.146	7.074	0.263
4.50	2.180	7.438	0.254
4.75	2.213	7.803	0.246
5.00	2.246	8.168	0.238

[a] Formulae: $V_{mr} = BV_{mp}$ [cf. Eq. (10)]; $P_{mr} = RCP_{am}$ [cf. Eq. (11)]; $(L/D)'_{max} = (D/R)(S_d/kA)^{1/2}$.

TABLE III

FACTORS FOR USE IN RANGE CALCULATIONS

F	$\ln \dfrac{1}{(1 - F)}$	$E = \left(\dfrac{1}{1 - F}\right)^{7/4}$
0.05	0.0513	1.09
0.10	0.105	1.20
0.15	0.163	1.33
0.20	0.223	1.48
0.25	0.288	1.65
0.30	0.357	1.87
0.35	0.431	2.13
0.40	0.511	2.44
0.45	0.593	2.85
0.50	0.693	3.36
0.55	0.799	4.04

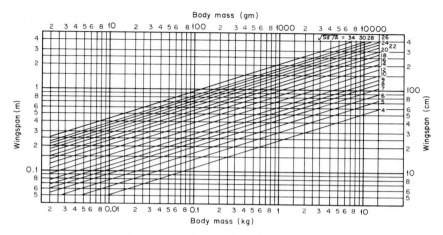

FIG. 5. Chart for finding $(S_d/A)^{1/2}$ from body mass and wingspan.

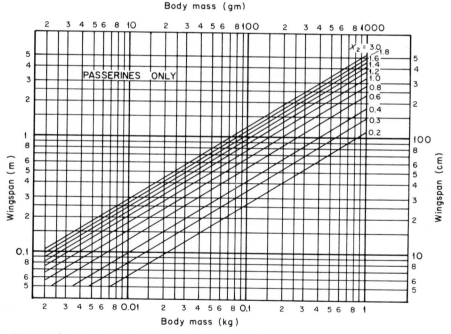

FIG. 6. Chart for finding X_2 from body mass and wingspan (for passerines only).

11. Proceed with steps 2–7, as in Method 1, but using M_2 instead of M_1 when consulting the charts (Figs. 5 and 6 or 7). This yields an estimate of the effective lift:drag ratio at the end of the flight.

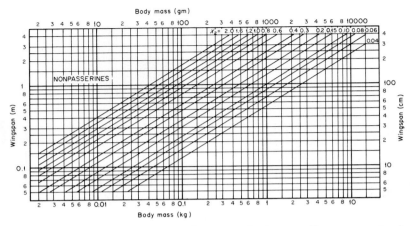

Fig. 7. Chart for finding X_2 from body mass and wingspan (for nonpasserines).

12. Divide the value of X_2 obtained in step 11 by the factor E obtained from Table III. This gives the value of X_2 at the beginning of the flight. Find the corresponding value of D from Table II.

13. Divide the previous value of $(S_d/A)^{1/2}$ by $(M_1/M_2)^{1/2}$. This gives the value of $(S_d/A)^{1/2}$ at the beginning of the flight.

14. Repeat step 7 with these new values, to get an estimate of $(L/D)'$ at takeoff.

15. Use the mean of the two values of $(L/D)'$ obtained in steps 11 and 14 to obtain the range as in step 9.

9. Effect of Wind on Range

The "range" considered so far is the air range and is not strongly dependent on the size or weight of the bird. However, larger birds on the whole fly faster than smaller ones and are therefore more affected by head- and tailwinds. The ratio of achieved range (over the ground) to air range in birds cruising at different speeds, with head- and tailwind components of different velocities, is plotted in Fig. 8.

Figure 8 assumes that the bird flies at a constant airspeed, regardless of the wind, but in practice the effect of wind on range may be modified somewhat, depending on whether the bird makes appropriate adjustments to its cruising speed.

The maximum range speed V_{mr} is the speed at which maximum air range is obtained. To obtain maximum range over the ground, the bird must maximize the ratio of ground speed to power. The power curve must therefore be replotted with ground speed as the abscissa, which

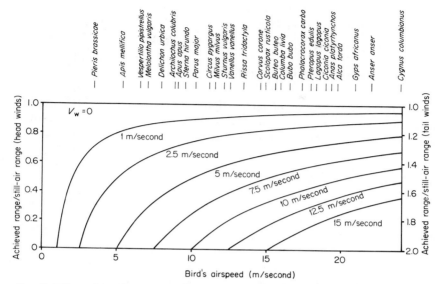

FIG. 8. Effect of head- and tail-wind components (V_w) on range, assuming that the same cruising airspeed is maintained, regardless of the wind strength. Some species of birds, bats, and insects are marked opposite their estimated maximum range speeds. (From Pennycuick, 1969.)

is done by simply shifting the y axis to the left for a tailwind and to the right for a headwind (Fig. 9). The optimum cruising speed is then found, as before, by drawing a tangent to the curve. When flying

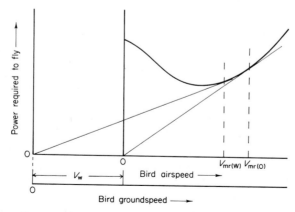

FIG. 9. When flying with a tailwind V_w, the maximum range speed $V_{mr(w)}$ is less than the maximum range speed in still air $V_{mr(0)}$. Similarly, speed must be increased above $V_{mr(0)}$ to obtain maximum range when flying against a headwind. (From Pennycuick, 1972a.)

against a headwind, the maximum range speed is higher than in still air, and conversely, speed must be reduced if there is a tailwind.

III. Limitations on Flight Performance

Several different kinds of factors limit the performance of flying birds and, indirectly, limit the kinds of birds that can evolve. A few of the more important of these will be considered in this section.

A. Limits of Wing-Beat Frequency

In general a bird flying along at some steady speed is free to choose its wing-beat frequency within a certain range. This range is limited by definite minimum and maximum wing-beat frequencies, which are determined by different factors and are affected in different ways by changes of scale. These limitations in turn exert an effect on the lower and upper limits of weight that are available to flying birds.

1. Upper Limit of Wing-Beat Frequency

Hill (1950) has argued that the upper limit of the frequency of vibration of any limb is set by the strengths of the muscles and tendons that accelerate and decelerate the limb at either end of its stroke. In a series of geometrically similar animals, the way in which the maximum oscillation frequency varies with the linear dimensions l of the animal can be deduced as follows.

The stress that can be borne by the muscles and their attachments is a property of the materials of which they are made and hence can be assumed constant. The force exerted by a muscle will therefore be proportional to the cross-sectional area of its attachment, and hence to the square of the length. The moment J that it exerts about the center of rotation of the proximal end of the limb will therefore vary with the length cubed:

$$J \propto l^3$$

The limb's moment of inertia I about its proximal end varies with the fifth power of the length:

$$I \propto l^5$$

so that the angular acceleration $\dot{\omega}$ that the muscle can impart to the limb varies inversely with the length squared:

$$\dot{\omega} = \frac{J}{I} \propto \frac{l^3}{l^5} = l^{-2}$$

If the wing is assumed to accelerate steadily through a fixed angle (not necessarily the whole stroke angle), then the time τ taken for the stroke is inversely proportional to the square root of the angular acceleration, and hence varies directly with the length

$$\tau \propto \dot{\omega}^{-1/2} \propto l$$

The maximum wing-beat frequency f_{max}, being inversely proportional to τ, varies inversely with the length:

$$f_{max} \propto 1/\tau \propto l^{-1} \tag{44}$$

This type of argument cannot be used to calculate the actual maximum wing-beat frequency in particular cases, for which detailed information about the mechanical properties of the parts of the particular bird would be required.

Proportionality (44) can be rewritten

$$f_{max} \propto W^{-1/3} \tag{44a}$$

where W is the weight. In this form, the proportionality indicates that if the logarithms of the maximum wing-beat frequencies of a number of different-sized birds are plotted against the logarithms of their respective weights, the points should fall on a straight line with a slope of $-\frac{1}{3}$ (Fig. 13). Although variations from this relationship can be caused by variations of wing planform or aspect ratio at any particular size, proportionality (44a) should represent the general trend over a wide range of weights.

2. Lower Limit of Wing-Beat Frequency

In this section, the generation of lift in hovering and slow forward flight will be assumed to follow the rules of steady-state aerodynamics; that is, the lift ΔL on an element of wing area ΔS is given by

$$\Delta L = \tfrac{1}{2}\rho V_r^2 \, \Delta S \, C_L$$

where V_r is the local air velocity relative to the wing, ρ is the air density, and C_L is the lift coefficient. In this case the minimum wing-beat frequency is determined by the weight and dimensions of the bird and by the maximum lift coefficient of its wings. In fast forward flight, however, the ratio of lift to profile drag of the wings, rather than their maximum lift coefficient, is the determining factor. For an account of the wing movements seen in slow and fast flapping flight, the reader is referred to Brown (1948, 1953).

Starting with the hovering case, consider a simplified bird with rectangular wings, of semispan $\frac{1}{2}b$ and chord c, which it flaps horizontally to and fro (Fig. 10). The problem is to find at what angular velocity ω each wing must be rotated in order to generate a lift force equal to half the weight. The simplifying assumptions of constant angular velocity, constant upward force throughout the cycle, and rectangular planform would lead to errors if used to calculate the absolute angular velocity in particular cases, but should not affect the conclusions as to the dependence of flapping frequency on weight, which is the limited objective of the present calculation.

Referring to Fig. 10, consider a strip of spanwise width δy, distant y from the shoulder joint. The relative airspeed V_y at the strip is the resultant of its horizontal velocity (ωy) due to flapping, and the vertical induced velocity V_i, so that

$$V_y = (\omega^2 y^2 + V_i{}^2)^{1/2}$$

Combining Eqs. (3) and (6), the induced velocity is

$$V_i = (2W/\rho \pi b^2)^{1/2} \tag{45}$$

where W is the weight. The lift δL generated on the strip is therefore

$$\delta L = \frac{1}{2} \rho C_L \left(\omega^2 y^2 + \frac{2W}{\rho \pi b^2} \right) c \, \delta y$$

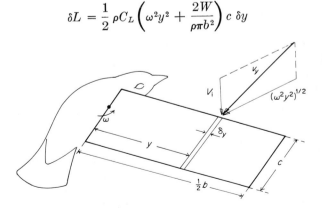

FIG. 10. Simplified bird with rectangular wings—hovering.

where ρ is the air density. Integrating this from the shoulder joint to the wing tip gives the lift L on one wing, which is assumed equal to half the weight

$$
\begin{aligned}
L = \frac{1}{2} W &= \frac{1}{2} \rho c C_L \int_0^{b/2} \left(\omega^2 y^2 + \frac{2W}{\rho \pi b^2} \right) dy \\
&= \frac{1}{2} \rho c C_L \left[\frac{1}{3} \omega^2 y^3 + \frac{2Wy}{\rho \pi b^2} \right]_0^{b/2} \\
&= \frac{1}{2} \rho c C_L \left(\frac{1}{24} \omega^2 b^3 + \frac{W}{\rho \pi b} \right)
\end{aligned}
$$

whence, by rearrangement

$$
\omega^2 = \frac{24W}{\rho} \left(\frac{1}{cC_L b^3} - \frac{1}{\pi b^4} \right) \tag{46}
$$

When considering the influence of the area and shape of the wing, it is more convenient to express this in terms of the wing area S (both wings), and the aspect ratio Λ, where $S = bc$ and $\Lambda = b/c = b^2/S$. Equation (46) then becomes

$$
\omega^2 = \frac{24W}{\rho S^2 \Lambda} \left(\frac{1}{C_L} - \frac{1}{\pi \Lambda} \right) \tag{47}
$$

If the weight, wing area, and aspect ratio are regarded as fixed, then substitution of the maximum lift coefficient in Eq. (47) defines the minimum wing-beat frequency, which will be proportional to the angular velocity ω, if the stroke angle is assumed constant. The effects on the minimum wing-beat frequency of varying the wing area or aspect ratio are apparent from Eq. (47). The effect of planform can be seen in a general way if the above calculation is repeated for a pointed, straight-tapered wing of the same area and aspect ratio as before, that is, one whose root chord is $2c$, and which tapers to a point at a semispan of $\frac{1}{2}b$. In this case ω^2 will be found to be twice as great as before, so that the minimum wing-beat frequency is increased by a factor of $\sqrt{2}$ as compared to that of the rectangular wing. Most bird wings are intermediate in shape between these two cases.

Where a low minimum wing-beat frequency is important, the wings need to be as broad at the tip as possible. Some insects, especially among the Lepidoptera, have wings with negative taper, that is, whose chord increases from the root to the tip, and this would result in the minimum wing-beat frequency being less than for a rectangular wing.

Turning now to the case of horizontal flight, with the bird traveling forward at some steady speed V, the minimum wing-beat frequency is generally lower than in hovering, and is determined by somewhat different considerations. First of all, the same approach as that used above for hovering will be applied to this situation, that is, the wing-beat frequency will be calculated by considering the relative air speed necessary to support the weight. The assumptions to be made now are the following.

1. The wing is rectangular, with semispan $\frac{1}{2}b$ and chord c as before.
2. The induced velocity is small compared to the forward speed and can be neglected when calculating the relative air speed at a point on the wing.
3. The wings flap vertically up and down (see Fig. 11).

The relative velocity V_y at a chordwise strip, distant y from the shoulder joint, is now

$$V_y = (\omega^2 y^2 + V^2)^{1/2}$$

so that the lift on the strip is

$$\delta L = \tfrac{1}{2}\rho C_L(\omega^2 y^2 + V^2)c\,\delta y$$

and the lift on the wing, assumed equal to half the weight, is

$$
\begin{aligned}
L = \tfrac{1}{2}W &= \tfrac{1}{2}\rho c C_L \int_0^{b/2} (\omega^2 y^2 + V^2)\,dy \\
&= \tfrac{1}{2}\rho c C_L[\tfrac{1}{3}\omega^2 y^3 + V^2 y]_0^{b/2} \\
&= \tfrac{1}{2}\rho c C_L(\tfrac{1}{24}\omega^2 b^3 + \tfrac{1}{2}V^2 b)
\end{aligned}
\tag{48}
$$

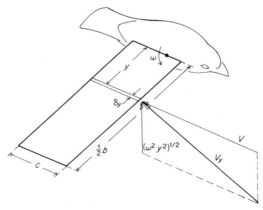

FIG. 11. Simplified bird with rectangular wings—horizontal flight.

Rearranging this yields

$$\omega^2 = \frac{24(W - \frac{1}{2}\rho V^2 S C_L)}{\rho C_L S^2 \Lambda} \tag{49}$$

where S is the wing area and Λ the aspect ratio, as before.

The right-hand term in the bracket of the numerator of this expression, $\frac{1}{2}\rho V^2 S C_L$, is the lift that would be developed by the wings in gliding flight at the forward speed V. Equation (49) suggests that if this lift exceeds the weight, there is no need to flap the wings at all. This is the case if V exceeds the stalling speed in gliding; that is, if

$$V > (2W/\rho S C_L)^{1/2}$$

when C_L is at its maximum value. This stalling speed is generally in the neighborhood of the minimum power speed, and most powered cruising flight takes place at speeds considerably above this. In this case, the minimum flapping frequency is determined by considerations other than the need to support the weight, and in fact depends upon the drag of the body and the lift : profile drag ratio of the wings.

In Fig. 12, a bird is seen from the side, flapping its wing directly downward. The relative air velocity V_r at some point on the wing is the resultant of an upward component V_f due to the downward motion of the wing, and the forward speed V. The angle between the aerodynamic reaction q on the wing strip and the resultant air velocity V_r always exceeds a right angle by an amount that depends on the ratio of the strip's profile drag to the lift developed by it. Because the drag of the

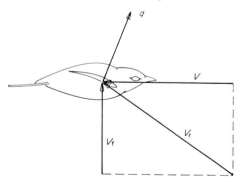

Fig. 12. At any particular point on the wing, the direction of the total reaction q always makes an angle of more than a right angle with the relative air flow V_r. Some points on the wing must have a forward component of q during the down-stroke, and this sets a lower limit (increasing with V) to the downward angular velocity of the wing.

body must be overcome, the reaction on at least some parts of the wing must have a forward component, and the flapping angular velocity must be great enough to allow the necessary forward force to be developed.

Since the velocity component due to flapping V_f, at any strip is proportional to its distance from the shoulder joint, the more distal parts of the wing make a greater contribution to the forward force than the more proximal parts. The actual value of the minimum flapping frequency at high forward speeds is rather complicated to predict, but one can see that if the body drag were independent of speed, the flapping angular velocity would have to increase in proportion to the speed. Since the body drag in fact increases about with the square of the speed, the product of flapping frequency and stroke angle must actually increase somewhat more rapidly than in direct proportion to the speed.

The dependence of the minimum wing-beat frequency on the weight in a series of geometrically similar birds can now be assessed for either hovering [Eq. (47)] or slow forward flight [Eq. (49)]. It can be seen that in either case

$$\omega_{min} \propto (W/S^2)^{1/2}$$

where ω_{min} is the minimum angular velocity needed to support the weight. Since the weight is proportional to the cube and the wing area to the square of any representative length, this proportionality can be rewritten

$$\omega_{min} \propto (l^3/l^4)^{1/2} = l^{-1/2} \qquad (50)$$

or alternatively,

$$\omega_{min} \propto W^{-1/6} \qquad (50a)$$

That is, the minimum angular velocity, and hence the minimum wing-beat frequency, is inversely proportional to the square root of the linear dimensions.

3. Lower and Upper Limits of Weight

It seems that the time required for a vertebrate locomotor muscle to be activated, to shorten through its work stroke, to lengthen again, and to be reset in readiness for the next contraction imposes a practical upper limit to the contraction frequency in the region of 80–100 sec^{-1}. Because of the trend expressed by proportionality (50a), this in turn imposes a lower limit on the weight, which appears to be around 0.02 N. Lower weights are possible in insects with fibrillar flight muscles, since these muscles can contract at higher frequencies than vertebrate skeletal muscle (Pringle, 1957).

Figure 13 is a log–log plot representing proportionalities (44a) and (50a). It can be seen that as the size of birds is increased, the maximum and minimum wing-beat frequencies converge, and there must be some weight above which flapping flight is impossible for this reason alone. It is possible that this is indeed the factor that sets the upper limit to the size of birds that are able to hover. The maximum size of very large birds that run to take off, however, is more probably limited by considerations of power output from the flight muscles, which are examined in the next section.

4. Lift Generation Depending on Nonsteady Aerodynamics

Weis-Fogh (1973) showed that the lift developed by flapping wings may be generated by mechanisms which depend on nonsteady aerodynamics and in which the assumptions underlying the calculation of ω_{min} above are not obeyed. The general principle is that certain types of accelerated motion may induce a transient circulation (and hence lift) greater than that which would be associated with the prevailing relative air velocity. Weis-Fogh's examples are based mainly on studies of insects, but he points out that the same mechanisms could, and probably do, apply in birds and bats also.

Weis-Fogh's first mechanism consists of a "clap" followed by a "fling." In the clap, the dorsal surfaces of the wings are pressed together above the back at the top of the upstroke. In the fling, the wings rotate so that the leading edges separate while the trailing edges remain in contact. This causes air to flow around the leading edges of the wings from the ventral to the dorsal surfaces, so setting up a strong circulation, which persists well into the downstroke. Weis-Fogh's description, which refers to a tiny Chalcid wasp, agrees closely with the wing motions seen in high-speed film of pigeons taking off and hovering.

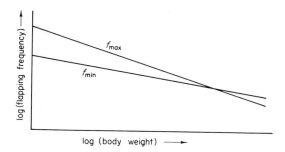

FIG. 13. Double logarithmic plot of minimum and maximum wing-beat frequencies (f_{min} and f_{max}) against body weight. This relationship would set an upper limit to the weight of flying birds if a limit were not already set by considerations of muscle power required and available.

The conventional idea of a lift coefficient does not really apply to this situation, since the lift is not directly related to the square of the relative airspeed. A symptom that something of the kind is taking place is that estimates of lift coefficient, calculated on steady-state assumptions, come out implausibly high. For example, Pennycuick (1968b) gives a maximum value of $C_L = 2.78$ in a hovering pigeon, whereas it is difficult to envisage any value much over half this being attained under steady-state conditions. It now seems that Weis-Fogh's "clap–fling" mechanism is the most likely explanation of this anomaly.

Weis-Fogh's second mechanism, the "flip" consists of an abrupt supination of the wing at the beginning of the upstroke and an abrupt pronation at the beginning of the downstroke. These motions also have the effect of establishing a strong circulation in the appropriate sense. Weis-Fogh says that this mechanism is important in hover-flies (Syrphinae) and in Odonata, being responsible for generating most of the lift in the former. One result of this is that these insects can hover with the wing-beat plane steeply inclined to the horizontal, whereas hovering bees and hummingbirds, which apparently depend mainly on steady-state lift generation, hover with the wing-beat plane nearly horizontal. Certain small passerines, such as sunbirds (Nectariniidae) and tits (Paridae), also hover with the wing-beat plane steeply inclined, and it is possible that they, too, may generate lift by some variant of the flip mechanism.

B. LIMITATIONS DUE TO MUSCLE POWER

1. Speed Range and Rate of Climb

A given amount of power available from the muscles, when combined with the curve of power required to fly versus speed, defines the maximum and minimum speeds, V_{max} and V_{min}, at which a bird can fly horizontally (Fig. 14). If the power available exceeds the power required

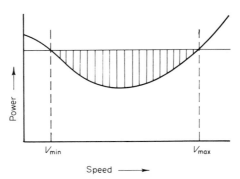

FIG. 14. Horizontal flight is possible at any speed at which the power available from the muscles (horizontal line) equals or exceeds the power required (curve).

to hover, then the bird can fly at any speed from zero up to its maximum speed, at which the power required just equals that available.

At intermediate speeds, the excess of power available P_a over power required P_r can be devoted to increasing the bird's potential energy by climbing. Thus if V_u is the greatest available upward vertical component of the bird's speed in straight, unaccelerated, but nonhorizontal flight, then

$$V_u = \frac{P_a - P_r}{W} \tag{51}$$

where P_r is a function of speed, expressed by the power curve.

The relationship between the maximum rate of climb (V_u) and the forward speed is plotted in Fig. 15. The maximum possible rate of climb (though not the steepest angle of climb) is obtained by flying at full power at the minimum power speed V_{mp}. Note that the bird can fly slower than V_{min}, or faster than V_{max}, provided that it is also descending (V_u negative). In this case, the power required to fly exceeds that available from the muscles, but the excess can be made good from gravity.

2. Sprint and Cruise Performance

Two types of "maximum power" available from the muscles have to be recognized in practice. In the first place, the maximum power P_{max} available in a brief burst of exertion is obtained when the muscle is shortening at its maximum tensile stress and speed of shortening, and is limited by the mechanical properties of the muscle and its attachments. The maximum power continuously available P_{ac}, however, is generally less than P_{max}, being limited by the rates at which fuel and oxygen can be supplied to the muscle by the blood system, and heat removed.

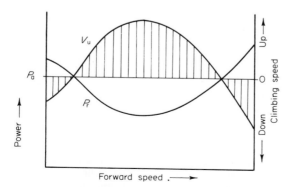

FIG. 15. Power required (P_r) and maximum rate of climb (V_u) plotted against forward speed, from Eq. (51). P_a is the power available from the muscles.

Both sorts of maximum power available generally lie higher, in relation to the curve of power required, in small than in large birds, for reasons that will be considered in the next section. In some hummingbirds (Trochilidae), such as *Calypte costae*, P_{ac} exceeds the power required to hover (Lasiewski, 1963), and such birds can hover continuously without incurring an oxygen debt (Fig. 16a). In the pigeon, which is probably representative of most small and medium-sized birds in this respect, P_{max} exceeds the power required to hover, but P_{ac} does not (Fig. 16b), so that the bird can take off by jumping directly into the air, but has to accelerate to some finite forward speed before it can fly continuously (Pennycuick, 1968b). Very large birds, such as the larger Old World vultures (Accipitridae, subfamily Aegypiinae) cannot hover even momentarily and have to accelerate by running along the ground before they can fly (Fig. 16c). The need for a take-off run can, however, be reduced or eliminated by taking off into wind and/or downhill, or by diving off a tree or cliff. It is doubtful whether the largest birds, such as the California Condor (*Gymnogyps californianus*) can fly continuously at all (Fig. 16d), and these birds spend nearly all their airborne time soaring (Koford, 1953), flapping their wings only at take-off and in emergencies.

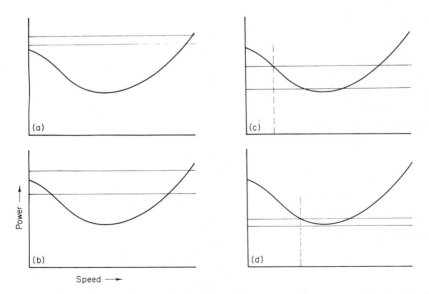

FIG. 16. Power required compared with power available in (a) hummingbird, (b) pigeon, (c) White-backed Vulture, and (d) California Condor. The upper horizontal line in each diagram is the maximum (sprint) power available, and the lower one maximum power continuously available. The vertical lines represent take-off speed, which must be attained by running. (From Pennycuick, 1968b.)

3. Effect of Scale on Power Required and Power Available

The reason the excess of power available over that required is less in larger than in smaller birds can be seen if one considers the way in which changes of scale affect these two power levels. The power required P_r to fly at some characteristic speed V (which might be the maximum range speed, for example) can be represented as the product of an average drag D' and the forward speed:

$$P_r = D'V \tag{52}$$

The effective lift:drag ratio (Eq. 36) should be independent of the weight, in a set of geometrically similar birds all flying at their respective maximum range speeds, so that D' is proportional to the average lift, and thus to the weight:

$$D' \propto W$$

The dependence of V_{mp} (and hence of V_{mr} also) on the weight can be deduced from Eq. (10), which shows that

$$V_{mp} \propto \frac{W^{1/2}}{S_d^{1/4}A^{1/4}} \propto \frac{W^{1/2}}{(W^{1/6})^2} = W^{1/6}$$

Hence, from Eq. (52)

$$P_r \propto W \times W^{1/6} = W^{7/6} \tag{53}$$

The power-to-weight ratio P_r/W must therefore increase with the one-sixth power of the weight. This argument has been known since the early days of aeronautics (von Helmholtz, 1874). It may be noted that, since the metabolic rate increases with approximately the two-thirds power of the weight, proportionality (53) implies that the ratio of the power required in flight to the metabolic rate is not constant, but is greater in large than in small birds.

Since it appears that the fraction of a bird's mass consisting of flight muscle does not vary in any systematic way with the total mass (Greene-walt, 1962), it follows that the specific power required P_{rs} (that is, the power required per unit mass of muscle) also increases with the one-sixth power of the weight (or of the mass):

$$P_{rs} \propto W^{1/6} \tag{54}$$

The maximum power available P_a from a bird's flight muscles is the product of the greatest work that can be done in one contraction when the muscles are shortening at their optimum speed multiplied by the frequency of contraction. The specific power available P_{as} (power available per unit mass of muscle) is the product of the specific work Q_s (work done in one contraction by unit mass of muscle) multiplied by the contraction frequency f:

$$P_{as} = Q_s f \qquad (55)$$

Q_s depends only on the bulk mechanical properties of the muscle tissue. The product of the tensile stress s which the muscle can develop and the proportion λ of its initial length through which it shortens gives the work done per unit volume in one contraction, and this divided by the density d of the muscle gives Q_s, thus:

$$Q_s = s\lambda/d \qquad (56)$$

so that

$$P_{as} = s\lambda f/d \qquad (57)$$

λ and d should be similar for all muscles of similar type. s can be regarded as constant so long as it refers to the contractile elements only, but the stress developed across the whole cross-sectional area of the fiber declines if the contents of the fiber are "diluted" with noncontractile material. This happens in a systematic way as the specific power output increases (and hence as animals get smaller), since a progressively increasing fraction of the muscle volume has to be devoted to mitochondria. This point is further considered below. At low contraction frequencies, this complication can be neglected, and s regarded as constant. The specific power output is then proportional to f, whose maximum value varies with the one-third power of the body weight [proportionality (44a)], so that

$$P_{as} \propto f_{max} \propto W^{-1/3} \qquad (58)$$

Comparison of proportionality (58) with proportionality (54) shows that whereas the power required from unit mass of muscle increases with increasing body weight, that available decreases. These opposing trends imply that there must be a well defined upper limit to the practicable weight of flying birds. In practice the limit seems to be in the neighborhood of 120 N, corresponding to a mass of about 12 kg.

It may be noted that rough numerical values can be assigned to all the quantities on the right-hand side of Eq. (57). According to Alexander

(1973), about 90 kN/m² would be typical of the stress exerted during shortening by a vertebrate locomotor muscle during vigorous exercise, and 0.25 would be a typical value for λ. The value of d is about 1060 kg/m³ for muscle. Hence, from Eq. (57)

$$P_{as} = 21f \qquad (59)$$

where P_{as} is in watts per kilogram of muscle.

The form in which Eq. (59) is expressed will serve to emphasize that specific power output in different animals varies over a very wide range because of differences in contraction frequency. It is meaningless to compare this quantity in widely different animals without taking any differences in contraction frequency into account. Two estimates from birds carrying out maximum-rate climbs will serve to illustrate the point. Pennycuick (1968b) estimated that the maximum specific power output in the domestic pigeon was 227 W/kg at a flapping frequency of 9.4 sec⁻¹. McGahan (1973b) estimated only 54 W/kg for the Andean Condor (*Vultur gryphus*) at a flapping frequency of 2.50 sec⁻¹. Expressed in the form of Eq. (59), these estimates become 24.1f and 21.6f W/kg, respectively.

At very high contraction frequencies, the specific power available is not as great as Eq. (59) would suggest because of the need to devote a progressively increasing fraction of the muscle volume to mitochondria. If we suppose that a volume σ of mitochondria is needed per unit of mechanical power output, then it is easily shown that

$$P_{as} = \frac{s\lambda f}{d(1 + \sigma s\lambda f)} \qquad (60)$$

where s is now the stress borne by the contractile elements only.

At low contraction frequencies, Eq. (60) approximates to Eq. (57), whereas at extremely high frequencies, P_{as}, instead of approaching infinity as Eq. (57) suggests, approaches instead a limiting value of $1/d\sigma$. As no reliable estimate of σ is available at present, this relationship cannot be given numerical expression. However, the departure from Eq. (57) is certainly large in the smaller insects, and it is probably substantial in hummingbirds and small passerines also.

Equation (59) should apply to more or less any vertebrate skeletal muscle, provided the specific power output is not too high. In insects with fibrillar flight muscles, the proportional shortening is only about one-tenth that of vertebrate muscles, and so the specific work is also only about a tenth as great. In spite of this, the specific power available in insects weighing less than about 0.1 N (i.e., nearly all insects) exceeds that of large birds because of their high wing-beat frequencies.

4. Disposable Load

All birds must be able to lift some weight in addition to the functioning parts of their bodies in order to carry food for themselves or their young. Raw food may be carried externally in the beak or feet or stowed internally in the crop, while fuel reserves are stored as fatty tissue distributed about the body. The addition of weight increases the power required to fly at any characteristic speed in proportion to the three-halves power of the total weight [Eq. (13)]. A bird's weight-lifting ability depends mainly on how much spare muscle power it has available over and above that needed to fly in the unladen condition.

The argument of the last section shows that at some limiting weight, somewhere over 120 N, no spare power at all is available, so that still larger flying birds are not possible. At some lower limiting weight, probably in the region of 60 N, just enough power is available to fly at the maximum range speed when unladen, but the addition of weight progressively narrows the gap between the maximum speed and the minimum power speed. At some much lower weight still, a bird can take on food or fat equal in weight to its unladen weight and still have enough power to fly at its maximum range speed. Pennycuick (1969) suggests that the heaviest birds able to do this should weigh about 7 or 8 N unladen, but this is somewhat conjectural.

The exact relation between unladen weight and weight-lifting ability is uncertain at present. It would be most valuable if falconers, or others with access to large performing birds, would investigate this matter. A few observations on the maximum weight that can be lifted by birds weighing between about 10 and 100 N would allow the cruising range of large birds in powered flight to be estimated with greater confidence than at present (see Section II,B,1), and would also fix the maximum weight of flying birds with greater precision.

It seems that small passerines preparing for long migratory flights are typically able to store fat up to about half their total weight (see, for example, Odum *et al.*, 1961). Fat loads much larger than this do not appear to occur and are probably impracticable for structural reasons.

IV. Gliding

A. GLIDING PERFORMANCE

1. Theory of the Glide Polar

The flight path of a bird or glider gliding at some constant speed V is always inclined below the horizontal. Its speed thus has a downward vertical component V_z, which is referred to as its sinking speed, or rate of sink, and is a function of V. The graph of sinking speed against forward speed is called the glide polar and is the most usual and convenient method of summarizing the gliding performance of a particular bird or glider.

The main features of the glide polar are shown in Fig. 17. This curve is the gliding analog of the power curve for horizontal flight (Fig. 3), but unlike the power curve, it starts at some minimum speed V_{min}, and does not extend down to zero speed. The value of V_{min} is determined by the maximum lift coefficient $C_{L_{max}}$, and by the weight W and wing area S:

$$V_{min} = \left(\frac{2W}{\rho C_{L_{max}} S}\right)^{1/2} \tag{61}$$

Somewhat higher than the minimum speed is the speed for minimum sink V_{ms}, at which V_z is a minimum. Higher still is the speed V_{bg} for best glide ratio, which can be found by drawing a tangent to the curve from the origin. The glide ratio is defined as the ratio V/V_z, and by flying at V_{bg} (in still air), the bird covers the greatest horizontal distance for a given loss of height.

The simplest interpretation of the glide polar regards it as the sum of two components, one due to the induced drag D_i and the other to the profile drag D_o, which in this interpretation is actually the sum of several different components of drag. The power input to the bird from gravity is equal to the product of the weight and the sinking speed, while the power expended is the product of the total drag times the forward speed.

$$W V_z = (D_i + D_0) V$$

and

$$V_z = (D_i + D_0) V/W \tag{62}$$

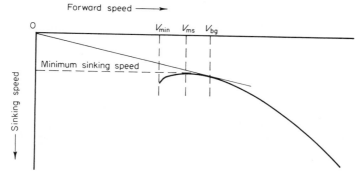

FIG. 17. The glide polar.

According to classical aeronautical theory, the induced drag D_i in a straight unaccelerated glide can be represented as

$$D_i = \frac{2kW^2}{\pi \rho V^2 b^2} \tag{63}$$

where b is the wing span. The factor k is an "induced drag factor" which would be equal to 1 in the ideal case of elliptical spanwise lift distribution, and is commonly found to be around 1.1 or 1.2 in airplane wings of reasonably efficient design.

The term "profile drag" is used in the crude analysis of glide polars to refer to the sum of all those components of drag that vary approximately with the square of the forward speed. It can be expressed in terms of a profile drag coefficient C_{D_0} such that

$$D_0 = \tfrac{1}{2} C_{D_0} \rho V^2 S \tag{64}$$

C_{D_0} is really the sum of the drag coefficients due to (1) the profile drag of the body, (2) skin friction, and (3) the profile drag of the wings. Whereas the first two of these components are more or less independent of speed, the profile drag coefficient of the wings varies with the lift coefficient and hence with speed, generally rising sharply at lift coefficients near the maximum (very low speeds). Thus, if C_{D_0} is defined as the drag coefficient of the whole bird at zero lift (as is usual), Eq. (64) tends to underestimate the profile drag at speeds near the minimum.

Combining Eqs. (62), (63), and (64), one can write the equation of the glide polar as

$$V_z = \frac{2kW}{\pi \rho b^2 V} + \frac{C_{D_0} \rho S V^3}{2W} \tag{65}$$

It should therefore be possible to fit a regression equation of the form

$$V_z = \beta/V + \gamma V^3 \tag{66}$$

through a set of experimental measurements of sinking speed at different forward speeds. Having found the constants β and γ empirically, estimates of k and C_{D_0} can be obtained, since comparison of Eq. (66) with Eq. (65) shows that

$$k = \beta \pi \rho b^2 / 2W \tag{67}$$

and

$$C_{D_0} = 2W\gamma/\rho S \tag{68}$$

Because of the variation of C_{D_0} with lift coefficient, the regression equation tends to give rather a poor fit to gliding measurements at low speeds (Fig. 18). The sharp rise of sinking speed in the region of the minimum forward speed, which is mostly due to a rise in C_{D_0}, is interpreted as being entirely due to the induced drag, so that the calculated value of k is too high.

The regression method of analyzing glide polars, which has been tried by Pennycuick (1971a), thus suffers from an inherent ambiguity in that it is not possible to separate the respective contributions of induced drag and varying wing profile drag to the total observed at low speeds. To resolve this difficulty, direct measurements are needed on actual bird wings of either the induced drag or the profile drag as a function of lift coefficient. These data are not available as yet, and because of this, experimental values of the induced drag factor k cannot yet be deduced from measured glide polars.

The intriguing question of whether or not the splayed primaries seen in many birds specializing in thermal soaring (vultures, storks, etc.) reduce the induced drag below the value for an elliptical lift distribution ($k = 1$) therefore remains open for the time being. Cone (1962b) has suggested on theoretical grounds that this should be possible, and Raspet (1960) claimed to have observed such an effect in the American Black Vulture (*Coragyps atratus*). Recently, however, Tucker and Parrott

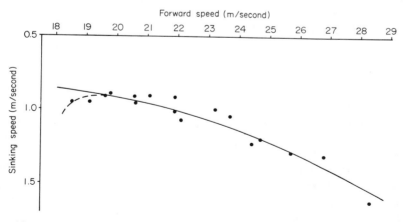

FIG. 18. Measured glide polar for ASK-14 motor glider (modified). The regression curve according to Eq. (66) (solid line) diverges from the truth (broken line) at speeds near the stall because of a rise in the profile drag coefficient of the wing. It may be noted that the performance of this aircraft in its normal (unmodified) configuration is considerably better than that shown. (From Pennycuick, 1971a.)

(1970), Parrott (1970), and Pennycuick (1971a) have suggested that, although Raspet's interpretation of his results may have been correct, his data were not sufficient to establish the point because of lack of reliable observations at low speed and also because of the inherent ambiguity in interpreting polars discussed above.

The simple theory of the glide polar expressed by Eq. (65) assumes that wing span and the wing area are fixed. In fact, a gliding bird, as it increases speed, reduces its wing span by flexing the carpal and elbow joints, and because this results in increased overlap of the flight feathers, the same movement also decreases the wing area. Tucker and Parrott (1970) found in a wind-tunnel study of the Laggar Falcon (*Falco juggr*) that the changes of wing area were proportional to those of the span; in other words, the mean chord remained constant, so that the aspect ratio was also proportional to the span.

The general nature of these changes of wing shape was described by Hankin (1913), and their effect on performance was interpreted in qualitative terms by Pennycuick and Webbe (1959). Reduction of the wing span produces an increase of induced drag, which is reflected in an increase in the first term of the right-hand side of Eq. (65). Simultaneous reduction of the wing area, however, reduces the second term because of a reduction of wing profile drag and skin friction.

Elaborating this argument, Tucker and Parrott (1970) showed that at any given speed there is a unique value of the wing span at which the total drag (and hence V_z) is at a minimum, and that this optimum wingspan decreases with increasing speed. The calculated optimum wingspan agreed well with the measured wing span of their falcon over a range of different gliding speeds.

2. Measurements of Gliding Performance

a. Maximum Lift Coefficient. The maximum lift coefficient in gliding flight has been measured for only three species of birds. The method in each case was to determine the minimum speed at which the bird could glide in a tilting wind tunnel. Tucker and Parrott (1970) obtained a value of 1.6 for the maximum lift coefficient of the Laggar Falcon. Pennycuick (1968a) recorded 1.3 for the pigeon, but this figure was based on the sum of the wing and tail areas, whereas the value of Tucker and Parrott was based on the wing area alone. The maximum lift coefficient of the pigeon, recalculated with reference to wing area alone, becomes 1.5.

Parrott (1970) obtained 1.1 for the American Black Vulture. This is a surprisingly low figure when one considers that glider wings equipped with NACA 6-series wing sections (chosen for low-drag rather than high-lift properties) can achieve a maximum lift coefficient of about

1.3 (Merklein, 1963), and so also can the forewing of the locust *Schisto-cerca gregaria* (Jensen, 1956). The lowest glide ratio available in Parrott's wind tunnel was 7.6, and it seems possible that his vulture would have been able to glide more slowly at a steeper gliding angle.

It is generally believed that the alula, which separates from the leading edge of the wing in most birds during the downstroke of flapping flight (especially at low speeds) serves to delay the stall over the rest of the wing and to increase the maximum lift coefficient. Nachtigall and Kempf (1971) mounted wings of the European Blackbird (*Turdus merula*), House Sparrow (*Passer domesticus*), and Mallard (*Anas platy-rhynchos*) in a wind tunnel to test this idea. On raising the alula, the stall was detectably delayed, and there was a small increase (up to 0.11) in the maximum lift coefficient. However, the maximum lift coefficients observed were very low (0.8–1.1). Also, the test speeds were low (5–6 m/second), and the conditions of the experiment seemed to correspond most closely to low-speed gliding flight, in which the alula is not normally raised. It remains possible that the effect of the alula in flapping flight is greater than these results suggest.

b. Measured Polars. Measured polars, obtained from wind-tunnel measurements, are presented in the three papers quoted in the last section. More approximate polars based on measurements of free-flying birds are given by Pennycuick (1960) for the Northern Fulmar (*Fulmarus glacialis*), and by Pennycuick (1971a) for the African White-backed Vulture (*Gyps africanus*). McGahan (1973a) measured speeds and gliding angles of Andean Condors gliding in the vicinity of the speed for best gliding angle ($C_L \approx 0.7$). He estimated a best glide ratio of 14 for this species, which agrees well with Pennycuick's (1971a) estimate of 15 for the African White-backed Vulture.

Tucker and Parrott (1970) plotted a number of measured or estimated polars for various birds and gliders on the same graph. Although this gives the most direct comparison, the result is a little confusing, because the polars of the birds with the lower wing loadings become compressed in comparison with those of the gliders. The effect of wing loading can be eliminated by dividing both the forward speed V and the sinking speed V_z by $(2W/\rho S)^{1/2}$. The effect of this is to plot $C_L^{-1/2}$ instead of V, and $C_D/C_L^{3/2}$ instead of V_z (Merklein, 1963). A selection of these dimensionless polars is plotted in Fig. 19.

B.   STABILITY AND CONTROL

A bird gliding at a constant speed must be in equilibrium; that is to say, the weight must be exactly balanced by the net aerodynamic force, and the line of action of this force must pass through the center

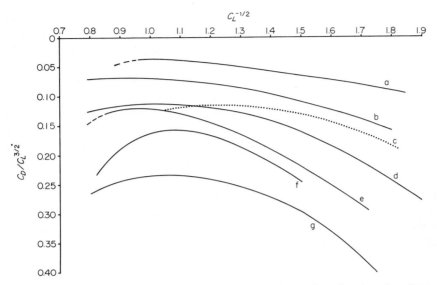

FIG. 19. Some measured or estimated glide polars expressed in dimensionless form:
(a) Schleicher ASK-14 powered sailplane (manufacturer's data); (b) White-backed
Vulture (*Gyps africanus*) (Pennycuick, 1971a); (c) (dotted) Black Vulture (*Coragyps atratus*) (Parrott, 1970); (d) Laggar Falcon (*Falco jugger*) (Tucker and Parrott, 1970); (e) Northern Fulmar (*Fulmarus glacialis*) (Pennycuick, 1960); (f)
Dog-faced Bat (*Rousettus aegyptiacus*) (Pennycuick, 1971c); (g) pigeon (Pennycuick, 1968a).

of gravity, so that there are no residual forces or moments. This does
not necessarily mean that the glide is stable. For stability, an additional
requirement has to be met, namely, that any disturbance from equilibrium must give rise to a force or moment that tends to restore equilibrium. Analyses of the stability of birds or aircraft are in practice almost
entirely concerned with moments, rather than with forces as such. Disturbances from equilibrium take the form of rotations of the bird, and
in a stable glide, any such rotations lead to the appearance of moments
tending to restore the original gliding attitude.

Rotations of a bird or aircraft can be resolved into components about
three axes mutually at right angles. Pitching is rotation about the pitch
axis, which runs horizontally and transversely through the center of
gravity, while the roll axis is horizontal and longitudinal, and the yaw
axis is vertical. These axes are regarded as fixed with respect to the
bird, as shown in Fig. 20, not with respect to the earth.

1. Pitch

The simplest method of providing both stability and control about
the pitching axis is to put a tailplane (a small auxiliary wing) on the

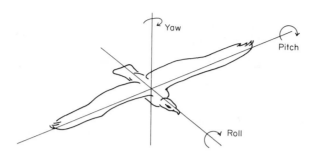

FIG. 20. Axes of rotation.

end of a rigid boom some distance behind the main wing, as in most airplanes (Fig. 21). A disturbance in pitch, say a rotation of the airplane in the nose-down sense, also rotates the tailplane, so inducing a downward increment of the force acting on it, and hence a nose-up pitching moment tending to reestablish the original attitude. The control moments needed to induce deliberate rotations in pitch can be produced by rotating the tailplane relative to the wing or by altering its profile shape.

Sturdy tails that appear capable of acting in the same way as an airplane tail are seen in some of the earliest fossil representatives of both birds and pterosaurs (*Archaeopteryx, Rhamphorhynchus*), but in later members of both groups, the area, and more especially the moment arm of the tail, was greatly reduced. In many late pterosaurs, the tail disappeared altogether as an aerodynamic surface, as it has in the modern fruit bats (Megachiroptera). Although most modern birds have an aero-

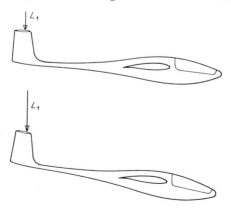

FIG. 21. Rotation of a conventional (tailed) aircraft in the nose-down sense also rotates the tailplane, so producing an increase in the downward force L_t acting on it. This in turn produces a nose-up pitching moment tending to restore the original attitude.

dynamically functional tail that is no doubt used to some extent in both stability and control, they are not dependent on it for either function and can glide quite well without it.

The main means of longitudinal (pitching) control in birds is fore-and-aft movement of the wings relative to the center of gravity. A nose-up pitching moment is produced by moving the wings forward, while moving them back gives a nose-down moment. With the wings forward, equilibrium is attained at a high angle of attack and hence at a low speed, while the swept-back wing position corresponds to a high-speed glide. The wing movements are largely effected at the carpal and meta-carpal joints and automatically produce the changes of wing span and wing area, whose effect on gliding performance was discussed above in Section IV,A,1 (Fig. 22).

Longitudinal stability in fast glides is probably achieved by the combi-nation of sweepback-with-washout. "Washout" means twisting of the wing so that the distal parts are rotated in the nose down sense relative

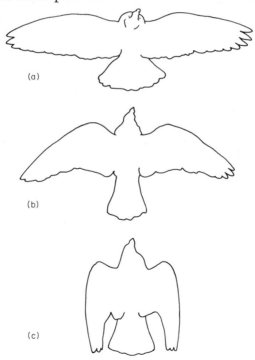

FIG. 22. A pigeon gliding in a wind tunnel at speeds of (a) 8.6 m/second, (b) 12.4 m/second, and (c) 22.1 m/second. The trim changes controlling speed are produced by fore-and-aft movements of the wings, which also result in a change of wing area over a range of 1.6:1 and of span over a range of 2.7:1. (From Penny-cuick, 1968a.)

to the proximal parts. With this arrangement, a swept-back wing achieves longitudinal stability in much the same way as a conventional airplane, with the wing tips performing the function of the tailplane.

At low speeds, when the wings are held straight out from the body, or even swept slightly forward, stability is thought to be attained by the downward deflection of the wing tips about an axis that is not parallel to the body axis (Fig. 23). Wing tips so deflected are known as "diffuser wing tips," and the reader is referred to Weyl (1945a,b) for an account of their action.

2. Roll and Yaw

Rolling moments for control purposes are easily produced by rotating one or both wings from the shoulder joint, so that the two wings have different angles of attack. It also seems to be possible for the manus to be twisted differentially through a small angle. When an extremely high rate of roll is required, one wing may be partially retracted by flexion of the elbow and wrist joints.

The term "stability" as applied to roll generally refers to rolling moments induced by sideslip, since a disturbance in roll leads to sideslip and can be corrected by such moments. A stable relationship of rolling moments to sideslip is produced both by sweepback and by dihedral (upward deflexion of the wings from the shoulders); the mechanisms involved are explained in textbooks of aeronautics, such as von Mises (1959) and Babister (1961). In certain display maneuvers, some birds— e.g., some pigeons (Columbidae) and auks (Alcidae)—may be seen gliding about with the wings set at an extremely large dihedral angle. In normal gliding, however, the extreme rolling stability thus produced would be an embarrassment. As speed is increased, the associated increase of sweepback would increase rolling stability, but this tendency

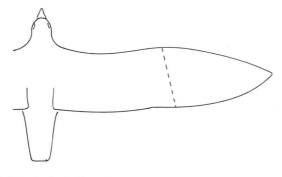

Fig. 23. A wing tip bent downward about an axis inclined in the way shown is called a diffuser wing tip. This arrangement contributes to both longitudinal and directional stability.

is counteracted by reducing the dihedral. Indeed, birds gliding at very high speeds normally exhibit negative dihedral, or anhedral (Pennycuick and Webbe, 1959).

The control of yaw is not well understood, but there appears to be no structure in birds that is adapted to be a yaw control as such. Instead, yawing moments are most probably produced by alterations of wing shape giving different amounts of drag on the two sides. The diffuser wing tips mentioned above should provide some stability in yaw as well as pitch.

C. AUXILIARY ORGANS OF FLIGHT

1. The Tail

Although the tail is not the primary means of longitudinal or directional control in modern birds, many gliding birds, especially kites (*Milvus* spp.) continually move their tails about in a way that suggests the correction of small errors of pitch and yaw. The primary function of the tail, however, seems to be to help maintain lift at very low speeds, both in gliding and flapping flight.

Most birds can vary their tail area by a factor of 3 to 4 by fanwise spreading of the rectrices, and the tail is normally spread in low-speed flight, especially when landing and taking off, and furled in fast flight. In slow flight, the spread tail forms an auxiliary surface behind and below the wings, which helps to suck the air over the main wing surface, so keeping the flow attached at higher angles of attack than would otherwise be possible. The maximum lift coefficient of the wing is increased by this, an effect that was demonstrated in the gliding pigeon by Pennycuick (1968a). In birds with long forked tails, such as the frigatebirds (Fregatidae) and most terns (Laridae, subfamily Sterninae), the spread tail forms, in effect, a long slotted flap behind the wings.

In slow flapping flight, the tail is moved up and down at the same frequency as the wings, and in hovering it appears to help divert the downwash from the wings into a vertically downward direction.

2. The Feet

In normal flight, the feet are carried either horizontally beneath the tail or tucked up under the flank feathers, with the ankle joint fully flexed, and in either of these positions they generate little or no drag. When lowered into the airstream, however, with the toes spread perpendicularly to the flow, the feet of land birds have a drag coefficient of 1.1–1.2 (Pennycuick, 1968a, 1971b) and thus form effective air brakes in those species that have reasonably large feet. The function of this air-brake action is usually not to slow the bird down (as is often thought), but to decrease the lift:drag ratio and hence steepen the angle of glide. This may be necessary when approaching to land or when trying to soar at a constant height in strong lift (Section V,A,1).

The air-brake action of the feet, which can be represented as an increase in C_{D_0} in Eq. (65), is most effective at high speeds. It has been suggested that in slow flight, in which the feet are less effective, steepening of the gliding angle is effected by changes of wing shape that increase k and hence augment the induced drag (Pennycuick, 1971b).

The webbed feet of water birds make the most effective air brakes because of their large area, and certain land birds, such as vultures of the genus *Gyps*, also have small webs between the toes which doubtless serve to augment the air-brake effect.

In some water birds with high wing loadings and very small tails, notably among the auks (Alcidae), the large webbed feet may also be spread at either side of the tail during slow flapping or gliding flight. In this position, their function appears to be to augment the tail area and thus improve the low-speed performance, although they can also be swung beneath the body and used as air brakes in the ordinary way.

V. Soaring

The term "soaring" covers any technique of flight by which energy is extracted from movements of the atmosphere and converted into potential or kinetic energy of the bird. Although some birds habitually soar with "power on," it is more usual to glide, and estimates of soaring capabilities are normally based on a bird's gliding performance.

Birds use soaring for one or both of two distinct functions. In the first place, it can be used as a means of staying airborne with reduced energy expenditure while patrolling in search of food, and second, cross-country soaring offers an alternative method of travel to the straightforward use of powered flight. It may be noted in passing that soaring birds are most probably the only animals (apart from people) that depend for their livelihood on a source of energy external to their own bodies.

A. METHODS OF SOARING

Some authors, notably Cone (1962a), have emphasized the distinction between static soaring, which depends on vertical movements of the atmosphere, and dynamic soaring, in which energy is extracted from variations in the horizontal wind. This distinction is quite clear in theory, but is an unsatisfactory basis for classification, because most practical soaring methods involve elements of both sorts. It is more convenient to classify soaring techniques according to the meteorological process that is exploited, as glider pilots do.

1. Slope Soaring

The simplest method of soaring is slope soaring, whereby the bird flies in the region of rising air (slope-lift) caused by the upward deflection of the wind over a slope (Fig. 24a). If the upward air velocity in this region equals or exceeds the bird's sinking speed, then the bird can maintain or increase its height above the ground without having to flap its wings. If the bird must fly at a forward speed greater than the wind speed in order to maintain a low enough sinking speed, then it is obliged to tack back and forth along the slope, aiming at an angle to the wind (Fig. 24b). In moderate to strong winds, however, many birds can make their airspeed equal to the wind speed, and can then hang motionless in the air above the slope. To maintain station vertically in this type of flight, the gliding angle must be suitably adjusted by the use of the feet (Section IV,C,2).

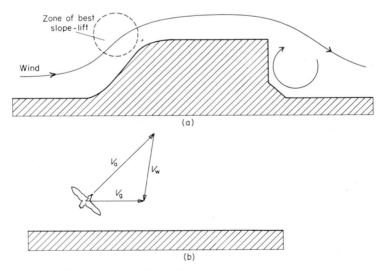

FIG. 24. (a) The best slope lift is found when the wind impinges on a smooth slope like that shown at left. Vertical cliffs can produce more complicated flow patterns, and sometimes give usable lift when facing downwind (right). (b) When a bird's airspeed V_a exceeds the windspeed V_w, it has to aim at an angle to the wind and tack along the slope at a groundspeed V_g.

Slope soaring, using ocean waves as slopes, is the main method of locomotion of most of the small and medium-sized petrels (Procellariiformes). On occasion, these birds even manage to slope-soar in zero wind, since a wave moving through stationary air displaces air upward in the same way as a stationary slope exposed to the wind.

One sometimes sees birds slope-soaring close in to a cliff facing *downwind*, especially if there is a sharp transition from the vertical face to horizontal ground at the top. In this case the air, flowing over the level surface to the cliff edge, cannot follow the right-angle change of direction of the ground, and instead of flowing vertically down the cliff face, follows a more gradual route to the lower level. A large vortex then forms in front of the cliff, with the air in contact with the face flowing upward (Fig. 24a). The same process in reverse can give rise to sink along a cliff face facing into wind.

On land, the opportunities for slope-soaring are limited to the vicinity of suitable hills and cliffs. All sea birds that nest on cliffs seem to spend a great deal of their time slope-soaring, although those with high wing loadings, such as the auks (Alcidae) and cormorants (Phalacrocoracidae) can only do so in strong winds. Most birds of prey take advantage of slope lift when patrolling in search of food, and so do gulls (Laridae, subfamily Larinae), which also use the slope lift along sea cliffs and buildings as a means of travel between their roosting and foraging areas. Although it is sometimes possible to slope-soar for distances running into hundreds of kilometers in areas where there are lines of hills, sea cliffs, or fault escarpments, slope-soaring cannot in general be considered a means of overland migration because of the restricted distribution of the lift.

2. Thermal Soaring

Thermal convection, when it occurs, is much more widely distributed than slope lift, although it tends to be poorly developed over certain types of surface, especially water. A general account of convection in the atmosphere is given by Ludlam and Scorer (1953), while the process is explained in relation to soaring by Wallington (1961). New information derived from radar studies has recently been reviewed by Konrad and Brennan (1971).

The basic requirement is an unstable vertical distribution of temperature in the atmosphere, caused either by differential heating from the bottom, or cooling from the top. The base of the atmosphere most commonly acquires heat by contact with a land surface heated by the sun, so that thermals tend to develop best in clear weather in low latitudes where this heating effect is strong. At night in clear conditions, however, the ground, and the air in contact with it, cools quickly by radiation, so that the formation of thermals is inhibited until some hours after sunrise, when solar heating once again breaks down this temperature inversion.

The individual units of thermal convection are vortices of limited extent known as "thermals." Thermals show an infinite variety of form and structure, but can be broadly classified into two main types, the dust-devil, or columnar type and the vortex-ring, or doughnut type (Fig. 25).

The dust-devil type of thermal consists of a rapidly rotating column of air with a zone of reduced pressure up the middle caused by centrifugal force. Friction with the ground causes air to flow inward at the bottom into the low-pressure area, so producing a vertical flow up the column. By circling in the zone of rising air a gliding bird can maintain or increase its height. Dust devils are triggered by solar heating of the ground. They are the predominant type of thermal near the ground when heating is intense and are often visible as whirling columns of dust. Except over intense and persistent sources of heat (such as grass fires), they usually penetrate only a few hundred meters above the ground and have lifetimes of only a few minutes. The strongest lift is found in the immediate vicinity of the whirling column, but weak lift and turbulence generally extend over a considerable area round about. The structure and behavior of dust devils have been studied by Ives (1947) and Williams (1948).

Thermals of the vortex-ring type may be initiated directly from heated ground, but when fully developed, they rise through the atmosphere as distinct bubbles unconnected with the ground. The lift is confined to a central core that is limited in extent both horizontally and vertically and is surrounded by a ring of descending air (sink). Vortex rings

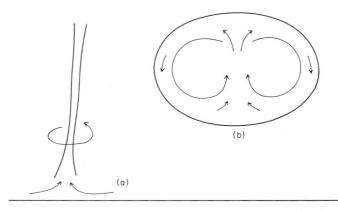

Fig. 25. Diagrams of the two main types of thermals: (a) dust devil, (b) vortex ring.

grow in size by entraining environmental air as they rise, and at heights of 2–3 km above the ground, very large cores 1–2 km across are often encountered.

When thermals penetrate above the condensation level, they are marked by blobs of cumulus cloud, which provide the soaring bird or pilot with a convenient means of locating them. If the air at cloud level is moist, however, the cumuli may spread out and coalesce, eventually inhibiting convection by cutting off the sunshine from the ground. This is known to glider pilots as "overconvection."

Strong convection leads to complex mixtures of different types of thermals. A common variant is for thermals to form into lines (thermal streets) along the wind direction, and they may even coalesce into roll vortices, giving continuous lines of lift. These are marked by "cloud streets" when they penetrate above the condensation level. Thermal streets can be exploited by flying straight along the line of thermals.

Isolated thermals, in the absence of streets, can be regarded as typically round in plan, having the strongest lift in the middle, and progressively weaker lift further from the center. The appropriate tactic for gaining height in such thermals is to fly in circles of as small a radius as possible, so as to remain in the strongest part of the lift. It is also possible to soar by flying in straight lines through randomly distributed, round thermals, if the speed is reduced while traversing lift and increased in sink. This method of soaring is more effective in birds of low than of high wing loading (Pennycuick, 1972b), and in birds with wing loadings below about 20 N/m², there would usually be little advantage in circling in thermals unless the lift was very weak (see also Section V,B,4).

Both columnar and vortex-ring thermals contain horizontal, radially convergent components of flow at the bottom, and the former contain circular tangential components as well. Either sort of horizontal flow can be exploited by suitable maneuvers (Pennycuick, 1971a). This could be considered a type of "dynamic" soaring, but it is impracticable to distinguish sharply between static and dynamic components in this case.

3. Gust Soaring

Random eddies, sometimes of considerable amplitude, are set up by the passage of the wind over the ground, especially if the latter is rough or wooded. This is felt by an observer on the ground as variations in the horizontal wind speed (gusts), but above the ground, the gusts have substantial vertical as well as horizontal components.

It is not generally possible to circle in up-gusts because of their transitory nature, but Pennycuick (1972b) records a case of Black Kites (*Milvus migrans*) apparently using the straight-line soaring technique (mentioned in the last section) to soar in random turbulence near the ground using mainly vertical gusts. Klemperer (1958) lists several methods of using the horizontal components of gusts (a form of dynamic soaring), and it is possible that the Bateleur Eagle (*Terathopius ecaudatus*) may use some such technique. Once again, the static and dynamic components of these methods, while distinct in principle, cannot readily be separated in practice.

Hendriks (1972) has made a detailed theoretical analysis of gust soaring. He concluded that under typical conditions sufficient energy for soaring should be available in the form of atmospheric turbulence, but he was not able to propose a practical flight procedure whereby the bird could actually extract this energy. The difficulty is that some characteristic of the time course of wind variations must be anticipated if the energy is to be used, but random gusts are by definition unpredictable.

4. Frontal Soaring

A front occurs when two air masses with different physical characteristics converge, giving rise to a narrow zone of lift along the convergence. The familiar cold fronts of temperate latitudes and sea-breeze fronts caused by the incursion of maritime air over a heated land surface are two types which can be used for soaring. They are favored by European Swifts (*Apus apus*), and in fact sea-breeze fronts over the British Isles were first identified from radar echoes of concentrations of swifts soaring in them.

5. Wave Soaring

Lee waves are standing waves that occur on the downwind sides of hills, and the highest climbs in gliders have been made in this type of lift. Harrison (1971) gives an account of the meteorology and technique of wave soaring. Barlee (1957) records the regular use by Northern Gannets (*Morus bassanus*) of small lee-wave systems set up by the offshore stacks on which they nest. Soaring birds that live in hilly areas no doubt use waves when opportunity offers, but there do not seem to have been any systematic investigations into this.

6. Wind-Gradient Soaring

According to Idrac (1924, 1925), albatrosses (Diomedeidae) use a form of dynamic soaring that depends on the wind gradient over the surface of the sea. The lowest layers of air are retarded by contact with the water, so that for the first few meters above the surface the wind strength increases fairly rapidly with increasing height. By gliding upward into wind, the albatross rises through layers of air coming progressively faster toward it, so that its airspeed tends to increase. As long as this increase offsets the deceleration due to its weight and drag, the bird can climb without losing airspeed. Eventually, the wind gradient becomes too weak to sustain any further gain of height, and the bird then turns and glides downwind, whereby it can gain energy on the descent as well as on the climb.

The principle of this technique was understood before that of any other method of soaring (Lord Rayleigh, 1883). It was analyzed by Walkden (1925), who calculated the minimum lift:drag ratio required. Cone (1964) analyzed the process in terms of the bird's ground speed, but this led to a very complicated analysis, whose conclusions are somewhat difficult to interpret. More recently, Wood (1973), and Hendriks (1972) have independently carried out analyses and computer simulations in terms of air speed, with a view to investigating the merits of various kinds of flight paths and establishing the conditions under which this type of soaring should be possible. Both concluded that it should be quite feasible for albatrosses to make use of this source of energy.

Even in this classic example of dynamic soaring, the situation is not quite so clear cut as might appear, since albatrosses, like the smaller Procellariiformes, also rely partly on the use of slope lift along waves.

It was pointed out by Klemperer (1958) that dynamic soaring should also be possible using the horizontal wind gradient set up when the wind blows parallel to a cliff face, but no examples of either birds or gliders using this technique seem to have been recorded.

B. Soaring Performance in Thermals

1. Climbing Performance

To climb well in a small thermal, it is necessary to fly in circles of small radius and at the same time to achieve a low sinking speed. To fly in circles, the bird must bank its wings at some angle to the horizontal. Steepening the angle of bank reduces the radius of turn, but at the expense of an increase in the sinking speed. At any particular angle of bank ϕ the radius of turn r is given by

$$r = \frac{W}{S} \frac{2}{\rho g C_L \sin \phi} \tag{69}$$

where g is the acceleration due to gravity. The minimum turning radius at this angle of bank is thus obtained by flying at the maximum lift coefficient, i.e., at the lowest possible forward speed. By flying faster at the same angle of bank, a larger turning radius and a different (not necessarily larger) sinking speed results, so that for any one angle of bank, a curve of sinking speed versus turning radius can be drawn. Figure 26 shows a family of such curves, with angle of bank as parameter. The envelope of this family is called the circling envelope, and shows the minimum sinking speed attainable, as a function of radius of turn. Families of curves such as that shown in Fig. 26 can be constructed directly from a straight-flight glide polar, and this has been done for the African White-backed Vulture by Pennycuick (1971a). For an explanation of the principles involved, the reader is referred to Welch *et al.* (1968) or Haubenhofer (1964).

The circling envelope determines the rate of climb that can be achieved in a given thermal. For this purpose, the thermal, which is assumed to be circular, is represented by its profile, which is a plot of upward air velocity versus distance from the center of the thermal. Goodhart (1965) pointed out that in general the maximum rate of climb is obtained when the slope of the circling envelope is equal and opposite to that of the thermal profile (Fig. 27). Thus, in very narrow thermals,

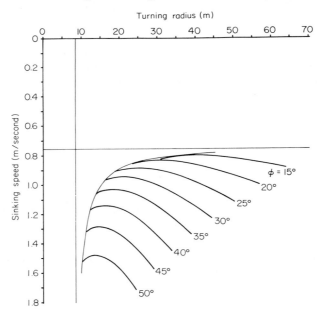

FIG. 26. Circling curves for the White-backed Vulture (*Gyps africanus*) (from Pennycuick, 1971a). At any particular angle of bank ϕ, increasing the airspeed increases the radius of turn. Each of the curves shown is for one angle of bank, with speed (not marked) increasing from left to right. The left-hand end of each curve is determined by the maximum lift coefficient. The envelope of these curves (circling envelope) defines the minimum sinking speed obtainable at any particular radius. The thin straight lines are its asymptotes, the horizontal one being the minimum sinking speed in straight flight.

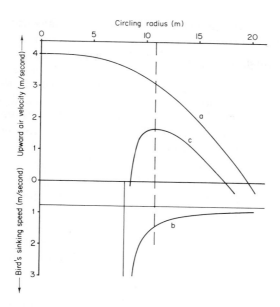

FIG. 27. Curve (a) is an arbitrary thermal profile, and (b) is a bird's circling envelope. Curve (c) is the bird's rate of climb, obtained by subtracting (b) from (a). The maximum possible rate of climb is obtained by circling at a radius such that the slopes of curves (a) and (b) are equal and opposite (Goodhart, 1965). The thin lines are the asymptotes of the circling envelope. (From Pennycuick, 1971a.)

in which the upward velocity is strong in the center but declines rapidly outward, it is necessary to circle tightly, at the expense of a high sinking speed, in order to stay in the lift, while in wider thermals a more gentle turn produces the best results.

It follows from Eq. (69) that at any given lift coefficient and angle of bank, the radius of turn is directly proportional to the wing loading. A low wing loading is therefore an asset to a bird that needs to remain aloft in narrow thermals, and the benefit is enhanced by the fact that the sinking speed varies with the square root of the wing loading, other things being equal.

From the point of view of food searching in vultures and birds of prey, the ability to remain airborne in the widest possible range of weather conditions is all important. It was observed long ago by Hankin (1913) that different species start soaring in the morning in the order of their wing loadings. This is because the first thermals to form are generally weak and of small diameter, and so usable only by birds with the lowest wing loadings. As Cone (1962a) pointed out, this factor must exert an appreciable selective pressure in favor of a low wing

loading (i.e., large wing area), especially in the more competitive scaven-
gers such as the vultures.

2. Cross-Country Performance

In the simplest form of thermal cross-country soaring, the bird gains
some height in a thermal, then glides off through dead air, covering
distance but losing height, until it contacts another thermal, when it
climbs up again, and so on (Fig. 28). The average cross-country speed
is given by the distance between thermals divided by the total time
taken for the climb plus the interthermal glide. The highest average
cross-country speed that can be attained with a given rate of climb
in thermals can be determined graphically from the glide polar by the
construction shown in Fig. 29.

This average speed is obtained only if the bird flies at the correct
optimum speed on the straight glides between thermals; this speed,
which is never less than V_{bg}, can also be read from Fig. 29. The higher
the achieved rate of climb, the faster the bird should fly between thermals if it is to maximize its cross-country speed. In practice, the optimum
interthermal speed is not highly critical, and glider pilots have found
that it is generally advantageous to fly at a somewhat lower speed than
the theoretical optimum in order to flatten the glide and so increase
the chance of striking a strong thermal for the next climb. The gliding
performance of such birds as vultures and storks is poor by glider standards, and one result of this is that the average cross-country speed
is highly sensitive to the rate of climb in thermals, but not very sensitive
to interthermal speed.

3. Effect of Wing Loading

It has been seen that a low wing loading, by minimizing the turning
radius and the sinking speed, leads to good climbing performance in

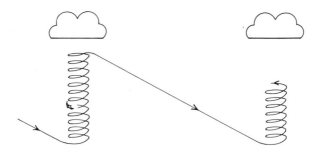

FIG. 28. Basic principle of cross-country flying using thermals.

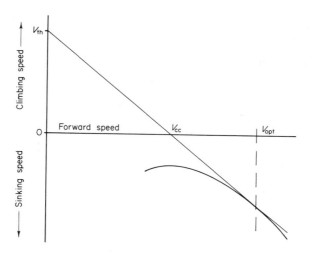

FIG. 29. Cross-country speed in thermal soaring. V_{th} is the achieved rate of climb in thermals (not the upward air velocity) and is plotted upward from zero on the same axis as the sinking speed. The optimum speed V_{opt} at which to glide between thermals is found by drawing a tangent to the polar from V_{th}. The tangent cuts the speed axis at the average cross-country speed V_{cc}. The cross-country speed corresponding to some interthermal speed other than V_{opt} can be found by the same construction, but will be less than V_{cc}. The reader is referred to Welch *et al.* (1968) for an explanation and discussion of this theorem.

thermals. On the straight interthermal glides, however, a low wing loading is a disadvantage, because the gliding angle at high speeds becomes excessively steep, and the cross-country speed that can be achieved with a given rate of climb in thermals is therefore low. Figure 30 shows the effect of doubling the weight on the polar of a glider of fixed wing area, and it can be seen from this why glider pilots often carry water ballast when strong convection is anticipated. If the thermals are wide and strong, doubling the circling radius and increasing the sinking speed by a factor of $\sqrt{2}$ has relatively little effect on the rate of climb, and this loss is more than offset by the increased speed obtained in interthermal glides. If the thermals are very narrow and/or weak, a low wing loading may be needed for a bird to soar at all, but when thermals are wide and strong, the birds with the highest wing loadings go the fastest; and gliders, which have still higher wing loadings, go faster than birds.

In vultures, the ability to stay aloft in weak lift is no doubt important, but cross-country performance is also important for those species that forage for food far from their nests, such as the African vultures of the genus *Gyps* (Pennycuick, 1972b). It was shown by Pennycuick

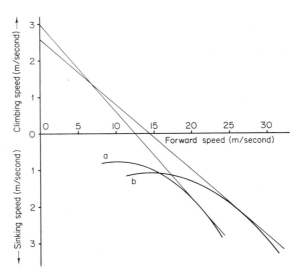

Fig. 30. Curve (a) is the polar estimated for the White-backed Vulture (*Gyps africanus*) by Pennycuick (1971a), based on a wing-loading of 76.5 N/m². Curve (b) is the same polar, recalculated for a wing loading of 153 N/m², which shows the effect on the polar if the vulture were to carry ballast equal to its own unladen weight. The addition of ballast increases the minimum sinking speed by 0.3 m/second in straight flight, and by 0.4 m/second when circling at an angle of bank of 25°. Thus, if the unladen vulture climbs at 3 m/second in thermals, the ballasted one will climb at 2.6 m/second, provided the thermals are wide. In spite of this, the ballasted vulture achieves a cross-country speed of 14.5 m/second, as compared to 12.7 m/second for the unladen one. If the thermals were very narrow (core radius <35 m), the ballasted vulture would lose out, because its circling radius is doubled.

(1971a) that these vultures should achieve a large gain in cross-country performance by adopting an albatross-like wing shape, at the expense of a very small loss in climbing performance (in fact, rate of climb should actually be better in all thermals except those with radii between 14 and 17 m). Cross-country speed is even more important for the White Stork (*Ciconia ciconia*), which migrates between Europe and Africa by thermal soaring, while the mere ability to stay airborne in thermals would seem to be relatively much less important in this species. Nevertheless, it, too, has wings of low aspect ratio (about 7.5), very similar to those of vultures. It is difficult to believe that the low-aspect-ratio wing of vultures and storks is an adaptation to soaring performance as such, and it is more probably adapted to some requirement of take-off and landing performance, which is more difficult to meet on land than at sea, as suggested by Cone (1962a). Exactly why low-aspect-ratio wings should be better in this respect is still unclear, however.

4. Straight-Line Soaring

It was pointed out by Nisbet (1962) that a bird flying straight through thermals, embedded in compensating sink, must fly more slowly through the thermals than through the sink if it is to gain any energy. Analyzing this method of soaring for those species whose polars could be estimated, Pennycuick (1972b) found that for optimum results they should traverse the sink at a speed about 1.6–1.8 times that at which they fly through the thermals. It was shown that the upward air velocity V_t needed in thermals, for height to be maintained, is related to the fraction a of the ground area covered by thermals by the relation

$$V_t = \frac{V_c V_d}{an(V_d - V_c)} \tag{70}$$

where V_c and V_d are the speeds at which the bird traverses lift and sink, respectively, and n is the glide ratio; V_c and V_d are chosen so that n is the same for both. The average cross-country speed $\bar{V}$ is

$$\bar{V} = \frac{V_c V_d}{V_c + a(V_d - V_c)} \tag{71}$$

Substitution of estimated speeds and glide ratios showed that for a glider (the Schleicher ASK-14) to soar in this way, the thermals would either have to be improbably strong (10 m/second), or else of normal strength but covering 65% of the area, which is also improbable. At the other end of the scale, a Bonelli's Warbler (*Phylloscopus bonelli*), if it had occasion to soar in this way, would be able to do so in thermals of 4 m/second covering 21% of the area. A Black Kite soaring in random turbulence (up-gusts 50% of area) would need a gust strength of about 2 m/second.

When straight-line soaring is possible at all, the cross-country speed achieved is about the same as in conventional thermal soaring, or a little faster. The conditions for its feasibility are harder to satisfy for the larger birds (because of their higher wing loadings), and they are therefore more often seen circling in thermals than the smaller ones.

5. Observed Cross-Country Performance and Behavior

Actual cross-country speeds achieved by large birds, including White Storks, Old World White Pelicans (*Pelecanus onocrotalus*), and various species of vultures (Aegypiinae) have been determined in East Africa by Pennycuick (1972b) by following birds about in a glider. In normal dry-season conditions, with rates of climb typically about 3 m/second or so, such birds achieve cross-country speeds around 45 km/hour, a result that is consistent with estimates of their glide polars (Section V,B,2).

An interesting difference in behavior was noted between the storks and pelicans on the one hand, which soar in flocks, and the vultures and eagles on the other, which do not. It was suggested that vultures, which soar as individuals, locate lift by reacting to visible weather signs in the same way as glider pilots. They are extremely successful at locating and using strong cores and thermal streets, and will make detours in order to take advantage of favorable soaring conditions. Storks and pelicans, however, use the flock to search for lift, by spreading out laterally on the interthermal glides. As soon as one part of the line happens to enter lift, the birds in that part are carried up in relation to the rest, and the others, on seeing this, quickly concentrate around them and start circling. By observing one another's progress, the individuals constituting the flock quickly maneuver into the strongest part of the lift. White Storks seem to pay rather little attention to weather signs and will hold a steady course into indifferent conditions, even when quite a small detour would take them into better weather.

C. ENERGETICS OF SOARING

It seems obvious that soaring on motionless wings is a less strenuous process than horizontal flapping flight. Soaring is not, however, effortless for any bird, since in addition to the requirements of basal metabolism, some power is consumed by the pectoralis muscles, which must hold the wings down in the horizontal position. In all soaring birds, regardless of their taxonomic position, the pectoralis muscle is divided into a large superficial and a small deep portion (Kuroda, 1961), and it seems most probable that the deep portion is a tonic muscle, specialized for holding the wings down in gliding flight. Birds that do not soar do not have the subdivided pectoralis muscle.

It appears that the power required to run the tonic muscles is of the same order of magnitude as the basal metabolic rate and should scale in the same way with changes in body mass. In large birds like vultures, where the metabolic power ratio (X_2) would be in the region of 0.07, a large saving in energy expenditure can be obtained by soaring. In small passerines, however, where X_2 will often exceed 1, a much smaller proportion of the total power expenditure can be saved.

In their recent study of the huge gliding pterosaur *Pteranodon,* Bramwell and Whitfield (1974) found that the shoulder joint could be mechanically locked in the gliding position, in a manner which would have significantly reduced the energy expenditure required from the tonic muscles. This feature is apparently unique to *Pteranodon*, and certainly no soaring bird is known to possess any such adaptation.

Large birds are also under stronger pressure than are small ones to use alternative sources of energy because of the difficulty they face in meeting the power requirements for flapping flight (Section III,B,3) and in lifting substantial loads of fuel (Section III,B,4). In opting for soaring, however, they have to accept the difficulties caused by the unpredictability of soaring weather, and its restricted distribution in time and space.

D. Migration by Soaring

Several kinds of large birds, notably White Storks, Black Storks (*Ciconia nigra*), and various species of eagles of the genus *Aquila*, which migrate annually between their breeding grounds in Europe and wintering areas south of the Sahara, make the journey mainly by thermal soaring. Because thermal soaring is not feasible over the sea, these soaring birds have to go round the east end of the Mediterranean, by the Bosporus and Suez (Rüppell, 1942), making a typical journey length of perhaps 7000 km between central Europe and equatorial Africa. Pennycuick (1972b) estimated, on the basis of an average cross-country speed of 45 km/hour and 6–8 hours of soaring per day, that a White Stork would need about 23 days for this journey. It should be possible to lift enough fat to last 48 days, so that the stork should theoretically be able to make the entire trip without feeding at all. If the stork were to fly under power, it would have to break the journey into at least five stages, stopping between each to replenish its fat reserves.

By comparison, a small passerine flying under power would require 7 days of continuous day-and-night flying to make the same journey in a straight line (5500 km), but would have to break this into at least two, and possibly three stages because of the need for refueling stops. Several days would probably be needed at each stop to replenish the fat reserves. A realistic estimate of the journey time might be 2 weeks, with no restrictions on choice of route. Radar observations indicate that passerines cross the Mediterranean and the Sahara on a very broad front, and take both obstacles in a single stage (Casement, 1966). If the warbler were to elect to soar like the storks, instead of going straight under power, it could do so, but would need 37 days for the journey.

The calculations on which these figures are based are, of course, very approximate, but they do show why large birds like storks are more or less obliged to soar on migration, whereas small passerines can more advantageously take the direct route and fly under power. The fact that they do so does not, however, necessarily mean that they do not also soar at the same time. "Soaring" in the general sense means flying in such a way as to extract energy from the atmosphere, irrespective of whether the wings are being flapped as well. It was pointed out by Nisbet (1962) that if passerines crossing the Sahara by day were to vary their airspeed when passing through thermals in the fashion indicated in Section V,B,4, this would result in a gain of energy to the bird that would appear as an increase in its cross-country speed, with no extra expenditure of energy.

VI. Loss of Flight

Birds differ from other flying vertebrates in that the development of their wing system has not placed any restrictions on the simultaneous evolution of the legs for other forms of locomotion, either on land or in the water, or for food manipulation. As nearly all birds perform well in two quite different forms of locomotion, one depending on the wings and the other on the legs, the needs of movement on land or in water seldom impose any appreciable selective pressure toward reduction of the power of flight. On the other hand, the pressure for its retention, as a fast, economical and versatile method of travel, is nearly always strong.

Some water birds, notably the auks (Alcidae), diving-petrels (Pelecanoididae), and cormorants (Phalacrocoracidae) use their wings both for flying and for swimming, and here some compromise on wing area is necessary. The extremely fast flight imposed on auks by their small wing area causes little difficulty at sea, but obliges them to use special kinds of landing and take-off maneuvers when they come ashore to breed. Where selective pressure for the retention of flight has been relaxed for some reason, the wings have been reduced to the optimum dimensions for swimming, and the power of flight has then been lost, as in the Great Auk (*Pinguinus impennis*), the penguins (Spheniscidae), and the Galapagos Flightless Cormorant (*Phalacrocorax* [*Nannopterum*] *harrisi*).

The power of flight can be lost in a more indirect way by selection for characteristics that are inconsistent with it. For instance, any trend toward increasing mass must either stop at about 12 kg, as in the largest bustards (Otididae), or else continue with the loss of flight, as in the ratites. No case is known of a bird having lost the power of flight through selection for small size, although such an occurrence is conceivable in view of the lower limit of mass set by the maximum possible wing-beat frequency (Section III,A,2). It is not very likely, however, as considerations of heat loss apparently set about the same lower limit.

On the whole, the extraordinary versatility of the avian dual locomotor system enables birds to retain the power of flight and at the same time to compete effectively with other animals on land or in the water.

VII. List of Symbols

A Equivalent flat-plate area of body

a Fraction of ground area covered by lift

B Ratio of V_{mr} to V_{mp}

b Wing span

C Factor for finding P_{mr}

C_{D_p} Wing profile drag coefficient

$C_{D_{pa}}$ Parasite drag coefficient

C_{D_0} Zero-lift drag coefficient

C_L Lift coefficient

$C_{L\,max}$ Maximum lift coefficient

c Wing chord

D Factor for finding $(L/D)'$

D' Average drag

D_{pa} Parasite drag

D_i Induced drag

D_o Composite profile drag (gliding)

d Density of muscle

E Correction factor for $(L/D)'$

e Energy content of fuel

F Fuel ratio

f Flapping frequency

f_{min} Minimum flapping frequency

f_{max} Maximum flapping frequency

g Acceleration due to gravity

I Wing moment of inertia

J Moment exerted by muscle

k Induced power (or drag) factor

L Lift on wing

L' Average lift

$(L/D)'$ Effective lift:drag ratio

$(L/D)'_{max}$ Maximum effective lift:drag ratio

$(L/D)'_{ult}$ Maximum effective lift:drag ratio for ideal bird

l Representative length

M Body mass

M_1 Body mass at start of flight

M_2 Body mass at end of flight

n Glide ratio

P Mechanical power

P_a Power available from muscles

P_{ac} Maximum power available continuously from muscles

P_{am} Absolute minimum power (ideal bird)

P_{as} Specific power available from muscles

P_{fwd} Profile power due to forward motion

P_{flap} Profile power due to flapping

P_i Induced power

P_{ih} Induced power in hovering

P_m Metabolic power (mechanical equivalent)

P_{min} Minimum power for real bird

P'_{min} Minimum power excluding metabolic power

P_{mr} Power required at maximum range speed

P_{max} Maximum (sprint) power available from muscles

P_{pa} Parasite power

P_{pr} Profile power

P_r Power required to fly

P_{rs} Specific power required to fly

Q_s Specific work (flight muscles)

q Reaction on wing

R Ventilation and circulation factor

Re Reynolds number

r Circling radius

S Wing area

S_d Disk area

S_{pa} Widest cross-sectional area of body

s Active stress developed by muscle

T Flight time

t Time

V Forward speed

V_{bg} Speed for best gliding angle

V_c Speed to fly through lift

V_d Speed to fly through sink

V_f Speed component due to flapping

V_i Induced velocity

V_{max} Maximum speed for horizontal flight

V_{min} Minimum speed for horizontal flight; minimum gliding speed

V_{mp} Minimum power speed

V_{mr} Maximum range speed

V_{ms} Speed for minimum sink

V_r Local relative airspeed

V_t Upward air velocity in thermals

V_u Rate of climb

V_y Local relative airspeed at spanwise station y

V_z Sinking speed

$\bar{V}$ Average speed

W Body weight

X_1 Profile power ratio

X_2 Metabolic power ratio

Y Range

y Spanwise distance from shoulder joint

α Empirical constant for calculating Π_m

β, γ Regression coefficients for glide polar

δ Empirical constant for calculating Π_m

η Mechanical efficiency

θ Descent angle

Λ Aspect ratio

λ	Proportional shorten-ing (strain) of muscle	τ	Time for downstroke
Π	Power input	ϕ	Angle of bank
Π_m	Basal metabolic rate	ω	Angular velocity of wing
ρ	Air density	ω_{min}	Minimum angular velocity of wing
σ	Inverse power den-sity of mitochondria	$\dot\omega$	Angular acceleration of wing

REFERENCES

Alexander, R. McN. (1973). Muscle performance in locomotion and other strenuous activities. In "Comparative Physiology" (L. Bolis, S. H. P. Maddrell, and K. Schmidt-Nielsen, eds.), pp. 1–21. North-Holland Publ., Amsterdam.

Babister, A. W. (1961). "Aircraft Stability and Control." Pergamon, Oxford.

Barlee, J. (1957). The soaring of gannets and other birds in standing waves. *Ibis* 99, 686–687.

Bernstein, M. H., Thomas, S. P., and Schmidt-Nielsen, K. (1973). Power input during flight of the Fish Crow *Corvus ossifragus*. *J. Exp. Biol.* 58, 401–410.

Bramwell, C. D. and Whitfield, G. R. (1974). Biomechanics of *Pteranodon*. *Phil. Trans. Roy. Soc. London, Ser. B* 267, 503–581.

Brown, R. H. J. (1948). The flight of birds: The flapping cycle of the pigeon. *J. Exp. Biol.* 25, 322–333.

Brown, R. H. J. (1953). The flight of birds. II. Wing function in relation to flight speed. *J. Exp. Biol.* 30, 90–103.

Brown, R. H. J. (1961). The power requirements of birds in flight. *Symp. Zool. Soc. London* 5, 95–99.

Casement, M. B. (1966). Migration across the Mediterranean observed by radar. *Ibis* 108, 461–491.

Cone, C. D. (1962a). Thermal soaring of birds. *Amer. Sci.* 50, 180–209.

Cone, C. D. (1962b). The theory of induced lift and minimum induced drag of nonplanar lifting systems. *NASA Tech. Rep.* R-139.

Cone, C. D. (1964). A mathematical analysis of the dynamic soaring flight of the albatross with ecological interpretations. *Va. Inst. Mar. Sci., Spec. Sci. Rep.* 50, 1–104.

Fullerton, Col. (1925). The flight of birds. *J. Roy. Aeronaut. Soc.* 29, 535–543.

Goldspink, G. (1965). Cytological basis of decrease in muscle strength during starvation. *Amer. J. Physiol.* 209, 100–104.

Goodhart, H. C. N. (1965). Glider performance: A new approach. *Organ. Sci. Tech. Vol Voile, Publ.* 8.

Greenewalt, C. H. (1962). Dimensional relationships for flying animals. *Smithson. Misc. Collect.* 144, No. 2.

Hankin, E. H. (1913). "Animal Flight: A Record of Observation." Iliffe, London.

Harrison, K. A. (1971). Wave soaring. *Sailplane & Gliding* 22, 2–9 and 92–100.

Haubenhofer, M. (1964). Die Mechanik des Kurvenfluges. *Schweiz. Aerorev.* 39, 561–565.

Hendriks, F. (1972). Dynamic soaring. Ph.D. Thesis, University of California, Los Angeles.

Hill, A. V. (1950). The dimensions of animals and their muscular dynamics. *Sci. Progr.* (*London*) **38**, 209–230.

Idrac, M. P. (1924). Etude théorique des manoeuvres des albatros par vent croissant avec l'altitude. *C. R. Acad. Sci.* **179**, 1136–1139.

Idrac, M. P. (1925). Experimental study of the "Soaring" of albatrosses. *Nature* (*London*) **115**, 532.

Ives, R. L. (1947). Behavior of dust devils. *Bull. Amer. Meteorol. Soc.* **28**, 168–174.

Jensen, M. (1956). Biology and physics of locust flight. III. The aerodynamics of locust flight. *Phil. Trans. Roy. Soc. London, Ser. B* **239**, 511–552.

Kendall, M. D., Ward, P., and Bacchus, S. (1973). A protein reserve in the pectoralis major flight muscle of *Quelea quelea. Ibis* **115**, 600–601.

Klemperer, W. B. (1958). A review of the theory of dynamic soaring. *Organ. Sci. Tech. Vol Voile, Publ.* **5**.

Koford, C. B. (1953). "The California Condor." Dover, New York.

Konrad, T. G., and Brennan, J. S. (1971). Radar observations of the convective process in clear air—A review. *Schweiz. Aerorev.* **46**, 425–427.

Kuroda, N. (1961). A note on the pectoral muscles of birds. *Auk* **78**, 261–263.

Lasiewski, R. C. (1963). Oxygen consumption of torpid, resting, active and flying hummingbirds. *Physiol. Zool.* **36**, 122–140.

Lasiewski, R. C., and Dawson, W. R. (1967). A re-examination of the relation between standard metabolic rate and body weight in birds. *Condor* **69**, 13–23.

Ludlam, F. H., and Scorer, R. S. (1953). Convection in the atmosphere. *Quart. J. Roy. Meteorol. Soc.* **79**, 317–341.

McGahan, J. (1973a). Gliding flight of the Andean condor in nature. *J. Exp. Biol.* **58**, 225–237.

McGahan, J. (1973b). Flapping flight of the Andean condor in nature. *J. Exp. Biol.* **58**, 239–253.

Merklein, H. J. (1963). Bestimmung aerodynamischer Beiwerte durch Flugmessungen an 12 Segelflugzeugen mit Brems- und Landeklappen. *Flugwiss. Forschungsanstalt Muenchen Ber.* No. 63.

Nachtigall, W., and Kempf, B. (1971). Vergleichende Untersuchungen zur flugbiologischen Funktion des Daumenfittichs (*Alula spuria*) bei Vögeln. *Z. Vergl. Physiol.* **71**, 326–341.

Nisbet, I. C. T. (1962). Thermal convection and trans-Saharan migration. *Ibis* **104**, 431.

Odum, E. P., Connell, C. E., and Stoddard, H. L. (1961). Flight energy and estimated flight ranges of some migratory birds. *Auk* **78**, 515–527.

Parrott, G. C. (1970). Aerodynamics of gliding flight of a Black Vulture *Coragyps atratus. J. Exp. Biol.* **53**, 363–374.

Pennycuick, C. J. (1960). Gliding flight of the fulmar petrel. *J. Exp. Biol.* **37**, 330–338.

Pennycuick, C. J. (1968a). A wind-tunnel study of gliding flight in the pigeon *Columba livia. J. Exp. Biol.* **49**, 509–526.

Pennycuick, C. J. (1968b). Power requirements for horizontal flight in the pigeon *Columba livia. J. Exp. Biol.* **49**, 527–555.

Pennycuick, C. J. (1969). The mechanics of bird migration. *Ibis* **111**, 525–556.

Pennycuick, C. J. (1971a). Gliding flight of the White-backed Vulture *Gyps africanus. J. Exp. Biol.* **55**, 13–38.

Pennycuick, C. J. (1971b). Control of gliding angle in Rüppell's Griffon Vulture *Gyps rüppellii. J. Exp. Biol.* **55**, 39–46.

Pennycuick, C. J. (1971c). Gliding flight of the dog-faced bat *Rousettus aegyptiacus* observed in a wind tunnel. *J. Exp. Biol.* **55**, 833–845.

Pennycuick, C. J. (1972a). "Animal Flight." Arnold, London.

Pennycuick, C. J. (1972b). Soaring behaviour and performance of some East African birds, observed from a motor-glider. *Ibis* 114, 178–218.

Pennycuick, C. J., and Webbe, D. (1959). Observations on the Fulmar in Spitsbergen. *Brit. Birds* 52, 321–332.

Pringle, J. W. S. (1957). "Insect Flight." Cambridge Univ. Press, London and New York.

Raspet, A. (1960). Biophysics of bird flight. *Science* 132, 191–200.

Rayleigh, Lord. (1883). The soaring of birds. *Nature (London)* 27, 534–535.

Rüegg, J. C. (1971). Smooth muscle tone. *Physiol. Rev.* 51, 201–248.

Rüppell, W. (1942). Versuch einer neuen Storchzugkarte. *Vogelzug* 13, 35–39.

Schmitz, F. W. (1960). "Aerodynamik des Flugmodells." Lange, Duisberg.

Shapiro, J. (1955). "Principles of Helicopter Engineering." Temple Press, London.

Tucker, V. A. (1971). Flight energetics in birds. *Amer. Zool.* 11, 115–124.

Tucker, V. A. (1972). Metabolism during flight in the laughing gull, *Larus atricilla*. *Amer. J. Physiol.* 222, 237–245.

Tucker, V. A. (1973). Bird metabolism during flight: Evaluation of a theory. *J. Exp. Biol.* 58, 689–709.

Tucker, V. A. and Parrott, G. C. (1970). Aerodynamics of gliding flight in a falcon and other birds. *J. Exp. Biol.* 52, 345–367.

von Helmholtz, H. (1874). Ein Theorem über geometrisch ähnliche Bewegungen flüssiger Körper, nebst Anwendung auf das Problem, Luftballons zu lenken. *Monatsber. Kgl. Preuss. Akad. Wiss. Berlin* pp. 501–514.

von Mises, R. (1959). "Theory of Flight." Dover, New York.

Walkden, S. L. (1925). Experimental study of the soaring of albatrosses. *Nature (London)* 116, 132–134.

Wallington, C. E. (1961). "Meteorology for Glider Pilots." Murray, London.

Weis-Fogh, T. (1952). Fat combustion and metabolic rate of flying locusts (*Schistocerca gregaria* Forskål). *Phil. Trans. Roy. Soc. London, Ser.* B 237, 1–36.

Weis-Fogh, T. (1956). Biology and physics of locust flight. II. Flight performance of the desert locust (*Schistocerca gregaria*). *Phil. Trans. Roy. Soc. London, Ser.* B 239, 459–510.

Weis-Fogh, T. (1972). Energetics of hovering flight in hummingbirds and in *Drosophila*. *J. Exp. Biol.* 56, 79–104.

Weis-Fogh, T. (1973). Quick estimates of flight fitness in hovering animals, including novel mechanisms for lift production. *J. Exp. Biol.* 59, 169–230.

Weis-Fogh, T., and Jensen, M. (1956). Biology and physics of locust flight. I. Basic principles in insect flight. A critical review. *Phil. Trans. Roy. Soc. London, Ser.* B 239, 415–458.

Welch, A., Welch, L., and Irving, F. G. (1968). "New Soaring Pilot." Murray, London.

Weyl, A. R. (1945a). Stability of tailless aeroplanes. *Aircr. Eng.* 17, 73–81.

Weyl, A. R. (1945b). Wing tips for tailless aeroplanes. *Aircr. Eng.* 17, 259–266.

Williams, N. R. (1948). Development of dust whirls and similar small scale vortices. *Bull. Amer. Meteorol. Soc.* 29, 106–117.

Wood, C. J. (1973). The flight of albatrosses (a computer simulation). *Ibis* 115, 244–256.

Chapter 2

MIGRATION: CONTROL AND METABOLIC PHYSIOLOGY

Peter Berthold

I. Introduction

Except for some types of irruptive movements, migration occurs twice annually, from the breeding grounds to the winter quarters and return. Thus, in a migratory species there must be a cyclic development of relatively precise control mechanisms stimulating and terminating migratory activity. Further, there must be a cyclic development of a distinct metabolic state that provides the energy requirements for migration. In the following sections, a survey of our knowledge of the control mechanisms as well as of the distinct metabolic conditions of bird migration will be presented.

Extensive investigations of physiological problems of bird migration did not begin until the nineteenth century (e.g., Brehm, 1828; Palmén, 1876; von Homeyer, 1881). In the history of our knowledge of the annual stimulation of migration (Section IV,A,1), Farner (1955) differentiated two periods—the period of observation (1825–1925, comprehensively and critically reviewed by von Lucanus, 1923; Wachs, 1926), and the period of experimental investigations, which began with experiments with Slate-colored Juncos (*Junco hyemalis*) by Rowan (1925). It should be emphasized initially that migration may have evolved several times in the evolution of modern birds (Section VI), and that different or even contradictory findings among different species or groups may be related to this.

II. Methods

Since the pioneer experiments of Rowan, the combination of observational and experimental approaches to the problems of migratory physiology is characteristic of the kind of experimental analysis performed today. Nowadays, it is useful and common to compare the physiology of migratory birds (species, races, populations, individual birds) and of nonmigratory birds, of typical and less typical migrants, of migratory birds before, during, and after migration and during molt and breeding periods. Comparisons such as these reveal characteristic features of migratory birds as well as correlations between migration and other annual events. By simultaneous observations of migrants and changes in the environment, correlations between migratory events and environmental factors become known, and by experimentation, cause-and-effect relationships of such correlations can be verified. On the other hand, predictions from experiments can be proved in the wild. Here, the use of optical equipment, Wilkinson apparatus (flight duration recorder), radar and miniature radio transmitters, and traps (for example, nets and weirs), as well as the naked eye, are all important in the study of annual and diurnal patterns of migration and nocturnal activity. In the so-called experimental migration (Farner, 1955), a planned treatment (displacement, castration, injection, implantation, keeping in various experimental conditions, etc.) is followed by release of experimental birds with attempts to obtain recoveries. Most important is the study of caged birds. Under natural and various artificial conditions and after different treatments, migratory activity (migratory restlessness, Section III), changes in body weight and fat deposition, metabolism (oxygen uptake, release of carbon dioxide, net caloric intake, utilization of foodstuffs, expenditure of energy, and water loss during flight), hormone levels, etc., can be observed and above all quantitatively measured.

Furthermore, caging allows continuous investigations of individual birds in long-term experiments. As a rule, most conclusive data were obtained in simultaneous laboratory and field studies.

III. Migratory Restlessness

Caged birds, in which a migratory flight is prevented, display migratory activity by jumping, whirling, and fluttering, the so-called migratory restlessness (*Zugunruhe*) (Figs. 1–3). This restlessness, often displayed in an intense, madness-like manner, has long been known to bird fanciers (e.g., Naumann, 1822; Brehm, 1828; Ekström, 1828). It is most conspicious in nocturnal migrants that normally show little or no nocturnal activity during nonmigratory periods (e.g., Farner, 1955; Dorka, 1966), but it is also recognizable in daytime migrants (e.g., Palmgren, 1949; Dolnik and Blyumental, 1967). Restlessness is quantitatively measurable in recording cages first introduced by Szymanski (1914) and first applied in important experiments on migratory restlessness by Wagner (1930). These cages are usually equipped with movable perches, electrical switches, selenium cells, and event recorders or with moist ink pads (for review, see, e.g., Farner and Mewaldt, 1953a; also, e.g., Eyster, 1952; Emlen and Emlen, 1966).

In numerous investigations for many species (for reviews, see, e.g., Palmgren, 1944, 1949; Farner, 1950, 1955; Helms, 1963; further, e.g., Weise, 1963; Hamilton, 1967; Gwinner, 1968a, 1972a; Berthold *et al.*,

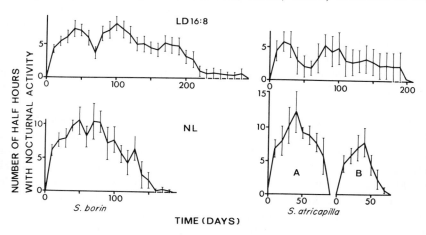

Fig. 1. Migratory restlessness of *Sylvia borin* and *Sylvia atricapilla* under constant conditions (LD 16:8) and under natural light conditions (NL). A = birds born early in the season (May), B = birds born late in the season (August). (From Berthold *et al.*, 1972b.)

1972b), it has been shown that migratory restlessness in caged birds often corresponds fairly closely in annual and daily patterns to natural migratory activity. However, migratory restlessness of the spring migration often persists throughout the breeding season, even during post-nuptial or post-juvenile molt, and the restlessness of the autumn migration may persist throughout the winter and even during pre-nuptial molt and gonadal development (for review, see, e.g., Helms, 1963; further, e.g., Weise, 1963; Gwinner, 1968b, 1972a; Berthold *et al.*, 1971b, and unpublished data) (Fig. 3). Some nonmigratory species investigated showed, as expected, no migratory restlessness (e.g., *Passer domesticus*, Eyster, 1952; *Nucifraga columbiana*, Farner, 1955), but others displayed at least in part some type of restlessness (e.g., *Zonotrichia leucophrys nuttalli*, Mewaldt *et al.*, 1968; Smith *et al.*, 1969; Gwinner, 1972a). Mewaldt *et al.* (1968) interpret the night activity of *Zonotrichia leucophrys nuttalli* as an atavistic remnant of ancestral migratory behavior. With the differences between migratory restlessness and natural migratory activity in mind, Helms (1963) warned that all nocturnal locomotor activity should not be necessarily regarded as migratory. Helms proposed the use of the neutral term "nocturnal activity" (*Nachtunruhe*) and called only the restlessness in the first period of more intense activity motivational migratory restlessness (*Zugunruhe, sensu stricto*); the restlessness in the later, more variable period of activity he termed merely "nocturnal activity" or "adaptational migratory restlessness" (*Zugunruhe, sensu lato*). Merkel (1956) called the prolonged nocturnal activity of the spring migration "*Sommerunruhe*" (summer restlessness), and Delvingt (1962) used the term "pseudorestlessness." Concerning the question as to whether migratory restlessness is similar or analogous physiologically to migratory behavior, there are, according to Farner (1955), reasons for a cautiously affirmative answer, but it is obviously impossible to be certain of the degree of similarity in general. Indeed, in many cases, there is a high degree of temporal agreement of natural migratory activity and nocturnal restlessness, but not in others. Therefore, generalizations are impossible, and the extent to which nocturnal activity can be said migratory must be demonstrated in each case. Nothing is known about a possible feedback relating actually covered distances to migratory activity (and fat deposition) (Sections IV,C and V,A,1), which could result in differences, in principle, between caged and wild birds, and very little is known to what extent the recording of activity may be influenced by changes of the type of movement in caged birds (e.g., Kramer, 1949; Farner and Mewaldt, 1953a; Wagner, 1961). Recently, McMillan *et al.* (1970) have shown in the White-throated Sparrow (*Zonotrichia albicollis*) that spring migratory restlessness is established by a circadian rhythmicity as suggested earlier by Wagner (1956b).

Gwinner (in Aschoff, 1967) proposed a model of two coupled oscillators, one producing daytime activity, the other nocturnal activity. The exact interrelations between daytime and nighttime activity must be further investigated.

IV. Controlling Factors and Control Mechanisms of Migration

A. THE SOURCES OF INFORMATION

1. Definition of Proximate and Ultimate Factors

With respect to causation in bird migration, two groups of environmental factors, proximate and ultimate, may be recognized (e.g., Baker, 1938; Thomson, 1950; Farner, 1955). "Ultimate factors . . . give a survival value to the adaptation of the bird's cycle to that of the environment . . . ; proximate (or 'immediate') factors . . . provide the actual timing that brings adaptation into play" (Thomson, 1950). According to Farner (1955), the ultimate factors are those that have made the migratory pattern useful to a species and have allowed it to survive by exerting a strong positive selection of those individuals that developed a certain hereditary migratory pattern and a negative selective influence on those individuals that failed to do so. The proximate factors include those that twice annually bring the migratory bird into actual migration. Separate factors may be involved in the development of a migratory disposition (*Zugdisposition*) (Groebbels, 1928) (Section V,A) as well as for the development of migratory activity (*Zugstimmung*) (Groebbels, 1928), or should this categorization be an oversimplification (Verwey, 1949), for the development of an "annual stimulus for migration" (Farner, 1950). It should be observed, as Farner (1955) pointed out, that the distinction between proximate and ultimate factors probably can only be made with respect to typical migrants (Section IV,A,2). In these species, the proximate factors operate to initiate migration in advance of the seasonal occurrence of the ultimate factors; in weather migrants (Section IV,A,2), movements occur in direct response to factors that may be ultimate and proximate at once.

2. Climate, Weather, and Food Supply

Climatic conditions and food supply were the main forces in the evolution of migration (e.g., Schüz, 1971). In this section, the direct influences of weather conditions and food supply as proximate factors on actual migration will be discussed. As early as in the 13th century, the Emperor Friedrich II (1964) noticed that in spring starlings, thrushes, finches and other species migrate "following food supply and heat." Through innumerable field observations, many correlations among weather conditions, food supply, and migration became known including such specific factors as temperature; moisture; dry and rainy seasons; atmospheric pressure; atmospheric humidity (especially fog); cloudiness; direction and velocity of wind; blanket of snow; freezing; water level of rivers; flowering, fruit, and seed times of plants; migrations of ants and termites; and savannah fires (for reviews, see, e.g., Bagg et al., 1950; Williams, 1950; Steinbacher, 1951; Drost, 1960; Lack, 1960, 1963, 1968; Immelmann, 1963; Nisbet et al., 1963; Bellrose, 1967, 1971; Eastwood, 1967; Mueller and Berger, 1967; Graber, 1968; Nisbet and Drury, 1968; Williamson, 1969; Schüz, 1971; further, e.g., Keskpaik, 1965; Sick, 1967; Curtis, 1969; Evans, 1969, 1970a; Owen, 1969; Parslow, 1969; Bruderer, 1970; Eriksson, 1970; Netterstrøm, 1970; Orr, 1970; Thelle, 1970; Thiollay, 1970; Gauthreaux, 1971). Most striking examples are mass arrivals in spring with increased temperature in the front of cyclonics and in autumn with cold air from polar regions and shifts in wind directions from south to north in the northern hemisphere, further rush migrations and reverse migrations as reactions to winter conditions move in, the agreement of isepeptises and isotherms (for review, see, e.g., Schüz, 1971), and annual variations of migration times, covered distances, and the extent of migration in partial migrants related to fluctuations in weather conditions and food supply.

In addition to field observations, several workers experimenting with many different species (for reviews, see, e.g., Farner, 1955; Kendeigh et al., 1960; Helms, 1963; further, Merkel, 1966) have been able to initiate or enhance nocturnal restlessness in autumn by lowering the temperature or restricting food, or inhibited it by increasing temperature and/or food. In spring or even in late winter, released or enhanced restlessness and fat deposition could be recorded with increased temperature. Stolt (1969) observed increasing activity with falling air pressure and vice versa in caged Emberiza and Carduelis. The displacement of European Starlings (Sturnus vulgaris) to a favorable wintering area inhibited further migration and unfavorable surroundings induced prolongation of the journey (Perdeck, 1964). As a rule, close correlations among various weather conditions, changes in food supply and migration characterize less typical migrants—early arrivals at or late-departing short-distance migrants from the breeding ground, partial migrants, weather migrants (Weigold, 1924); außenweltbedingte Zugvögel (Putzig, 1938a); weniger ausgeprägte Zugvögel (Dorka, 1966); rush migrants. Such correlations are weak in typical migrants (Thomson, 1949)—late arriving, early departing long-distance migrants, instinct migrants (Weigold, 1924); innenweltbedingte Zugvögel (Putzig, 1938a); echte trekvogels (van Oordt, 1943, 1960); introverse trekvogels (Vleugel, 1948); migrateurs par excellence (Leclerqc and Delvingt, 1960); ausgeprägte Zugvögel (Dorka, 1966). It has long been well known that in typical migrants arrival at breeding grounds depends on a specific time in spring rather than on various weather and food conditions ("calendar birds") and that autumn migration occurs before these conditions worsen (e.g., von Homeyer, 1881; Gätke, 1891). Typical migrants also display migratory restlessness relatively independently of environmental temperature, the amount of food supply, etc. (e.g., Naumann, 1822; Schüz, 1952; Wagner, 1961). Concerning the extent of the influences of weather and food supply on actual migration, there is obviously an extended scale from extreme weather migrants to extreme instinct migrants (e.g., Schüz, 1952; Farner, 1955).

Unfortunately, from correlations of weather conditions, food supply, and migration, obtained by field observations, the actually effective factors rarely, if ever, become known because of the complexity of these factors (e.g., Bruderer, 1967). Even when these factors are experimentally revealed, it is presently difficult, if not impossible, to decide whether such factors play a modifying (inhibitory or accelerating), a primary stimulatory role, or exert a synchronizing effect (Section IV,A,4). Furthermore, it is difficult to be certain to what extent nocturnal restlessness released by experimental variations of environmental temperature, food supply, etc., can be called "migratory" (Section III). At present, a primary stimulatory effect of weather and/or food conditions on migration appears most evident only in facultative migrants (e.g., rush migrants). As we know from a few species (*Melospiza melodia,* Nice, 1937; *Mimus polyglottos,* Brackbill, 1956; see also Thomas, 1934; Kessel, 1953, for *Sturnus vulgaris*) even one and the same individual bird migrates in one year but not in the other. Displacements showed similar findings for populations as mentioned for individual birds. For example, English Mallards (*Anas platyrhynchos*), normally resident, migrated when hatched in the colder Finland and Prussia [in part without social attachment to native birds (Välikangas, 1933; Putzig, 1938c)]. The normally migrating Canada Goose (*Branta canadensis*), displaced to Europe, is still migrating in Sweden but stopped migration abruptly in the milder England (e.g., Ringleben, 1956). The Song Thrush (*Turdus philomelos*), among others, became resident when introduced to Australia and elsewhere (e.g., Wagner and Schildmacher, 1937). These observations obviously demonstrate a potential migratory instinct in all individual birds of a population, and this instinct only becomes effective with distinct meteorological and/or nutritional conditions (e.g., Schüz, 1952). Presently, a participation of genetic factors cannot be excluded (e.g., Lack, 1944; Milne and Robertson, 1965). Unfortunately, nothing is known about the migratory disposition (Section V,A) of individual birds with different migratory behavior in different years.

Whether irruptions (for reviews, see, e.g., Rudebeck, 1950; Svärdson,

1957; Ulfstrand, 1963; Schüz, 1971) in general are controlled primarily by food supply (e.g., starvation migration) (Höglund and Lansgren, 1968), population density and social factors (Section IV,B,5), or both (imbalance between food supply and population level) (Cornwallis and Townsend, 1968) is not yet clear. The entire field of meteorological and nutritional influences on actual migration requires further investigation, especially experimental work.

3. Photoperiod

Among all known environmental factors, seasonal changes in day length appear with the greatest regularity from year to year and thus bear the most precise relationship to the annual seasons. Because it is the most useful environmental source of predictive information, photoperiod long ago was assumed to play an important role in the control of bird migration (e.g., the poet Runeberg, 1874; von Homeyer, 1881; Seebohm, 1888). Rowan (since 1925) first demonstrated an influence of day length on actual bird migration. In his pioneer experiments (Section IV,B,2), Rowan subjected Slate-colored Juncos (*Junco h. hyemalis*) and Common Crows (*Corvus brachyrhynchos*) to an artificially increasing day length in winter and when subsequently released, these birds showed precocious migratory movements. These results were confirmed by Wolfson (1942) with several California subspecies of *Junco hyemalis*. Following the investigations of Rowan, many experiments utilizing caged light-treated birds of many different species showed that, at least in short- and middle-distance migrants, hyperphagia, body weight increase, and fat deposition (migratory disposition, Section V,A), as well as nocturnal restlessness, could be evoked already in late autumn, winter, or early spring (before the time of spring migration) with artificially prolonged day lengths (Fig. 2). Nonmigrants showed no comparable fat storage when subjected to long day lengths (for reviews, see, e.g., Farner, 1955, 1966; Dolnik and Blyumental, 1964; King and Farner, 1965). In alternating long and short day periods, Wolfson (1954) observed five periods of fat deposition (correlated with long days) within 369 days in Slate-colored Juncos. In general, the photoperiodically evoked phenomena in migrants closely resembled that of normal migration. Emlen (1969) has demonstrated that photoperiodically manipulated Indigo Buntings (*Passerina cyanea*) in autumnal conditions oriented southward, those in spring conditions, northward. Hence, photoperiod appears to be an important factor in the control of actual migration in many species. For Gambel's Sparrow (*Zonotrichia leucophrys gambelii*), in which the photoperiodic control of annual cycles has been most extensively investigated, Farner and King proposed the following model: In the vernal migratory period, fat deposition and migratory activity are direct photo-

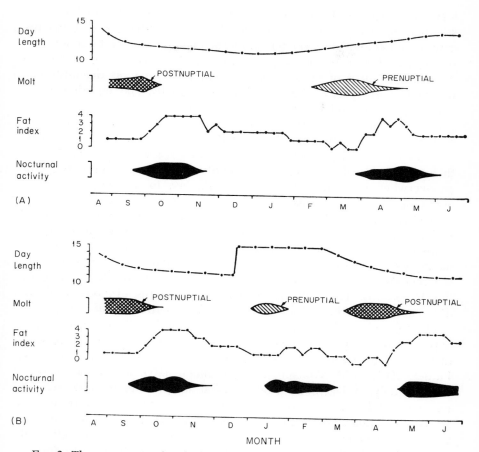

FIG. 2. The occurrence of molt, fat deposition, and nocturnal activity in *Passerina cyanea*. (A) Control birds on a natural photoperiod. (B) Experimental birds exposed to an accelerated photoperiod regimen. (From Emlen, 1969. Copyright 1969 by the American Association for the Advancement of Science.)

periodically induced by increasing vernal day length (direct long day effects); in the autumn migratory period both phenomena occur as indirect long-day effects. In the latter, the increasing vernal photoperiod acts as a remote timer. Thus, spring and autumn migration in this species are believed to be completely and exclusively controlled by photoperiod (for reviews, see King, 1963; Farner, 1964, 1966). Evans (1970a) pointed out that a remote timer set in spring could hardly be responsible for timing the migration of young birds in autumn. Recently, King (1968, 1970) observed in Gambel's Sparrow that cycles of fat deposition persist for at least one year with essentially normal frequency and intensity in conditions previously assumed to maintain an indefinite photorefractory state (LD 20:4) and in conditions customarily regarded as non-stimulatory (LD 8:16). At present, it must remain open, if in this case the photoregulation is effected by the annual initiation of an autonomous or continuously controlled sequence that is completed in one year or by entrainment of endogenous circannual rhythms, as demonstrated in some other species (Section IV,A,4). Actually, the question as to whether photoperiod in the control of migration in any species has a primary stimulatory effect also must remain open. It is not unlikely that photoperiod may generally exert its effect by entrainment of more or less distinct endogenous circannual rhythms (Section IV,A,4). Nevertheless, because of its temporal reliability, photoperiod certainly plays an overwhelming role, if only as a *Zeitgeber*, in the control of bird migration (for more details, see Section IV,A,4).

Wolfson (1947, 1970) proposed the hypothesis, newly discussed by Evans (1970a), that the summation of day lengths could be important in the control of migration in transequatorial migrants. Long-term experiments with Garden Warblers (*Sylvia borin*), kept three years in three different constant light–dark ratios (10:14, 12:12, 16:8) do not support this hypothesis. The different constant day lengths did not result in differences in the timing of migratory events (Berthold *et al.*, 1972a).

The physiological transduction of photoperiodic effects involves the hypothalamohypophyseal system and the use of photosensitive phases in the endogenous circadian rhythmicity of the bird (for reviews, see Farner, 1966, 1970; Wolfson, 1966; Hamner, 1968; among others).

Repeatedly, it has been proposed that in autumn, with decreasing environmental temperature, an increasing energy requirement for thermoregulation and decreasing opportunity for food intake cause a deterioration of energy balance because of shortening day length and consequent reduction in time available for daily food intake. In spring, on the other hand, with increasing environmental temperature, a reduction of caloric loss in thermoregulation, and with increasing opportunity for caloric intake, a lengthening photoperiod was assumed to take part in creating a more favorable metabolic situation because of an increasing daily period for food intake. In part, these metabolic situations were thought to lead directly to migratory activity in spring as a device for physiological dissipation of excess energy (productive energy) (Kendeigh, 1934; for reviews, see Weissmann, 1878; Farner, 1955; Kendeigh *et al.*, 1960; Merkel, 1966; among others). In Gambel's Sparrow, however, King (1961a) has demonstrated by the use of a fractionated photoperiod that light energy directly influences the metabolic status of this species and that increased feeding time appears to be insignificant. In the Canada Goose, Williams (1965) found that northward migration in spring does not involve a more favorable energy balance, so that factors other than energy economy must be responsible for the northward migration of this species. Farner (1955) drew attention to the fact that the course of expenditure of assumed excess or productive energy must be directed by specific mechanisms to migratory, molt, or reproductive energy. The exact role of photoperiod as a permissive factor in the control of migration is still unknown.

Finally, the influence of light intensity on bird migration requires mentioning. On one hand, field observations indicated intensified migration under a full moon, but on the other hand, they argued for the opposite (for review, see Gwinner, 1967a). In caged birds, low intensity of light is a necessary prerequisite for optimal display of migratory restlessness (e.g., Besserer and Drost, 1935; Wagner, 1961; Helms, 1963). Gwinner (1967a) observed a higher mean migratory restlessness in European Robins (*Erithacus rubecula*) and European Redstarts (*Phoenicurus phoenicurus*) during full moon phases than new moon phases. The completion of restlessness at night was strongly correlated with moonset. Brown and Mewaldt (1968) found in *Zonotrichia* in some birds a slight but inconsistent trend for more locomotor activity in moonlit nights and a significant positive correlation between both the height and brightness of the moon and the quantity of nocturnal activity. These observations indicate a modifying effect of light intensity on migratory activity. The same may be true in the observation of Wallraff (1966) that artificial stars exert a stimulatory effect on migratory restlessness (see, also, Merkel and Wiltschko, 1966).

4. Circannual Rhythmicity

Since the earliest days of research on problems of bird migration, internal timers were assumed in migratory birds. Von Pernau as far back as 1702 postulated a "hidden urge for migration" (in the bird itself) responsible for the onset of autumnal migratory activity. The assumption of endogenous control mechanisms in migration essentially was initiated and reinforced by three facts: first, typical migrants depart to their winter quarters long before adverse environmental conditions appear (Section IV,A,2); second, birds wintering in equatorial areas generally depart at the same time of the season, although the environmental factors at the winter quarters are fairly constant or fluctuate relatively irregularly; and third, the application of models on the control of migration to transequatorial migrants is difficult based solely on photoperiod (for reviews, see, e.g., Aschoff, 1955; van Oordt, 1959; Marshall, 1961; Gwinner, 1972a,b). Although not conclusive, the first experimental evidence for the control of migration by an endogenous circannual rhythmicity was that of Merkel (1963) on European Robins and Whitethroats (*Sylvia communis*), and that of Zimmerman (1966) on Dickcissels

(*Spiza americana*). The first and the most convincing evidence is that of Gwinner (1967b, 1968b; for reviews, see Farner, 1970; Gwinner, 1971a, 1972b) on Willow Warblers (*Phylloscopus trochilus*) and Wood Warblers (*Phylloscopus sibilatrix*) and that of Berthold *et al.* (1971b) on Garden Warblers and Blackcaps (*Sylvia atricapilla*).* Willow Warblers, Garden Warblers, and Blackcaps, hand-reared and kept under natural day length until summer, autumn, or winter but then transferred to constant conditions (light–dark ratio 10:14, 12:12, and 16:8) for 28–36 months, showed thereafter, at least in part and in so far as investigated, clear circannual cycles of nocturnal restlessness, changes in body weight, molt, and gonadal development (Fig. 3). Nocturnal restlessness and fat deposition occurred spontaneously and as a rule alternated with molt and gonadal development as expected. The same has been observed in Garden Warblers and Blackcaps already hand-reared in constant conditions (Berthold *et al.*, 1972a). The latter observation and the fact that many birds were transferred to constant conditions before 1 year old prove that the endogenous circannual rhythmicity is inborn.

In all warbler species investigated, the onset of fat deposition and the development of nocturnal restlessness depend on the date of birth of the birds, an observation also reported in part in the Scarlet Rosefinch (*Carpodacus erythrinus*) by Promptov (1949). On the other hand, as demonstrated for Garden Warbler and Blackcap, both phenomena are also influenced by the season. As a rule, fat deposition and nocturnal restlessness occur relatively sooner in late-born birds compared to early broods (Section V,C) (P. Berthold, E. Gwinner, and H. Klein, unpublished data).

Comparative studies on migratory restlessness under natural as well as constant artificial conditions for a number of closely related species that migrate different distances in the fall have shown that the long-distance migrants develop more restlessness than the short- or middle-distance migrants or almost-resident species (for review, see Gwinner, 1972a; Berthold *et al.*, 1972b; Berthold, 1973). Gwinner (1968b, 1972a) quantitatively compared the migratory restlessness displayed by caged Willow Warblers and Chiffchaffs (*Phylloscopus collybita*) during the first fall migration with the actual autumnal migratory activity of banded and recovered wild birds of both species. He calculated the theoretical distances traveled by the experimental birds during autumn migration and found that the birds would have reached their respective winter quarters under the assumption they had traveled along their normal migratory routes. In six species of the genus *Sylvia* with different migratory habits,

* The investigations of Bernhoft-Osa (1945) are not conclusive as misstated by Salomonsen (1967). The experimental *Sylvia atricapilla* was not kept under constant environmental conditions but in a living room.

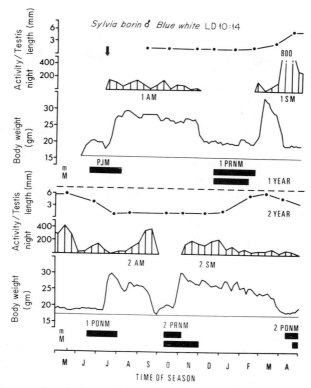

Fig. 3. Circannual periodicity of testis length, nocturnal restlessness, body weight changes and molt in a *Sylvia borin* in LD 10:14. The bird was born 29 May 1968 and transferred to constant conditions in July 1968 (↓). AM = autumn migration; SM = spring migration; PJM = postjuvenile molt; PRNM = prenuptial molt; PONM = postnuptial molt; m = molt of body feathers; M = molt of wing and tail feathers. Top: events in the first year. Bottom: events in the second year. (From Berthold *et al.*, 1971b.)

Berthold (1973) has demonstrated that the ratios of the amount of migratory activity of caged birds and the average migration journey of free-living conspecifics of these species correlate quite well. These results highly support the hypotheses first proposed by von Lucanus (1923) and Stresemann (1934) that a temporal program of migration is organized species-specifically in such a way that just enough migratory activity during the migratory season is produced as required to reach the goal migrating along a fixed route. Some displacement and detention experiments with several species (*Corvus corone cornix*, Rüppell and Schüz, 1948; *Sturnus vulgaris*, Perdeck, 1958; and *Anas discors*, Bellrose, 1958) are at least compatible with this hypothesis (Gwinner, 1972a). In the released birds in which migration time had more or less passed, the covered distances, in general, were not as long as normally. Evans (1968) criticized the hypothesis of a direction-plus-time basis in favor to that of a direction-plus-distance basis with respect to the ignorance of the effect of wind velocity on the speed of bird flight. At present, however, the hypothesis of an endogenous temporal mechanism for finding the winter quarters appears most convincing compared with those involving an innate knowledge of environmental features of the winter quarters or of the distance to travel (see, also, e.g., Sauer and Sauer, 1959; Wallraff, 1960a,b; Evans, 1966, 1968; for review, see Gwinner, 1972a). It must remain open as to whether the endogenous time mechanism is significantly different between typical and less typical migrants and in the first fall migration and in following migrations. Furthermore, its accuracy and the extent to which it is assisted by other mechanisms is not clear. *Sylvia* species, for example, exhibit longer restlessness under long day lengths and less under a short day length compared to natural day lengths (Berthold *et al.*, 1972b; Fig. 1). The recurrence of trans-Saharan and American trans-Gulf migrants in winter quarters (e.g., Elgood *et al.*, 1966; Nickell, 1968; Moreau, 1969a), for instance, obviously requires additional factors of spatial and perhaps also of temporal orientation besides an endogenous time mechanism.

In *Phylloscopus* (Gwinner, 1968a, 1972a) and possibly also in *Sylvia* (Berthold *et al.*, 1972b), the development, duration, total amount, and maximal values of migratory restlessness are subject to higher interindividual variability in less typical short- and middle-distance migrants than in more typical long-distance migrants. From these results, it was concluded that migration in the latter group is more rigidly endogenously controlled than in the former group. This assumption is supported by other experimental data as well as by field observations. The migratory restlessness can be affected much more by variations of environmental factors in less typical migrants than in typical migrants (Section IV,A,2). In typical migrants, the annual patterns of trapped and observed passage migrants are more symmetrical or left-skewed; in less typical migrants, they are more right-skewed. The former may best be interpreted as representative of an endogenous control mechanism of migration, the latter essentially as an effect of environmental factors. Only in less typical migrants do unfavorable environmental conditions, especially late in the season, cause mass migration resulting in right-skewed patterns (Berthold and Dorka, 1969; see, also, Preston, 1966). Furthermore, there is more variability in the diurnal patterns of migration throughout the season and in the annual patterns from year to year in less typical migrants than in typical migrants (e.g., Dorka, 1966; Berthold and Dorka, 1969).

According to Preston (1966), migratory birds are better synchronized in spring than in autumn. From the study of annual patterns of migration (Berthold and Dorka, 1969) as well as from several experiments and field observations (e.g., Curry-Lindahl, 1963; Merkel, 1963; Lofts, 1964; King, 1968, 1970; Schwab, 1971; for reviews, see Gwinner, 1971a, and Chapter 4), it must be concluded that the number of birds in which migration is controlled by endogenous circannual rhythmicity may be considerable. Much more information, however, is needed to decide the question whether this rhythmicity is basic in the control of bird migration.

Endogenous circannual rhythms, deviating more or less from the solar calendar, require that Zeitgeber synchronize the endogenous rhythms with environmental cycles by influencing the phase of the rhythm (e.g., Aschoff, 1958, 1965; Hoffmann, 1969). It has been suggested that seasonal changes in day length provide the predominating Zeitgeber for circannual rhythms with respect to the high reliability of photoperiod (e.g., Aschoff, 1955) (Section IV,A,3). Indeed, there is experimental evidence for this assumption (Gwinner, 1971a; Berthold et al., 1971a, 1972b). Gwinner (1971a) assumed that Phylloscopus react mainly to the continuous action of photoperiod and that the sensitivity to one and the same day length is different at different phases of the annual cycle. Results from Sylvia species indicate that the transition from one day length to another is important (differential effect, Berthold et al., 1972a). At present, the knowledge of the mechanisms of photoperiodic synchronization is speculative and the matter needs further investigation. In this connection, it should be pointed out that it appears not unlikely that in different populations of the same species (e.g., in Phylloscopus trochilus, Sylvia borin, Hirundo rustica), which breed in extreme northern latitudes and winter far south of the equator, or breed at lower latitudes and winter in equatorial areas, whole-year preferences of longer or shorter day lengths, respectively, may have evolved with respect to the mechanisms of synchronization. Population differences at the time of migration and in the covered distances not genetic in origin may be the result of interactions of circannual rhythms and different annual photoperiodic patterns. These problems and the question whether other environmental factors than photoperiod act as Zeitgeber on circannual rhythms producing migratory disposition and migratory activity (e.g., Immelmann, 1963) need further investigation.

B. COMPONENTS OF A CONTROL SYSTEM

1. Hypothalamus, Pituitary Gland, Pineal Gland

The pituitary as a fundamental regulatory organ in the endocrine system was often suggested to play a primary stimulatory role in bird migration (e.g., Rowan, 1946; for reviews, see Farner, 1955; Dorst, 1956; among others). Wolfson (1945) injected in *Junco hyemalis* subspecies gonadotropic preparations from pregnant mare serum, a chorionic gonadotropic preparation, and Antuitrin G (for the most part somatotropic hormone, but also containing small amounts of thyrotropic and gonadotropic hormones). Only in the latter case did fat deposition occur. Merkel (1937, 1938, 1940) and Putzig (1938c) observed, at least in some cases, a simulated migratory restlessness in several species treated with thyrotropic hormone preparations. For several reasons, including the small numbers of birds involved, the results obtained and their relationship to actual migration must be regarded as suggestive rather than conclusive (Farner, 1955). More recently, George and Naik (1964a) observed a degranulation of the thyrotrope cells of the pituitary in Rose-colored Starlings (*Sturnus roseus*) a few days prior to migration. In the same species, an increased output of adrenocorticotropic hormone prior to migration is assumed (Naik and George, 1963). In *Zonotrichia leucophrys gambelii* and *Zonotrichia albicollis,* prolactin experimentally administered, induces migratory behavior and fat deposition that resemble in magnitude and temporal pattern that of the premigratory periods. Gonadotropin is synergistic with prolactin in promoting an increase in lipid reserves and corticosterone in the development of migratory restlessness. In the domestic pigeon, prolactin markedly accelerates lipogenesis in the liver. Pituitary prolactin levels in wild *Zonotrichia leucophrys gambelii* are highest during migratory periods (Meier and Farner, 1964; for reviews, see Farner *et al.*, 1969; King, 1970). Consequently, prolactin and the anterior pituitary apparently play an important role in the control of bird migration. Since prolactin secretion in birds is controlled by a prolactin-releasing factor (PRF) of the hypothalamus (Assenmacher and Baylé, 1964; Kragt and Meites, 1965), a primary stimulatory role can be attributed to the hypothalamus rather than to the pituitary. Electron microscopic observations and estimates of enzyme activities indeed have established the importance of the hypothalamus in the physiological transduction of photoperiodic effects (e.g., Farner *et al.*, 1964). According to George and Naik (1956) and John and George (1967a) the release of neurosecretory material of the hypothalamus before migration may act as a trigger for migration. Numerous hypothalamic lesions in different species showed different effects on fattening and nocturnal restlessness (e.g., Wilson, 1965; Kuenzel and Helms, 1967, 1970; Farner *et al.*, 1969).

At present, it must remain open as to whether the "clock" of circannual rhythms (Section IV,A,4) is located in the hypothalamus. If circannual rhythmicity is based on a multioscillatory system, as suggested by the apparent dissociation of several normally coupled circannual rhythms (Gwinner, 1968b; Berthold *et al.*, 1972a), the "clock" may be located at different centers simultaneously. In this connection it is interesting that Underwood and Menaker (1970) produced some evidence that the avian pineal may be involved in the circadian periodicity in locomotor activity in lizards. On the other hand, pinealectomy causes no detectable changes in gonadal cycles in birds (e.g., Donham and Wilson, 1970). More information is needed to clarify the exact role of the centers discussed here in bird migration.

2. Gonads

As early as the eighteenth century, the gonads were thought to control bird migration (Jenner, 1824). The first evidence supporting such an assumption was that of Rowan's (1925, 1931) pioneer experiments. In Slate-colored Juncos and Common Crows subjected in winter to artificially increasing day length, the testes increased in size as in spring. When light-treated birds were released in midwinter, many disappeared and some moved in a direction typical of normal migration in spring, whereas controls with undeveloped gonads and light-treated castrates remained in the vicinity or moved in a direction typical of normal migration in autumn. These results were confirmed in part with other *Junco* races by Wolfson (1942). Rowan first proposed that spring migration is stimulated by hormones secreted by the developing gonads and that autumn migration is caused by gonad regression (for review, see Rowan, 1931; for revision in part, see Rowan, 1946). With respect to numerous parallelisms between gonadal development and migratory behavior in different species, races, populations, and intraspecific sex and age classes, several authors before and after Rowan suggested similar hypotheses. From numerous older experiments with injections and implantations of sex and gonadotropic hormones in different species, it was concluded that the gonads and their hormones play a primary stimulatory or at least a modifying role in bird migration (for reviews, see, e.g., Farner, 1955; Berthold, 1971).

In contrast to a simple hypothesis involving gonadal control of migration are several arguments and experimental data, including hypotheses based exclusively on correlations not necessarily involving cause-and-effect relationships (Farner, 1950). Briefly, some earlier experiments with injections of hormones are inconclusive because of the small numbers of birds involved and the use of unphysiological dosages of hormones (e.g., Thomson, 1936). Furthermore, castrates of several species migrated or displayed migratory behavior (for reviews, see, e.g., Putzig, 1939a; Farner, 1950; and below). There are also marked differences in migratory behavior in different species, races, populations, and even in siblings with similar gonadal cycles. In some species, development of migratory restlessness could be observed before essential development of the gonads (e.g., Putzig, 1938d; Farner and Mewaldt, 1953b; Farner, 1955). For many species, without any sex differences in migratory behavior, identical effects of androgens and estrogens should be assumed (Farner, 1950). According to Wolfson (1945) gonadotropic preparations failed to cause fat deposition. In several species, there is an inhibitory effect of migration on gonadal development (Berthold, 1969; for more details and for reviews, see, e.g., Farner, 1950, 1955; Berthold, 1971).

Highly instructive were investigations with caged castrates of several species (e.g., Millar, 1960; Lofts and Marshall, 1961; Morton and Mewaldt, 1962; King and Farner, 1963), recently reviewed and continued by Weise (1967). Castrations performed close to a migratory period had only a moderate, if any, effect on premigratory fat deposition and migratory behavior; castrations performed some months before a migratory period effectively suppressed the development of both phenomena in White-throated Sparrows (Fig. 4). Weise (1967) suggested that this suppression is effected by a reduction in prolactin release caused by lack of a negative feedback mechanism of gonadal hormones on the hypothalamohypophyseal system. Stetson and Erickson (1972) and Gwinner et al. (1971) also report on an eliminated fat deposition in castrated Gambel's Sparrows and in Golden-crowned Sparrows (*Zonotrichia atricapilla*) and Gambel's Sparrows in which testicular growth was prevented by reduction of extraocular light perception.

At present, the available evidence supports the view that gonads and sex hormones do not play a primary regulatory role in bird migration in general. However, the questions as to whether the gonads are directly or indirectly involved in a primary control of migration in some species and to what extent and how they exert a general effect on fat deposition (e.g., Rautenberg, 1957; Brenner, 1965; Lofts et al., 1966; Weise, 1967; Stetson and Erickson, 1972; Gernandt, 1969) and on migratory activity (e.g., Weise, 1967), as in White-throated Sparrows (Weise, 1967) require further investigation. Caged birds of several species that attained breeding conditions showed no summer restlessness (Section III). From these observations it was concluded that high blood titers of sex hormones may prevent migratory activity (e.g., Merkel, 1960; Helms, 1963; see, also, Wagner, 1956a; Creutz, 1961). This problem also needs further investigation.

3. Thyroid Gland

In his extensive investigations on the possible role of the thyroid gland in bird migration, Putzig (1937, 1938a) observed, as a rule, a great variability of thyroid activity in birds in a similar migratory state, even within the same species. In some species (e.g., limicoline birds, Putzig, 1938a; Rose-colored Starlings, George and Naik, 1964a), there are certain

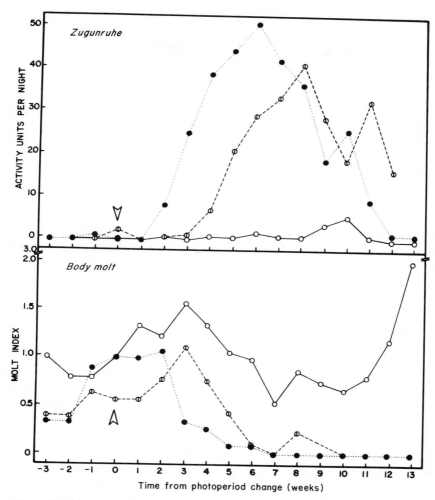

FIG. 4. Weekly means of migratory restlessness and body molt in prestimulus castration experiments in *Zonotrichia albicollis*. Arrows indicate date of change from short to long photoperiod. (From Weise, 1967.)

correlations between thyroid and migratory activity, but not in others (e.g., Laridae, Accipitridae, Putzig, 1937; and Gambel's Sparrows, Wilson and Farner, 1960). In different species, nocturnal restlessness could be stimulated, modified, or interrupted, and body weight could be increased or decreased by the application of thyroid hormone, thyroid extracts, and thyrotropic preparations and hormones (e.g., Wagner, 1930; Merkel, 1938, 1960; Putzig, 1938a; Schildmacher and Rautenberg, 1952; for review, see, e.g., Farner, 1955). But considering the small numbers of birds involved and the variability of results, even with similar dosages of hormones, the investigations reported can only be suggestive. Furthermore, according to Thomson (1936) and Farner (1955), the simulated restlessness may not necessarily represent true migratory behavior. From a critical viewpoint, the thyroid gland can hardly be assumed to play a primary stimulatory role in bird migration. To what extent thyroid hormone may be involved in part in the control of migratory activity and in the migratory fat metabolism (Schildmacher, 1963; George and Naik, 1964a; John and George, 1967b) for the present remains undecided.

4. Pancreas, Suprarenal Gland

According to George and Naik (1964b), premigratory fat deposition and the lower blood-glucose level just before migration in Rose-colored Starlings are attributed to the increased activity of the islet cells of the pancreas and increase in insulin production. However, in Gambel's Sparrows, no correlation between islet activity and the degree of migratory fat deposition could be found (Epple and Farner, 1967). According to Goodridge (1964), insulin does not accelerate lipid synthesis in Gambel's Sparrows during migratory periods. It was further hypothesized that the output of glucagon falls, allowing lipid synthesis to proceed at a more rapid rate during migratory periods. In Chaffinches (*Fringilla coelebs*), injections of small doses of insulin depressed locomotor activity, whereas those of hydrocortisone increased it abruptly. Both substances did not significantly influence fat deposition and feeding (Dolnik and Blyumental, 1967). In Rose-colored Starlings, White Wagtails (*Motacilla alba*), and Yellow Wagtails (*Motacilla flava*), an increase in adrenaline, noradrenaline, and corticoids, possibly augmenting muscular efficiency, was observed in the migratory period (Naik and George, 1963, 1964, 1965; John, 1966; John and George, 1967c). In thrushes, the injection of adrenaline was found not to affect nocturnal restlessness (Bergman, 1950). Meier *et al.* (1965) demonstrated that adrenocortical hormones are synergistic with prolactin in increasing nocturnal restlessness in Gambel's Sparrows. It should be added that in the Common Eider (*Somateria mollissima*), Gorman and Milne (1971) found heavy fat deposition (before egg-laying) at a time of increased interrenal activity and assumed thereby an induced hyperphagia, which in turn prevented the catabolic effects of high glucocorticoid production.

From the results reported, it must be concluded that both the pancreas and suprarenal gland are involved in the control of migration. Their exact role, however, is not yet clear. With respect to other glands and hormones it should be recalled that Rowan (1946) suggested that the entire physiology of the animal may be involved in the stimulation of migration.

5. Social Factors

Wallace (1876), ignorant of the periodicity and spontaneity of migratory restlessness (Section IV,A,4), still assumed that in caged birds it was only an expression of social excitement. Often, the typical call notes of nocturnal migrants were thought to be involved in the stimulation of migratory activity in birds in migratory condition, holding together migrating flocks and keeping and correcting the migratory direction (e.g., Hamilton, 1962; Drost, 1963; Gwinner, 1971b). In caged Bobolinks (*Dolichonyx oryzivorus*), Hamilton (1962) observed that nocturnal call notes, recorded and played back, increased the nocturnal restlessness of birds in migratory condition or stimulated them to fly into the lid of the apparatus. Hamilton suggested that the call notes of migrants aloft induce grounded birds to fly up. In some social migrants, e.g., White Stork (*Ciconia ciconia*), European Crane (*Grus grus*), hand-reared birds may fail to depart in autumn or to leave their caged parents. This is thought to be due to the lack of a social stimulus for migration (e.g., Schüz, 1971). On the other hand, hand-reared storks may develop migratory restlessness (Broekhuysen, 1970). Herring Gulls (*Larus argentatus*), normally resident in Britain, were extensively migratory when hatched and reared by migratory Lesser Black-backed Gulls (*Larus fuscus*) (Harris, 1970). Possibly, the young birds followed their foster parents. In the control of density-dependent invasions (Section IV,A,2), a psychologically acting pressure factor has been assumed (Berndt and Henss, 1967). The evidence available strengthens the opinion that social factors are somehow involved in the control or at least in the synchronization of migration, chiefly in the control of migratory activity, but their exact role is rather obscure.

C.   Relationships between Migratory Disposition and Migratory Activity

It is possible that either the development of migratory disposition may by itself culminate in the release of migratory behavior or that once a state of migratory disposition is attained, migratory behavior may be released by an additional stimulus (e.g., Farner, 1955). In several species, migratory restlessness develops in some individuals before the deposition of fat reserves (e.g., Putzig, 1939b; King and Farner, 1963; Wagner, 1963; Dolnik, 1970a). Lofts *et al.* (1963) demonstrated migratory restlessness in Bramblings (*Fringilla montifringilla*), in which premigratory fat deposition experimentally was prevented by starvation. A Garden Warbler displayed periodically migratory restlessness through 3 years under constant conditions without considerable changes in body weight (P. Berthold, E. Gwinner, and H. Klein, unpublished data). In *Phylloscopus* and *Sylvia* species, desynchronizations of the circannual rhythms of fat deposition and restlessness occurred (Gwinner, 1968b; Berthold *et al.*, 1972a). In displacement experiments (for reviews, see Schüz, 1952; Gwinner, 1971b; among others), birds of various species covered considerable distances immediately after release without specific "premigratory" fat depots. The evidence reported so far indicates that premigratory fat deposition is not a necessary prerequisite for the development of migratory activity. The mechanisms of development of both phenomena, therefore, obviously are independent as previously concluded by King and Farner (1963). However, despite the apparent independence in the initiation of fat deposition and migratory activity, each process influences the other. Commonly, intense migratory restlessness is correlated with large fat reserves and vice versa (e.g., Merkel, 1960; Engels, 1962; Weise, 1963; Gifford and Odum, 1965; Evans, 1968) (Fig. 5). According to West *et al.* (1968), migratory flocks of Lapland Longspurs (*Calcarius lapponicus*) probably are composed of fat birds in synchronous migratory condition. In the Chaffinch, among other species (Dolnik and Blyumental, 1967; Dolnik, 1970a; Czaja-Topinska, 1969), a wavelike alternation of migratory and feeding activity in the migratory period is typical and can be influenced by subcutaneous injections of fat. Stopover periods of lean birds during migration for replenishment of fat reserves are

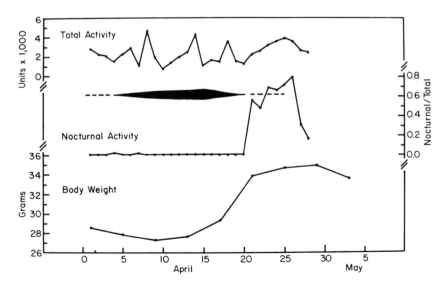

Fig. 5. Molt, activity, and body weight in a *Zonotrichia leucophrys gambelii* exposed outdoors to natural weather and photoperiod in spring. (From King and Farner, 1963.)

well known in many species, and they are obviously longer in lean than in fat birds (e.g., Stresemann, 1944; Dolnik and Blyumental, 1967; Page and Salvadori, 1969; Pearson, 1970; King, 1972).

If, as suggested (e.g., Vleugel, 1948; Schüz, 1952) certain species or populations of rush migrants may be in migratory condition throughout the winter, migratory activity, occurring irregularly and failing in some years, must be stimulated by additional factors. In such facultative and weather migrants (Section IV,A,2), the migratory disposition could be involved as a permissive factor.

V. Metabolic Physiology

A. THE MIGRATORY DISPOSITION

It is obvious now that each migratory period, as a rule, is preceded by or associated with the development of a specific physiological state leading finally to a general metabolic condition in which there is sufficient energy available for migration (for reviews, see Farner, 1955; Dolnik and Blyumental, 1964; King, 1972; among others). The most striking phenomenon of this migratory disposition (Groebbels, 1928) is the deposition of fat reserves that has long been well known to bird fanciers (e.g., Naumann, 1822). Fat metabolism and further typical features of the migratory disposition will be discussed in the following sections.

1. Fat Metabolism

Hyperlipogenesis and fat deposition associated with migration is now well established in more than forty families of birds (for reviews, see, e.g., Dolnik and Blyumental, 1964; King, 1972; further, Mascher, 1966; Fry et al., 1970; McNeil, 1970; Thomas 1970; Thomas and Dartnall 1970). It is world-wide in distribution, also occurring in intratropical migrants and summer migrants; exceptions are rare and scarcely investigated. Comparable fat deposition does not occur in resident species, races, and populations (nonmigrating individual birds of partial migrants?), but seems to be maintained at least in part in domestic geese (for reviews, see, e.g., Dolnik and Blyumental, 1964; King and Farner, 1965; King, 1972; further e.g., Wachs 1926; Helms and Smythe 1969; Fry, 1971). In some species, there are no sex, age, or racial differences in fat deposition—e.g., *Passerculus sandwichensis, Parus major, Spizella arborea* (respectively, Connell et al., 1960; Czaja-Topinska, 1969; Helms and Symthe, 1969)—but in others, males increase in weight sooner than females—e.g., *Zonotrichia albicollis* (Odum and Major, 1956)—or juveniles appear to gain weight more slowly or to a lesser extent than adults—e.g., *Dendroica striata, Turdus naumanni* (respectively, Nisbet *et al.*, 1963; Kuroda, 1964). Fattening by captive birds is often a good reflection of the event as it occurs in the free-living population, but there are exceptions. Sometimes an exaggeration of the magnitude of the reserves accumulated and/or a prolongation of the fattening period can be observed (e.g., Odum and Major, 1956; King and Farner, 1959; King, 1972; Gifford and Odum, 1965). The migratory fat reserves are stored as triglycerides in many parts of the body, the heart being an exception (e.g., Odum and Perkinson, 1951; Vallyathan and George, 1964). In Gambel's Sparrow, among others, cutaneous and subcutaneous adipose tissue account for 35–50% of the total carcass lipid. These lipids are stored in fifteen distinct fat organs and during the migratory fattening, coalesce into an essentially continuous sheath around the torso, exposing only the medial part of the pectoral muscles. The storage of fat in visceral depots also increases sharply during migratory fattening, and the level of liver and pectoral muscle lipids approximately doubles (for review, see, e.g., King and Farner, 1965). Adipose tissue in migratory birds may function on the basis of a "tank" principle, in which storage and mobilization involves only the addition or subtraction of dry lipid (Odum *et al.*, 1964), but see also Section V,A,4. It has generally been found in various species that the quantity of unsaturated fatty acids exceeds by about two to three times the quantity of saturated acids. Oleic, linoleic, and palmitic acids were the major components, linoleic acid ranging up to 39% of the total. Ostensibly, the fatty acid composition is similar in migratory and nonmigratory birds, as well as during migratory and nonmigratory periods, and relative composition of depot lipid is related to basic species differences and to varying dietary composition rather than to different stages of migratory fattening (for review, see, e.g., King and Farner, 1965; further, e.g., Hicks, 1967; Nakamura, 1969; Bower and Helms, 1968; Dolnik, 1970a).

As shown by several authors for several species, the principal mechanism in fat deposition, in some species obviously the only one, is an active hyperphagia. In Ortolan Buntings (*Emberiza hortulana*) (Wallgren, 1954), European Robins and Whitethroats (Merkel, 1958), and Gambel's Sparrows (King, 1961b), no decrease of the standard metabolic rate during fattening could be observed. However, Pohl (1971) found a lower rate of oxygen consumption during migratory periods compared with nonmigratory periods in Bramblings, but only in birds kept under quasinatural conditions outdoors and not under constant conditions. As a rule, the total daily locomotor activity is not consistently reduced during this period (Fig. 5), and the efficiency of digestion, absorption, and assimilation is stable or increases only slightly during fattening. These events, therefore, cannot in a substantial way spare calories for lipogenesis as formerly often assumed (for reviews, see, e.g., King and

Farner, 1965; King, 1970, 1972; further Zimmerman, 1965). Furthermore there is no important change in the daily feeding pattern during fattening periods, and increased daily feeding time is not a necessary prerequisite for fat deposition, at least in Gambel's Sparrow (King, 1961a; Morton, 1967; for review, see King, 1970). Fat deposition is also not affected by high loads of the blood parasite *Haemoproteus* (Ashford, 1971). However, according to Vallyathan and George (1964), Rosen (1966), George and Chandra-Bose (1967) and Dolnik *et al.* (1969), an increased expenditure of glycogen for muscular work in short flights and for thermoregulation could help to spare fat for fattening in the premigratory period in at least some species. During hyperphagia, food intake is increased up to about 40%, and completion of fat deposition in small passerines may require only 4–10 days with a fattening rate of about 0.1–1.5 gm/day. Typical during the migratory disposition is the ability to rapidly restore fat reserves when depleted in a matter of days. This suggests an adaptive hyperphagia and a negative feedback relationship between the level of stored fat and the rate of food intake. Among others, results obtained in experiments with lipid injections in migratory finches support this suggestion. The exact control of hyperphagia, however, and the question as to whether hypothalamic "hunger" and "satiety" systems are involved as in the rat (Leibowitz, 1970) are still unknown. For hypotheses on primary and secondary hyperphagia, see Farner *et al.* (1969); for reviews, see Dolnik and Blyumental (1964); King and Farner (1965); Farner *et al.* (1969); Dolnik (1970a); also, e.g., Dolnik (1968); Fry *et al.* (1970).

There is some evidence from field studies and experimental investigations that relatively short day lengths (in late summer and autumn and during overcast periods compared with cloudless periods) accelerate (autumnal) premigratory fat deposition (Gifford and Odum, 1965; P. Berthold, E. Gwinner, and H. Klein, unpublished data). That might be very useful especially for birds hatched very late in the season.

Obviously, depot fat is synthesized from carbohydrate, in some species possibly principally at night (in *Zonotrichia leucophrys gambelii* mostly during daytime; D. S. Farner, personal communication), and the major site for synthesis of fatty acids and for the primary regulation of lipogenesis in birds seems to be the liver (e.g., Fisher and Bartlett, 1957; Merkel, 1958; Rosen, 1966; George and Chandra-Bose, 1967; Evans, 1970a; for review, see Farner *et al.*, 1969). In migratory birds, fatty acids are directly used in the pectoral muscles (e.g., Farner *et al.*, 1961; George and Berger, 1966).

This section should be closed with an interesting calculation of Moreau and Dolp (1970) for small passerine birds, especially *Phoenicurus*: "If the weight of the feathers and the skeleton is deducted from the lean dry weight, there remains only 1.8–2.4 g of tissue that is physiologically active in the context of migration. This exiguous basis accumulates about five times its own weight in fat."

2. Carbohydrate Metabolism

During migratory fattening a considerable reduction in lean body mass found in several species has been interpreted as an adaptive reduction in ballast that reduces the energy cost of migratory flight (e.g., Dolnik, 1963; Dolnik and Blyumental, 1964, 1967; King et al., 1965). In addition, a reduction of protein and/or water content in body mass (Sections V,A,3 and 4), a diminished carbohydrate metabolism during migratory fattening (interruption or reduction of glycogen accumulation in the liver, pectoral muscles, and other muscles, to a mean maximal carbohydrate reserve in Zonotrichia, e.g., of only 0.2 kcal/bird) has been observed in about a dozen European species (e.g., Dolnik et al., 1965; Dolnik and Blyumental, 1967; Dolnik, 1970a) and in the North American Zonotrichia leucophrys gambelii (e.g., Farner et al., 1961). In Indian birds (Sturnus roseus, Motacilla flava, Naik, 1963; John and George, 1966) and in the central Asian Passer domesticus bactrianus (Dolnik, 1970b), however, a considerable increase in the glycogen level in that period has been described. Furthermore, in Zonotrichia leucophrys gambelii, the autumnal migrants have higher levels of hepatic and pectoral muscle glycogen than spring migrants. At present, it must remain open whether the different findings can be attributed to species differences or represent different states of the migratory disposition. According to Evans (1970a), the differences may be caused by variations in the extent of molt immediately before migration, and carbohydrate metabolism obviously predominates during molt (e.g., Rosen, 1966; Dolnik and Blyumental, 1967).

3. Protein Metabolism

In Sturnus roseus, Naik (1963) observed a reduction of liver protein content during premigratory fattening. In Passer domesticus bactrianus, accumulation of protein prior to migration has been described (Dolnik, 1971b). Evans (1969) and Fry et al. (1970) suggested that an increase in lean body weight during fattening in some Palaearctic migrants is caused by an increase in size of the flight muscles. Kuroda (1964) also assumed a slight weight gain in heart and lung in migrating Naumann's Thrushes (Turdus naumanni). In Common Redpolls (Acanthis flammea), Evans (1970b) suggested that muscle protein was used up during migratory flight. For the present, the data available do not permit any generalization concerning the protein metabolism in migratory birds.

4. Water Economy and Temperature Regulation

As in lean (dry) body weight (Section V,A,2 and 3), there are also contradictory findings with respect to the water content during migratory fattening. In several species, a fairly constant or relatively decreased water content (decrease in water per milligram fat tissue) has been observed (e.g., McGreal and Farner, 1956; Naik, 1963; Odum *et al.*, 1964; Helms and Smythe, 1969; Dolnik, 1970b). Some data even indicate the existence of a negative correlation between the fat index and the water index in migratory birds (McNeil, 1970). In other species, however, it was noted that water content increases during fattening (Fry *et al.*, 1970; Dolnik, 1970b). In the latter, a hypertrophy of the pectoral muscles was suggested to account entirely for the increase in water fractions (Fry *et al.*, 1970). Moreau and Dolp (1970) drew attention to the considerable variation of water content within individual species. Generalizations concerning the water content of migratory birds may become possible if individual organs are compared separately.

With regard to transoceanic and transdesert migratory flights that require migrants to fly nonstop for 40–60 hours (e.g., Moreau, 1961, 1966; Johnston and McFarlane, 1967), the important question arises whether fuel, water content, or both limit the flight range. By experimental investigations of three species (domestic pigeon, *Anas rubripes*, and *Melopsittacus undulatus*), it was shown that the respiratory water losses during flight exceed the metabolic water production considerably. The discrepancy will be even greater if there is also a substantial cutaneous water loss (for reviews, see Hart and Berger, 1972; Bartholomew, 1972). Hart and Berger, therefore, assume that dehydration in migratory birds could be as important in limiting flight range as the energy reserves (see also, e.g., Yapp, 1956). Evidence from field observations concerning this problem is scarce, but it should be noted that in American trans-Gulf migrants, for example, only a slight dehydration was found even though fat indices were extremely low (Johnston, 1968; Child, 1969; see, also, Nisbet *et al.*, 1963; Dolnik, 1969). According to Dolnik (1969), water loss during flight is not high, since only about 15% of heat is dispersed by respiratory evaporation. At present, it must remain open as to whether the results obtained by the experimental investigations of the three species mentioned above may be applied to typical migrants flying nonstop across oceans and deserts, or whether typical migrants may have evolved special means of ensuring water homeostasis or water sparing during long nonstop flights.

With respect to this problem, finally, attention must be drawn to the observations that water loss decreases with decreasing environmental temperature (*Anas rubripes, Melopsittacus undulatus,* Hart and Berger,

1972) and that body temperature in *Passer domesticus* at a simulated altitude of 6100 m decreases 2°C (Tucker, 1968). In this context, the interesting question arises as to whether high-altitude flights may help to spare water during long nonstop flights. Unfortunately, the preferred altitudes of migrants during long flights, particularly, when crossing deserts, are unknown, although there are some records of migrating birds at very high altitudes (e.g., Besson, 1969; Harrison, 1969). It is also assumed that nocturnal migration may have evolved in part to prevent excessive water loss and hyperthermy (e.g., Dorka, 1966). In the domestic pigeon, a control of the rate of heat loss in flight may involve vasomotor regulation combined with slight venting of the feathers to allow some through-passage of the air stream (Hart and Berger, 1972).

According to Turček (1966), migrants tend to have less plumage mass than nonmigrants. The middle-distance migrant *Sylvia atricapilla* travels to its relatively cold winter quarters with a first adult plumage that is heavier than in the long-distance migrant *Sylvia borin* on the way to its tropical winter quarters. This difference is interpreted chiefly in terms of adaptation to the climatic conditions of the winter quarters and to shorter and longer flights, respectively (Berthold and Berthold, 1971). Finally, in several taxa, resident birds have darker plumages than migrants and short-distance migrants darker than long-distance migrants (e.g., Black Wheatear, *Oenanthe leucura,* and other *Oenanthe* species; Black Redstart, *Phoenicurus ochruros* and European Redstart, *Phoenicurus phoenicurus*). This color difference may be interpreted as an adaptation to environmental temperatures experienced by each species in its winter quarters. It is well known that dark birds have a greater ability to absorb and use solar radiation for thermoregulatory purposes than white birds (Heppner, 1970; Lustick, 1971). Until more is known, discussions on water economy and temperature regulation in migratory birds must be speculative rather than conclusive.

5. Behavior

The migratory disposition is characterized by behavioral as well as physiological peculiarities. In many species, including solitary species, there is a more or less distinct tendency to flock together during premigratory periods and migration itself. Flocking behavior may result in a better orientation (e.g., Gwinner, 1971b), an easier discovery of food supplies (both supported by the use of special calls) (Section IV,B,5), and in the Hirundinidae may lead to a higher survival value of torpid birds. Furthermore, in birds in migratory disposition, there is a tendency to sleep more lightly (e.g., Farner, 1955), and the diel pattern of activity changes considerably. Most striking is the development of nocturnal activity in birds that normally are completely or almost exclusively active in daytime. Nocturnal migratory activity is often interpreted as a means of sparing daytime that can be used for feeding (e.g., Stresemann, 1934), of protecting migrants from raptors (for discussion, see Dorka, 1966), and of utilizing better wind conditions (e.g., Berthold, 1971) (for other explanations, see Section V,A,4). Furthermore in migratory disposition, there is often a reduction of the total daytime activity, a disappearance of the late afternoon peak of activity and the appearance of a characteristic pause in the evening (*Einschlafpause,* Bergman, 1941), and a cessation of feeding before the onset of nocturnal activity (e.g., Palmgren, 1949; King and Farner, 1965; Morton, 1967; Dolnik, 1970a; Klein *et al.,* 1971). These changes in the pattern of activity in part may help to spare energy. As Dorka (1966) pointed out, short-distance migrants often migrate in the early morning hours when feeding is most unprofitable, and partial migrants (e.g., rush migrants) may be able to start migration during the day as well as at night. Many birds must rest and feed in unusual habitats during migration. Parnell (1969), however, has shown that Parulidae during migration tend to feed and rest most often in habitats similar to those in which they nest.

In many insect feeders, there is a remarkable eating of such food as berries, fruit, olives, and seed before migration (e.g., Dolnik and Blyumental, 1964; Hintz and Dyer, 1970). Most interesting is the observation of Fry (1967) that the White-throated Bee-Eater (*Merops albicollis*) in its winter quarters includes the lipid-rich epicarp of the oil palm in its diet, a behavior possibly related to premigratory fattening. However, according to Odum and Major (1956), the amount of fat in the diet of *Zonotrichia* has little effect on fat deposition. The consumption of berries, fruit, etc., during fattening is believed to promote fat deposition by offering carbohydrates necessary for fattening (Section V,A,1) (see, e.g., Dolnik and Blyumental, 1964; Vallyathan and George, 1964; Orr, 1970). It is not known whether seasonal changes in the food in migrants are primarily controlled by the food supply or by endogenous factors.

B. ADAPTABILITY OF ENERGY RESERVES

1. Adaptations to Distances

Small nonmigrants and migrants not in migratory disposition have lipid contents of only about 3–5% of the live weight (e.g., Ward, 1964, 1969; King and Farner, 1965). Migratory fat deposition, then, is adaptive to the distances that must be covered; small long-distance migrants (e.g., hummingbirds, small passerines) that cross such ecological barriers as the Gulf of Mexico, the Mediterranean, and the Sahara in long nonstop flights deposit fat on the average up to 30–47% of the live weight (the maximum is slightly more than 50%). The fat depots in short- and middle-distance and short-flight migrants only average about 13–25% (e.g., Odum *et al.*, 1961; Ward, 1963, 1964; Dolnik and Blyumental, 1964; Nakamura and Ishizawa, 1965; King and Farner, 1965; King, 1972; Helms and Smythe, 1969; Moreau, 1969b; Moreau and Dolp, 1970; Fry *et al.*, 1970). Furthermore, the maximum fat reserves in long-distance migrants are obviously directly related to long nonstop flights. As shown for several species, long-distance migrants begin migration with low to moderate fat reserves that increase during migration; the maximum level is attained at or near points where long nonstop flights are undertaken (e.g., Odum *et al.*, 1961; Caldwell *et al.*, 1963; King and Farner, 1965). To begin migration with relatively low fat reserves, even before additional fat accumulation, seems not to be uncommon (e.g., Odum *et al.*, 1961; Johnston, 1964; King and Farner, 1965; Yarbrough and Johnston, 1965; Yarbrough, 1970; Helms and Smythe, 1969), but there are also species (e.g., *Elaenia parvirostris,* McNeil and Carrera de Itriago, 1968) that show a considerable increase in the amount of fat before the beginning of migration.

Finally, attention should be drawn to some interesting aspects of the interactions of body weight, fuel reserve, and flight velocity in migratory birds. In larger birds, the relative fat reserves are considerably less than in small birds. In larger limicoline birds, the average fat reserves do not exceed about 20% of the live weight even before long nonstop flights (e.g., Johnston and McFarlane, 1967; Pearson *et al.*, 1970; Thomas and Dartnall, 1970). According to Schaefer (1968), the power/weight ratio decreases slowly from smaller to larger birds, so that waders should be able to cross ecological barriers with only about half the proportional fat deposit of small birds. On the other hand, only birds weighing up to about 750 gm (fat free) may be able to double their fat-free mass by fat deposition and still have sufficient power to fly at the maximum-range speed (Pennycuick, 1969). With respect to this calculation, the advantage of fat deposition in comparison with carbohydrate accumulation is self-evident, since 1 gm of fat is equivalent to more than 9 kcal and 1 gm carbohydrate or protein only to about 4 kcal. Fat deposition during migration may allow, at least in smaller birds, sufficient energy storage without affecting prejudicially maximum-range speed. Some large birds, however, are able to minimize dependence on fuel reserves by thermal soaring (Pennycuick, 1969, 1970; King, 1972). According to Pennycuick, addition of fat (extra weight) in migratory birds increases the flight velocity, because the maximum-range velocity varies in direct proportion to the square root of body weight. Indeed, there is evidence at least in part from field observations that increased body weight due to fat deposition results in an acceleration in the velocity of migration (e.g., Blyumental, 1968; for review, see King, 1972). Hence, maximal fuel load in long-distance migrants not only has the advantage of providing sufficient energy but also allows the birds to cross unfavorable ecological barriers with relatively high velocity. For more details of the interactions of body weight, flight speed, altitude and wind velocity, see Pennycuick (1969, 1970), Blyumental and Dolnik (1970), King (1972), and Chapter 1.

2. Adaptations to Migratory Seasons

In a variety of short- and middle-distance migrants (e.g., *Zonotrichia, Erithacus rubecula, Fringilla coelebs*), lower energy reserves have been observed in the fall migratory period than in the spring migratory period (for reviews, see, e.g., King et al., 1963; King, 1972). This difference is probably correlated with weather and velocity of migration. The more extensive fat deposition in spring may serve as an emergency reserve during inclement weather and poor feeding conditions on the migratory route and within the breeding grounds, provides for relatively high caloric demands of the relatively rapid vernal migration, and permits a more rapid vernal migration by carrying a high fuel load (Section V,B,1) (e.g., King et al., 1963, 1965; King and Wales, 1965). In Gambel's Sparrows, the levels of hepatic and pectoral muscle glycogen are much higher in autumn than in spring, and the existence of at least quantitatively different metabolic adaptations for the accumulation of reserves during spring and fall migratory seasons are suggested (King et al., 1963). Late-arriving long distance migrants (e.g., Garden Warbler, P. Berthold, E. Gwinner, and H. Klein, unpublished data) may have already exhausted their fat reserves at arrival. McNeil (1969) showed that several species of shorebirds are fatter in autumn and suggests that this is associated with a migration route that is longer in autumn than in spring. There are further reports on different as well as equal amounts of fat reserves in spring and autumn migratory period (for review, see, e.g., Berthold, 1971; King, 1972). As many of the data on which these reports are based were obtained from captive birds or relatively small samples in part including different populations, general conclusions must remain speculative until more is known.

3. Energy Calculations

Physiologists have often tried to calculate flight ranges of migratory birds, partly to make more credible and intelligible the nonstop flights across oceans and deserts. Calculations based on the decline in body weight and fat content of birds killed at the same point at different times during one night; on the differences in weight, fat, and time of migrating birds examined at different geographic locations during a migratory season; on weight loss of caged birds displaying nocturnal restlessness; and on measurement of oxygen consumption and carbon dioxide production of experimental birds yielded following results. In small birds with fat reserves of about 40% of the live weight, the range of nonstop flights with an average consumption of about 0.1 kcal/gm/hour may exceed 2500 km and 100 hours; in shore birds, it may reach 10,000 km, and in hummingbirds, 1000 km (for reviews, see, e.g., Nisbet 1963, 1967; King and Farner, 1965; Bernis, 1966; Raveling and LeFebvre, 1967; Farner, 1970; Hart and Berger, 1972; Tucker, 1971; also, Lasiewski, 1962; Johnston and McFarlane, 1967; Helms, 1968; Hussell 1969). At present, the degree of accuracy of single calculations remains unknown (e.g., Hart and Berger, 1972), but considered together, the calculations are at least compatible with the evidence from field observations on flight range and energy consumption. According to these calculations, even very small birds with maximum fat reserves as a rule are easily capable of crossing the largest ecological barriers of the world and still have sufficient energy to fly hundreds of kilometers before reaching or after crossing a barrier (e.g., Odum *et al.*, 1964). On the winter grounds, some species arrive with large quantities of fat, while others have already exhausted all fat reserves (e.g., Rogers and Odum, 1966; McNeil and Carrera de Itriago, 1968; Nisbet, 1968). In order to obtain reliable calculations of maximum flight range in single species and individual birds, more investigations and more precise methods are necessary.

C. INTERACTIONS OF MIGRATION AND OTHER ANNUAL PROCESSES

The annual cycle of migratory birds is chiefly characterized by the sequence of three events: migration, reproduction, and molt. The caloric demands of all these processes are great, and therefore, the events must harmonize economically.

1. Migration and Molt

Intact wings are a necessary prerequisite at least for long-distance migration, and therefore, molting of wing and tail feathers and migration are usually mutually exclusive (Stresemann and Stresemann, 1966; for exception, see Traylor, 1970). With regard to the time available within the breeding grounds, different populations of the same species (e.g., *Charadrius hiaticula*) may molt wing and tail feathers before or after autumn migration. In other species, e.g., *Streptopelia turtur,* molt may be interrupted by migration (Stresemann and Stresemann, 1966). Molt of body feathers and migration more often occur simultaneously, but peak intensities of both events normally are not coincident (e.g., Berthold *et al.*, 1971a; Fig. 5). With respect to the interactions of molt and migration, some further general rules may be stated. In migratory species, races, populations, and individual birds(?) molt occurs earlier in the season and at an earlier age and is of shorter duration than in resident birds. The same holds true for late and early migrants, respectively. Probably, early migrants such as *Sylvia borin* begin acceleration of preparation for migration while still in the egg (Berthold *et al.*, 1971a). Molt begins earlier than usual when there is an early completion, or nonbreeding, or a failure in breeding, which may be adaptive to migration (e.g., Keast, 1968; Newton, 1968; Snow, 1969; Baggott, 1970). Furthermore, there are very interesting adaptations in birds hatched relatively late. It is well known that in several bird species molt begins at an earlier age and has a shorter duration in individuals born relatively late in the season (for review, see Berthold *et al.*, 1971a). In *Sylvia atricapilla* and *Sylvia borin*, it was found, as a rule, that juvenile and adult plumage, migratory disposition, and migratory restlessness develop relatively earlier and more rapidly the later the birds are hatched in the season. Furthermore, the later these birds are hatched, the greater is the extent of overlap of consecutive events. This calendar reaction permits birds hatched very late to be ready in time for migration (Berthold *et al.*, 1971a, 1972b, and unpublished data). On the other hand, *Sylvia atricapilla* hatched very late may not migrate as far as birds hatched earlier during autumn migration. Thus, they may be subjected to relatively shorter day lengths in their winter quarters, which may result in a further acceleration of annual events culminating in similar arrival times of both early and late hatched birds to the breeding grounds the following spring.

2. Migration and Reproduction

As an adaptation to the short time that they are present on the breeding grounds, late-arriving and early-departing typical migrants, as a rule, are single-brooded and show less autumnal courtship display than less typical migrants and resident birds (e.g., Kipp, 1943). Often, rapid gonadal development does not occur before the end of homing migration, and for several species, it was shown that the time of gonadal development is dependent at least in part on the time of migration (e.g., Lofts, 1962; for review, see Berthold, 1969). In the Rose-colored Starling, there is evidence for a suspension of spermatogenesis during migration, perhaps effected by an excessive release of TSH (Naik and George, 1964). Very late breeding may delay migration to some extent (e.g., Salomonsen, 1967), but it also may happen that young birds are deserted by their parents only when it is time to migrate. As shown in this short section, there is considerable evidence that molt and reproductive activity are adaptive to migration, particularly since neither overlap with migration.

VI. Concluding Remarks

The review of the literature concerning the physiology of bird migration presented so far on one hand illustrates the relatively large amount, on the other hand, the paucity of knowledge in different fields of investigation. Fat metabolism, for example, is relatively well understood in migratory birds, but for most other aspects, generalizations are rather suggestive than conclusive. It should be noted that our collective knowledge has been obtained from relatively few groups of birds, chiefly New World and Old World finches, warblers and thrushes, some starlings, and waders. The migratory physiology of the vast majority of migrants remains unknown, and even with respect to fat metabolism, only relatively little is known about nonpasserines and large birds. Furthermore, it should be recalled that it is not unlikely that migratory behavior may have evolved several or many times in the course of the evolution of modern birds and that we may actually be confronted with many cases of divergent and convergent evolution (e.g., Farner, 1955). Only with this concluding remark in mind should the generalizations and hypotheses presented be cautiously accepted.

ACKNOWLEDGMENTS

For very useful criticism and suggestions, and for polishing my English, I am very grateful to Professor Donald S. Farner, Seattle, and Professor James R. King and John D. Chilgren, Pullman.

REFERENCES

Aschoff, J. (1955). Jahresperiodik der Fortpflanzung bei Warmblütern. *Stud. Gen.* **8**, 742–776.

Aschoff, J. (1958). Tierische Periodik unter dem Einfluss von Zeitgebern. *Z. Tierpsychol.* **15**, 1–30.

Aschoff, J. (1965). Response curves in circadian periodicities. *In* "Circadian Clocks" (J. Aschoff, ed.), pp. 95–111. North-Holland Publ., Amsterdam.

Aschoff, J. (1967). Circadian rhythms in birds. *Proc. Int. Ornithol. Congr., 14th, 1966* pp. 81–105.

Ashford, R. W. (1971). Blood parasites and migratory fat at Lake Chad. *Ibis* **113**, 100–101.

Assenmacher, I., and Baylé, J. D. (1964). La sécrétion prolactinique du pigeon en réponse à différents traitements. *C.R. Soc. Biol.* **158**, 255.

Bagg, A. M., Gunn, W. W. H., Miller, D. S., Nichols, J. T., Smith, W., and Wolfarth, F. P. (1950). Barometric pressure-patterns and spring bird migration. *Wilson Bull.* **62**, 5–19.

Baggott, G. K. (1970). The timing of the moults of the Pied Wagtail. *Bird Study* **17**, 45–46.

Baker, J. R. (1938). The evolution of breeding seasons. *In* "Evolution: Essays on Aspects of Evolutionary Biology" (G. R. de Beer, ed.), pp. 161–177. Oxford, Univ. Press, London and New York.

Bartholomew, G. A. (1972). The water economy of seed-eating birds that survive without drinking. *Proc. Int. Ornithol. Congr., 15th, 1970* pp. 237–254.

Bellrose, F. C. (1958). The orientation of displaced waterfowl. *Wilson Bull.* **70**, 20–40.

Bellrose, F. C. (1967). Radar in orientation research. *Proc. Int. Ornithol. Congr., 14th, 1966* pp. 281–309.

Bellrose, F. C. (1971). The distribution of nocturnal migrants in the air space. *Auk* **88**, 397–424.

Bergman, G. (1941). Der Frühlingszug von *Clangula hyemalis* (L.) und *Oidemia nigra* (L.) bei Helsingfors. *Ornis Fenn.* **18**, 1–26.

Bergman, G. (1950). Zur Zugphysiologie der Drosseln (*Turdus pilaris* L., *T. merula* und *T. musicus* L.). Über Blutzuckerspiegel, Körpertemperatur und Einwirkung von Adrenalin. *Comment Biol.* **10**, 1–20.

Berndt, R., and Henss, M. (1967). Die Kohlmeise, *Parus major*, als Invasionsvogel. *Vogelwarte* **24**, 17–37.

Bernhoft-Osa, A. (1945). Bidrag til munkens—*Sylvia atricapilla* (L)—biologi i bur. *Stavanger Mus. Årsh.* pp. 130–137.

Bernis, F. (1966). "Migración en Aves." Soc. Española de Ornitol., Madrid.

Berthold, P. (1969). Über Populationsunterschiede im Gonadenzyklus europäischer *Sturnus vulgaris*, *Fringilla coelebs*, *Erithacus rubecula* und *Phylloscopus collybita* und deren Ursachen. *Zool. Jahrb. Syst. Oekol. Geogr. Tiere* **96**, 491–557.

Berthold, P. (1971). Physiologie des Vogelzugs. *In* "Grundriss der Vogelzugskunde" (E. Schüz, ed.), pp. 257–299. Parey, Berlin.

Berthold, P. (1973). Relationships between migratory restlessness and migration distance in six *Sylvia* species. *Ibis* **115**, 594–599.

Berthold, P., and Berthold, H. (1971). Über jahreszeitliche Änderungen der Kleingefiederquantität in Beziehung zum Winterquartier bei *Sylvia atricapilla* und *S. borin. Vogelwarte* **26**, 160–164.

Berthold, P., and Dorka, V. (1969). Vergleich und Deutung von jahreszeitlichen Wegzugs-Zugmustern ausgeprägter und weniger ausgeprägter Zugvögel. *Vogelwarte* **25**, 121–129.

Berthold, P., Gwinner, E., and Klein, H. (1971a). Vergleichende Untersuchung der Jugendentwicklung eines ausgeprägten Zugvogels, *Sylvia borin*, und eines weniger ausgeprägten Zugvogels, *S. atricapilla. Vogelwarte* **25**, 297–331.

Berthold, P., Gwinner, E., and Klein, H. (1971b). Circannuale Periodik bei Grasmücken. *Experientia* **27**, 399.

Berthold, P., Gwinner, E., and Klein, H. (1972a). Circannuale Periodik bei Grasmücken. I. Periodik des Körpergewichtes, der Mauser und der Nachtunruhe bei *Sylvia atricapilla* und *S. borin* unter verschiedenen konstanten Bedingungen. *J. Ornithol.* **113**, 170–190.

Berthold, P., Gwinner, E., Klein, H., and Westrich, P. (1972b). Beziehungen zwischen Zugunruhe und Zugablauf bei Garten- und Mönchsgrasmücke (*Sylvia borin* und *S. atricapilla*). *Z. Tierpsychol.* **30**, 26–35.

Besserer, J., and Drost, R. (1935). Ein Beitrag zum Kapitel "Vogelzug und Elektrizität." *Vogelzug* **6**, 1–5.

Besson, J. (1969). Migration de bergeronnettes printanières *Motacilla flava* à très haute altitude. *Oiseaux* **30**, 23.

Blyumental, T. I. (1968). Dependence of the flight velocity of some passerines on fat formation and the reserves of energy (in the light of the data for ringing recoveries). *Commun. Baltic Comm. Study Bird Migration* **5**, 146–154.

Blyumental, T. I., and Dolnik, V. R. (1970). Body weight, wing length, fat depots and flight in birds. *Zool. Zh.* **49**, 1069–1072.

Bower, E. B., and Helms, C. W. (1968). Seasonal variation in fatty acids of the Slate-colored Junco (*Junco hyemalis*). *Physiol. Zool.* **41**, 157–168.

Brackbill, H. (1956). Unstable migratory behavior in a Mockingbird. *Bird-Banding* **27**, 128.

Brehm, C. L. (1828). Der Zug der Vögel. *Isis* **21**, 912–922.

Brenner, F. J. (1965). The influence of testosterone on the pattern of fat deposition in birds. *J. Sci. Lab., Denison Univ.* **46**, 131–136.

Broekhuysen, G. J. (1970). White Storks breeding in the most southern part of the wintering quarters and the behaviour of their young. *Abstr. Int. Ornithol. Congr., 15th, 1970* p. 75.

Brown, I. L., and Mewaldt, L. R. (1968). Behavior of sparrows of the genus *Zonotrichia*, in orientation cages during the lunar cycle. *Z. Tierpsychol.* **25**, 668–700.

Bruderer, B. (1967). Zur Witterungsabhängigkeit des Herbstzuges im Jura. *Ornithol. Beob.* **64**, 57–90.

Bruderer, B. (1970). Radar studies on spring migration in northern Switzerland. *Abstr. Int. Ornithol. Congr., 15th, 1970* pp. 75–76.

Caldwell, L. D., Odum, E. P., and Marshall, S. G. (1963). Comparison of fat levels in migrating birds killed at a central Michigan and a Florida Gulf coast television tower. *Wilson Bull.* **75**, 428–434.

Child, G. I. (1969). A study of nonfat weights in migrating Swainsons's Thrushes (*Hylocichla ustulata*). *Auk* **86**, 327–338.

Connell, C. E., Odum, E. P., and Kale, H. (1960). Fat-free weights of birds. *Auk* **77**, 1–9.

Cornwallis, R. K., and Townsend, A. D. (1968). Waxwings in Britain and Europe during 1965/66. *Brit. Birds* **61**, 97–118.

Creutz, G. (1961). Nochmals: Freilassung von Bergfinken (*Fringilla montifringilla*) nach der Zugzeit. *Vogelwarte* **21**, 53–54.

Curry-Lindahl, K. (1963). Molt, body weights, gonadal development, and migration in *Motacilla flava*. *Proc. Int. Ornithol. Congr., 13th, 1962* pp. 960–973.

Curtis, S. G. (1969). Spring migration and weather at Madison, Wisconsin. *Wilson Bull.* **81**, 235–245.

Czaja-Topinska, J. (1969). Migration dynamics and changes in fat deposition in the Great Tit, *Parus major* L. *Acta. Ornithol., Warsz.* **11**, 357–378.

Delvingt, W. (1962). Régulation des migrations aviennes. *Gerfaut* **52**, 644–652.

Dolnik, V. R. (1963). Bioenergeticheskie adaptatsii k migratsii u vorobinykh. *Tezisy Dokl. Pyatoi Pribaltiiskoi Ornithol. Konf., Tartu, 1963* pp. 61–63.

Dolnik, V. R. (1968). The role of fat depot in metabolism regulation in birds behaviour during migration. *Zool. Zh.* **47**, 1205–1216.

Dolnik, V. R. (1969). Bioenergetics of the flying bird. *Zh. Obshch. Biol.* **30**, 273–291.

Dolnik, V. R. (1970a). Fat metabolism and bird migration. *Colloq. Int. Cent. Nat. Rech. Sci.* **172**, 351–364.

Dolnik, V. R. (1970b). Physiology of migratory state: A comparison of migratory and sedentary races. *Abstr. Int. Ornithol. Congr., 15th, 1970* pp. 14–15.

Dolnik, V. R., and Blyumental, T. I. (1964). The bioenergetics of bird migration. *Successes Mod. Biol.* **58**, 280–301.

Dolnik, V. R., and Blyumental, T. I. (1967). Autumnal premigratory and migratory periods in the Chaffinch (*Fringilla coelebs coelebs*) and some other temperate-zone birds. *Condor* **69**, 435–468.

Dolnik, V. R., Dobrynina, I. N., and Blyumental, T. I. (1965). Seasonal variations in the metabolism of carbohydrates and fats and their influence on the migration behaviour of the Chaffinch. *Commun. Baltic Comm. Study Bird Migration* **3**, 171–182.

Dolnik, V. R., Keskpaik, J., and Gavrilov, V. P. (1969). Bioenergetics of autumnal pre-migratory period of the Chaffinch. *Commun. Baltic Comm. Study Bird Migration* **6**, 125–146.

Donham, R. S., and Wilson, F. E. (1970). Photorefractoriness in pinealectomized Harris' Sparrows. *Condor* **72**, 101–102.

Dorka, V. (1966). Das jahres- und tageszeitliche Zugmuster von Kurz- und Langstreckenziehern nach Beobachtungen auf den Alpenpässen Cou/Bretolet. *Ornithol. Beob.* **63**, 165–223.

Dorst, J. (1956). "Les migrations des oiseaux." Payot, Paris.

Drost, R. (1960). Über den nächtlichen Zug auf Helgoland. *Proc. Int. Ornithol. Congr., 12th, 1958* pp. 178–192.

Drost, R. (1963). Zur Frage der Bedeutung nächtlicher Zugrufe. *Vogelwarte* **22**, 23–26.

Eastwood, E. (1967). "Radar Ornithology." Methuen, London.

Ekström, C. U. (1828). Zerstreute Bemerkungen über schwedische Zugvögel. *Isis* **21**, 696–707.

Elgood, J. H., Sharland, R. E., and Ward, P. (1966). Palaearctic migrants in Nigeria. *Ibis* **108**, 84–116.

Emlen, S. T. (1969). Bird migration: Influence of physiological state upon celestial orientation. *Science* **165**, 716–718.

Emlen, S. T., and Emlen, J. T. (1966). A technique for recording migratory orientation of captive birds. *Auk* **83**, 361–367.

Engels, W. L. (1962). Migratory restlessness in caged Bobolinks (*Dolichonyx oryzivorus*), a transequatorial migrant. *Biol. Bull.* **123**, 542–554.

Epple, A., and Farner, D. S. (1967). The pancreatic islets of the White-crowned Sparrow, *Zonotrichia leucophrys gambelii,* during its annual cycle and under experimental conditions. *Z. Zellforsch. Mikrosk. Anat.* **79**, 185–197.

Ericksson, K. (1970). Wintering and migration ecology of Siskin (*Carduelis spinus*) and Redpoll (*C. flammea*). *Abstr. Int. Ornithol. Congr., 15th, 1970* pp. 96–97.

Evans, P. R. (1966). An approach to the analysis of visible migration and a comparison with radar observation. *Ardea* **54**, 14–44.

Evans, P. R. (1968). Reorientation of night migrants after displacement by the wind. *Brit. Birds* **61**, 281–303.

Evans, P. R. (1969). Ecological aspects of migration, and pre-migratory fat deposition in Lesser Redpoll, *Carduelis flammea cabaret*. *Condor* **71**, 316–330.

Evans, P. R. (1970a). Timing mechanisms and the physiology of bird migration. *Sci. Progr.* (*London*) **58**, 263–275.

Evans, P. R. (1970b). The physiology and nutrition of Lesser Redpoll during moult and autumn migration. *Abstr. Int. Ornithol. Congr., 15th, 1970* pp. 97–98.

Eyster, M. B. (1952). Mechanically recorded nocturnal unrest in captive songbirds. *Abstr. Pap., A.O.U. 70th Stated Meet.* p. 3.

Farner, D. S. (1950). The annual stimulus for migration. *Condor* **52**, 104–122.

Farner, D. S. (1955). The annual stimulus for migration: Experimental and physiologic aspects. In "Recent Studies of Avian Biology" (A. Wolfson, ed.), pp. 198–237. Univ. of Illinois Press, Urbana.

Farner, D. S. (1964). The photoperiodic control of reproductive cycles in birds. Amer. Sci. 52, 137–156.

Farner, D. S. (1966). Über die photoperiodische Steuerung der Jahreszyklen bei Zugvögeln. Biol. Rundsch. 4, 228–241.

Farner, D. S. (1970). Predictive functions in the control of annual cycles. Environ. Res. 3, 119–131.

Farner, D. S., and Mewaldt, L. R. (1953a). The recording of diurnal activity patterns in caged birds. Bird-Banding 24, 55–65.

Farner, D. S., and Mewaldt, L. R. (1953b). The relative roles of diurnal periods of activity and diurnal photoperiod in gonadal activation in male Zonotrichia leucophrys gambelii (Nuttall). Experientia 9, 219–221.

Farner, D. S., Oksche, A., Kamemoto, F. I., King, J. R., and Cheney, H. E. (1961). A comparison of the effect of long daily photoperiods on the pattern of energy storage in migratory and non-migratory finches. Comp. Biochem. Physiol. 2, 125–142.

Farner, D. S., Kobayashi, H., Oksche, A., and Kawashima, S. (1964). Proteinase and acid-phosphatase activities in relation to the function of the hypothalamo-hypophysial neurosecretory systems of photostimulated and of dehydrated White-crowned Sparrows. Progr. Brain Res. 5, 147–156.

Farner, D. S., King, J. R., and Stetson, M. H. (1969). The control of fat metabolism in migratory birds. In "Progress in Endocrinology" (C. Gual, ed.), Int. Congr. Ser. No. 184, pp. 152–157. Excerpta Med. Found., Amsterdam.

Fisher, H. I., and Bartlett, L. M. (1957). Diurnal cycles in liver weights in birds. Condor 59, 364–372.

Friedrich II. (1964). "Über die Kunst mit Vögeln zu jagen." Insel-Verlag, Frankfurt am Main.

Fry, C. H. (1967). Lipid levels in an intra-tropical migrant. Ibis 109, 118–120.

Fry, C. H. (1971). Migration, moult and weights of birds in northern Guinea savanna in Nigeria and Ghana. Ostrich Suppl. 8, 239–263.

Fry, C. H., Ash, J. S., and Ferguson-Lees, I. J. (1970). Spring-weights of some palaearctic migrants at Lake Chad. Ibis 112, 58–82.

Gätke, H. (1891). "Die Vogelwarte Helgoland." Meyer, Braunschweig.

Gauthreaux, S. A. (1971). A radar and direct visual study of passerine spring migration in southern Louisiana. Auk 88, 343–365.

George, J. C., and Berger, A. J. (1966). "Avian Myology." Academic Press, New York.

George, J. C., and Chandra-Bose, D. A. (1967). Diurnal changes in glycogen and fat levels in the pectoralis of the migratory starling Sturnus roseus. Pavo 5, 1–8.

George, J. C., and Naik, D. V. (1956). The hypothalamo-hypophysial neurosecretory system of the migratory starling, Sturnus roseus. J. Anim. Morphol. Physiol. 12, 42–56.

George, J. C., and Naik, D. V. (1964a). Cyclic changes in the thyroid of the migratory starling, Sturnus roseus (Linnaeus). Pavo 2, 37–49.

George, J. C., and Naik, D. V. (1964b). Cyclic histological and histochemical changes in the pancreas in relation to blood glucose levels in the migratory starling, Sturnus roseus (Linnaeus). Pavo 2, 88–95.

Gernandt, D. (1969). Über den Energiehaushalt der Zugvögel. Ber. Offenbach. Ver. Naturk. 37, 19–23.

Gifford, C. E., and Odum, E. P. (1965). Bioenergetics of lipid deposition in the Bobolink, a transequatorial migrant. *Condor* 67, 383–403.

Goodridge, A. G. (1964). The effect of insulin, glucagon and prolactin on lipid synthesis and related metabolic activity in migratory and non-migratory finches. *Comp. Biochem. Physiol.* 13, 1–26.

Gorman, M. L., and Milne, H. (1971). Seasonal changes in the adrenal steroid tissue of the Common Eider *Somateria mollissima* and its relation to organic metabolism in normal and oil-polluted birds. *Ibis* 113, 218–228.

Graber, R. R. (1968). Nocturnal migration in Illinois — different point of view. *Wilson Bull.* 80, 36–71.

Groebbels, F. (1928). Zur Physiologie des Vogelzuges. *Verh. Ornithol. Ges. Bayern* 18, 44–74.

Gwinner, E. (1967a). Wirkung des Mondlichts auf die Nachtaktivität von Zugvögeln. *Experientia* 23, 227–238.

Gwinner, E. (1967b). Circannuale Periodik der Mauser und der Zugunruhe bei einem Vogel. *Naturwissenschaffen* 54, 447.

Gwinner, E. (1968a). Artspezifische Muster der Zugunruhe bei Laubsängern und ihre mögliche Bedeutung für die Beendigung des Zuges im Winterquartier. *Z. Tierpsychol.* 25, 843–853.

Gwinner, E. (1968b). Circannuale Periodik als Grundlage des jahreszeitlichen Funktionswandels bei Zugvögeln. Untersuchungen am Fitis (*Phylloscopus trochilus*) und am Waldlaubsänger (*P. sibilatrix*). *J. Ornithol.* 109, 70–95.

Gwinner, E. (1971a). A comparative study of circannual rhythms in warblers. *In* "Biochronometry" (M. Menaker, ed.), pp. 405–427. National Academy of Science, Washington.

Gwinner, E. (1971b). Orientierung. *In* "Grundriss der Vogelzugskunde" (E. Schüz, ed.), pp. 299–348. Parey, Berlin.

Gwinner, E. (1972a). Endogenous timing factors in bird migration. *In* "Animal Orientation and Navigation" (S. R. Galler, K. Schmidt-Koenig, G. J. Jacobs, and R. E. Belleville, eds.), pp. 321–338. NASA, Washington.

Gwinner, E. (1972b). Adaptive functions of circannual rhythms in warblers. *Proc. Int. Ornithol. Congr., 15th, 1970* pp. 218–236.

Gwinner, E., Turek, F., and Smith, S. D. (1971). Extraocular light perception in three photoperiodic responses of the White-crowned Sparrow (*Zonotrichia leucophrys*) and the Golden-crowned Sparrow (*Zonotrichia atricapilla*). *Z. Vergl. Physiol.* 75, 323–331.

Hamilton, W. J., III (1962). Evidence concerning the function of nocturnal call notes of migratory birds. *Condor* 64, 390–401.

Hamilton, W. J., III (1967). Temporal pattern of nocturnal activity of a caged Fox Sparrow. *Anim. Behav.* 15, 527–533.

Hamner, W. M. (1968). The photorefractory period of the House Finch. *Ecology* 49, 211–227.

Harris, M. P. (1970). Abnormal migration and hybridization of *Larus argentatus* and *L. fuscus* after interspecies fostering experiments. *Ibis* 112, 488–498.

Harrison, J. (1969). The altitude of a migrating Shoveler. *Bull. Brit. Ornithol. Club* 89, 72.

Hart, J. S., and Berger, M. (1972). Energetics, water economy and temperature regulation during flight. *Proc. Int. Ornithol. Congr., 15th, 1970* pp. 189–199.

Helms, C. W. (1963). Annual cycle and Zugunruhe in birds. *Proc. Int. Ornithol. Congr., 13th, 1962* pp. 925–939.

Helms, C. W. (1968). Food, fat, and feathers, *Amer. Zool.* 8, 151–167.

Helms, C. W., and Smythe, R. B. (1969). Variation in major body components of the Tree Sparrow (*Spizella arborea*) sampled within the winter range. *Wilson Bull.* 81, 280–292.

Heppner, F. (1970). The metabolic significance of differential absorption of radiant energy by black and white birds. *Condor* 72, 50–59.

Hicks, D. L. (1967). Adipose tissue composition and cell size in fall migratory thrushes (Turdidae). *Condor* 69, 387–399.

Hintz, J. V., and Dyer, M. I. (1970). Daily rhythm and seasonal change in the summer diet of adult Red-winged Blackbirds. *J. Wildl. Managem.* 34, 789–799.

Hoffman, K. (1969). Die relative Wirksamkeit von Zeitgebern. *Oecologia* 3, 184–206.

Höglund, N. H., and Lansgren, E. (1968). The Great Grey Owl and its prey in Sweden. *Viltrevy* 5, 364–421.

Hussell, D. J. T. (1969). Weight loss of birds during nocturnal migration. *Auk* 86, 75–83.

Immelmann, K. (1963). Tierische Jahresperiodik in ökologischer Sicht. *Zool. Jahrb. Syst. Oekol. Geogr. Tiere* 91, 91–200.

Jenner, G. C. (1824). Some observations on the migration of birds. *Phil. Trans. Roy. Soc. London* 1, 11–44.

John, T. M. (1966). A histochemical study of adrenal corticoids in the pre- and post-migratory phases in the migratory wagtails, *Motacilla alba* and *M. flava*. *Pavo* 4, 9–14.

John, T. M., and George, J. C. (1966). Seasonal variations in the glycogen and fat contents of the liver and the pectoralis muscle of migratory wagtails. *Pavo* 4, 58–64.

John, T. M., and George, J. C. (1967a). Cyclic histochemical changes in the hypo-thalamo-hypophyseal neurosecretory system of the migratory wagtails, *Motacilla alba* and *M. flava*. *J. Anim. Morphol. Physiol.* 14, 216–222.

John, T. M., and George, J. C. (1967b). Certain cyclic changes in the thyroid and parathyroid glands of migratory wagtails. *Pavo* 5, 19–28.

John, T. M., and George, J. C. (1967c). Seasonal variations in the levels of tyrosine and phenylalanine in the adrenal of the migratory starling, *Sturnus roseus*. *Pavo* 5, 47–51.

Johnston, D. W. (1964). Ecological aspects of lipid deposition in some postbreeding arctic birds. *Ecology* 45, 848–852.

Johnston, D. W. (1968). Body characteristics of Palm Warblers following an over-water flight. *Auk* 85, 13–18.

Johnston, D. W., and McFarlane, R. W. (1967). Migration and bioenergetics of flight in the Pacific Golden Plover. *Condor* 69, 156–168.

Keast, A. (1968). Moult in birds of the Australian dry country relative to rainfall and breeding. *J. Zool.* 155, 185–200.

Kendeigh, S. C. (1934). The role of environment in the life of birds. *Ecol. Monogr.* 4, 299–417.

Kendeigh, S. C., West, G. C., and Cox, G. W. (1960). Annual stimulus for spring migration in birds. *Anim. Behav.* 8, 180–185.

Keskpaik, J. (1965). The interpretation of the prolongation of the spring migration. *Commun. Baltic Comm. Study Bird Migration* 3, 163–168.

Kessel, B. (1953). Distribution and migration of the European Starling in northern America. *Condor* 55, 49–67.

King, J. R. (1961a). On the regulation of vernal premigratory fattening in the White-crowned Sparrow. *Physiol. Zool.* 34, 145–157.

King, J. R. (1961b). The bioenergetics of vernal premigratory fat deposition in the White-crowned Sparrow. *Condor* **63**, 128–142.

King, J. R. (1963). Autumnal migratory-fat deposition in the White-crowned Sparrow. *Proc. Int. Ornithol. Congr., 13th, 1962* pp. 940–949.

King, J. R. (1968). Cycles of fat deposition and molt in White-crowned Sparrows in constant environmental conditions. *Comp. Biochem. Physiol.* **24**, 827–837.

King, J. R. (1970). Photoregulation of food intake and fat metabolism in relation to avian sexual cycles. *Colloq. Int. Cent. Nat. Rech. Sci.* **172**, 365–385.

King, J. R. (1972). Adaptive periodic fat storage by birds. *Proc. Int. Ornithol. Congr., 15th, 1970* pp. 200–217.

King, J. R., and Farner, D. S. (1959). Premigratory changes in body weight and fat in wild and captive White-crowned Sparrows. *Condor* **61**, 315–324.

King, J. R., and Farner, D. S. (1963). The relationship of fat deposition to Zugunruhe and migration. *Condor* **65**, 200–223.

King, J. R., and Farner, D. S. (1965). Studies of fat deposition in migratory birds. *Ann. N. Y. Acad. Sci.* **131**, 422–440.

King, J. R., and Wales, E. E. (1965). Photoperiodic regulation of testicular metamorphosis and fat deposition in three taxa of Rosy Finches. *Physiol. Zool.* **38**, 49–68.

King, J. R., Barker, S., and Farner, D. S. (1963). A comparison of energy reserves during the autumnal and vernal migratory periods in the White-crowned Sparrow, *Zonotrichia leucophrys gambelii. Ecology* **44**, 513–521.

King, J. R., Farner, D. S., and Morton, M. L. (1965). The lipid reserves of White-crowned Sparrows on the breeding ground in central Alaska. *Auk* **82**, 236–252.

Kipp, F. (1943). Beziehungen zwischen dem Zug und der Brutbiologie der Vögel. *J. Ornithol.* **91**, 144–153.

Klein, H., Berthold, P., and Gwinner, E. (1971). Vergleichende Untersuchung der tageszeitlichen Aktivität gekäfigter und freilebender *Sylvia atricapilla* und S. *borin. Oecologia* **8**, 218–222.

Kragt, C. L., and Meites, J. (1965). Stimulation of pigeon pituitary prolactin release by pigeon hypothalamic extract in vitro. *Endocrinology* **76**, 1169.

Kramer, G. (1949). Über Richtungstendenzen bei der nächtlichen Zugunruhe gekäfigter Vögel. *In* "Ornithologie als biologische Wissenschaft" (E. Mayr and E. Schüz, eds.), pp. 269–283. Winter, Heidelberg.

Kuenzel, W. J., and Helms, C. W. (1967). Obesity produced in a migratory bird by hypothalamic lesions. *BioScience* **17**, 395.

Kuenzel, W. J., and Helms, C. W. (1970). Hyperphagia, polydipsia, and other effects of hypothalamic lesions in the White-throated Sparrow, *Zonotrichia albicollis. Condor* **72**, 66–75.

Kuroda, N. H. (1964). Analysis of variations by sex, age, and season of body weight, fat and some body parts in the Dusky Thrush, wintering in Japan: A preliminary study. *Misc. Rep. Yamashina Inst. Ornithol.* **4**, 90–104.

Lack, D. (1944). The problem of partial migration. *Brit. Birds* **37**, 122–130 and 143–150.

Lack, D. (1960). The influence of weather on passerine migration, a review. *Auk* **77**, 171–209.

Lack, D. (1963). Weather factors initiating migration. *Proc. Int. Ornithol. Congr., 13th, 1962* pp. 412–414.

Lack, D. (1968). Bird migration and natural selection. *Oikos* **19**, 1–9.

Lasiewski, R. C. (1962). The energetics of migrating hummingbirds. *Condor* **64**, 324.

Leclercq, J., and Delvingt, W. (1960). "Les migrations des oiseaux." Labor. Zool. Gen. Inst. Agron. Etat Gembloux, Gembloux.

Leibowitz, S. F. (1970). Hypothalamic β-adrenergic "satiety" system antagonizes an α-adrenergic "hunger" system in the rat. Nature (London) 226, 963–964.

Lofts, B. (1962). Cyclic changes in the interstitial and spermatogenetic tissue of migratory waders "wintering" in Africa. Proc. Zool. Soc. London 138, 405–412.

Lofts, B. (1964). Evidence of an autonomous reproductive rhythm in an equatorial bird (Quelea quelea). Nature (London) 201, 523–524.

Lofts, B., and Marshall, A. J. (1961). Zugunruhe activity in castrated Bramblings Fringilla montifringilla. Ibis 103, 189–194.

Lofts, B., Marshall, A. J., and Wolfson, A. (1963). The experimental demonstration of pre-migration activity in the absence of fat deposition in birds. Ibis 105, 99–105.

Lofts, B., Murton, R. K., and Westwood, N. J. (1966). Gonadal cycles and the evolution of breeding seasons in British Columbidae. J. Zool. 150, 249–272.

Lustick, S. (1971). Plumage color and energetics. Condor 73, 121–122.

McGreal, R. D., and Farner, D. S. (1956). Premigratory fat deposition in the Gambel White-crowned Sparrow: Some morphologic and chemical observations. Northwest Sci. 30, 12–23.

McMillan, J. P., Gauthreaux, S. A., and Helms, C. W. (1970). Spring migratory restlessness in caged birds: a circadian rhythm. BioScience 20, 1259–1260.

McNeil, R. (1969). La détermination du contenu lipidique et la capacité de vol chez quelques espèces d'oiseaux de rivage (Charadriidae et Scolopacidae). Can. J. Zool. 47, 525–536.

McNeil, R. (1970). Hivernage et estivage d'oiseaux aquatiques nord-Américains dans le nord-est du Venezuela (mue, accumulation de graisse, capacité de vol et routes de migration). Oiseau Rev. Fr. Ornithol. 40, 185–302.

McNeil, R., and Carrera de Itriago, M. (1968). Fat deposition in the Scissor-tailed Flycatcher (Muscivora t. tyrannus) and the Small-billed Elaenia (Elaenia parvirostris) during the austral migratory period in northern Venezuela. Can. J. Zool. 46, 123–128.

Marshall, A. J. (1961). Breeding seasons and migration. In "Biology and Comparative Physiology in Birds" (A. J. Marshall, ed.), Vol. 2, pp. 307–339. Academic Press, New York.

Mascher, J. W. (1966). Weight variations in resting Dunlins (Calidris a. alpina). Bird-Banding 37, 1–34.

Meier, A. H., and Farner, D. S. (1964). A possible endocrine basis for premigratory fattening in Zonotrichia leucophrys gambelii. Gen. Comp. Endocrinol. 4, 584–595.

Meier, A. H., Farner, D. S., and King, J. R. (1965). A possible endocrine basis for migratory behavior in Zonotrichia leucophrys gambelii. Anim. Behav. 13, 453–465.

Merkel, F. W. (1937). Zur Physiologie des Vogelzugtriebes. Zool. Anz. 117, 297–308.

Merkel, F. W. (1938). Zur Physiologie der Zugunruhe bei Vögeln. Ber. Ver. Schles. Ornithol. 25, 1–72.

Merkel, F. W. (1940). Neuere Untersuchungen über die Ursachen des Vogelzugtriebes. Natur Volk 70, 167–178.

Merkel, F. W. (1956). Untersuchungen über tages- und jahresperiodische Aktivitätsänderungen bei gekäfigten Zugvögeln. Z. Tierpsychol. 13, 278–301.

Merkel, F. W. (1958). Untersuchungen über tages- und jahresperiodische Änderungen im Energiehaushalt gekäfigter Zugvögel. Z. Vergl. Physiol. 41, 154–178.

Merkel, F. W. (1960). Stoffwechselvorgänge regeln den Wandertrieb der Zugvögel. Umschau 60, 243–246.

Merkel, F. W. (1963). Long-term effects of constant photoperiods on European Robins and Whitethroats. *Proc. Int. Ornithol. Congr., 13th, 1962* pp. 950–959.

Merkel, F. W. (1966). The sequence of events leading to migratory restlessness. *Ostrich Suppl.* 6, 239–248.

Merkel, F. W., and Wiltschko, W. (1966). Nächtliche Zugunruhe und Zugorientierung bei Kleinvögeln. *Zool. Anz. Suppl.* 29, 356–361.

Mewaldt, L. R., Kibby, S. S., and Morton, M. L. (1968). Comparative biology of Pacific coastal White-crowned Sparrows. *Condor* 70, 14–30.

Millar, J. B. (1960). Migratory behavior of the White-throated Sparrow, *Zonotrichia albicollis*, at Madison, Wisconsin. Ph.D. Dissertation, University of Wisconsin, Madison.

Milne, H., and Robertson, F. W. (1965). Polymorphismus in egg albumen protein and behaviour in the Eider Duck. *Nature (London)* 205, 367–369.

Moreau, R. E. (1961). Problems of Mediterranean-Saharan migration. *Ibis* 103a, 373–427 and 580–623.

Moreau, R. E. (1966). "The Bird Faunas of Africa and its Islands." Academic Press, New York.

Moreau, R. E. (1969a). The recurrence in winter quarters (Ortstreue) of trans-Saharan migrants. *Bird Study* 16, 108–110.

Moreau, R. E. (1969b). Comparative weights of trans-Saharan migrants at intermediate points. *Ibis* 111, 621–624.

Moreau, R. E., and Dolp, R. M. (1970). Fat, water, weights and wing-lengths of autumn migrants in transit on the northwest coast of Egypt. *Ibis* 112, 209–228.

Morton, M. L. (1967). Diurnal feeding patterns in White-crowned Sparrows, *Zonotrichia leucophrys gambelii. Condor* 69, 491–512.

Morton, M. L., and Mewaldt, L. R. (1962). Some effects of castration on a migratory sparrow (*Zonotrichia atricapilla*). *Physiol. Zool.* 35, 237–247.

Mueller, H. C., and Berger, D. D. (1967). Fall migration of Sharp-shinned Hawks. *Wilson Bull.* 79, 397–415.

Naik, D. V. (1963). Seasonal variation in the metabolites of the liver of the Rosy Pastor, *Sturnus roseus* (Linnaeus). *Pavo* 1, 44–47.

Naik, D. V., and George, J. C. (1963). Histochemical demonstration of increased corticoid level in the adrenal of *Sturnus roseus* (Linnaeus) towards the migratory phase. *Pavo* 1, 103–105.

Naik, D. V., and George, J. C. (1964). Certain cyclic histological changes in the testis of the migratory starling, *Sturnus roseus* (Linnaeus). *Pavo* 2, 48–54.

Naik, D. V., and George, J. C. (1965). Certain cyclic changes in the histology and histochemistry of the adrenal in the migratory starling, *Sturnus roseus* (Linnaeus). *Pavo* 3, 121–130.

Nakamura, T. (1969). Physiological and ecological studies on bird migration with special reference to body lipids. *Tori* 19, 87–108.

Nakamura, T., and Ishizawa, J. (1965). Studies on the migration of *Locustella fasciolata.* II. Duration of migration, flock formation and physiology. *Misc. Rep. Yamashina Inst. Ornithol.* 4, 217–220.

Naumann, J. A. (1822). "Naturgeschichte der Vögel Deutschlands." Fleischer, Leipzig.

Netterstrøm, B. (1970). Efterårstraekket af Islandsk Ryle (*Calidris canutus*) i Vestjylland. *Dan. Ornithol. Foren. Tidsskr.* 64, 223–228.

Newton, I. (1968). The moulting seasons of some finches and buntings. *Bird Study* 15, 84–92.

Nice, M. M. (1937). Studies in the life history of the Song Sparrow. *Trans. Linn. Soc. N. Y.* 4, 61–247.

Nickell, W. P. (1968). Return of northern migrants to tropical winter quarters and banded birds recovered in the United States. *Bird-Banding* **39**, 107–116.

Nisbet, I. C. T. (1963). Weight-loss during migration. Part II. Review of other estimates. *Bird-Banding* **34**, 139–159.

Nisbet, I. C. T. (1967). Aerodynamic theories of flight versus physiological theories. *Bird-Banding* **38**, 306–308.

Nisbet, I. C. T. (1968). Weights of birds in Malaya. *Ibis* **110**, 352–354.

Nisbet, I. C. T., and Drury, W. H. (1968). Short-term effects of weather on bird migration: A field study using multivariate statistics. *Anim. Behav.* **16**, 496–530.

Nisbet, I. C. T., Drury, W. H., and Baird, J. (1963). Weight loss during migration. Part I. Deposition and consumption of fat by the Blackpoll Warbler *Dendroica striata*. *Bird-Banding* **34**, 107–138.

Odum, E. P., and Major, J. C. (1956). The effect of diet on photoperiod-induced lipid deposition in the White-throated Sparrow. *Condor* **58**, 222–228.

Odum, E. P., and Perkinson, J. D. (1951). Relation of lipid metabolism to migration in birds: seasonal variation in body lipids of the migratory White-throated Sparrow. *Physiol. Zool.* **24**, 216–230.

Odum, E. P., Connell, C. E., and Stoddard, H. L. (1961). Flight energy and estimated flight ranges of some migratory birds. *Auk* **78**, 515–527.

Odum, E. P., Rogers, D. T., and Hicks, D. L. (1964). Homeostasis of the nonfat components of migrating birds. *Science* **143**, 1037–1039.

Orr, R. T. (1970). "Animals in Migration." MacMillan, New York.

Owen, D. F. (1969). The migration of the Yellow Wagtail from the equator. *Ardea* **57**, 77–85.

Page, G., and Salvadori, A. (1969). Weight changes of Semipalmated and Least Sandpipers pausing during autumn migration. *Ont. Bird-Banding* **5**, 52–58.

Palmén, J. A. (1876). "Über die Zugstrassen der Vögel." Engelmann, Leipzig.

Palmgren, P. (1944). Studien über die Tagesrhythmik gekäfigter Zugvögel. *Z. Tierpsychol.* **6**, 44–86.

Palmgren, P. (1949). On the diurnal rhythm of activity and rest in birds. *Ibis* **91**, 561–576.

Parnell, J. F. (1969). Habitat relations of the Parulidae during spring migration. *Auk* **86**, 505–521.

Parslow, J. (1969). The migration of passerine night migrants across the English channel studied by radar. *Ibis* **111**, 48–79.

Pearson, D. J. (1970). Weights of Red-backed Shrikes on autumn passage in Uganda. *Ibis* **112**, 114–115.

Pearson, D. J., Phillips, J. H., and Backhurst, G. C. (1970). Weights of some palaearctic waders wintering in Kenya. *Ibis* **112**, 199–208.

Pennycuick, C. J. (1969). The mechanics of bird migration. *Ibis* **111**, 525–556.

Pennycuick, C. J. (1970). Energetics of thermal soaring in birds. *Abstr. Int. Ornithol. Congr., 15th, 1970* p. 173.

Perdeck, A. C. (1958). Two types of orientation in migrating Starlings, *Sturnus vulgaris* L., and Chaffinches, *Fringilla coelebs* L., as revealed by displacement experiments. *Ardea* **46**, 1–37.

Perdeck, A. C. (1964). An experiment on the ending of autumn migration in Starlings. *Ardea* **52**, 133–139.

Pohl, H. (1971). Seasonal variation in metabolic functions of Bramblings. *Ibis* **113**, 185–193.

Preston, F. W. (1966). The mathematical representation of migration. *Ecology* **47**, 375–392.

Promptov, A. N. (1949). Sezonnye migratsii ptits kak biofiziologicheskaia problema. *Izv. Akad. Nauk SSSR, Ser. Biol.* pp. 30–39.

Putzig, P. (1937). Von der Beziehung des Zugablaufs zum Inkretdrüsensystem. *Vogelzug* 8, 116–130.

Putzig, P. (1938a). Der Frühwegzug des Kiebitzes. *J. Ornithol.* 86, 123–165.

Putzig, P. (1938b). Beobachtungen über Zugunruhe beim Rotkehlchen. *Vogelzug* 9, 10–14.

Putzig, P. (1938c). Über das Zugverhalten umgesiedelter englischer Stockenten (*Anas p. platyrhyncha*). *Vogelzug* 9, 139–145.

Putzig, P. (1938d). Die Triebkräfte des Vogelzugs. *Umschau* 42, 866–869.

Putzig, P. (1939a). Keimdrüsen und Heimzug. *Ber. Ver. Schles. Ornithol.* 24, 36–41.

Putzig, P. (1939b). Beiträge zur Stoffwechselphysiologie des Zugvogels. *Vogelzug* 10, 139–154.

Rautenberg, W. (1957). Vergleichende Untersuchungen über den Energiehaushalt des Bergfinken (*Fringilla montifringilla* L.) und des Haussperlings (*Passer domesticus* L.). *J. Ornithol.* 98, 36–64.

Raveling, D. G., and LeFebvre, E. A. (1967). Energy metabolism and theoretical flight range of birds. *Bird-Banding* 37, 97–113.

Ringleben, H. (1956). Kanadagänse (*Branta canadensis*) in Deutschland. *Ornithol. Mitt., Göttingen* 8, 185–187.

Rogers, D. T., and Odum, E. P. (1966). A study of autumnal postmigrant weights and vernal fattening of North American migrants in the tropics. *Wilson Bull.* 78, 415–433.

Rosen, E. (1966). A seasonal analysis of fatty acids in the Wood Thrush (*Hylocichla mustelina*). *Diss. Abstr. B* 27, 2200.

Rowan, W. (1925). Relation of light to bird migration and developmental changes. *Nature (London)* 115, 494–495.

Rowan, W. (1931). "The Riddle of Migration." Williams & Wilkins, Baltimore, Maryland.

Rowan, W. (1946). Experiments in bird migration. *Trans. Roy. Soc. Can., Sect.* 5, 3, 123–135.

Rudebeck, G. (1950). Studies on bird migration. *Vår Fågelvärld* Suppl. 1, 1–148.

Runeberg, J. L. (1874). Notes and news. *Academy* 6, 262.

Rüppell, W., and Schüz, E. (1948). Ergebnis der Verfrachtung von Nebelkrähen (*Corvus corone cornix*) während des Wegzuges. *Vogelwarte* 15, 30–36.

Salomonsen, F. (1967). "Fugletraekket og dets gåder." Munksgaard, Copenhagen.

Sauer, F., and Sauer, E. (1959). Nächtliche Zugorientierung europäischer Vögel in Südwestafrika. *Vogelwarte* 20, 4–31.

Schaefer, G. W. (1968). Energy requirement of migratory flight. *Ibis* 110, 413–414.

Schildmacher, H. (1963). Neuere Gesichtspunkte zur Physiologie des Vogelzuges. *Beitr. Vogelkunde* 9, 87–97.

Schildmacher, H., and Rautenberg, W. (1952). Untersuchungen zur hormonalen Regulierung des Fettwerdens der Zugvögel im Frühjahr. *Biol. Zentralbl.* 71, 397–405.

Schüz, E. (1952). "Vom Vogelzug." Schöps, Frankfurt am Main.

Schüz, E. (1971). "Grundriss der Vogelzugskunde." Parey, Berlin.

Schwab, R. G. (1971). Circannual testicular periodicity in the European Starling in the absence of photoperiodic change. *In* "Biochronometry" (M. Menaker, ed.), pp. 428–447. National Academy of Science, Washington.

Seebohm, H. (1888). "The geographical distribution of the family Charadriidae." Sotheran, London.

Sick, H. (1967). Hochwasserbedingte Vogelwanderungen in den neuweltlichen Tropen. *Vogelwarte* **24**, 1–6.

Smith, R. W., Brown, I. L., and Mewaldt, L. R. (1969). Annual activity patterns of caged non-migratory White-crowned Sparrows. *Wilson Bull.* **81**, 419–440.

Snow, D. W. (1969). The moult of British thrushes and chats. *Bird Study* **16**, 115–129.

Steinbacher, J. (1951). "Vogelzug." Kramer, Frankfurt am Main.

Stetson, M. H., and Erickson, J. E. (1972). Hormonal control of photoperiodically induced fat deposition in White-crowned Sparrows. *Gen. Comp. Endocrinol.* **19**, 355–362.

Stolt, B.-O. (1969). Temperature and air pressure experiments on activity in passerine birds with notes on seasonal and circadian rhythms. *Zool. Bidr. Uppsala* **38**, 175–231.

Stresemann, E. (1934). Aves. *In* "Handbuch der Zoologie" (T. Krumbach, ed.), Vol. 7. de Gruyter, Berlin.

Stresemann, E. (1944). Der zeitliche Ablauf des Frühjahrszuges beim Kappenammer, *Emberiza melanocephala* Scop. *Ornithol. Monatsber.* **52**, 85–92.

Stresemann, E. (1951). "Die Entwicklung der Ornithologie von Aristoteles bis zur Gegenwart." Peters, Berlin.

Stresemann, E., and Stresemann, V. (1966). Die Mauser der Vögel. *J. Ornithol.* **107**, Suppl.

Svärdson, G. (1957). The "invasion" type of bird migration. *Brit. Birds* **50**, 314–343.

Szymanski, J. S. (1914). Eine Methode zur Untersuchung der Ruhe- und Aktivitäts-perioden bei Tieren. *Pflügers Arch. Gesamte Physiol. Menschen Tiere* **158**, 343–385.

Thelle, T. (1970). Traekket af Strandskade (*Haematopus ostralegus*) fra Vestnorge til Vadehavet. *Dan. Ornithol. Foren. Tidsskr.* **64**, 229–247.

Thiollay, J. M. (1970). L'exploitation par les oiseaux des essaimages de Fourmis et Termites dans une zone de contact Savane-Forêt en Côte-d'Ivoire. *Alauda* **38**, 225–273.

Thomas, D. G. (1970). Wader migration across Australia. *Emu* **70**, 145–154.

Thomas, D. G., and Dartnall, A. J. (1970). Pre-migratory deposition of fat in the Red-necked Stint. *Emu* **70**, 87.

Thomas, E. S. (1934). A study of Starlings banded at Columbus, Ohio. *Bird-Banding* **5**, 118–128.

Thomson, A. L. (1936). Recent progress in the study of bird-migration: A review of the literature, 1926–1935. *Ibis* **6**, 472–530.

Thomson, A. L. (1949). "Bird Migration." Witherby, London.

Thomson, A. L. (1950). Factors determining the breeding seasons of birds. *Ibis* **92**, 173–184.

Traylor, M. A. (1970). Molt and migration in *Cinnyricinclus leucogaster*. *Abstr. Int. Ornithol. Congr., 15th, 1970* pp. 218–219.

Tucker, V. A. (1968). Respiratory physiology of House Sparrows in relation to high-altitude flight. *J. Exp. Biol.* **48**, 55–66.

Tucker, V. A. (1971). Flight energetics in birds. *Amer. Zool.* **11**, 115–124.

Turček, F. J. (1966). On plumage quantity in birds. *Ekol. Polska A* **14**, 617–633.

Ulfstrand, S. (1963). Ecological aspects of irruptive bird migration in northwestern Europe. *Proc. Int. Ornithol. Congr., 13th, 1962* pp. 780–794.

Underwood, H., and Menaker, M. (1970). Extraretinal light perception: Entrainment of the biological clock controlling lizard locomotor activity. *Science* **170**, 190–192.

Välikangas, I. (1933). Finnische Zugvögel aus englischen Eiern. *Vogelzug* **4**, 159–166.
Vallyathan, N. V., and George, J. C. (1964). Glycogen content and phosphorylase activity in the breast muscle of *Sturnus roseus*. *Pavo* **2**, 55–60.
van Oordt, G. J. (1943). "Vogeltrek." Brill, Leiden.
van Oordt, G. J. (1959). The reaction of the gonads to lengthening days in northern birds, migrating far beyond the equator. *Ostrich Suppl.* **2**, 342–345.
van Oordt, G. J. (1960). "Vogeltrek." Brill, Leiden.
Verwey, J. (1949). Migration in birds and fishes. *Bijdr. Dierk.* **28**, 477–504.
Vleugel, D. A. (1948). Enkele waarnemingen over "vorstflucht" en "randtrek" in het Sloe-Schengengebied tijdens de winters van 1935/1936 and 1936/1937. *Ardea* **36**, 143–162.
von Homeyer, E. F. (1881). "Die Wanderungen der Vögel mit Rücksicht auf die Züge der Säugethiere, Fische und Insecten." Grieben, Leipzig.
von Lucanus, F. (1923). "Die Rätsel des Vogelzuges." Beyer & Mann, Langensalza.
von Pernau, F. A. (1702). "Unterricht, Was mit dem lieblichen Geschöpff, denen Vögeln, auch ausser dem Fang, nur durch Ergründung deren Eigenschaften und Zahmmachung oder anderer Abrichtung man sich vor Lust und Zeitvertreib machen könne." (Cited from Stresemann, 1951.)
Wachs, H. (1926). Die Wanderungen der Vögel. *Ergeb. Biol.* **1**, 479–637.
Wagner, H. O. (1930). Über Jahres- und Tagesrhythmus bei Zugvögeln. *Z. Vergl. Physiol.* **12**, 703–724.
Wagner, H. O. (1956a). Die Bedeutung von Umweltfaktoren und Geschlechtshormonen für den Jahresrhythmus der Zugvögel. *Z. Vergl. Physiol.* **38**, 355–369.
Wagner, H. O. (1956b). Über Jahres- und Tagesrhythmus bei Zugvögeln. *Z. Tierpsychol.* **13**, 82–92.
Wagner, H. O. (1961). Beziehungen zwischen dem Keimdrüsenhormon Testosteron und dem Verhalten von Vögeln in Zugstimmung. *Z. Tierpsychol.* **18**, 302–319.
Wagner, H. O. (1963). Beziehungen zwischen Körpergewicht und Zugintensität der Vögel. *Z. Morphol. Ökol. Tiere* **53**, 152–165.
Wagner, H. O., and Schildmacher, H. (1937). Über die Abhängigkeit des Einsetzens der nächtlichen Zugunruhe verfrachteter Vögel von der geographischen Breite. *Vogelzug* **8**, 18–19.
Wallace, A. R. (1876). "Die geographische Verbreitung der Tiere." Meyer, Dresden.
Wallgren, H. (1954). Energy metabolism of two species of the genus *Emberiza* as correlated with distribution and migration. *Acta. Zool. Fenn.* **84**, 1–110.
Wallraff, H. G. (1960a). Können Grasmücken mithilfe des Sternenhimmels navigieren? *Z. Tierpsychol.* **17**, 165–177.
Wallraff, H. G. (1960b). Does celestial navigation exist in animals? *Cold Spring Harbor Symp. Quant. Biol.* **25**, 451–461.
Wallraff, H. G. (1966). Versuche zur Frage der gerichteten Nachtzugaktivität von gekäfigten Singvögeln. *Zool. Anz.* Suppl. **29**, 338–356.
Ward, P. (1963). Lipid levels in birds preparing to cross the Sahara. *Ibis* **105**, 109–111.
Ward, P. (1964). The fat reserves of Yellow Wagtails, *Motacilla flava*, wintering in southwest Nigeria. *Ibis* **106**, 370–375.
Ward, P. (1969). Seasonal and diurnal changes in the fat content of an equatorial bird. *Physiol. Zool.* **42**, 85–95.
Weigold, H. (1924). VII. Bericht der Vogelwarte der Staatl. Biologischen Anstalt auf Helgoland. *J. Ornithol.* **72**, 17–68.
Weise, C. M. (1963). Annual physiological cycles in capitve birds of differing migratory habits. *Proc. Int. Ornithol. Congr., 13th, 1962* pp. 983–993.

Weise, C. M. (1967). Castration and spring migration in the White-throated Sparrow. *Condor* **69**, 49–68.

Weissmann, A. (1878). Das Wandern der Vögel. *Samml. Gemeinverst. Wiss. Vortr.* H. 291.

West, G. C., Peyton, L. J., and Irving, L. (1968). Analysis of spring migration of Lapland Longspurs to Alaska. *Auk* **85**, 639–653.

Williams, G. G. (1950). Weather and spring migration. *Auk* **67**, 52–65.

Williams, J. E. (1965). Energy requirements of the Canada Goose in relation to distribution and migration. *Diss. Abstr.* **26**, 110.

Williamson, K. (1969). Weather systems and bird movements. *Quart. J. Roy. Meteorol. Soc.* **95**, 414–423.

Wilson, A. C., and Farner, D. S. (1960). The annual cycle of thyroid activity in White-crowned Sparrows of eastern Washington. *Condor* **62**, 414–425.

Wilson, F. E. (1965). The effects of hypothalamic lesions on the photoperiodic testicular response in White-crowned Sparrows. Thesis, Washington State University, Pullman.

Wolfson, A. (1942). Regulation of spring migration in juncos. *Condor* **44**, 237–263.

Wolfson, A. (1945). The role of the pituitary, fat deposition, and body weight in bird migration. *Condor* **47**, 95–127.

Wolfson, A. (1947). Fat deposition as a response to photoperiodism in migratory birds. *Anat. Rec.* **99**, 95.

Wolfson, A. (1954). Production of repeated gonadal, fat, and molt cycles within one year in the junco and White-crowned Sparrow by manipulation of day length. *J. Exp. Zool.* **125**, 353–376.

Wolfson, A. (1966). Environmental and neuroendocrine regulation of annual gonadal and migratory behavior in birds. *Recent Progr. Horm. Res.* **22**, 177–244.

Wolfson, A. (1970). Light and darkness and circadian rhythms in the regulation of the annual reproductive cycle in birds. *Colloq. Int. Cent. Nat. Rech. Sci.* **172**, 93–119.

Yapp, W. B. (1956). Two physiological considerations in bird migration. *Wilson Bull.* **68**, 312–319.

Yarbrough, C. G. (1970). Summer lipid levels of some subarctic birds. *Auk* **87**, 100–110.

Yarbrough, C. G., and Johnston, D. W. (1965). Lipid deposition in wintering and premigratory Myrtle Warblers. *Wilson Bull.* **77**, 175–191.

Zimmerman, J. L. (1965). Digestive efficiency and premigratory obesity in the Dickcissel. *Auk* **82**, 278–279.

Zimmerman, J. L. (1966). Effects of extended tropical photoperiod and temperature on the Dickcissel. *Condor* **68**, 377–387.

ADDITIONAL REFERENCES

Since the closing date of the manuscript for this chapter, July 1971, a considerable number of papers on the physiology of bird migration has been published. A selection of these—not discussed in this paper, but for information—is listed below.

Able, K. P. (1973). The role of weather variables and flight direction in determining the magnitude of nocturnal bird migration. *Ecology* **54**, 1031–1041.

Alerstam, T., Lindgren, A., Nilsson, S. G., and Ulfstrand, S. (1973). Nocturnal passerine migration and cold front passages in autumn—A combined radar and field study. *Ornis Scand.* **4**, 103–111.

Anonymous (1971). Ekologiceskie i fiziologiceskie aspekty pereletov ptic. *Nauka,* p. 244.

Berger, M., Hart, J. S., and Roy, O. Z. (1971). Respiratory water and heat loss of the Black Duck during flight and different temperatures. *Canad. J. Zool.* 49, 767–774.

Berthold, P. (1974a). Circannuale Periodik bei Grasmücken (*Sylvia*) III. Periodik der Mauser, der Nachtunruhe und des Körpergewichtes bei mediterranen Arten mit unterschiedlichem Zugverhalten. *J. Ornithol.* 115, 251–272.

Berthold, P. (1974b). In "Annual Biological Clocks" (E. T. Pengelley, ed.), pp. 55–94. Academic Press, New York.

Berthold, P., and Berthold, H. (1973). Jahreszeitliche Änderungen der Nahrungspräferenz und deren Bedeutung bei einem Zugvogel. *Naturwiss.* 60, 391–392.

Blyumental, T. I. (1973). In "Bird Migrations" (B. E. Bykhovskii, ed.), pp. 125–218. Wiley, New York.

Blyumental, T. I. (1974). Latitudinal geographical differences in the body weight of some passerines during summer and autumn movement. *Mater. Conf. Study Conserv. Migratory Birds Baltic Basin, Tallinn,* 1974 pp. 7–8.

Bolshakov, K. V. (1974). Social behaviour of birds during nocturnal migration. *Mater. Conf. Study Conserv. Migratory Birds Baltic Basin, Tallinn,* 1974 pp. 8–10.

Bolshakov, K. V., and Rezvyj, S. P. (1972). Predstartovoe i startovoe povedenie nekotorych vidov ptic s nocnym ritmom migracionnoj aktivnosti. *Tezisy Dokladov 8 Pribaltijskoj Ornitologiceskoj Konferencii, Tallinn,* 1972 pp. 22–24.

Bruderer, B. (1972). Zur Flughöhe ziehender Vögel. *Vögel der Heimat* 43, 27.

Bruderer, B. (1973). Fluggeschwindigkeit und Energiehaushalt ziehender Vögel. *Vögel der Heimat* 43, 134–136.

Caldwell, L. D. (1973). Fatty acids of migrating birds. *Comp. Biochem. Physiol.* 44B, 493–497.

Davydov, A. F. (1973). Sezonnye izmenenija energeticeskogo obmena i termoreguljacii u osedlych i migrirujuscich vorobinych ptic. *Ekologija* pp. 42–49.

Diamond, A. W., and Smith, R. W. (1973). Returns and survival of banded warblers wintering in Jamaica. *Bird-Banding* 44, 221–224.

Dingle, H., and Khamala, C. P. M. (1972). Seasonal changes in insect abundance and biomass in an East African grassland with reference to breeding and migration in birds. *Ardea* 60, 216–221.

Dobrynina, I. N. (1972). Sostojanie gipotalamo-gipofizarnoj nejrosekretornoj sistemy zjablika v period osennej migracii. *Tezisy Dokladov 8 Pribaltijskoj Ornitologiceskoj Konferencii, Tallinn,* 1972 pp. 36–37.

Dobrynina, I. N. (1974). Godovye cikly aktivnosti gipotalamiceskoj nejrosekretornoj sistemy migrirujuscego i osedlogo vidov ptic. *Zool. Zhurnal* 53, 96–103.

Dolnik, V. R. (1971a). Sravnenie energeticeskikh raskhodov na migratsiyu i zimovky u ptits. *Ekologija* 2, 88–89.

Dolnik, V. R. (1971b). Endokrinnaya sistema i sezonnye biologicheskie yavleniya u ptits. *Uspekhi Sovremennoi Biol.* 71, 412–427.

Dolnik, V. R. (1971c). Cirkade ritmy lokomotornoj aktivnosti pitanija i potreblenija kisloroda u zjablika (*Fringilla coelebs*) v migracionnyj period. *Zool. Zhurnal* 50, 1835–1842.

Dolnik, V. R. (1972a). Izmenenie sostava zirovogo tela pri migracionnom otlozenii zira u *Fringilla coelebs. Dokl. Akad. Nauk SSSR* 206, 247–249.

Dolnik, V. R. (1972b). In "Productivity, Population Dynamics and Systematics of Granivorous Birds" (S. C. Kendeigh and J. Pinowski, eds.), pp. 103–109. Polish Sci. Publ., Warschau.

Dolnik, V. R. (1973). Diurnal and seasonal cycles of the blood sugar in sedentary and migrating birds. *Zool. Zhurnal* 52, 94–103.

Dolnik, V. R. (1974a). The study of the physiological basis of the wavelike migration of some passerines through the Baltic area. *Mater. Conf. Study Conserv. Migratory Birds Baltic Basin, Tallinn,* 1974. pp. 19–21.

Dolnik, V. R. (1974b). Annual cycles of migratory fat deposition, sexual activity and moult in Chaffinches (*Fringilla coelebs*) under constant photoperiodic conditions. *Zhurnal Obscei Biol.* **35**, 543.

Dolnik, V. R., Gavrilov, V. P., and Dyachenko, V. P. (1974). Bioenergetics and regulation of autumnal premigratory period in Chaffinch (*Fringilla coelebs coelebs* L.). *Trudy Zool. Inst. Akad. Nauk SSSR, Leningrad* **55**, 62–100.

Dowsett, R. J., and Fry, C. H. (1971). Weight losses of trans-Sahara migrants. *Ibis* **113**, 531–533.

Eriksson, K. (1971). Irruption and wintering ecology of the great Spotted Woodpecker *Dendrocopos major. Ornis Fennica* **48**, 69–76.

Farner, D. S. (1970). Some glimpses of comparative avian physiology. *Fed. Proc.* **29**, 1649–1663.

Fogden, M. P. L. (1972). Premigratory dehydration in the Reed Warbler *Acrocephalus scirpaceus* and water as a factor limiting range. *Ibis* **114**, 548–552.

Forsyth, B. J., and James, D. (1971). Springtime movements of transient nocturnally migrating land birds in the Gulf coastal bend region of Texas. *Condor* **73**, 193–207.

Fry, C. H., Ferguson-Lees, I. J., and Dowsett, R. J. (1972). Flight muscle hypertrophy and ecophysiological variation of Yellow Wagtail *Motacilla flava* races at Lake Chad. *J. Zool.* **167**, 293–306.

Gauthreaux, S. A. (1971). Nocturnal songbird migration. *Nature (London)* **230**, 580.

Gauthreaux, S. A. (1972). Behavioral responses of migrant birds to daylight and darkness: A radar and direct visual study. *Wilson Bull.* **84**, 136–148.

Gustafson, T., Lindkvist, B., and Kristiansson, K. (1973). New method for measuring the flight altitude of birds. *Nature (London)* **244**, 112–113.

Gwinner, E. (1974). Endogenous temporal control of migratory restlessness in warblers. *Naturwiss.* **61**, 405.

Hebrard, J. J. (1971). The nightly initiation of passerine migration in spring: a direct visual study. *Ibis* **113**, 8–18.

Högstedt, G., and Persson, Ch. (1971). Phänologie und Überwinterung der über Falsterbo ziehenden Rotkehlchen (*Erithacus rubecula*). *Vogelwarte* **26**, 86–98.

Hummel, D. (1973). Die Leistungersparnis beim Verbandsflug. *J. Ornithol.* **114**, 259–282.

Hyytiä, K., and Vikberg, P. (1973). Autumn migration and moult of the Spotted Flycatcher *Muscicapa striata* and the Pied Flycatcher *Ficedula hypoleuca* at the Signilskär bird station. *Ornis Fennica* **50**, 134–143.

Jablonkevic, M. L. (1972a). Sostav zirnych kislot podkoznych zirovych depo zjablika v migracionnyj period. *Tezisy Dokladov 8 Pribaltijskoj Ornitologiceskoj Konferencii, Tallinn,* 1972 pp. 112–113.

Jablonkevic, M. L. (1972b). Sostavzirnych kislot v podkoznych zirovych depo u migrirujuscich i nemigrirujuscich ptic. *Doklady Akad. Nauk SSSR* **206**, 1465–1467.

Jablonkevic, M. L. (1974). Diurnal and seasonal cycles of plasma-free fatty acids in migrating and sedentary birds. *Zool. Zhurnal* **53**, 87–95.

Johnston, D. W. (1973). Cytological and chemical adaptations of fat deposition in migratory birds. *Condor* **75**, 108–113.

King, J. R., and Farner, D. S. (1974). *In* "Chronobiology" (L. E. Scheving, F. Halberg, and J. E. Pauly, eds.), pp. 625–629. Igaku Shoin Ltd., Tokyo.

Lewis, R. A., and Farner, D. S. (1973). Temperature modulation of photoperiodically induced vernal phenomena in White-crowned Sparrows (*Zonotrichia leucophrys*). *Condor* **75**, 279–286.

Lyuleeva, D. S. (1973). *In* "Bird Migrations" (B. E. Bykhovskii, ed.), pp. 219–272. Wiley, New York.

Lyuleeva, D. S. (1974). Annual rhythm of the migrational movements of the Swift and the House Martin. *Mater. Conf. Study Conserv. Migratory Birds Baltic Basin, Tallinn,* 1974 pp. 45–46.

McNeil, R., and Cadieux, F. (1972). Fat content and flight-range capabilities of some adult spring and fall migrant North American shore birds in relation to migration routes on the Atlantic coast. *Naturaliste Canad.* **99**, 589–606.

Moreau, R. E. (1972). "The Palaearctic-African bird migration systems." Academic Press, New York.

Morton, M. L., and Liebman, H. A. (1974). Seasonal variations in fatty acids of a migratory bird with and without a controlled diet. *Comp. Biochem. Physiol.* **48**, 329–335.

Newton, I. (1970). *In* "Animal Populations in Relation to Their Food Resources" (A. Watson, ed.). Symposium No. 10. pp. 337–353. British Ecological Society, London.

Oniki, Y. (1971). Wandering interspecific flocks in relation to ant-following birds at Belém, Brazil. *Condor* **73**, 372–374.

Page, G., and Middleton, A. L. A. (1972). Fat deposition during autumn migration in the semipalmated Sandpiper. *Bird-Banding* **43**, 85–96.

Petersen, F. D. (1972). Weight-changes at Hesselø in night migrating passerines due to time of day, season and environmental factors. *Dansk Ornithol. Foren. Tidsskr.* **66**, 97–107.

Pilo, B., and George, J. C. (1970). Serum tyrosine level as an index of thyroidal activity in a migratory starling. *J. Anim. Morphol. Physiol.* **17**, 26–36.

Postnikov, S. N. (1970). Geograficeskie razlicipa v ernergeticeskich rezervach i chode osennej migracii obyknovennoj cecetki. *Uc. zap. Perm. Gos. Ped. Int.* **99**, 107–111.

Rabøl, J., and Petersen, F. D. (1973). Lengths of resting time in various night-migrating passerines at Hesselø, Southern Kattegat, Denmark. *Ornis Scand.* **4**, 33–46.

Robiller, F., and Lauterbach, H. (1973). *In-vitro*-Funktionsuntersuchungen der Schilddrüsen von Girlitzen (*Serinus serinus*) während der Zugzeiten. *Naturwiss.* **60**, 522.

Safriel, U. N., Abramski, Z., and Boren, N. (1974). Interspecific competition among migrants en route. *Abstr. XVI Intern. Ornithol. Congr., Canberra,* 1974, pp. 115.

Shumakov, M. E., and Vinogradova, N. V. (1971). The formation and duration of the state of autumn migration in young and adult Chaffinches of the Kushsk population. *Analyzer Systems Orientation Birds* pp. 115–116.

Shumakov, M. E., Vinogradova, N. V., and Payevsky, V. A. (1972). Moult and migratory state in reared Chaffinch *Fringilla coelebs coelebs*. *Zool. Zurnal* **51**, 113–118.

Sinclair, A. R. E. (1974). The effect of food supply on the breeding of resident bird species and the movement of palaearctic migrants in a tropical African Savannah. *Abstr. XVI Intern. Ornithol. Congr., Canberra,* 1974, pp. 116.

Södergren, A., and Ulfstrand, S. (1972). DDT and PCB relocate when caged Robins use fat reserves. *Ambio* **1**, 36–40.

Stetson, M. H. (1971). Neuroendocrine control of photoperiodically induced fat deposition in White-crowned Sparrows. *J. Exper. Zool.* **176**, 409–413.

Stolt, B.-O. (1970). Ekologiska skillnader i höstflyttande tättingars uppträdande på Uppsala-slätten. *Vår Fågelvärld* **29**, 13–23.

Thiollay, J. M. (1973). Ecologie de migrateurs tropicaux dans une zone préforestière de Côte d'Ivoire. *Terre et Vie* **27**, 268–296.

Ulfstrand, S., Södergren, A., and Rabøl, J. (1971). Effect of PCB on nocturnal activity in caged Robins, *Erithacus rubecula*. *Nature* (*London*) **231**, 467–468.

D'Yachenko, V. P. (1972). Prolactin content of the hypophysis in the Chaffinch *Fringilla coelebs* during the autumn migratory period. *Dokl. Akad. Nauk. SSSR* **206**, 211–212.

Chapter 3

MIGRATION: ORIENTATION
AND NAVIGATION

Stephen T. Emlen

I. Introduction

For centuries, people have observed birds departing from north temperate areas each fall and returning the following spring. Worldwide programs of banding and recovery have largely answered the question of "where do migrants go?" Birds migrate great distances, often many thousands of kilometers, and with great regularity and accuracy. This information raises other questions of fundamental interest and importance to biologists and laymen alike. How have birds adapted to withstand the high energy cost of long-distance flight? And how have they solved the navigational problems involved in traveling over vast stretches of unfamiliar terrain between specific breeding and wintering locations? To give an idea of the types of orientational capabilities required for migratory journeys, a small amount of background information on the general behavior of migrants and typical features of migratory flight is outlined below.

The migratory habit is extremely widespread among birds. Fully two-thirds of the species breeding in the northern United States travel south for the winter. Distances covered typically average 1000–3000 km each autumn with a similar return trip in the spring, but one-way trips of 4000–6000 km are not uncommon (e.g., Dorst, 1962). Some shorebirds cover round-trip distances exceeding 20,000 km on their migrations between arctic breeding areas and southern hemisphere overwintering locations.

A migratory journey is usually broken into a series of short, discrete flights. Most songbirds are nocturnal travelers that probably fly singly or in widely spaced, loose groups. They initiate a flight 30 minutes to 1 hour after sunset and fly continuously throughout most of the night. Depending upon the flight speed of the bird and the direction and speed of the winds aloft, a bird may cover 300–600 km on each night of traveling. An individual usually does not fly each night, however, and a 3000 km trip may require 3 or 4 weeks.

Diurnal migrants, in contrast to their nocturnal counterparts, are more prone to fly in flocks and to follow along topographic leading lines. But they, too, travel along their migratory route in a series of discrete one-day flights. Longer flights in which birds travel continuously throughout the 24 hour period commonly occur among some shorebirds and waterfowl. Among passerines, such flights occur when long stretches of inhospitable terrain are crossed. These include the autumnal over-water flights from the east coasts of North America to South America, flights crossing the Gulf of Mexico during both migration seasons, and the trans-Sahara voyage of many European migrants to and from central and southern Africa.

Recaptures of banded birds attest to the navigational accuracy of most migrants. Adults routinely return to breed in the same general location where they bred in previous seasons. Evidence is accumulating that a similar fidelity occurs at the wintering localities; individual birds return year after year to the same specific area where they spent the previous winter.

Young birds show little evidence of possessing predetermined destinations or specific goals for migration (see Section III). Rather, they appear to adopt particular locations, perhaps by becoming imprinted to certain features of the local environment (Perdeck, 1958; Löhrl, 1959; Ralph and Mewaldt, 1975).

One major unknown of migratory behavior concerns the accuracy of birds while en route. This is particularly true in the case of nocturnal migrants. Do birds follow precise pathways that take them to the same intermediate locations year after year? Are experienced birds cognizant of any destinations or goals along the way? Unfortunately, neither banding returns nor experimental results provide sufficient information to answer these questions. What can be said is that many birds undertake long-distance migratory trips, and that they return year after year to rather precise geographic locations at both the breeding and the wintering ends of these journeys. Such travels require considerable navigational skills. In this chapter, I present a personalized overview of our current knowledge of these skills and of the knowns and unknowns of migratory orientation.

II. Philosophy of Orientation Research

Although biologists have been fascinated by migratory behavior for many decades, most of our current knowledge of navigation mechanisms has been obtained in the past 25 years. During most of this period, investigations centered on the development of all-inclusive theories to explain bird navigation. Underlying such studies was the assumption that one orientational cue or one mechanism of usage of a particular cue would provide the key to understanding all aspects of navigation behavior. Such a philosophy, I feel, is wholly inappropriate for this field of research.

The environment contains numerous cues that could give directional information to a migrating bird. Natural selection should favor the development of abilities to make use of all such information. By pooling several inputs, a bird could increase the accuracy of its directional decision. A system involving multiple cues could also provide checks and balances, enabling the bird to maintain its orienting ability in cases where one cue is lacking or provides equivocal information.

Not all sources of directional information are equally useful to the migrant. Some cues undoubtedly yield more accurate information than others. Some might be available throughout the entire route, whereas others would be useful only at specific geographic locations. Some might be available regardless of flight conditions, while others might be functional only under optimal meteorological situations. The probable existence of a hierarchy of redundant cues makes obsolete the search for "the" mechanism of migratory orientation. In all probability, a differential weighting of several directional cues occurs. These weightings should be expected to vary not only between species but, for any individual bird, with changing meteorological conditions and at different points along the migratory path (Emlen, 1971). Recent studies are confirming this prediction. They suggest that there are changes in the relative importance of different cues used during the spring and fall (S. T. Emlen and N. J. Demong, unpublished observations) and by birds of different ages (Emlen, 1970a, 1972; S. T. Emlen, N. J. Demong, and E. Gwinner, unpublished observations; Keeton and Gobert, 1970; Keeton, 1972).

The belief that most birds utilize multiple cues, some of them repetitive, necessitates changes in our evaluations and interpretations of previous studies. No longer should evidence for the importance of one directional cue be taken to imply, even indirectly, a lack of importance for an alternate cue. Redundancy of information was seldom considered in early experimental designs, and results must be reinterpreted accordingly. For example, let us assume that a bird can use two types of information, A and B, to determine a migration direction. Now suppose I block or modify sensory input to system A, yet find that the bird continues to orient properly. I would not be justified in concluding that system A was unimportant for orientation or that the bird did not have the capability of orienting using system A. The bird might simply have selected its direction using its alternate system, B. The experiment shows only that system A is not essential for proper orientation. This is very different from the conclusion that system A is not or cannot be used for direction determination.

Consider the following case. Several authors have attempted to study the role of magnetic information in orientation by attaching miniature magnets to the wings or bodies of birds and noting their abilities to home. Although the magnetic field surrounding the birds was seriously disrupted, homing ability generally was unimpaired and most authors concluded that the birds could neither detect nor use magnetic cues (Gordon, 1948; Matthews, 1951b; von Riper and Kalmbach, 1952). Unfortunately, almost all of these experiments were conducted either close to the home loft, on clear sunny days, or both. If we grant the birds the ability of using either familiar landmarks or information from the sun for orientational purposes, then the original conclusions of these studies must be reevaluated. No disorientation would have been expected in the presence of either of these alternate sources of directional information. The results did not provide any evidence for the use of geomagnetic cues, but neither did they rule it out.

Similar caution must be exercised in evaluating studies of inertial navigation in which semicircular canals have been severed. Experimentally treated birds performed as well as controls, but again, all releases were made in the close vicinity of home (Sobol, 1930; cited in Matthews, 1968; Huizinger, 1935) or under clear sunny skies (Wallraff, 1965a).

Unfortunately, the list of experiments where redundant cues were not taken into consideration is enormous.

I am the first to acknowledge the difficulties of designing experiments that control for repetitive cues. Often the best that can be done is to affect more than one cue simultaneously. But, at the very least, we must take this redundancy into account when interpreting experimental results.

We must also be careful not to overgeneralize among species. Migratory behavior has evolved independently many times among birds. Birds migrate for different reasons, and the paths taken, the distances traveled, and the hardships encountered en route vary greatly from species to species. The resulting selection pressures must differ accordingly.

Finally, I propose that caution be exercised in extrapolating the results from studies of homing to explanations of orientation during migration. Earlier workers assumed that the two processes were identical and that each could be used as an experimental tool for studying the other. But, as will be seen in the next section, there is little evidence that most migrants navigate to specific goals with the versatility and accuracy shown by homing pigeons. It is possible that the typical migrant switches to a homing-type process only during the terminal phases of its migratory travels. Throughout much of the trip, many migrants might use simple compass orientation. Migrants are faced with the additional problem of maintaining a given direction for periods of many hours, even days. During these long flights, the availability of different cues may change. Celestial cues may be obscured by clouds, features of the terrain may be enveloped by ground fog, the direction of the wind might change unexpectedly, or the bird might pass over a strong magnetic anomaly. The advantage of having redundancy built into the orientational systems is obvious. The orientational repertoire of a migrant, therefore, should be expected to include a number of simpler systems allowing for the *maintenance* of a given flight direction.

The problems faced by a pigeon homing to its loft and a migrant returning to its former breeding site at the end of its long migration may be identical. But the behavior of the migrant on the remainder of its journey may be quite different. At the very least, the weighting of various sources of directional information should be expected to be different in these two circumstances.

The realization that migrants probably use numerous cues and that different species may use them in different ways should not be considered as discouraging or slowing the advance to an understanding of orientational systems. The effect is exactly the opposite. Today there is an increased willingness to consider both a great variety of types of cues and a variety of ways in which they might be integrated. Alternative hypotheses frequently are viewed as reinforcing rather than competitive. The problem for the ornithologist interested in navigation is no longer searching for *the* mechanism of orientation. It is studying any and all components of the orientation systems and trying to decipher how these are integrated into a navigational complex that determines the directional behavior of the bird at any place in time or space.

III. Navigational Capabilities of Migrants

Before discussing the types of cues that may provide a migrant with directional information, mention should be made of the actual orientational capabilities exhibited by birds during the course of their migratory travels. In spite of the large number of studies of migration, surprisingly little is known about the accuracy or versatility of birds' navigational abilities. Do they know their geographic location at all times? Are they actually navigating toward a particular geographic goal, or do they rely on a system of simpler compass orientations during much of the migratory flight? Can they determine when they have been blown off course by strong winds, and do they correct or compensate for such displacement?

Many years ago, Griffin (1955) defined three levels of orientation ability. The first, which he called Type I, is a simple piloting based upon familiar landmarks (visual or otherwise). His second level, Type II, describes a bird's ability to fly in a particular compass direction without reference to landmarks. Type III, commonly called true navigation, refers to the ability to orient toward a specific goal from a variety of unfamiliar areas. As Schmidt-Koenig (1965) points out, reference to a goal could be maintained either by means of a complex inertial system ("reverse displacement navigation") or by reference to some grid of coordinates provided by celestial and/or geophysical cues ("bi-coordinate navigation"). Such true navigation has been shown convincingly in homing experiments for only three groups of birds—oceanic seabirds (Lack and Lockley, 1938; Matthews, 1953b, 1964; Kenyon and Rice, 1958; Billings, 1968), swallows (Rüppell, 1934, 1936, 1937; Southern, 1959, 1968), and homing pigeons (Matthews, 1951b; Wallraff, 1967, 1970; Keeton, 1974).

Many authors have assumed that migratory orientation falls under category III because individual birds cover thousands of kilometers and yet return to highly specific breeding and wintering locations. In actuality, there is very little convincing evidence for the widespread occurrence of Type III orientation among migrants.

A. DISPLACEMENT EXPERIMENTS WITH FREE-FLYING BIRDS

One way of studying orientational capabilities is to perform displacement experiments. Birds are captured as they migrate through an area and are transported and released at a different geographic location. If the displaced birds reorient back toward the original capture location or on a corrected course toward the normal breeding (in spring) or wintering (in autumn) area, it would imply that they had sensed and correctly compensated for the displacement.

Results from this type of experiment generally are dependent upon reports of recovered banded birds. Unfortunately, recovery rates are extremely low and are strongly biased by the uneven distribution of persons interested in capturing and reporting such birds. An additional error would be introduced if the displaced individuals showed any tendency to be influenced by, or to travel with, local migrants at the release location. Finally, one should consider only those recoveries occurring during the migration season in which the displacement was performed. Even taking this precaution, however, one must realize that an individual bird could travel over a tremendous area during the period before recapture; one cannot assume that the bird moved in a direct line from release to recapture location.

In spite of these potential drawbacks, displacement experiments have provided valuable information on the strategies of direction finding of migrants. The most extensive studies have been those of Perdeck (1958, 1967), who transported almost 15,000 European Starlings (*Sturnus vulgaris*) from Holland to release locations in Switzerland and Spain. From previous banding studies, Perdeck knew that the migrant starlings passing through The Hague each autumn were from a population that bred in eastern Holland, northern Germany, Denmark, and Poland. This population generally migrated west-southwest to wintering areas in western Holland, southern England, Belgium, and northern France (Fig. 1). In his initial experiments, 11,000 birds were displaced from The Hague approximately 750 km south-southeast and released at three locations in Switzerland. Of the 354 recaptures reported, 131 satisfied Perdeck's requirements of being recovered at a distance of more than 50 km from the release site and being recovered in the same autumn or in the following winter season. I have replotted these data in Fig. 2, and it is apparent that a dichotomy exists between the behavior of transported juvenile and adult starlings. The young birds on their first migratory flight continued to travel in the west-southwest compass direction even though this took them far from the population's normal wintering areas. There is no indication that the birds detected the displacement or corrected for it in any way. Rather, they traveled along a preferred compass direction (Type II orientation). Additional recoveries revealed that many immature birds adopted the new wintering areas in southern France and Spain and even returned in later years (Perdeck, 1958, 1967).

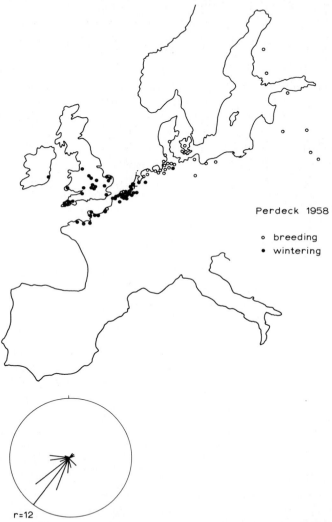

Perdeck 1958

o breeding
• wintering

r=12

FIG. 1. Top: Recapture locations of breeding (open circles) and wintering (solid circles) of European Starlings banded during autumn passage through The Hague. Banding location is designated by a plus (+) sign. Bottom: Vector diagram of the directions of winter recapture locations relative to The Hague. The data are grouped into 15° sectors and plotted on a proportionality basis with the radius equaling the greatest number of recaptures in any one sector. The number that this represents is presented at the lower left of the diagram. In all such figures, 0° or 360° represents geographic north. (Data from Perdeck, 1958.)

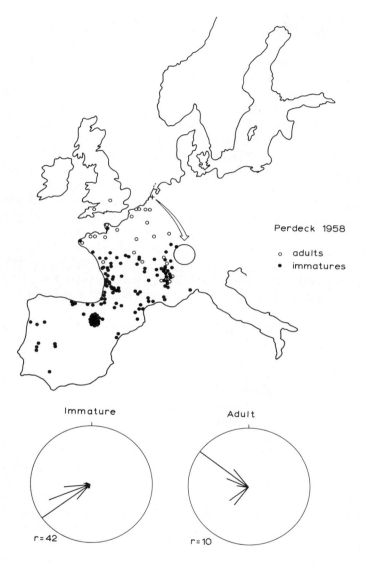

Fig. 2. Top: Recapture locations of adult (open circles) and immature (solid circles) European Starlings displaced from The Hague to three locations in Switzerland during autumn migration. Bottom: Vector diagram of the directions of winter recaptures of immature (left) and adult (right) European Starlings plotted relative to the release sites in Switzerland. (Data from Perdeck, 1958.)

The recaptures of adult displaced starlings showed a weakly bimodal distribution. One group of recaptures ($N = 8$) was to the west-south-west, suggesting a continuation along the previous compass direction.

But a second group ($N = 19$) was distributed to the north-northwest, and 13 of these birds returned successfully to the normal wintering area.

An additional 3600 starlings, also captured migrating through The Hague in the autumn, were transported to Barcelona, Spain (Perdeck, 1967). The recapture locations and directional vector diagrams for their recoveries are replotted in Fig. 3. Again there is a difference in the behavior of juvenile and adult birds. Young starlings continued to fly along the west-southwest compass bearing, seemingly unaware of the 1300 km displacement. The 12 recaptures of adults, however, support the findings of the Swiss experiments. Although there is a considerable spread in recovery locations, most were from the northwest.

One could be hypercritical and point out that the overwintering area for this population of starlings is not northwest, but north of Barcelona (Fig. 1). Only two birds successfully returned to this area. But the combined evidence from Perdeck's experiments clearly shows that most adult starlings did not continue in their "normal" migration directions. Rather, they traveled toward (and often reached) their previous wintering areas. This implies both a recognition of and a correction for the geographic displacement.

In contrast, first-flight migrants showed no change in behavior after transplantation. Either they did not detect the displacements or they were unable to reorient to compensate for them. In either case, Type III navigation capabilities were lacking.

Schüz (1949) conducted analogous experiments with White Storks (*Ciconia ciconia*). These storks breed over much of Europe and migrate considerable distances to winter throughout southern Africa. Individuals from populations breeding from central Germany eastward migrate in a southeasterly direction, passing around the eastern end of the Mediterranean on their way south. Storks from the western population take a southwesterly course, crossing the Mediterranean at the Strait of Gibraltar. Schüz captured young storks from the eastern population and released them in western Germany after the migratory departure of the local storks. Although they were in an area where the normal migration direction would have been to the southwest, recoveries indicated that the displaced young birds continued on a southeasterly compass course. Again we have the suggestion of a simple Type II directional capability.

Rüppell (1944) and Rüppell and Schüz (1948) performed similar displacements with Hooded Crows (*Corvus corone cornix*) during a spring

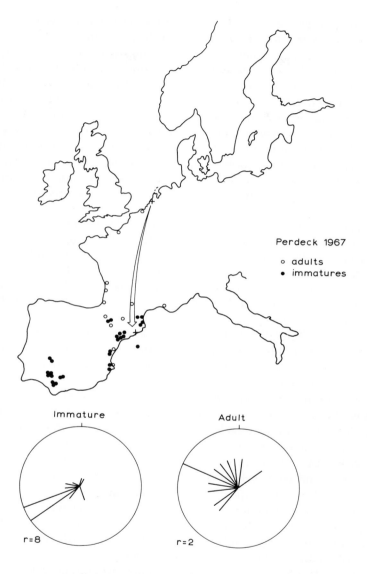

FIG. 3. Top: Recapture locations of adult (open circle) and immature (solid circle) European Starlings displaced from The Hague to Barcelona, Spain. Bottom: Vector diagram of directions of winter recapture locations of immature (left) and adult (right) European Starlings plotted relative to the release site at Barcelona. (Data from Perdeck, 1967.)

migration. Banding studies showed that the birds passing through Rossitten, on the Baltic Coast, migrated northeast to reach breeding areas in northern Poland, southern Finland, and northwestern U.S.S.R. Approximately 900 birds were transported from Rossitten west to central Denmark (770 km) or west-southwest to western Germany (960 km). Recaptures indicated that the dislocated birds continued to migrate to the northeast, failing to compensate and reorient back toward the normal breeding area. Recoveries from succeeding years showed that the birds adopted new, more westerly, breeding locations and even overwintered at points west of their normal wintering quarters. We must tentatively conclude that Hooded Crows, even on their second migration trip, were not goal-orienting but were following a Type II northeast–southwest compass course.

Occasionally, a natural experiment occurs in which migrants are blown off their normal course by adverse winds. Many autumn nocturnal migrants of Scandinavian origin leave Norway on a south-southwest flight that takes them over southern England and on to western France and Spain. Evans (1968) believes that such birds normally cross over the eastern coast of Britain at points south of Yorkshire. If, in crossing the North Sea, birds encounter stiff easterly winds, they may be blown off course and be forced to make landfall along the northeastern coast of Britain. Do such birds continue migrating to the south-southwest or do they reorient to the south and south-southeast, thereby returning to their normal migration route?

Birds captured at Northumberland (north of Yorkshire) and believed to have been displaced westward were tested in circular orientation cages. The data are extremely sparse, but many birds did display south or southeast tendencies (Evans, 1968). I hesitate to draw firm conclusions from these data, because no control tests were run with birds not presumed to be off course. But, at the very least, the results are consistent with a hypothesis of compensatory reorientation.

Evans next compared the autumn and winter recovery locations of European Redstarts (*Phoenicurus phoenicurus*) and Pied Flycatchers (*Ficedula hypoleuca*) banded north of Yorkshire with those netted on the coast south of Yorkshire. The recoveries from the northern or "displaced" birds ($N = 10$) showed the same geographic distribution as the recoveries from "normal" birds ($N = 25$), leading Evans to hypothesize that these two species possess the ability to correct for displacement.

Several words of caution are necessary before interpreting these results. Some workers question both the assumptions that all birds of these species on the British coast are of Scandinavian origin, and that those north of Yorkshire necessarily represent individuals that have been blown significantly off course (Evans, 1968, p. 30). Even if these assumptions are valid, we must ask ourselves what would be the expected distribution of band recoveries if those blown off course did not correct for displacement? Presumably, they would continue south-southwest on a course that would take them out over the Atlantic. Such birds would never be detected. Hence, a difference in recapture locations between the two groups of birds would not be expected regardless of their orientation behavior. Evans' case must rest on a comparison of the numbers or percentages of recaptures. These numbers are quite small and could be explained by alternative hypotheses requiring little in the way of complex navigation capabilities. Scandinavian migrants crossing the North Sea might have the tendency to turn to the east (toward the coast of France and Spain) at dawn after flying long periods over open water. Or they might demonstrate the "dawn ascent" phenomenon of climbing in altitude and turning toward the coast upon finding themselves at sea following a night's migration (Baird and Nisbet, 1960; Myres, 1964; Richardson, 1972). Simple strategies such as these could account for many of the recaptures, and they would seem clearly adaptive for passerines that frequently encounter easterly winds while traveling along the Atlantic coast of Europe.

Finally, radar studies of passerine migration in autumn show that the dominant direction of travel in England is to the south-southeast (Evans, 1966). Consequently, any tendency for displaced Scandinavian transients to associate with British migrants could also explain the apparent "reorientation" of Evans' birds. In view of these considerations, I must regard Evans' data as exciting, but his conclusions as unproven.

B. DISPLACEMENT EXPERIMENTS WITH CAGED MIGRANTS

One way to avoid the problems associated with banding-recapture studies is to study the orientation behavior of caged migrants following artificial displacement. Very few such studies have been undertaken and the preliminary results are somewhat confusing. Yet mention will be made of these studies since they constitute a powerful experimental approach that should be extended in the near future.

Russian investigators (Dolnik and Shumakov, 1967) working at Rybachii on the Nehrung Peninsula (longitude 21°E, latitude 55°N; near the Rossitten station used by Rüppell) transported four species of passerines over 3800 km to Dushanbey (69°E, 38°N) and almost 8300 km to Khabarodsk (135°E, 48°N). Two diurnal species, Chaffinches (*Fringilla coelebs*) and European Starlings, were selected as typical short-distance migrants, while two others, Scarlet Rosefinches (*Carpodacus erythrinus*) and Barred Warblers (*Sylvia nisoria*), were taken as examples of long-distance travelers. The birds were displaced in the autumn of

1964 and placed in circular cages exposed to the outdoor sky. The results are difficult to interpret, since all the data obtained from a particular species are presented in pooled form. This pooling includes an unknown number of individual birds, and diurnal as well as nocturnal activity. The latter is especially unfortunate, since it is probable that different mechanisms of orientation are used by day and by night. It is also unfortunate that the birds lived in the circular cages and were constantly exposed to the solar day at the displacement location. This means that the birds were in the process of resynchronizing their biological time senses to local time during the week the experimental data were being collected. Since all results are pooled, it is not possible to factor out the effects of clock-shifting and reentrainment, nocturnal versus diurnal activity, or the relative weighting of the data by different individual birds. Nevertheless, the pooled vector diagrams indicate that the two short-distance diurnal migrants either failed to correct for displacement or showed an overall low level of oriented behavior. The two long-distance nocturnal migrants, however, did shift their mean bearings from southeast at Rybachii to a more bimodal northeast-southwest axis at Khabarodsk. The results from the experiments with Barred Warblers are redrawn in Fig. 4. Since the birds had been transported approxi-

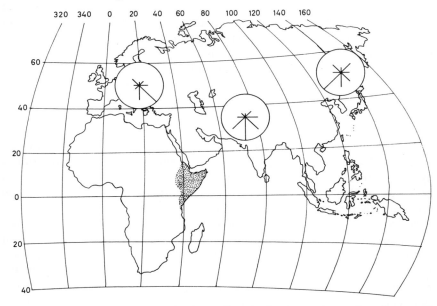

FIG. 4. Orientation of caged Barred Warblers (*Sylvia nisoria*) tested in the autumn at Rybachii (21°E, 55°N) and after artificial displacement to Dushanbey (69°E, 38°N) and Khabarodsk (135°E, 48°N). Stippled area denotes winter quarters. From Dolnik and Shumakov (1967). See text for full explanation.

mately 115° of longitude to the east, these findings are suggestive of a reorientation toward a wintering location and are so interpreted by Dolnik and Shumakov (1967). However, the same shifts in orientation (50° clockwise at Dushanbey and 115° clockwise at Khabarodsk) would be expected if the birds possessed not a goal-directed orientation (Type III), but rather a simple, celestial compass (Type II) incorporating a biological time sense to compensate for the clockwise motion of the sun or of stars located in the southern sky (see Sections IV,B and C). Unfortunately, the experiments were not designed or presented in a way that allows the reader to distinguish between these two very different interpretations.

One of the participants in the original study, R. L. Potapov, extended the study the following autumn by clock-shifting Garden Warblers (*Sylvia borin*), Scarlet Rosefinches, and Barred Warblers prior to their displacement. He presents his data in the same pooled form and includes results obtained when the birds were clock-shifted to different times as well as their behavior after they had been allowed to reset to local conditions at the points of displacement. Potapov (1966) states, however, that the clock-shifting had no effect upon the directional choices of the birds. The results that are given completely contradict those of Dolnik and Shumakov (1967). No evidence was obtained that the birds either detected or corrected for their eastward displacement. For each species, the overall orientation was similar regardless of test location. Barred Warblers, for example, continued to aim to the southeast at both Dushanbey and Khabarodsk. From this Potapov (1966, p. 892) concludes that his test species do not possess true navigational capabilities. Instead, he hypothesizes they use a simple star compass (Type II) and that this mechanism of orientation is independent of the biological clock.

Taken together, Dolnik, Potapov, and Shumakov must have accumulated exactly the types of data that are required to answer the questions posed in this section. Their sample sizes are large; their displacements are impressive. But as long as these data are presented in such an abbreviated, even cryptic, format, it will be impossible to interpret the results fully or to understand the opposing conclusions of the two studies. I can only plead that the data be reanalyzed and published in full detail. This reanalysis should treat nocturnal and diurnal activity separately and should factor out the effects of displacement, clock-shifting, and reentrainment.

Another experiment with displaced migrants is Hamilton's study of the nocturnal orientation of caged Bobolinks (*Dolichonyx oryzivorus*). Birds were obtained from a breeding population in North Dakota, transported to San Francisco, California, and allowed to resynchronize their activities with local time. When tested in the autumn under the natural night sky in California, the birds showed a southeasterly orientation, parallel to the normal direction taken by birds in North Dakota. No evidence of a directional shift to compensate for displacement was observed.

One adult female Bobolink was treated somewhat differently (Hamilton, 1962c). It was housed in a separate room where the light–dark cycle was kept in phase with North Dakota time. The bird was first exposed to the San Francisco sky in late August when its *Zugunruhe* orientation was tested. Interestingly, this individual took up a northeastward direction, approximating the direction to its breeding area in North Dakota. Why a bird in autumnal migratory condition should orient toward its summer home is unclear. During the successive two nights, presumably as the bird's chronometer synchronized with local, California time the directional preference shifted through east to the typical southsoutheast. This directional shift of approximately 100° is greater than would be expected if the bird were using a time-compensated compass (Type II). Unfortunately, no additional data are available because the bird escaped prior to its fourth test.

Jørgen Rabøl (1969, 1970, 1972) has recently initiated a series of short-distance displacement experiments with migrants in Denmark. He has transported several species of European warblers distances of 5° to 8° of longitude and latitude. Although his results are quite variable, mention will be made of one test that did provide a suggestion of reorientation (Rabøl, 1969). At Blavand, on the west coast of Denmark, migrant Garden Warblers typically exhibited a southward directional preference when tested in orientation cages. Rabøl transported a group of warblers 500 km east to Ottenby, off the southeast coast of Sweden, and tested them on the night of their arrival. Although the sample was small, the pooled mean direction of those birds that showed oriented behavior did shift westward (mean direction = 209°; $N = 5$, at Ottenby; mean direction = 165°, $N = 11$ at Blavand). By the second and fourth nights (it rained on night three), the pooled mean direction had shifted back eastward to south (184°; $N = 6$). These results could be interpreted to mean that some birds corrected for displacement, but only as long as their chronometers remained on Blavand time. Within 24 hours their internal clocks had reset to Ottenby conditions and the ability to detect the eastward displacement had disappeared.

If the caged warblers possessed a Type II compass system that incorporated time-of-day information, we also would have expected a clockwise shift in orientation. This shift would be equal to the longitudinal displacement, in this instance, 8°. The magnitude of Rabøl's shift (44°) suggests an awareness of geographic displacement, since it exceeds that predicted for the compass system alone. Additional tests with Garden Warblers, Willow Warblers (*Phylloscopus trochilus*), and European Redstarts have yielded more equivocal results (Rabøl, 1969, 1970).

C. DISCUSSION

When a bird is "naturally" displaced during a migratory flight, the causal agent is wind and the distance blown usually is small. Migrants don't often find themselves thousands or even hundreds of kilometers off route. When natural displacements do occur, migrants might be able to detect them directly, while in flight, either through reference to visual markers on the ground (Section IV,D,2) or by the exposure to strong winds and storm conditions aloft. Natural selection should operate to insure adequate correction for *these* errors, not necessarily for abilities to detect and compensate for displacements of thousands of kilometers. For small-scale wind displacements, changes of orientation strategy such as (1) reversing the direction of flight the following morning, (2) turning into the direction of the prevailing wind, or (3) reorienting toward a visible coastline, might be sufficient corrective measures. All three behaviors have been reported for songbird migrants on mornings following nights of strong wind or storm activity. All three might get the birds back on course; yet none presupposes geographic location detection or a Type III navigational capability.

Sections III,A and B reveal the meager state of our knowledge concerning the actual navigational capabilities of most migrants and point to the need for additional studies. From the information at hand, it is by no means clear that most migrants are using a bicoordinate navigational grid or are goal orienting. The orientation behavior of many migrants may be fundamentally different from that of goal-directed homing behavior found in shearwaters, petrels, swallows, or pigeons. Some might possess the capability of homing to a learned goal but not invoke this strategy until the final phases of the trip. A bird programmed only to fly in a preferred compass direction could still reach the general vicinity of its wintering quarters if it selectively timed its flights to avoid adverse winds. The accuracy of such a bird would be a direct function of (1) its degree of selectivity in flying with different winds, (2) its accuracy in predicting winds aloft, and (3) its ability to correct for wind drift (see Section IV,E).

Physiological studies have shown that the metabolic cost of flight is extremely high (Tucker, 1971; Berger, and Hart, 1974). Field observations suggest that birds often travel with a low margin of safety. Songbirds making landfall after a long, overwater flight in unfavorable winds frequently are so exhausted that they can be picked up by hand. Individ-

uals are often completely devoid of subcutaneous fat reserves, and some
have even metabolized part of their own flight muscles as a last source
of fuel (e.g., Rogers and Odum, 1966).

Birds can conserve energy by taking advantage of specific weather
situations, i.e., by initiating a journey only when wind conditions are
favorable for the preferred direction of flight. By flying with tail winds
and by taking advantage of vertical air currents, a migrant can both
reduce the energetic cost of flight and increase its potential flight range.

We should expect birds to follow migratory routes that allow a maximum utilization of the various prevailing winds occurring in different
geographical areas. This leads to the interesting hypothesis that *the
most advantageous route from an energetic viewpoint might be an indirect one*, following a path that is neither shortest in distance nor directly goal oriented.

Bellrose and Graber (1963) used this logic to explain their radar
observations of bird migration over central Illinois. They noted that
the flight paths of small birds were predominantly to the east of north
in the spring and to the east of south in the autumn. The absence
of reciprocal flight directions in the two seasons led them to speculate
that many species follow elliptical clockwise routes between breeding
areas in the north and wintering areas in Central and South America.
By following such an indirect path, the birds could take advantage
of the high frequency of westerly winds found at northern latitudes
and, more importantly, could utilize the westward drift provided by
the easterlies that predominate at more southerly latitudes.

A similar explanation may underlie the long-distance flights of passerines and shorebirds that depart the east coast of North America on
a nonstop trip to the West Indies or South America. Radar observations
indicate that birds leave the coasts of the Canadian Maritimes and New
England on southward and south-southeastward bearings (Drury and
Keith, 1962; Drury and Nisbet, 1964; Richardson, 1972, 1974, 1975;
Williams *et al.*, 1974). The courses of many of these birds would,
if extended, pass east of the coast of South America and take the birds
to Africa. Recent autumn radar observations in the Antilles (Hilditch
et al., 1973) and on Puerto Rico (Richardson, 1974, 1975) detected
a moderate volume of birds approaching from the north and north-
northeast. Many of these bearings, if extended backward, would suggest a European starting point. The most logical explanation is that
these birds are also following a clockwise elliptical course, shifting their
bearing more westward as they encounter the prevailing easterly trade
winds at low latitudes. We must hypothesize that the energetic savings
gained by using the low latitude easterly winds more than offsets the
increased mileage of this route to South America.

Peter Evans (1966) further stressed the importance of energetic considerations in the evolution of migratory routes. Banding studies of autumn migrants in Britain show that many species overwinter to the south-southwest in Spain. Yet, birds observed by radar generally depart from Britain on a preferred southeast or south-southeast bearing. Evans hypothesized that this indirect route had evolved to take advantage of the prevailing upper air winds and to minimize the chance of being blown out over the Atlantic.

In cases such as these, it might be selectively advantageous for a bird to orient along a preferred compass bearing (Type II orientation) even though it possessed the capabilities of true goal navigation. In the later stages of migration, such birds might change behavior and locate their specific wintering areas by a variety of means. These could include random search, recognition of familiar topographic landmarks, use of cues emanating directly from the final destination itself, or the use of intermediate cues providing bicoordinate grid information.

At this stage, our knowledge is meager enough to allow a considerable spread of speculation about the orientation strategies of migrants. In fact, a whole spectrum of hypotheses has been advanced in recent years. At one end of this spectrum are Sidney Gauthreaux and Kenneth Able (Gauthreaux and Able, 1970; Gauthreaux, 1972b; Able, 1972, 1973, 1974a; but see Able, 1974b) who believe that passerines in the southeastern United States have no strong directional preferences at all, but merely fly downwind. By this hypothesis, a migratory course would be rather circuitous. It would follow a modified random walk model with a strong bias imposed by the relative frequencies and temporal sequence of winds from different directions, and by the degree of selectivity in the departure behavior of the birds. In an intermediate view, William Cochran (personal communication) believes that thrushes possess a Type II compass capability but do not compensate for wind drift. Consequently, their direction of flight on any given night is merely the vector sum of the wind direction and speed plus their predetermined compass heading and flight speed. Other workers, including William Drury and Ian Nisbet (1964; Nisbet and Drury, 1967) and Frank Bellrose (1963), have proposed that birds not only have a preferred compass bearing but also an accurate ability to control for wind drift aloft (see Section IV,E). Discussing songbird migrants in Massachusetts, Nisbet and Drury state: "they appear to preserve their tracks so precisely that it is unnecessary to suppose that they use any form of bicoordinate navigation" (1967, p. 184).

At the other end of the spectrum, many authors hypothesize that migrants use a bicoordinate navigation system based upon celestial and/or geophysical cues (see Yeagley, 1947; Matthews, 1951a, 1968;

Sauer, 1957; Sections IV,B and C). This view reaches its most sophisticated form in the recent proposals of Jørgen Rabøl (1969, 1970, 1972), who suggests that a migrant is capable of detecting displacement and reorienting toward a goal by means of a bicoordinate system of star navigation. Additionally, he proposes that the coordinates of this goal change through time in an internally programmed sequence that brings the bird to its geographic destination at the end of the migration period.

At this point, it is too soon to look for generalizations about the navigational capabilities of birds. I have tried to point out that migrants undoubtedly make use of a variety of cues, some providing simple, others complex, directional information. In all probability, a whole array of navigational strategies exists. Their study will provide the ornithologist with an exciting challenge in the years ahead.

IV. Cues Used in Direction Finding

In spite of our ignorance about their actual migratory behavior, birds do reach specific breeding and wintering destinations with great regularity. The remainder of this chapter will be devoted to a survey of the types of directional cues available in the environment and the kinds of information that birds seem capable of extracting from them.

A. The Importance of Topographic Landmarks

It is well known that many diurnal migrants are influenced by general topographic features beneath them. Some regularly follow coastlines, avoiding flights over large bodies of water. Others appear to follow river valleys or to make use of updrafts that occur along mountain ridges. All of these landmark effects cause a concentration of diurnal migrants along specific "flight corridors" and are responsible for the mass flights at such famous birding localities as Point Pelee on the Ontario shore of Lake Erie, Cape May at the southern tip of New Jersey, and Hawk Mountain in northeastern Pennsylvania.

Piloting using familiar topographic landmarks is also widely believed to be of importance in the guidance process, particularly in the final stages of migration. This seems logical, since birds have quite good visual acuity and have been shown to be capable of recognizing salient features of the terrain and of retaining this information for long periods of time.

Most migration occurs at considerable heights above the surface of the earth and this has important consequences for the perception of the visual landscape beneath. The higher a bird flies, the greater the breadth of its field of view. For a bird at 200 feet (60 m), the visual horizon is approximately 30 km away, giving a breadth of vision of 60 km (Hochbaum, 1955). For a bird migrating at 2000 feet (610 m), the landscape stretches 100 km in all directions (visibility conditions permitting). Together with this increase in field of view, there is a decrease in the angular velocity of the apparent ground movement. Hochbaum (1955) correctly noted that this would give the migrant more time for visual inspection of the landscape below.

In spite of these facts, there are obvious drawbacks to basing an orientation system exclusively on topographic cues. Specific landmarks must be learned and thus are not available for first-flight migrants that travel separately or become separated from experienced adults. This system is useful only for traversing areas that have already become familiar as a result of previous flights. If a bird were blown off course or made an incorrect turn along the way, it would leave its zone of familiarity and, with it, the possibility of landmark orientation. Recognition and retention of landmark information becomes increasingly difficult as the distance covered in migration increases, and as the terrain covered becomes more homogeneous.

Natural selection should favor a reliance upon cues that can provide directional and/or geographical information and are available over large areas of the earth's surface. Landmark recognition probably is of considerable importance at specific areas along the migratory route—in finding familiar stopover points and in locating the final destination. But I would expect landmarks to take a secondary, supplemental role in the direction finding process en route.

Waterfowl show one of the strongest tendencies to return to traditional stopover locations during migration. Since ducks and geese typically travel in family groups, information about the migratory route and the guideposts along it could easily be transmitted from generation to generation. Not surprisingly, waterfowl provide some of the best examples of landmark orientation. Hochbaum (1955) and Bellrose (1967b) mention the importance of landscape features both at the breeding grounds and in the vicinities of migratory staging areas. "In addition to the use of landmarks for local orientation, large rivers, lakes and similar features of the landscape are used by diurnally migrating waterfowl as guidelines and guideposts" (Bellrose, 1967b, p. 95). Yet, even among waterfowl, other orientational cues play the major role in determining the direction and course of the migratory flights (Bellrose, 1958, 1963, 1964, 1967a, 1971).

In contrast, there is a growing body of evidence from radar studies that birds migrating at night generally ignore even the most obvious landscape features beneath them. This has been reported almost uniformly for waterfowl and waders (shorebirds), as well as for nocturnally migrating songbirds. It holds true in a variety of geographic locations ranging from southeastern Canada and the New England coast of North America (Drury and Keith, 1962; Drury and Nisbet, 1964; Nisbet and Drury, 1967; Richardson, 1971, 1972, 1974), the Great Lakes area (W. W. Gunn, unpublished observations) through the northern Mississippi Valley (Bellrose, 1964, 1967b) to the southeastern Gulf area of the United States (Gauthreaux, 1971), and the west coast of the Gulf of Mexico (Bellrose, 1967a, 1971). Similar reports come from a variety of locations in Europe (Lack, 1962; Gehring, 1963; Bergman and Donner, 1964; Evans, 1966; Eastwood, 1967).

Features of the landscape are less distinct on very dark nights. If migrants relied heavily upon these cues, they might be expected to compensate for the decreased visibility by flying at lower altitudes. Yet there is no indication that nocturnal migrants fly lower under overcast or on nights with no moon (Bellrose, 1971).

All of this implies that although a great deal of potential information is available in the form of major topographic cues, it is not of critical importance to most birds migrating overhead at night. Since the major features of the terrain are visible on all but the darkest of nights, it is difficult to avoid concluding that under most circumstances landmarks are openly ignored by the majority of nocturnal migrants.

Even in homing studies, there is a growing consensus that pilotage based on learned landmarks becomes crucial only in the immediate vicinity of the final destination. While tracking homing pigeons by airplane, Michener and Walcott (1966, 1967) and Walcott and Michener (1967) found that birds approaching home lofts in eastern Massachusetts did not turn and fly directly toward the loft location until they came within sight of a tall building in the Boston area. By plotting the track location at which pigeons made their final correction toward home, they estimated that landmark guidance occurred only within 15 to 20 km of the loft. Pigeons that approached the home area but whose paths were in error by more than this distance continued on past the loft, giving no indication of landmark recognition. Michener and Walcott further report that pigeons routinely flew over areas where they previously had flown many times without showing any apparent sign of recognition.

Wagner (1972) followed pigeons by helicopter as they homed to lofts in Switzerland. The mountains and valleys provided striking topographic features, and his birds often departed in directions dictated by the terrain. They were particularly prone to follow along valleys, provided that the discrepancy between the orientation of the valley and the direction of home was not great. But the pigeons changed strategies shortly after release and turned toward the home location.

Several investigators have shown that when pigeons whose internal clocks have been shifted are released at a familiar site, they ignore the landscape. Their initial bearings are significantly and predictably deflected from homeward. Similar results are obtained even when releases are made within a kilometer or two of the home loft [Graue, 1963; Matthews, 1968; Alexander (in Keeton, 1974)]. As long as the actual home loft is hidden from view, the clock-shifted birds choose deflected bearings even though they are within the area in which they have taken their daily exercise flights all their lives. Schmidt-Koenig (1972, p. 281) and Alexander (in Keeton, 1974) even report instances where the pigeons took shifted bearings when the loft building was in direct view.

Most recently, Schlichte and Schmidt-Koenig (1971), Schmidt-Koenig and Schlichte (1972), and Schlichte (1973) have equipped pigeons with miniature translucent contact lenses that prevent resolution of images at distances greater than about three meters. The position of the sun can still be localized but recognition of topographic landmarks becomes impossible. Yet such birds, when released 15 and 130 km from the loft, showed a homeward initial orientation that was quite comparable to control birds wearing transparent lenses. Even more astonishing, some individuals with translucent lenses still in place fluttered down and landed in the immediate vicinity of the home loft itself. When birds wearing frosted lenses were equipped with miniature radios and tracked by airplane (Schmidt-Koenig and Walcott, 1973), they showed typical homeward paths, except that they often failed to locate the loft and fluttered to the ground prematurely, terminating flights at distances several kilometers from home. In a few instances, a bird "missed" the loft, flying past at a distance of a few kilometers; but, shortly after the overflights, the birds reversed direction and returned toward the loft. These findings imply that pigeons can pinpoint the location of their home loft to within a kilometer or two in the absence of visual landscape cues.

While these studies do not imply that familiar landmarks cannot play an important role in orientation, they do provide strong evidence that they are rarely essential and that alternative cues generally take precedence in the guidance systems of both homing and migrating birds.

B. The Use of the Sun

1. Evidence of Sun Orientation

Early workers (Schneider, 1906; Rüppell, 1944) mentioned the possibility that migrants might use the sun, but it was the late Gustav Kramer who provided the first experimental evidence of its importance as a directional cue. He found that caged European Starlings exhibited a spontaneous migratory restlessness when placed outdoors in a circular cage (Kramer, 1950, 1951). This activity was oriented in an appropriate migratory direction in both the spring and fall as long as the sun was visible (Fig. 5a). Kramer then placed his cage inside a small, six-windowed pavilion that obscured all landmarks from view. The birds' orientation continued under clear skies but deteriorated under overcast conditions when the sun was not visible (Fig. 5b). This test was repeated and the results were confirmed by Perdeck (1957) a few years later. Further, the directional choices of the starlings were altered in a predictable way when Kramer (1951) changed the apparent position of the sun by using mirrors to deflect sunlight entering the pavilion (Fig. 5c and d).

These experiments implied that starlings could select a direction by orienting at a particular angle relative to the sun. But to use the sun

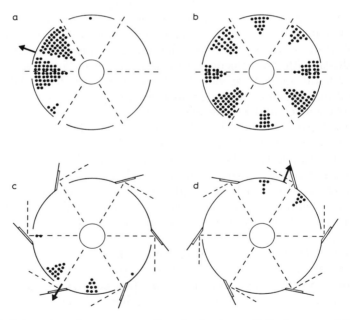

FIG. 5. Orientation of spontaneous diurnal migratory activity in a caged European Starling under various conditions of sun exposure. Bird was tested outdoors in a pavilion with six windows during the spring migration season. (a) Behavior under clear skies. (b) Behavior under total overcast, when the sun was not visible. (c) Behavior when the image of the sun was deflected 90° counterclockwise by means of mirrors. (d) Behavior when the image of the sun was deflected 90° clockwise by mirrors. Each dot represents 10 seconds of fluttering activity. Dotted lines show incidence of light from the sky. Solid arrow denotes mean direction of activity. (Redrawn from Kramer, 1951.)

as a cue during a long migratory flight, a bird must allow for the apparent motion of the sun across the sky; it must change its angle of orientation relative to the sun's position in a manner that correctly compensates for the earth's rotation.

To study this, Kramer and his co-workers developed the technique of training a bird to find food in a particular compass direction in a circular cage (Kramer and von St. Paul, 1950). Starlings, pigeons, and Western Meadowlarks (*Sturnella neglecta*) were trained during one part of the day and then tested at other times when the sun was in very different positions. Most birds continued to orient in their training directions (Kramer and von St. Paul, 1950; Kramer, 1952; von St. Paul, 1956). This implies that the birds had adjusted correctly for the sun's passage through the sky. The birds would even orient to an artificial sun consisting of a 250 W light bulb. If this source was held stationary, they shifted the angle of their movements through the day in such a way as to compensate for the nonexistent motion of this artificial sun (Kramer, 1953).

The most powerful technique for studying the role of a chronometer or time sense in sun orientation is the so-called clock-shift experiment. If a bird is exposed to a light–dark cycle whose onset and termination are shifted earlier or later than local time, it quickly resynchronizes its behavioral and physiological rhythms to be in phase with the new day–night regimen. This makes it possible to study the use of the sun by a bird whose biological clock is out of phase with local time. The rate of apparent motion of the sun is approximately 15° per hour. This leads to the prediction that a bird whose time sense has been retarded by 6 hours should take up an inappropriate orientation angle from the sun and make a clockwise error of approximately 90°. Hoffmann (1954), in a classic study, confirmed these predictions and provided convincing evidence of the integration of the time sense with sun orientation. Using European Starlings trained to find food in a particular compass direction, he was able to shift the directional choice of the birds at will by means of appropriate shifting of their biological clocks. As seen in Fig. 6, the starlings showed the predicted shift of 15° per hour when their clocks were slowed by 6 hours; when allowed to resynchronize with local time, their directional choices shifted back 90° counterclockwise.

After confirming Hoffmann's results using caged homing pigeons, Klaus Schmidt-Koenig (1958, 1960, 1961) studied the importance of sun orientation in free-flying birds. When clock-shifted pigeons were released at a variety of locations, they made large predictable errors in their selection of departure directions. Later studies have yielded similar results (Graue, 1963; Keeton, 1969).

The use of a sun compass has also been found among migratory waterfowl. Several species of ducks captured in Illinois showed a tendency

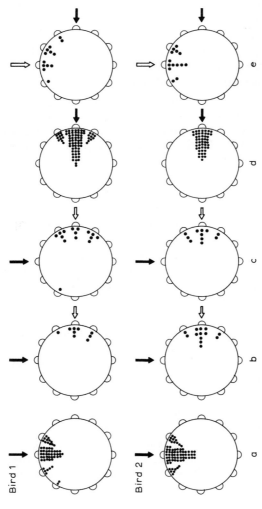

FIG. 6. Directional choices of two European Starlings trained to find food in a particular compass direction. From left to right a, tests during the training time in the natural day; b, tests following the artificial shifting of the birds' clocks 6 hours behind local time; c, tests after 3–28 days in constant conditions following the shift; d, results during training sessions in the artificial (shifted) day; e, results 8–17 days after reentrainment to the natural day. Each dot represents one critical choice (without reward); the black arrows indicate the training directions; the open arrows show the expected direction if the clock is used in sun-compass orientation. (Redrawn from Hoffmann, 1954.)

to depart in a specific northwesterly direction when released (Bellrose, 1958, 1963). This direction did not appear to be migratory and its meaning is unclear. Yet, whatever its biological significance, the direction is taken with reference to celestial cues. Orientation continued under clear skies by both day and night but disappeared under conditions of total overcast (Bellrose, 1958, 1963). Matthews (1961, 1963) made analogous observations on a sedentary population of Mallards (*Anas platyrhynchos*) in Britain. He extended the study to include clock-shifted ducks and found that their departure bearings were deflected in the predicted way when they were released under sunny skies (see Fig. 16).

Even species that are primarily nocturnal migrants seem capable of sun orientation. Von St. Paul (1953) placed two such species, the Red-backed Shrike (*Lanius collurio*) and the Barred Warbler, in orientation cages under the daytime sky and noted that each species took up its appropriate training direction.

2. Models of Sun Orientation

In the 20 years since Kramer first demonstrated the importance of the sun for orientation of the European Starling, sun-compass capabilities have been found to be widespread throughout the animal kingdom. The use of solar cues has now been documented in such diverse groups as decapod crustaceans, amphipods, arachnids, and insects, as well as in most classes of vertebrates. The Cold Spring Harbor Symposium volume of 1960 provides a good starting point for the reader interested in pursuing this topic in more detail.

Experiments such as those discussed above define a sun-compass mechanism in which the azimuth position of the sun is integrated with a biological time sense. But such a mechanism does not necessarily provide either the migrating or homing bird with sufficient information to orient toward its destination. Rather, the sun compass merely enables a bird to maintain a bearing once that bearing has been selected. Kramer was acutely aware of this distinction when, in 1953, he suggested that navigation be considered a two-step process. In the first, or map, step the bird determines its location relative to some goal, while in the second, or compass, step it takes up and maintains the appropriate bearing to that goal.

In principle, the sun could provide the necessary map information as well as serving as a compass. Matthews (1951a, 1951b, 1953a, 1955) was the first to propose a theory of complete bicoordinate navigation by means of the sun. He suggested that a displaced bird might determine its latitudinal displacement by observing the sun's movement in arc and extrapolating that arc to its zenith (noon) position. Comparison of the noon altitude at the new location with the remembered noon altitude of the sun at home would indicate whether the bird was north or south of its destination. Similarly, a bird could determine its longitudinal displacement by noting the sun's position along its arc and interpreting it with reference to a chronometer set on home (not local) time. If the bird is east of home, the sun will appear to have moved too far along its path; if west, the sun will not be far enough along its arc. With this information, the bird could take up the appropriate direction of flight to reach its destination.

An alternative model of bicoordinate sun navigation was proposed by Pennycuick (1960), who suggested that birds compare the sun's altitude and rate of change of altitude at their location with remembered values for the same time of day at their destination. The lines of equal values of these two parameters of the sun's position form a spatial grid that would also allow determination of distance and direction of displacement.

3. Theoretical Considerations

Although bicoordinate solar theories are attractive, both theoretical consideration and experimental evidence have failed to support them, and they have not gained wide acceptance. The theories demand extraordinarily accurate measurement of the sun's altitude, azimuth, and angular motion by a flying bird whose movement would result in changing parallactic errors in using landmarks as fixed reference points. Furthermore, the theory requires great precision in the bird's internal clock. A time change of one hour translates into a 15° change in the sun's position or, alternatively, to an east–west displacement of 15° of longitude. This means that a clock accurate to only ±10 minutes would be geographically accurate to only ±200 km (at a latitude of 40°N). A timing error of as little as 2–3 minutes might preclude the use of true sun navigation at distances of less than 40–50 km from home. Chronometers with this degree of accuracy have yet to be demonstrated in birds (Adler, 1963; Meyer, 1964; Miselis and Walcott, 1970).

Models of bicoordinate sun navigation assume that when a bird is displaced, its time sense remains in phase with the local time of the area from which it was displaced. But studies of behavioral and physiological circadian rhythms uniformly show that such rhythms readily and rapidly re-entrain (resynchronize) to new local conditions. Birds can be clock-shifted by as much as 6 hours in a matter of 2–6 days (Hoffman 1954; Schmidt-Koenig, 1958). This makes it difficult to believe that a migrant retains a memory of the local time of its overwintering location after migrating a thousand or more kilometers and spending several months on its breeding territory (and vice versa).

There are also difficulties associated with a migrant's use of a circadian clock to determine longitudinal displacement en route. The principal agents of displacement for migrating birds are crosswinds. Displacement distances of more than 200–300 km are probably extremely rare. These distances correspond to time differences of only 10–15 minutes, and it is logical to assume that the migrant would have resynchronized its time sense by this amount within a day or two. In other words, if a migrant is blown off course but remains in its new location for more than 24 hours, it may well have lost the ability to detect its displacement.

I would further suggest that migrants might be faced with a "latitude clock-shift" problem. The time of spring migration coincides with a time of increasing day lengths in the northern hemisphere. Birds traveling north compound this day-length change, since the days are longer at higher latitudes. If a bird took off from the Gulf Coast of the United States in early May and flew north all night, covering a distance of 550 km, dawn would find it at a locality where the day length had "suddenly" increased by 20 minutes. Sunrise would occur 10 minutes earlier than expected. Although the bird had flown due north, the sunrise information would be equivalent to an eastward displacement of some 200 km. How a migrant copes with this problem, if indeed it is a problem, is totally unknown.

For any model of bicoordinate sun navigation to be operative, the migrant must either correct for its displacement immediately and return to its original longitude before its chronometer has had the opportunity to reset, or else the bird must possess a second, stable time sense that is resistant to changes in environmental cues. To date, evidence for the existence of such a stable clock is lacking. One final possibility remains. If we view navigation as a two-step process, the migrant could perform the map step immediately following its displacement and determine a new, corrected preferred direction at that time. To take up this new bearing at some later time, the bird need only refer to the sun for compass (Type II) information. Ironically, this can be done accurately only *after* the chronometer has resynchronized with local time. This should be investigated by transporting birds longitudinally and comparing their preferred orientation before and after being allowed to resynchronize to the new local time.

4. Experimental Evidence

Aside from these theoretical considerations, a growing body of empirical results argues against Type III models of sun navigation and in favor of a much simpler use of solar information.

First, and most important, the results of many clock-shift experiments are contrary to the predictions of a bicoordinate model. Let us consider the example of a homing pigeon whose chronometer has been advanced by 6 hours. The pigeon now is released at 9:00 AM local time at a location 60 km north of the home loft. According to a bicoordinate model, the bird should interpret the sun's position with reference to its internal time sense, which indicates it is 3:00 PM. The sun is not as far along in arc as it would be at the home location, and it is rising rather than falling. The bird would interpret this time shift as a geographic displacement of 90° of longitude to the west. The 60 km north displacement would be inconsequential compared with the westward displacement of over 7000 km, and the bird would depart toward "home" to the east. If the bird had been displaced 60 km south, it would make little difference, since this distance also is minuscule compared with the apparent 90° westward displacement. This bird also would depart in an eastward direction (Fig. 7.)

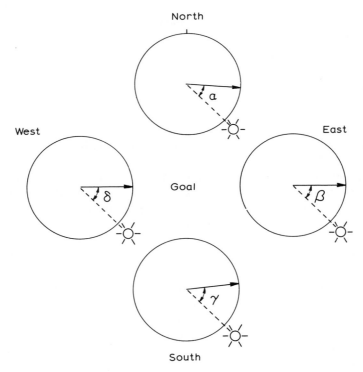

FIG. 7. Predicted orientation of clock-shifted birds according to Matthew's sun navigation hypothesis. The birds are released from the north, east, south, and west of home in the mid-morning. Their internal clocks have been reset by 6 hours and indicate mid-afternoon. Regardless of local release location, the birds should interpret the clock-shift as a massive displacement (90° of longitude; 7000 kilometers) to the west, and reorient toward "home" in the east. (See text for full explanation.)

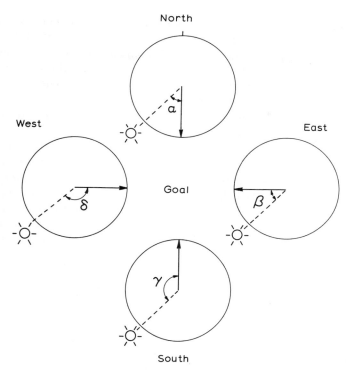

FIG. 8. Correct homeward orientation of birds released in the mid-afternoon when their internal clocks are in synchrony with local time.

If, on the other hand, the bird obtained its map information from some alternative source and merely used the sun as a compass, a very different set of predicted departure directions would be generated. When released 60 km north of home, the bird would determine (by unknown means) that home was to the south and it would take up a southward bearing relative to the sun's position. Under normal conditions, the sun is in the southwestern sky in the midafternoon, and south is approximately 45° counterclockwise from the sun (angle α in Fig. 8). In our experimental situation, the bird's chronometer indicates it is midafternoon, whereas in actuality, it is 9:00 AM. The pigeon will make a predictable error and fly off at angle α relative to the sun. Since the sun is rising in the southeastern sky, this results in an eastward departure, a shift of 90° counterclockwise from home (Fig. 9). Similar calculations for releases conducted east, south, or west of home all give the same predicted shift of 90° counterclockwise from homeward (Figs. 8 and 9).

Experiments now can be designed with clock-shifted birds in which opposite results are expected according to the two types of sun orientation hypotheses. In such experiments, it has been shown repeatedly that pigeons choose bearings consistent with a simple sun-compass (Type II) model (Schmidt-Koenig, 1958, 1960, 1961; Graue, 1963; Keeton, 1969). To dramatize this point, I enlisted the cooperation of my colleague William Keeton. Clock-shifted pigeons were released in the four cardinal compass directions from the Cornell lofts. Their initial departure bearings are presented in Fig. 10. The results confirm the use of the sun as a simple compass, as can be seen by comparing the predicted directions of Fig. 9 with the actual bearings of Fig. 10.

A second way of testing sun navigation hypotheses is by means of slight clock shifts involving time changes of an hour or less. According to a bicoordinate theory, a time error of 1 hour should be interpreted as a displacement of 15° of longitude along an east–west line. As mentioned previously, at a latitude of approximately 40°N, this is equivalent to a displacement of approximately 1200 km. Similarly, a 20 minute shift is equivalent to a 400 km displacement, and a 5 minute shift to

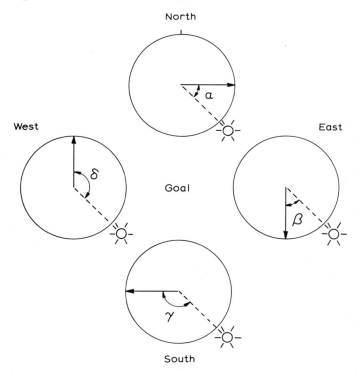

FIG. 9. Predicted orientation of clock-shifted birds according to a sun-compass model. The birds determine the direction home by some other means, then take up the appropriate compass bearing with reference to the position of the sun. The angles are as in Fig. 8, but the resulting directions are shifted 90° counter-clockwise since the sun is in the mid-morning position. (See text for full explanation.)

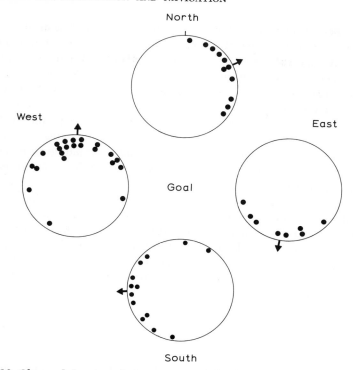

FIG. 10. Observed departure bearings of pigeons whose clocks have been advanced by 6 hours and then released 30 to 80 kilometers to the north, east, south, and west of the Cornell lofts. Solid arrows denote mean directions.

approximately 100 km. Thus, minor shifts in the time sense should give rise to major changes in predicted "homeward" bearings. If the sun is being used merely as a compass, however, the predicted deflection from the true homeward bearing should only approximate 15° per hour. Consequently, clock-shifts of 20 or 5 minutes should produce changes in departure bearings so slight as to be almost undetectable.

Two such series of experiments have been performed to date (Walcott and Michener, 1971; Schmidt-Koenig, 1972). Although the results are somewhat meager, no large-scale deflections in departure bearings were observed, and both authors concluded that the sun was being used solely as a compass.

Other experiments have attempted to dissect out the latitudinal component of bicoordinate models by confining pigeons out of sight of the sun at the time of the autumnal equinox. At this time of year, the sun's day-to-day changes in zenith altitude are considerable, and a bird that has been prevented from viewing the sun for a week or more should detect a significant decrease in the height of the sun at noon. According to Matthews' theory, this should indicate to the bird that it is north of home. It now becomes possible through a careful manipulation of time and distances to transport a pigeon to the south of its home loft and still have the sun's altitude lower at the release location than it was when *last viewed* at home. Matthews (1953a) performed such an experiment and found that the "deprived" pigeons departed to the south, away from home, in accordance with the hypothesis. Attempts to repeat these observations, however, have yielded negative results; deprivation of a view of the sun has not caused any significant change in homeward bearings (Rawson and Rawson, 1955; Kramer, 1955, 1957; Hoffmann, 1958; Keeton, 1970).

To summarize, a large body of evidence demonstrates that the sun is an extremely important directional cue for diurnally migrating and homing birds. Support for the idea that the sun provides accurate information about geographic location or the precise direction toward a predetermined goal remains unconvincing. Rather, the general use of the sun appears to be as a dominant cue for maintaining a bearing that is determined by some other means.

C. THE USE OF STELLAR CUES

1. Evidence of Star Orientation

Since the 1930's, physiologists have known that when birds are held in captivity for long periods of time they often exhibit intense activity at night (Palmgren, 1937, 1938, 1949a,b; Merkel, 1938, 1956). This activity seems almost entirely restricted to species that migrate at night; further, the seasonal timing of this behavior coincides with the periods of spring and fall migration.

Hypothesizing that this nocturnal restlessness, or *Zugunruhe*, is a direct expression of migratory behavior, Gustav Kramer decided to study its orientation. He placed Blackcap Warblers (*Sylvia atricapilla*) and Redbacked Shrikes outdoors under the night sky and recorded their spontaneous locomotor activity in a circular cage. He discovered that these species oriented in a particular direction (Kramer, 1949, 1951). As long as the sky was clear, orientation was consistent from night to night (although it was not in the predicted direction of migration). Activity decreased in magnitude and lost its directionality when the sky clouded over.

Franz Sauer (1957) extended these findings and reported that Blackcaps, Garden Warblers, and Lesser Whitethroats (*Sylvia curruca*) took up appropriate southerly directions when tested in cages outdoors under the autumn night sky. As long as the brightest stars were visible, the migratory direction was maintained; but if the stars were hidden by dense overcast, the birds circled randomly or ceased nocturnal activity altogether. Two Blackcaps were tested in the spring, and they both selected north-northeast bearings, in keeping with the predicted direction of spring migration.

Since this early study, star orientation has been reported for a wide variety of bird species ranging from waterfowl (Bellrose, 1958, 1963; Hamilton, 1962b; Matthews, 1961, 1963) and shorebirds (Sauer, 1963; S. T. Emlen, unpublished observations) to a large number of songbirds (Sauer and Sauer, 1960; Mewaldt and Rose, 1960; Mewaldt *et al.*, 1964; Hamilton, 1962a, 1966; Shumakov, 1965, 1967a; Emlen, 1967a,c; Dolnik and Shumakov, 1967; Sokolov, 1970; Rabøl, 1969, 1970, 1972; C. J. Ralph, personal communication; P. De Sante, personal communication). Few studies, however, do more than demonstrate that a caged bird can select its direction under starry skies or, perhaps, that orientational accuracy decreases under overcast.

Sauer (1957) was the first to investigate detailed mechanisms of star orientation. To do this, he exposed warblers to the artificial skies of a small planetarium. When he adjusted the star projector so that it accurately duplicated the sky outdoors, the birds again took up their migratory directions. Under autumn skies, Lesser Whitethroats selected a southeasterly direction, while a Garden Warbler and a Blackcap oriented to the southwest. Only a single Blackcap was studied under a spring sky in the planetarium; it adopted a northeasterly bearing in agreement with this species' migratory route in this season. By means of appropriate control experiments, Sauer was able to demonstrate that his birds definitely were responding to information provided in the artificial starry sky and not just to visual or acoustical artifacts that are so commonplace in planetarium settings. When he shifted the star projector so that the position of stellar north was reversed 180°, the birds shifted their orientation accordingly.

I have also conducted planetarium studies (Emlen, 1967a,b) on a North American migrant, the Indigo Bunting (*Passerina cyanea*). These birds selected southerly bearings in the autumn when they were tested under the natural night sky in small, circular cages. Individuals maintained approximately the same direction when they were brought indoors and tested under the artificial autumn skies of a large planetarium (Fig. 11). These birds also took up appropriate north to northeasterly directions in the spring when they were exposed to low-latitude spring skies. Results of control experiments were similar to those of Sauer; the buntings reversed their orientation when the direction of stellar north was reversed, and nocturnal activity either became random or ceased when the artificial stars were turned off and the planetarium dome was diffusely and dimly illuminated (Emlen, 1967a).

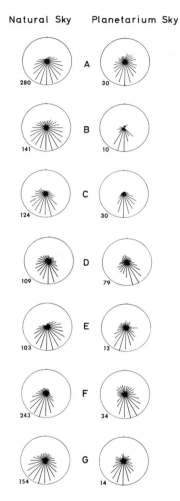

FIG. 11. *Zugunruhe* orientation of caged Indigo Buntings (*Passerina cyanea*). Left: Behavior when tested outdoors under clear night skies. Right: Behavior when tested in a large planetarium under artificial skies set to simulate outdoor conditions. Vector diagrams are plotted on a proportionality basis with the radius equaling the greatest amount of activity in any one 15° sector. The number that this represents is presented at the lower left of each diagram. (From Emlen, 1967a.)

The finding that a caged migrant will take up its migratory direction when in the confines of a planetarium has provided the ornithologist with a powerful experimental tool. By appropriate manipulations of the artificial sky, the investigator can tease out the different components of star orientation mechanisms.

2. Models of Star Orientation

Theoretically, sufficient information is available in the night sky to allow bicoordinate navigation. In a manner analogous to the sun navigation model described above (Section IV,B,2), a migrant could compare the altitude and azimuth position of a selected star or group of stars with their remembered positions at some goal. When coupled with an extremely accurate and stable time sense, such information could tell the bird its distance and direction of displacement.

Alternatively, stellar cues might provide the migrant with only compass (Type II) information. There are two basic ways of obtaining such directional information. In the first, a migrant could select a specific star or group of stars and orient by flying at a particular angle from them. This would require the integration of an internal time sense, since the bird could not maintain a straight course unless it changed its angle of orientation relative to the selected stars to compensate for their apparent motion. I wish to emphasize that the requirements for such a system to be operative at night are much more demanding than would be necessary for diurnal, solar orientation. Different stars are visible above the horizon at different times of night and at different seasons. Instead of one obvious celestial object, many potential cues are available in the night sky and the migrant must be able to locate consistently the specific one or ones of importance. This presumably requires some form of pattern recognition. Additionally, the necessary rate of compensation differs depending upon the position of the star or stars being used. Celestial motion is an apparent motion produced by the earth's rotation of one revolution each 24 hours. All stars appear to move with an *angular* velocity of 15° per hour. But the *linear* velocity of a star located near Polaris (the North Star), which moves through a very small arc, is quite small compared to that of a star located near the celestial equator. Consequently, if a bird were to use several star groupings located in different parts of the sky, it would have to have several compensation rates at its disposal and be able to pair specific stars and rates of movement appropriately. Finally, the actual direction of this compensation would have to vary, being clockwise for stars located in the northern sky and counterclockwise for those to the south (see Fig. 12).

Without time compensation, no single star (with the exception of Polaris) could provide sufficient information to allow a bird to maintain a given direction through time. In the second model, the bird would make use of additional, configurational, information available in the night sky and use patterns of stars to locate directional reference points. For example, the Big Dipper, Ursa Major, is localizable because of the characteristic spatial pattern of its component stars. By visually extending

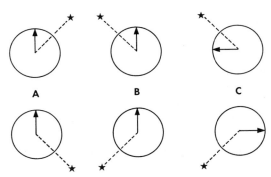

Fig. 12. Expected shift in orientation following resetting of the internal clock six hours behind local time, and assuming compensation for motion of a star located in the northern sky (upper diagrams), or in the southern sky (lower diagrams). (A) represents normal orientation early in the evening when the critical star is located to the NE or SE respectively. Later in the night, in (B), orientation continues although the stars have rotated to a more westerly position. In (C), the bird's time sense has been shifted to coincide with time (A) while the actual time and, hence, the position of the stars, coincides with situation (B). Thus, the direction of the predicted error is dependent upon the north-south location of the critical star. (After Emlen, 1967b.)

the pouring line of the dipper, one can find Polaris and hence geographic north. The Ursa Major pattern moves across the sky, but its shape remains constant and it preserves its relationship to Polaris. Since each star bears a fixed geometric relationship with all others, it is theoretically possible for a bird to locate a directional reference point from an infinite number of configurations. The principal difference between this model and the one presented above is that a compass direction can be located as long as the important star patterns are visible, regardless of any knowledge of the time of night, the season, or the geographic location.

3. Clock-Shift Experiments

These two alternative hypotheses can be separated by means of clock-shift experiments. The usual procedure is to confine birds for several days to a week under an altered light–dark regime. With a planetarium, however, one can create a situation in which the bird's internal time sense is out of phase with astronomical time without subjecting the bird to the trauma of confinement. The star projector can be altered to present a sky that is advanced or retarded from local time. If the chronometer is an integral component of the star orientation system, predictable shifts in migratory orientation should result.

A demonstration that the chronometer is important for nocturnal orientation would not, by itself, allow us to distinguish a bicoordinate navigation system from a time-compensated star compass similar to the sun compass discussed previously (Section IV,B). To separate these possibilities, we must examine the direction and magnitude of the resulting shifts in the bird's orientation. This can also provide valuable information about the portion of the sky and, thus, the critical stars used in orientation. If nocturnal migrants are responding solely to configurational information in the night sky, no directional change would be expected under the clock-shift situation. Note, however, that without integration of temporal information, a migrant could not obtain bicoordinate information from the stars alone.

Two different questions thus are posed by planetarium clock-shift experiments. First, is the chronometer an integral part of star orientation? And, second, does the system provide sufficient information for direct goal orientation or for reorientation following displacement? To date, only a few experiments have been performed, and different species performed differently. Because of their theoretical importance, these studies will be discussed in some detail.

The Sauers studied the behavior of three individual warblers—one Lesser Whitethroat (Sauer, 1957) and two Blackcaps (Sauer and Sauer, 1960)—under clock-shifted planetarium skies. Their technique was to observe the birds directly from underneath the circular orientation cage and to record the length of time the birds spent fluttering their wings while aiming in a particular direction. Periods without wing quivering were not considered *Zugunruhe* and were omitted from the analyses.

The Lesser Whitethroat was studied in the autumn of 1956 and oriented to the south-southwest when tested under a "normal" planetarium sky. When the projector was manipulated to give a sky that was advanced relative to local time (equivalent to the bird's chronometer being slow) the bird altered its directional behavior. When the discrepancy between local and planetarium time was 1 hour, the whitethroat headed southwest; when the discrepancy was $1\frac{1}{2}$ hours, it aimed to the west; and when the difference was 3–5 hours, it either oriented west-northwest or was random. When the sky's position was retarded from local time (equivalent to the bird's chronometer running fast) the results were less clear cut. The whitethroat continued to orient south-southeast in directions not significantly different from normal under skies phase shifted 1 and 3 hours (Sauer, 1957; Wallraff, 1960a,b).

An incorrect planetarium sky could be interpreted in one of three ways: either the bird's chronometer is wrong, the season has changed, or the bird has been displaced geographically to the location where the experimental sky would be in phase with local time. Sauer (1957) interpreted his experiments in terms of projected geographic displacements and considered his results as support for a hypothesis of bicoordinate stellar navigation. His advanced skies would be "equivalent" to instantaneous relocations 13° to 73° of longitude to the east. The whitethroat's westerly shifts in orientation under these skies are in accord with such a hypothesis, as is the fact that the amount of the shift exceeds the 15° per hour predictable on the basis of a time-compensated star compass alone. But the results under the retarded skies fail to provide evidence for the incorporation of a chronometer at all, much less for a bicoordinate system. These skies were equivalent to geographic displacements of 17° and 51° of longitude to the west. Yet the whitethroat showed no significant tendency to compensate with a more easterly orientation.

In the autumn of 1958, Sauer and Sauer (1960) repeated planetarium clock-shift experiments with two Blackcap Warblers. Under a normal planetarium sky, both individuals displayed a southwesterly directional preference. When exposed to skies advanced from local time, one individual (no. 632) changed its behavior in a manner similar to that of the whitethroat. Its orientation shifted more westward (clockwise) as the sky was altered from 1 to 5 hours out of phase with local time. (I have recalculated, pooled, and replotted the original data for this bird in Fig. 13.) The Sauers interpreted these results as confirmation of their Type III star navigation hypothesis. I feel, however, that the magnitude of the Blackcap's directional shifts is not significantly different from 15° per hour. Consequently, this bird's behavior is also consistent with a time-compensated star compass model. The direction and amount of change further implies the use of stars located in the southern sky, close to the celestial equator.

When the planetarium sky was set behind local time, the Blackcap's behavior was highly variable. With skies 1, 3, and 6 hours behind local time (equivalent to the chronometer being advanced equivalent amounts), orientation did shift to the south and southeast, although the degree of scatter increased considerably. In the first two cases, the amount of shift was more than would be expected from a star compass alone, while in the last case, it was somewhat less than such a model would predict. Interspersed with these results, the Blackcap showed a weak southwesterly orientation (no change from normal) under a sky 2 hours retarded, and its activity was random under skies deviating 4 and 5 hours from local time.

In one final experiment, another Blackcap (No. 652) was observed for $2\frac{1}{2}$ minutes while exposed to a sky 3 hours behind local time. The bird did shift its behavior to the southeast, but the importance of such a short observation period is questionable.

All in all, the results of these experiments are difficult to interpret and have been the cause of considerable controversy (see Wallraff, 1960a,b). Personally, I feel that they provide a strong suggestion of the use of a time sense in star orientation. As support for a model

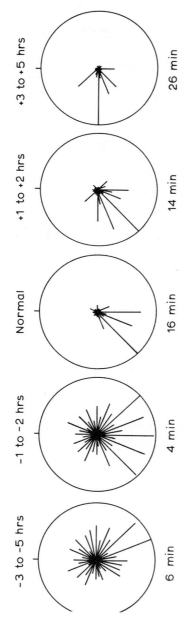

FIG. 13. *Zugunruhe* behavior of a Blackcap (*Sylvia atricapilla*) (No. 632) under planetarium skies adjusted to be out-of-phase with local time. Data were taken from Sauer and Sauer (1960, Figure 5) and translated into amounts of time that the bird fluttered its wings while aiming toward different sectors of the cage. Several experiments have been pooled and replotted in this figure. The number shown below each diagram represents the total amount of time (rounded to the nearest minute) spent in "active *Zugunruhe*" under that planetarium condition. The actual groupings used were the following: for tests under skies retarded 3 to 5 hours, (Sauer and Sauer, 1960) Figure 5t, a, and v; for −1 to −3 hours, Figure 5w and x; for "normal" skies, Figure 5y and z; for +1 to +2 hours, Figure 5a, b; and for skies advanced 3 to 5 hours, Figure 5d and e.

of bicoordinate star navigation (Sauer, 1957, 1958, 1961; Sauer and Sauer, 1960), I find them less convincing.

My studies (Emlen, 1967b) suggest a very different orientational strategy for Indigo Buntings. Pilot experiments conducted during the autumn migration season under temporally altered planetarium skies failed to produce any deflection in orientation (S. T. Emlen, 1967b; unpublished observations). I performed a more extensive series of experiments under clock-shifted skies in the spring of 1965. Eight different buntings were active for a total of 39 experiments, each approximately 2 hours in duration. The north to northeast orientation of these birds under planetarium-normal conditions was compared with their behavior when exposed to artificial skies 3, 6, and 12 hours out of phase with local time.

All birds tended to maintain their normal spring migratory directions. The results from one of the individuals whose orientation was most concentrated are given in Fig. 14. Even if the birds were relying upon stellar cues located very near Polaris with slow rates of movement, one would expect slight, consistent, changes in direction. But, although variations in orientation did occur, they were neither consistently clockwise nor counterclockwise in direction. In Fig. 15, I have replotted the deviations between the mean bearings from individual clock-shift tests and the mean directions obtained under planetarium-normal conditions. In the case of the 3-hour shifts (which, according to a bicoordinate model, would be indicative of a longitudinal displacement of 45° or approximately 3600 km), the mean bearings are remarkably close to the normal spring migration directions. The same is true for the results obtained under the 6-hour retarded and 12-hour advanced skies. Only in the case of the 6-hour advance (chronometer 6 hours behind local time) is there a slight suggestion of a counterclockwise shift in orientation. This difference is not statistically significant, and the magnitude (21°) is too minor to be considered as compensatory for an eastward displacement of over 7000 km. Rather, if these results were to be interpreted as evidence for the integration of a clock mechanism at all, they would imply a star compass based upon stars with slow rates of movement and located in the northern sky.

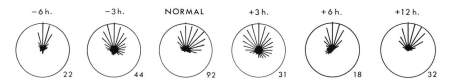

FIG. 14. *Zugunruhe* orientation of a caged Indigo Bunting exposed to planetarium skies advanced and retarded from local time. Vector diagrams are plotted as in Fig. 11. (Redrawn from Emlen, 1967b.)

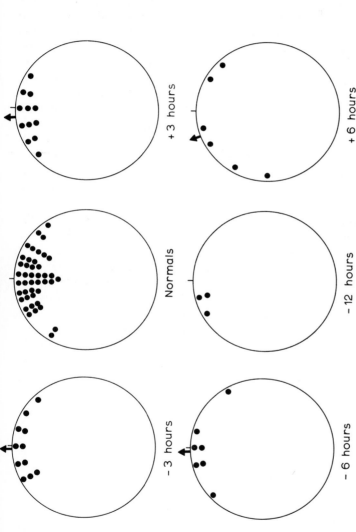

FIG. 15. Mean directions taken by caged Indigo Buntings under temporally altered planetarium skies. Results from 39 clockshift and 43 control (planetarium "normal") experiments, each of two hours duration and representing eight individual buntings are pooled in these diagrams. Each dot represents the mean direction from one such experiment. The data are plotted in 15° sectors with 360° representing the mean direction under unmanipulated planetarium conditions. The "normal" diagram gives an indication of the consistency of orientation from night to night. The remaining vector diagrams (with overall mean directions given by the black arrows) show any shift in orientation caused by resetting the planetarium sky 3, 6, and 12 hours out-of-phase with local time. No significant directional changes occurred. (Recalculated from Emlen, 1967b, and unpublished data.)

Since the buntings maintained their north to northeastward orientation under phase-shifted skies, I hypothesized (Emlen, 1967b) that temporal compensation for stellar motion was *not* an essential component in the migratory orientation mechanism of this species. Indigo Buntings presumably can determine their migratory direction from configurational information provided by the patterning of stars. This would give them a directional reference and hence a compass capability.

White-throated Sparrows (*Zonotrichia albicollis*) may also use the patterning of stars to determine direction. Preliminary studies involving radar tracking of free-flying clock-shifted birds failed to yield consistent deflections in migratory bearings (S. T. Emlen and N. J. Demong, unpublished observations). Gauthreaux (1969) also studied the directional preferences of nine individual White-throated Sparrows that he caged for extended periods of time under stationary skies in a small planetarium. With one exception, the birds maintained a stable migratory direction throughout the night, failing to compensate for the expected rotation of the sky.

Finally, Matthews (1963), in his studies of the orientation of Mallards, found that clock-shifted birds released in the daytime exhibited typical deflections in their departure bearings (Fig. 16, top). However, when similarly treated individuals taken from the same population were released at night under starry skies, they continued to orient northwest in the same direction as the controls (Fig. 16, bottom). This led Matthews to state (1963, p. 426), "we are forced to the conclusion that the time element does not enter into star-compass orientation, that measurement of the movement of certain stars in azimuth is not the concern of these birds. From this it does not seem possible to escape the implication that the birds are finding their compass direction by reference to the orientation of star *patterns*, possibly using the same constellations as we do ourselves."

Only one attempt has been made to study which star patterns are of particular importance to migrating birds. Using Indigo Buntings, I conducted a series of experiments in which selected stellar configurations or portions of the planetarium sky were blocked from view (Emlen, 1967b). My initial finding was that most major star groups could be removed one at a time with no discernable effect on the buntings' behavior. When I removed entire sections of the sky, however, orientation deteriorated considerably. A general tendency emerged for most buntings to rely upon the northern, circumpolar area of the sky within about 35° of Polaris.

Of equal importance, perhaps, were two corollary findings. First, there appears to be considerable redundancy in the pattern recognition process of buntings. Several individuals were disoriented when the circumpolar

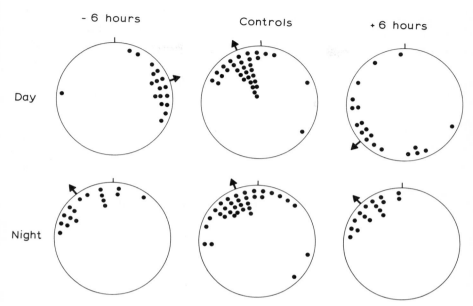

FIG. 16. Effect of clock-shifting upon the departure directions selected by Mallards (*Anas platyrhynchos*) when released in the daytime (top) and at night (bottom) near Slimbridge, England. Arrows denote mean bearings. (Redrawn from Matthews, 1963.)

portion of the sky was blocked but oriented properly when major groups of stars within this zone were removed individually. If a bird is familiar with a particular portion of the night sky, removal of a group of stars might merely force it to rely upon some alternative constellation. Because of this redundancy, continued orientation in the absence of a particular star pattern cannot be interpreted as evidence for the unimportance of that pattern. All one could state would be that sufficient directional information remained in its absence.

Second, different individual buntings were relying upon different cues in the night sky. This variability is inconsistent with any hypothesis of a predetermined, genetically fixed "star map" of constellations and suggests that the maturation of star orientation abilities is a complex, individualistic process. [For a fuller discussion of the ontogenetic development of star orientation capabilities, the reader is referred to Emlen (1969b, 1970a, 1972) and Wallraff (1972).]

4. Seasonal Changes in Orientation

If some birds do not possess a celestial map and compass capability, what factors determine the directions they select as their preferred compass bearings? And what is responsible for the major reversal of migratory directions between spring and fall? Does stellar information play a role in either process? The first of these questions is largely unanswered; the second is only beginning to receive attention.

The stars that are present in an autumn night sky are very different from those available in the spring. The slight inequality of the length of the solar and sidereal days produces a seasonal change in the temporal positions of stars. Because of these seasonal differences, a migrant might possess a specific northward directional response to the stellar stimuli present in a spring night sky and a different, southerly, response to the stellar stimuli of the autumn sky. This hypothesis has been tested with European warblers (Sauer, 1957, 1961; Sauer and Sauer, 1960) and Indigo Buntings (Emlen, 1967b), again with differing results.

When a planetarium sky is adjusted to be 6 hours behind local time, the new sky simulates the conditions that will be present at the same location and same time of night but at a season 3 months later. While performing clock-shift experiments, the Sauers noted that the Lesser Whitethroat and a Blackcap frequently showed random activity under skies phase shifted from 5 to 10 hours. [The actual conditions under which disorientation resulted were the following: for the Lesser Whitethroat, under skies 7 hours retarded and 4 hours advanced from local time; for the Blackcap, under skies 2, 4, 5, and 7 hours retarded and 6, 7, 8, 9, 10, and 11 hours advanced from local time.] Since these skies are those of summer and winter, they hypothesized that the warblers were unable to orient because the stellar cues essential for navigation are only present at the actual times of migration. Further, when the warblers (one Garden Warbler and three Blackcaps) were confronted with a planetarium sky typical of the opposite migration season, they exhibited "conflict" behavior, concentrating their locomotor activity to the north and south, along an axis parallel to the line of migratory flight.

The behavior of Indigo Buntings under skies adjusted by 6 and 12 hours was very different. Even with the sky advanced 12 hours, the celestial equivalent of the opposite migration season, the buntings continued to orient in their typical migratory direction, to the northeast (Figs. 14 and 15). The visual stimuli typical of the opposite migration season failed to evoke any change in directional response. This, coupled with the finding that the northern circumpolar area of the sky was most important during *both* spring and fall migration seasons (S. T. Emlen, 1967b; unpublished observations), raised the possibility that Indigo Buntings may use the same stellar cues during both their northbound and southbound travels. Perhaps it is not the visual features of the spring or fall sky per se, but rather some feature of the bird's physiologi-

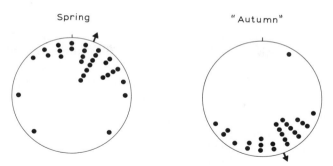

FIG. 17. *Zugunruhe* orientation of Indigo Buntings artificially brought into the physiological states appropriate for spring (left) and autumn (right) migration. Birds from both groups were tested simultaneously, in May, under a spring planetarium sky. Each dot represents the mean direction from one experiment. The results from 6 and 5 birds, respectively, are pooled in these diagrams. Mean direction in spring = 22°; probability by Rayleigh test = 0.000. Mean direction of "autumn" birds = 157°; Rayleigh probability = 0.000. (Replotted from Emlen, 1969a.)

cal state that dictates the preferred migratory direction. By appropriate photoperiod manipulations, I brought two groups of male Indigo Buntings into spring and autumn migratory conditions synchronously (Emlen, 1969a). When their directional preferences were tested under identical spring planetarium skies, the two groups showed different behavior. Birds that physiologically were anticipating a spring migration oriented to the north-northeast; those in autumnal condition headed south-southeast (Fig. 17). Important differences must exist between the physiological states of migration in the spring and autumn, and I hypothesized that seasonal changes in the hormonal state of the birds would be found to play a key role underlying the seasonal reversals of preferred migration directions (Emlen, 1969a). Recent studies by Martin and Meier (1973) with White-throated Sparrows support this prediction. They report being able to reverse the polarity of orientation in caged sparrows by administering exogenous prolactin and corticosterone at different times of day.

5. Discussion

Planetarium experiments suggest that at least two different strategies of star orientation may have evolved among nocturnal migrants. As additional studies are conducted, still other mechanisms may emerge.

The Indigo Bunting is a moderate-distance migrant that breeds throughout much of the eastern United States and winters in the Bahamas, southern Mexico, and Central America. For most individuals, the journey is probably 1500–2200 km in length. Even the southernmost extension of the species' wintering range lies well north of the equator

(10°N latitude). For such a migrant, use of the northern circumpolar area of the sky makes good sense. These stars are constantly above the horizon and hence are visible at all seasons and along the entirety of the migration route.

But consider the problems of a transequatorial migrant that travels far into the southern hemisphere each winter. Stars close to the northern celestial pole sink below the horizon and are no longer visible to guide the bird on the southern portion of its journey. Star orientation by pattern recognition alone might be more difficult for such long-distance migrants. Presumably they would have to be able to use several patterns of stars having different declination values. If a north reference point is to be extrapolated from star patterns alone, this extrapolation becomes less accurate as the angular distance between the critical stars and Polaris increases. The stars that might be optimal from the standpoint of visibility at all latitudes along the flight path would be those located in the southern sky, close to the celestial equator. This is precisely the area of the sky where the rate of apparent motion of stars approximates that of the sun. Consequently, it might be advantageous for these birds to employ some form of clock-compensated star orientation system, using the same rate and direction of compensation that they presumably use for sun-compass orientation during the day.

All this will remain speculative until many more comparative studies are performed. But it is interesting in this respect that several species mentioned as providing hints of time-compensated nocturnal orientation [Blackcaps, Whitethroats, Barred Warblers, Scarlet Rosefinches, and Lesser Golden Plovers (*Pluvialis dominica*)] migrate farther and to lower latitudes than do Indigo Buntings or White-throated Sparrows.

D. BEHAVIOR UNDER OVERCAST SKIES

As evidence of the importance of celestial cues accumulated in the 1950s and 1960s, it produced a bandwagon effect, and claims soon began appearing in the popular literature that scientists had solved the mysteries of bird navigation. But the bandwagon was short-lived. However attractive the hypotheses of true celestial navigation might be, experimental support was still lacking. Directional orientation by means of a celestial compass alone cannot fully explain migratory behavior; such orientational systems can just as well enable a bird to select and maintain an inappropriate bearing as a proper one. Current knowledge of celestial orientation simply does not provide answers to the rather fundamental questions: What factors operate in the initial selection of one direction rather than another? What causes young, inexperienced birds to select an appropriate direction for their first migration?

1. Orientation under Cloud

While the attention of experimentalists was focussing on laboratory studies of celestial orientation, other ornithologists were studying broad patterns of migration by means of surveillance radar. Early radar investigations concentrated on describing the directional patterns of migration at different geographic localities and on assessing the volume of migration and the behavior of migrants under different meteorological conditions (see Eastwood, 1967).

It soon became apparent that the presence of overcast skies had only a slight, if any, inhibitory effect upon the volume of migration aloft. There are reports of birds changing direction and thereby avoiding or skirting around areas of cloud (e.g., Williams et al., 1972; Richardson, 1974), and Bellrose and Graber (1963) believed that passerine migrants changed altitude on many overcast nights, apparently attempting to climb above the cloud layer if it was sufficiently low. But, in general, radar studies have shown that migrants continue to fly along straight well-oriented paths under totally overcast skies. Both the mean direction of flight and, in many instances, the angular spread of individual tracks are comparable under clear and cloudy conditions. To date, documentation of orientation under complete cloud cover has been reported for birds migrating over the flat agricultural lands of Illinois (Hassler et al., 1963; Bellrose and Graber, 1963), inland in the Canadian Maritimes and upstate New York (Richardson, 1971, 1974; Griffin, 1972, 1973), and along the Atlantic Coast from Nova Scotia (Richardson, 1971, 1974) to Massachusetts (Drury and Nisbet, 1964; Nisbet and Drury, 1967) and Virginia (Williams et al., 1972; S. T. Emlen and N. J. Demong, unpublished observations), and the southeastern United States, including Georgia on the Atlantic Coast (S. A. Gauthreaux, Jr., personal communication) and Lousiana on the northern coast of the Gulf of Mexico (Gauthreaux, 1971; Able, 1974a). Similar observations have been made in Europe, both in Britain (Evans, 1966, 1968, 1972) and in Switzerland (Steidinger, 1968). Taken together, these observations have been made both in the daytime and at night and include shorebirds and waterfowl as well as passerines.

Vector diagrams of the orientation of nocturnal migrants passing over a radar site in New Brunswick are presented in Fig. 18. These data, generously provided by W. John Richardson, clearly show the birds maintaining an accurate northeastward orientation on both clear and totally overcast nights. Figure 19 is a photograph of the display of a height-finding radar taken on the night of May 12, 1971. It shows that the birds whose orientation is depicted in Fig. 18 were flying at altitudes well below the layer of overcast.

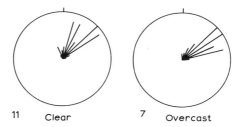

FIG. 18. Flight directions of nocturnal migrants near St. Margarets, New Brunswick, as determined by radar analysis. Left: Orientation under clear skies on the night of 11 May 1971. Right: Orientation under total overcast during the following night of 12 May 1971. Surface winds on 12 May were from the south-southwest at 13 knots. Diagrams plotted on a proportionality basis with the greatest number of birds in any one 10° sector shown to the left of the figure. (From Richardson, 1974.)

Instances of disorientation are rarely observed on radar, but where they have been reported, they have invariably been associated with conditions of fog, total overcast, or both (Lack and Eastwood, 1962; Drury and Nisbet, 1964; Bellrose, 1967a). But the growing list of studies cited above demonstrates that overcast per se does not result in disorientation. This suggests that although overcast skies obscure the sun and stars and therefore eliminate one system of direction finding, most migrants possess alternate guidance systems that are at least sufficient to enable the birds to *maintain* their flight directions. Only when these alternative cues are simultaneously eliminated does disorientation result.

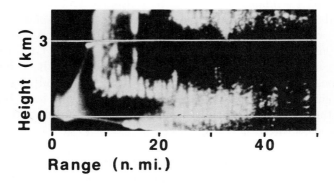

FIG. 19. Height distribution of birds and clouds on the night of 12 May 1971. Photograph shows the display of a high-power nodding-height finder radar located at St. Margarets, New Brunswick. Cloud echo occurs at 4000 meters while the bird echoes are concentrated well below this, near 1000 meters. Echoes within 15 nautical miles of the radar site are obscured by ground clutter. (From Richardson, 1974.)

2. Use of Visual Reference Markers

What sources of information might a migrant use to maintain its directional track in the absence of celestial input? The most obvious would be to use visual features of the landscape as points of reference. The bird would merely note the location of some visual marker on the ground below that is ahead of it along the desired line of flight. As long as two or more such markers remained visible, a migrant could easily maintain its course by successively piloting toward first one reference point and then another. In this way the bird could detect any slight displacement, turning, or wind drift and correct its heading to keep the same resultant track. This would require neither a familiarity with the landscape beneath nor an unrealistic degree of visual acuity.

On the basis of several hundred hours of night flying in a small aircraft at the altitudes where birds migrate, Bellrose (1971) reports that several features of the landscape are visible on even the darkest of nights. Landscape features, such as hills, mountains, and land–water interfaces, were the most prominent. On clear nights when the moon was in the sky, all but the smaller details of the landscape were detectable. On dark, overcast nights, topographic features were much less distinct, but major features, such as rivers, could be distinguished from 300 m and the shoreline of the Atlantic Ocean was clearly visible from at least 600 m. Artificial lights produced by human civilization could also provide optimal reference points for nocturnal migrants. Considering the apparent ease of maintaining a direction through the use of visual reference markers, it is not surprising that oriented flight continues under occluded skies.

3. Flight in Cloud

But what happens if visual contact with the ground is obscured by low overcast or fog? Can birds continue to orient properly in the absence of all visual cues? This question is of considerable importance and has led to a concerted search for cases in which nocturnal migrants are flying between cloud layers or in clouds. Understandably, birds do not often fly under these conditions; and when they do, ornithologists are rarely on hand or prepared to record all of the information necessary to document the event. This documentation requires an accurate assessment of the altitude of the flying birds, the horizontal extent of the total overcast conditions, and the height of the bottoms and tops of the cloud layers. Because of the difficulty of obtaining such information, no single instance yet compiled provides entirely convincing evidence that birds can maintain normal tracks while traveling through cloud.

However, several cases are extremely suggestive, and if one pools the available evidence it is difficult to avoid the conclusion that at least some birds are capable of maintaining fairly straight tracks while flying without visual contact with either celestial cues above or landscape cues below.

Bellrose and Graber (1963) provided the first suggestion of orientation in cloud. On the night of 27 May 1960, northern Illinois was covered by heavy cloud that extended from 300 m above ground to an estimated top of 1200 or 1500 m. The height distribution of the birds migrating aloft indicated that most were enveloped by the clouds. The mean direction of the radar targets was to the north-northeast, appropriate for spring migration through Illinois. However, the spread of tracks was quite large, with standard deviations ranging from 44° to 82° for samples of birds migrating at different altitudes. On 15 October 1960, clouds again extended from 200 m to an estimated 1050 m above ground, while birds were distributed up to 1500 m with a slight increase in density in the 300 m above the presumed cloud tops. Approximately half of the birds were flying within the cloud while the remainder were traveling above it. The flight directions of birds at all altitudes were well oriented to the south. No differences were apparent between the behavior of birds traveling in and above the overcast (standard deviation ranged from 20° to 33°).

The second report of possible orientation in overcast comes from Williams et al. (1971, 1972) who used NASA tracking radars at Wallops Island, Virginia, to obtain long, detailed tracks of individual bird targets. All targets are presumed to be shorebirds or waterfowl on the basis of their air speeds. On the night of 2 April 1969, heavy fog or low clouds covered the sky at an altitude of about 185 m. Using the radar in an RHI mode, the Williamses located an upper cloud layer at approximately 5000 m. They state that "surface observations and satellite photographs before and after our tracking indicated that this was a totally opaque cloud layer." Unfortunately, the surface observations taken at Wallops cannot be relied upon, since all five of the birds tracked that night were followed at distances of from 20 to 80 km out to sea. Furthermore, the resolution of satellite photographs is insufficient to rule out the possibilities of even sizable gaps in the upper layer of cloud. The assumption that a coastal fog would have extended without gaps for a distance of 80 km out to sea is also questionable. Nevertheless, the birds certainly were flying in conditions of heavy cloud. Of the five targets, one was traveling downwind to the east-southeast, while three tracks were oriented northeast to east-northeast, the typical direction for spring migration at this location. The last bird was descending in altitude and maintaining a constant heading while its track curved counterclockwise toward land. It is difficult to interpret these tracks considering the uncertainty of the meteorological information. But, with one exception, the birds were maintaining straight tracks and three were traveling in the appropriate migration direction.

Griffin (1972, 1973) used a low power, 3 cm, tracking radar to plot the courses of individual migrants. By specifically seeking out conditions of thick overcast, he was able to obtain accurate information on short segments (up to 5 km) of tracks of targets moving in or near cloud. His data also provide us with the most reliable information available concerning the specific nature and extent of the clouds.

His most impressive evidence of orientation in overcast comes from the night of 16 May 1970 at a location in upstate New York. The radar site was enveloped by fog, and available evidence (dew point depression information from evening radiosondes 150 km to the north-northeast and south-southeast, respectively, as well as surface observations from a variety of nearby locations) indicated the continuous presence of cloud to an altitude of approximately 2000 m. Twenty-one birds were tracked, all within this cloud zone. The behavior of the targets appeared perfectly normal; they were traveling downwind and all tracks were within 45° of a mean direction to the north-northeast. Furthermore, the tracks were straight.

On 23 April 1970, birds were tracked as they passed over New York City. Meteorological information indicated the presence of continuous cloud from a few hundred meters up to at least 1000 m and possibly much higher. The birds tracked were flying in directions ranging from 315° through 0° to 123°, although the mean direction was still to the north-northeast. More interestingly, many of these tracks were extremely circuitous, the birds apparently looping or zigzagging as they traveled. "Clearly, many of these birds were making extensive but small scale deviations from a perfectly straight track, while nevertheless maintaining a consistent direction of progress" (Griffin, 1972, p. 178). Only about one-third of the targets showed the straight, north to northeastward track typical of migration under clear skies.

I interpret the tracks obtained on the night of 30 April 1970, as suggesting a dichotomy in the behavior of migrants traveling inside and out of cloud. Neighboring airports reported fog and low-lying clouds, and radiosonde data suggested a well-defined cloud top at 500 m. Birds were tracked at a variety of altitudes although only four individuals definitely can be placed below the cloud tops. Each of these four tracks was quite nonlinear and showed twists, turns, or zigzags. The overall direction of three tracks was to the north-northwest to north, while the fourth curved to the southeast. The twenty-five tracks that were above 500 m in height (hence, presumably out of the cloud layer) were distinctly straighter than the four low tracks. They were also tightly clumped around a mean direction to the northeast.

I would summarize the available evidence as follows: (1) most birds avoid flying through thick layers of cloud; (2) when confronted with this situation, many birds seem able to maintain a general direction of travel; (3) the accuracy of orientation is considerably decreased; (4) some birds exhibit a totally different flying strategy, twisting and looping their way along their course.

4. Selection versus Maintenance of Direction

An important distinction must be made between *maintaining* a direction taken up previously, either on the ground prior to departure or at times during the flight when visual cues might be visible through breaks in cloud, and *selecting* a correct course under or in total overcast. The former requires much less navigational versatility. For example, if a bird could localize discrete and repetitious sound sources on the ground, it might use them as acoustical reference points in a manner analogous to the use of visual markers (D'Arms and Griffin, 1972). The call notes given by many species of nocturnal migrants could, under some conditions, provide an additional indirect source of directional information. If the altitudinal distribution of birds aloft extended above or below the layers of overcast (as in the cases described by Bellrose and Graber, 1963), the birds within the cloud might hear the call notes of birds flying at different altitudes where additional directional cues were available. By comparing the locations of successive call notes, a bird might be able to extrapolate and follow the flight direction of an unseen companion (Griffin, 1969; Steidinger, 1972).

Other possible nonvisual guidance systems might include flying at a constant angle relative to the direction of the wind (see Section IV,E), using some crude form of inertial dead-reckoning system (see Section IV,F), or determining direction from cues provided by the earth's magnetic field (Section IV,G). Of these five possible means of maintaining direction (and undoubtedly there are others), only geomagnetism, a complex form of inertial guidance, or an accurate prediction of winds aloft could be used for selecting a migratory direction.

One way of learning whether most migrants are able to *select* an appropriate direction of flight in the absence of visual cues is to observe departure behavior under conditions of total overcast. While on the ground, a bird presumably integrates the various types of directional information at its disposal. If the sky is clear, these might include the position of the sun, and perhaps most importantly, the position of sunset as well as the early evening location of the stars. The topographic features of the immediate surroundings could serve as reference points.

Wind could be an important additional cue that might be integrated into this system. If discrete, cumulus clouds were present in the sky, the migrant presumably could determine the direction of the wind (at least at the altitude of the clouds) relative to its celestial and topographic framework without ever leaving the ground. If no clouds were present, the bird theoretically could still determine the wind direction by circling and noting changes in its ground speed using topographic features as reference points. Geomagnetic information might also be incorporated into this selection process.

If celestial information is removed, do the additional cues provide sufficient information for selection of the migratory direction? While studying autumn migration on the southeast coast of Louisiana, Hebrard (1972) conducted a "natural" experiment when a cold front brought low, continuous overcast to his study area. The clouds arrived in mid-afternoon on 13 October 1969 and persisted without a break until the morning of 15 October. Birds departing under the total overcast on the night of 13 October did so after a 2 hour deprivation of solar information and without access to stellar cues (clouds arrived at 3:15 PM; sunset was at 5:32 PM). Yet these birds (passerines) took off downwind in meaningful migration directions (see Fig. 20A). Birds departing under the overcast on the night of 14 October had been unable to view either the sun or the stars for over 24 hours. The orientation of these birds was quite poor, although the statistical mean direction remained significantly oriented to the south (Fig. 20B).

One interpretation of these results is that some birds are able to "remember" the position of the sun or the extrapolated position of sunset by using landscape cues as reference markers. In this way, the sun could provide critical compass information even if it were absent at the time of departure; the bird would select its bearing with reference to topographic features that had taken on directional significance. Such transferring of information onto a topographic reference system should

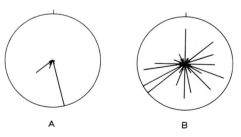

A   B

FIG. 20. Orientation of passerine migrants observed crossing through a ceilometer beam on the overcast nights of 13 October (A) and 14 October 1969 (B). See text for discussion. (Redrawn from Hebrard, 1972.)

become more difficult the longer the primary cue is unavailable. The accuracy should also decrease whenever the bird changes its position significantly, since this introduces a parallax error onto the landmark cue.

Many authors have reported normal migratory departure under 10/10 overcast (Bellrose and Graber, 1963; Drury and Nisbet, 1964; Evans, 1966, 1968; Nisbet and Drury, 1967; Bellrose, 1971; Gauthreaux, 1971; W. J. Richardson, 1974). Such cases deserve additional study and re-analysis, giving consideration to the time intervals between disappear-ance of the sun and oriented migratory departure.

A second way of distinguishing between the capabilities of maintaining and selecting directions under overcast involves bringing the departure behavior of free-flying migrants under experimental control. We have recently devised techniques for artificially releasing birds into the air space and following them with NASA tracking radars as they make their directional decisions and depart on migration (Emlen, 1974; S. T. Emlen and N. J. Demong, unpublished observations). In this way the experimenters can choose the meteorological conditions for depar-ture. The birds can be deprived of the opportunity to integrate sunset information with other directional cues. For the first time, it becomes possible to experimentally manipulate the orientational systems of a free-flying migrant. We are just beginning the analysis of our results from three seasons of radar tracking; hence, it would be unwise to weight our preliminary findings too heavily. However, we have been able to release White-throated Sparrows deep within cloud during the spring migration season. The spread in departure directions suggests that this species, at least, does *not* seem able to select its normal northeastward course under these conditions. Further, the detailed tracks showed multi-ple loops and zigzags very similar to those reported by Griffin (1972) when he was tracking unidentified bird echoes flying through clouds and suggestive, we feel, of a shift to an alternative flight strategy.

Finally, I feel that valuable information about possible nonvisual orien-tation systems can be obtained from detailed analyses of the spread of track directions obtained from surveillance radar studies. A bird that is maintaining a predetermined course will make flight errors that should accumulate through time (unless it possesses a complex inertial system). This should result in an increasing scatter between the directions of individual migrants. Comparisons of angular deviations from nights when birds are believed to be flying in clear skies, under cloud, and within cloud would be extremely valuable as would analyses of the changes in angular deviation occurring between the beginning and the end of such migratory flights. Information of this sort should be forthcoming from Richardson (1974), who is conducting analyses of this type on several seasons of radar data of migration over the Canadian Maritimes.

In summary, there are very few data available from free-flying migrants pertinent to the important question of direction *determination* in the absence of both celestial and topographic cues. What data there are seem more reconcilable with the hypothesis that most migrants can only *maintain* their flight course under these conditions.

In contrast to the scanty evidence from migrants, Keeton (1969) has reported well-oriented homing behavior in pigeons under total cloud. Birds from the Cornell lofts were given routine exercise flights and short training flights on overcast days so they became accustomed to flying under these conditions. Both control and clock-shifted birds then were taken to release sites 48 km north and 33 km east of the home loft. Under sunny skies, the vanishing bearings of control pigeons were oriented roughly homeward, while those of clock-shifted birds were deflected approximately 90° (Section IV,B). Under total overcast, however, both groups of birds oriented homeward.

To exclude the possibility of pilotage by familiar landmark cues, two additional tests were performed at a site 160 km east. Both clock-shifted and control birds departed toward home when released under opaque cloud. The data from all control birds used in these experiments have been pooled and replotted in Fig. 21. The similarity in orientation under clear and cloudy skies caused Keeton to conclude that "the sun is used as a compass when it is available, but . . . the pigeon navigation system contains sufficient redundancy to make accurate orientation possible in the absence of both the sun and familiar landmarks" (Keeton, 1969, p. 922). Although these findings conflict with many earlier reports, they have recently been corroborated by Baldaccini *et al.* (1971), Wagner (1972, discussion following his paper), and Walcott and Green (1974). W. T. Keeton believes that the discrepancies between his and previous

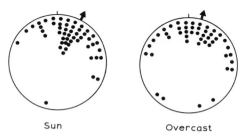

Sun   Overcast

FIG. 21. Departure bearings of homing pigeons released under clear and cloudy skies at distances of 33 to 160 kilometers from the home loft. The results from several experiments are pooled with 360° representing the home direction. With the sun visible, the mean direction, ϕ is 27°; probability by Rayleigh test = 0.000; probability by "v" test with home the expected direction = 0.000. Under overcast, $\phi = 13.5°$, Rayleigh probability = 0.000 and "v" probability = 0.000. (Recalculated from Keeton, 1969.)

studies may relate not to the navigational capabilities of the pigeons involved but rather to their motivation to fly under suboptimal conditions (personal communication).

Keeton's results again raise the possibility of differences in orientation strategies, if not in capabilities, between pigeons and migrants. The pigeons are homing to a loft from which they have just been transported. The migrants are traveling to a destination where they have not resided for many months, if at all.

These findings of migrants traveling through, and pigeons homing under, cloud are of extreme theoretical importance. They raise intriguing questions about nonvisual guidance systems in animals. The search for these alternative sources of directional information stands as one of the most exciting challenges in the field of avian orientation today.

E. The Importance of Meteorological Cues

Birds have evolved into excellent meteorologists. Much of their lives is spent in the air space, and during migratory flights they are often aloft at considerable heights for long periods of time. This is especially true for long-distance travelers, particularly those making nonstop flights over inhospitable terrain such as the Sahara Desert or portions of the Atlantic and Pacific Oceans. A strong selective premium must be placed on the ability to predict the weather correctly, particularly the winds aloft.

For decades, it has been known that birds are selective in timing their migration departures. Both ground observations and radar studies have shown that the bulk of migration at north temperate latitudes occurs on a small percentage of the days and nights of the migration season. The times of these dense migrations are quite predictable and coincide with "favorable" meteorological conditions. Stated overly simply, large autumn movements generally occur on the east side of a high-pressure cell following the passage of a cold front, while large spring flights concentrate on the west side of a high pressure area and ahead of an advancing low with its associated cold front (for reviews, see Drury and Keith, 1962; Richardson, 1971, 1972).

Until recently, weather factors were viewed primarily in a negative light, as possible sources of navigation error or disaster. Migrants might be caught up in storms and be blown far off course; they might encounter stiff cross or head winds and fail to reach their destinations; they might find themselves in fog or heavy rain and lose their way. The ornithological literature is filled with notes of severe exhaustion or mass mortality of migrants that misread meteorological information and encountered snowstorms, overtook cold fronts, or battled headwinds aloft. The birds' selectivity in choosing a time of departure was considered a safeguard against going aloft under hazardous conditions.

In reality, the migrant is probably seeking optimal flight conditions as well as avoiding dangerous ones. The metabolic cost of flight is extremely high, and most migrants lay down sizable fat deposits that serve as energy stores for long-distance flights. If a migrant could correctly predict the winds aloft and initiate a flight only when those winds approximated a tailwind situation, the energetic savings would be immense. A small passerine with an air speed of 20 knots flying with a 20-knot tail wind would effectively double its flight range without significantly increasing the energy cost of the trip. This leads to a selective advantage not just to avoid unfavorable weather conditions, but to postpone departures until optimal or near optimal winds are present. Significantly, most recent radar studies show a marked tendency both for songbirds to migrate downwind, and for migration to be heaviest when the winds are in the predicted migratory direction.

Some migrants show considerable accuracy in interpreting wind information. On nights when wind directions differ at different altitudes, birds often concentrate in those height strata most favorable for migratory travel (Bellrose, 1967a; Blokpoel, 1970; Richardson, 1971; Bruderer and Steidinger, 1972; Gauthreaux, 1972a; Steidinger, 1972). Other studies have raised the interesting possibility that some migrants adjust their air speed as a function of the wind velocity, resulting in the maintenance of a more constant ground speed (Bellrose, 1967a; Bruderer, 1971; Emlen, 1974). If migrants are this closely attuned to interpreting the winds, might they not use them more directly, as a source of directional information?

Gauthreaux and Able (1970; Gauthreaux, 1972b; Able, 1972, 1973, 1974a) feel that wind is the dominant directional cue for many songbirds moving through Georgia and coastal Louisiana. Their observations show birds moving almost directly downwind regardless of the wind direction. This indicates, they feel, that many passerine migrants have only weakly developed directional preferences and will ride the prevailing winds rather than attempt to fly on a predetermined course.

Alternatively, songbirds might be so precise in choosing their flight conditions, that the only species going aloft on a particular night are those whose preferred directions are approximately downwind. According to this model, the changes in mean flight direction observed by Gauthreaux and Able (1970) could be due partly to different species of migrants being aloft on different nights. Evans (1966), Nisbet and Drury (1967), Lack (1969), and Richardson (1972, 1974) each advance this hypothesis to explain similar, but less pronounced, correlations between mean flight and wind directions.

The upper winds present in most areas are primarily dependent upon the location of high- and low-pressure centers. Within any given sector of such a pressure cell, wind direction is quite predictable. Evidence suggests that many birds have evolved the ability to identify and respond to synoptic weather situations favorable for migration. Many spring migrants, for example, will depart on the east side of a low pressure cell, presumably using their full complement of guidance systems to select their flight directions. Suppose, however, that all principal cues were unavailable, and the birds had no direct means of determining their migration directions. The recent increase in temperature and humidity, the drop in barometric pressure, and the buildup of stratus clouds all would indicate the passage of a warm front (and hence a low pressure system to the west). If, under these stimulatory conditions, the migrants that departed adopted a strategy of flying downwind, they would usually travel on a course at least approximating the general migration direction. The predictability of the upper winds from meteorological information available on the surface could make this a crude means of selecting as well as maintaining a direction in the absence of celestial, magnetic, or inertial information.

What happens when a migrant departs on a preferred direction that is not directly downwind? If the bird has a northward heading but the winds are from the east, it will travel along the resultant vector of the two, to the northwest. Its exact direction will depend upon its air speed relative to the speed and direction of the wind. To reach a destination to the north, therefore, the bird must change its heading and aim somewhat east of north. Only by flying at some angle relative to the wind and changing this angle as the wind direction shifts could the bird compensate for wind drift.

Do migrants detect and correct for such drift? This has been a major question since the advent of radar ornithology two decades ago. Different workers have voiced conflicting opinions and there is no general consensus (e.g., Gauthreaux and Able, 1970; Evans, 1970). Much of the disagreement might be due to different species responding differently to drift or to individual species changing their flight strategies depending upon the strength and inappropriateness of the wind.

Many excellent studies provide strong evidence that at least some migrants do correct for drift, although such corrections may not be complete (Drury and Keith, 1962; Drury and Nisbet, 1964; Evans, 1966, 1972; Bellrose, 1967a; Nisbet and Drury, 1967; Lack, 1969; Bruderer and Steidinger, 1972; Gauthreaux, 1972b; Steidinger, 1972; Williams *et al.*, 1972). This evidence is based primarily on comparisons of the mean and spread of tracks and headings on nights with different wind conditions. Tracks generally show less scatter than headings. This implies that the birds alter their headings in different winds, thereby maintaining more constant tracks. When the wind direction shifts during the course of a single night, Drury and Nisbet (1964), Evans (1966), and Nisbet and Drury (1967), but not Gauthreaux and Able (1970) report that passerine migrants compensate for the wind change and keep the same mean track.

Small-scale wind drift cannot be detected using a celestial or magnetic compass sense alone. Detection requires a crude integration between compass information and some measure of forward progress along the ground (presumably by means of visual reference markers). Once this integration has taken place and the bird has adjusted its heading appropriately, the track could be maintained simply by flying at a constant angle to the wind.

Vleugel (1954, 1959, 1962) has repeatedly stressed that birds could maintain their bearings by using the wind. He proposed that nocturnal migrants use the position of sunset to select their flight direction and then maintain it through the night by keeping their angle to the wind constant. This seems plausible for cases where wind direction can be noted with reference to features of the landscape. But what happens when visual contact with the ground is eliminated by low clouds or fog?

Under these conditions, a bird traveling in a totally homogeneous air flow would have no means of detecting the direction or speed of the wind. But air flow near the surface of the earth is seldom homogeneous. Friction resulting from the passage of air over the ground produces patterns of turbulence that are related to the direction of the wind. Similar patterns may be produced whenever there is wind shear (Stewart, 1956, cited in Griffin, 1969). Nisbet (1955) pointed out that this turbulence takes the form of a series of gusts that should be detectable to a bird as sudden changes in velocity. These gusts have a characteristic structure consisting, at the onset, of a sharp increase of wind velocity in the direction of the mean wind stream, followed by a much more gradual decrease (Nisbet, 1955, p. 558). Consequently, if a bird were flying at right angles to the wind direction, it would be buffeted most strongly on its upwind side. By sampling the turbulent structure of the air a bird might be able to determine the downwind direction or to fly at a particular angle relative to the wind without visual reference to the ground. Bellrose (1967a,b) believes that such patterns of wind turbulence provide the necessary information enabling migrants to maintain their courses when traveling under or in thick cloud.

If many migrants possess this capability, then cases of disorientation should be even rarer. Presumably migrants do not take off unless sufficient directional information is available for the selection of a preferred direction. Disorientation should occur only when a variety of cues become unavailable after the initiation of a normal flight. Such conditions might include overcast skies (loss of celestial information), ground fog or low clouds (loss of visual contact with the ground), *and* relatively uniform air flow prohibiting the use of wind turbulence.

Massive disorientation has rarely been observed on radar. When recorded it has been associated with total overcast, low visibility, and the presence of rain and/or fog (Lack and Eastwood, 1962; Drury and Nisbet, 1964; Bellrose, 1967a). Bellrose (1967a, p. 307) argues that rain tends to reduce whatever gust structure is present, while fog often indicates a general lack of air turbulence. Drury and Nisbet (1964) make the additional interesting point that the same weather situation is frequently associated with mass bird mortality at television towers and with conspicuous calling of nocturnal migrants.

F. INERTIAL ORIENTATION HYPOTHESES

The possibility that birds orient their flights by means of inertial navigation has been reviewed by Barlow (1964, 1966). Basically, such models require that the animal accurately sense both angular and rectilinear motion. By integrating information from all changes in flight speed and direction, a bird could, in theory, always know its position relative to an arbitrary starting point. For cases of homing orientation, this starting point presumably would be the breeding territory or, for pigeons, the home loft. Thus, a pigeon that accurately detected and integrated the accelerations produced by each twist and turn of its outgoing path (as it was transported to a release site) should be able to select the homeward bearing without resort to visual or geophysical cues.

Numerous attempts have been made to eliminate inertial cues from homing experiments. Birds have been rotated or spun during transit to their release sites with no significant decrease in performance (Rüppel, 1936; Griffin, 1940; Matthews, 1951b). Other investigators, the most recent being Walcott and Schmidt-Koenig (1973), transported birds under deep anesthesia, again with no noticeable effect upon homing success. Wallraff has bisected the horizontal semicircular canal (1965a) and extirpated the cochlea and lagena (1971) in pigeons and reported normal orientation and homing success when they were released approximately 150 km distant at several different release locations. Most of these experiments, however, were performed on clear days when the sun was visible as a possible alternative cue. Keeton and Money (in Keeton, 1974) conducted tests on pigeons with the sacculi removed. When released at unfamiliar sites under conditions of total overcast, performance was unimpaired and both test and control birds oriented homeward.

Errors in inertial systems occur if the animal unknowingly enters a gradual turn that is below the threshold of detection by the vestibular system. Measured thresholds for angular and linear acceleration are approximately $0.2°$ sec^{-2} and 6 cm sec^{-2}, respectively (Barlow, 1964). These levels of sensitivity are insufficient for accurate position localization. The acceleration-sensing elements of the inertial systems of aircraft are commonly 10^3–10^4 times this precise (Barlow, 1964).

The case of long-distance migration poses an additional constraint upon inertial navigation hypotheses. To navigate to a migration destination, a bird would have to rely upon information obtained and integrated many months earlier, on the previous migratory flight. For the immature bird embarking on its first migration, such information does not even exist.

Drury and Nisbet (1964) have postulated that a much simpler form of inertial dead reckoning might help explain a bird's ability to maintain its bearing during periods when alternate cues are unavailable. If absolute position is not required, the demands on the vestibular apparatus are greatly reduced. If birds have to rely upon this crude inertial system for only short periods of time and can double-check their position and or direction by frequent reference to other cues, the requirements become even less stringent. Whether migrants actually use such a simplified form of inertial guidance is unknown. Unfortunately, there have been very few experimental tests of inertial hypotheses and even fewer are under way at the present time. Consequently, this remains as a theoretically appealing and yet inadequately studied area of animal orientation.

G. POSSIBLE USE OF GEOMAGNETISM

1. Introduction

The hypothesis that the earth's magnetic field plays an important role in bird orientation has had a cyclical history of acceptance since first being proposed over a century ago. The first experimental support was provided by Yeagley (1947) who initially reported that attaching miniature magnets to the carpal joints of the wings of pigeons had a detrimental effect on the birds' homing ability. These findings received wide attention and stimulated considerable thought. Much of this thought was critical, however, and Yeagley's ideas were seriously challenged on theoretical grounds (Davis, 1948; deVries, 1948; Slepian, 1948; Varian, 1948; Wilkinson, 1949). Later attempts to repeat the experiments, including one by Yeagley himself, failed to produce positive results (Gordon, 1948; Matthews, 1951b; Yeagley, 1951; von Riper and Kalmbach, 1952). Following this surge of negative reports, the field lay dormant for nearly a decade. It was not until the mid-1960's that positive evidence obtained from new techniques began to restimulate serious discussion of magnetic orientation.

Theoretically, a tremendous amount of navigational information could be obtained from geomagnetic cues. A bird might learn the magnetic "topography" of an area and orient by magnetic piloting or by following familiar magnetic contours of the terrain. Alternatively, a bird might localize the direction of the magnetic pole and use it as a reference for compass orientation. Geomagnetic cues could also form part of the basis of a bicoordinate navigation system. Yeagley originally proposed that birds determine latitude from the Coriolis force while obtaining longitude information from the strength of the vertical component of the magnetic field. An earlier model (Viguier, 1882, cited in Matthews, 1968) proposed that the bird measures both the intensity and the dip angle of the earth's field. Over some portions of the earth's surface, the lines of equal intensity and equal dip angle intersect and form a potential navigational grid; in other localities, these lines are parallel and such a model would be inoperative. Hybrid models can also be envisaged. A bird might obtain crude latitude information from the inclination angle while receiving longitude information from the integration of its chronometer with the position of the sun, to mention just one possibility.

Can birds detect magnetic fields, and are they sensitive to intensities of the magnitude of the earth's field (approximately 0.5 G)? To be useful for orientational purposes, sensitivities should be considerably in excess of this value, since birds presumably would need to respond to minor differences in the strength or direction of different components of the magnetic field. Physicists are quick to point out that the intensity of the earth's field at any location is not constant through time but shows both daily and seasonal fluctuations. Superimposed on these changes are irregular perturbations caused primarily by changes in solar storm activity. It is against this backdrop of unpredictably fluctuating magnetic fields that a bird would have to make its measurements and comparisons if it were to use geomagnetic information for navigational purposes.

2. Conditioning Experiments

What is the evidence that birds can detect magnetic fields at all? Several workers have tried to condition birds to respond to magnetic stimuli. Griffin (1952) and G. Bartholomew (personal communication) each attempted unsuccessfully to train homing pigeons to strong magnetic fields. Meyer and Lambe (1966) reported negative results in attempts to condition pigeons to detect slight changes in the intensity of an artificial field whose strength approximated that of the earth. I also obtained negative results trying to train Indigo Buntings to respond to changes in the direction of a magnetic field of 0.55 G (Emlen, 1970b). Griffin (1952), Yeagley (cited in Griffin, 1952), and Orgel and Smith (1954) all exposed pigeons to oscillating magnetic fields. The bird remained stationary while magnetic lines of force moved through it, thereby inducing an electromagnetic force considerably greater than might be experienced by a free-flying bird. Once again, all experiments failed to produce positive results.

One must be extremely cautious, however, in the interpretation of negative results from conditioning experiments. Psychologists correctly stress the importance of coupling a particular conditioned stimulus with an appropriate behavioral response. While this makes sense ethologically, it is difficult to predict what type of behavior would be optimally associated with presentation of a magnetic stimulus. To date, most studies have used some form of locomotor activity as the conditioned response. Pigeons have moved through T-mazes, walked across electric grids, or pecked at standard response keys, while the Indigo Buntings jumped from the floor of their cage.

Autonomic conditioning provides an alternative and, in all probability, a much more sensitive means of studying magnetic detection. A physiological response such as heart rate is monitored while the animal is presented with test stimuli. Each stimulus presentation is immediately followed by an electrical shock. If the animal can detect the stimulus its cardiac rhythm should show an involuntary increase at each stimulus presentation in anticipation of the forthcoming shock. Using this approach, Reille (1968) reported initial success in conditioning pigeons to magnetic stimuli. Positive responses were greater to oscillating fields of 0.80 G (frequencies 300–500 Hz) than to a continuous, static field of the same strength. These exciting data spurred several laboratories to try to corroborate the findings. None has been successful. Kreithen (Kreithen and Keeton, 1974) has, perhaps, performed the most extensive series of replicate tests to date. Using ninety-seven pigeons from the Cornell loft, he found no indication that responses to magnetic fields were significantly different from control levels.

3. Zugunruhe Experiments

Several recent behavioral studies have linked magnetic effects more directly with migratory activity and orientation. El'Darov and Kholodov (1964) exposed caged migrants [European Bullfinches (*Pyrrhula pyrrhula*), Greenfinches (*Carduelis chloris*), Coal Tits (*Parus ater*), Red Crossbills (*Loxia curvirostra*), and Chaffinches (*Fringilla coelebs*)] to an artificial magnetic field produced by a large pair of Helmholtz coils.

When the field was increased from 0.6 to 1.7 G, the locomotor activity of the birds increased significantly. A similar finding was reported by Shumakov (1967a,b), who studied the behavior of five species of passerine migrants at the site of a strong magnetic anomaly. First, the activity and orientation of the birds was tested in circular cages at a location on the Baltic. Then the birds were transported to the center of the Kursk magnetic anomaly (located in the town of Gubkin), where the field intensity is significantly increased and the direction of magnetic north is altered by 60°. When tested at this location under overcast skies, the birds failed to show any orientation. However, their level of general activity was double or triple that observed on the Baltic. The biological significance of these findings is unclear, but they could suggest a sensitivity and responsiveness to magnetic information.

William Southern (1971, 1972) has obtained indirect evidence that geomagnetism affects gull orientation. He placed juvenile Ring-billed Gulls (*Larus delawarensis*) in outdoor, circular arenas measuring 2.4 or 9 m in diameter and noted a significant tendency for birds to walk to the southeast. This direction persisted under both clear and overcast skies and was present in gull chicks ranging in age from 2 to 20 days (Southern, 1969). Southern interprets this orientation as evidence for a spontaneous preference for what will later become the autumn migration direction.

Next, Southern grouped all of his tests according to the intensity of magnetic disturbance present during the 3-hour period closest to the time of each experiment. These fluctuations in the earth's magnetic field are primarily a result of differential solar activity and are classified by the United States Coast and Geodetic Survey on a 0–9 scale of K values. A disturbance is quantified by the amount of change that it produces in the intensity of the earth's magnetic field. The K value represents the greatest amount of change (measured in gammas, with $1 \gamma = 10^{-5}$ G) occurring during a 3-hour period. Indices of 1–4 signify low amounts of magnetic storm activity; 5 refers to moderate storm activity; and values of 6–9 indicate increasingly severe disturbances.

When the accuracy of the gull chicks' orientation was analyzed as a function of K values, the southeast orientation was strong with K indices of 0, 1, or 2, became weak at 3, and disappeared at $K = 4$, 5, and 7 (Fig. 22). The behavior under moderately severe storm activity ($K = 5$ and 7) showed an intriguing tendency for reorientation in the opposite direction, to the north.

Southern concludes that magnetic disturbances (or some factor correlated with magnetic disturbances) are responsible for the observed deterioration of gull orientation. If replicable, his findings would suggest an amazing sensitivity to minor fluctuations in the magnetic field. Orien-

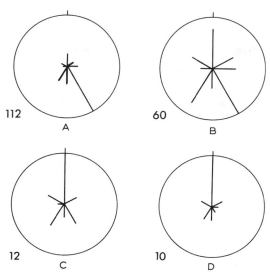

Fig. 22. Walking orientation of Ring-billed Gull (*Larus delawarensis*) chicks tested in a circular arena. Data are grouped according to the amount of magnetic disturbance present at the time of the experiments. (A) Orientation during periods of very low magnetic storm activity. K indices = 0, 1 and 2. Mean direction = 155°; Rayleigh probability <0.01. (B) Orientation when K indices were 3 or 4. Mean direction = 152°; $p < 0.01$. (C) Orientation during moderate storm activity, K index = 5. Mean direction = 321°; $p > 0.05$. (D) Orientation under moderately severe magnetic disturbance, K = 7. Mean direction = 360°; $p > 0.05$. Vector diagrams are drawn on the proportionality basis with the radius equaling the greatest number of gull chicks choosing a particular 30° sector. The number this represents is given adjacent to each diagram. (Regrouped, calculated and plotted from data of Southern, 1971.)

tation broke down at K values between 3 and 4, values signifying a disruption of 20–70 γ. This is equivalent to approximately 1/1000 of the intensity of the normal earth's field.

In a series of important papers, F. Merkel and W. and R. Wiltschko have hypothesized that European Robins (*Erithacus rubecula*) actively use the geomagnetic field to determine their migratory direction. In initial tests, caged robins spontaneously oriented their nocturnal activity in the appropriate seasonal direction, even when placed in a room isolated from "normal visual environmental cues" (Merkel and Fromme, 1958; Fromme, 1961; Merkel *et al.*, 1964). If the birds were placed in a large steel chamber that greatly reduced the total intensity of the magnetic field (from 0.41 G to 0.14 G) this orientation ability disappeared (Fromme, 1961; Merkel and Wiltschko. 1965; Wiltschko, 1968). Later tests showed that migratory activity became random if the magnetic field surrounding the cage was artificially decreased to values below

0.34 G or increased to values greater than 0.68 G in intensity (Wiltschko, 1968). Interestingly, if the robins were allowed to adjust to the abnormal magnetic intensities by living under these test conditions for a period of 3 days or more, they redeveloped the ability to orient in the appropriate directions (Merkel and Wiltschko, 1965; Wiltschko, 1968). This led Wiltschko (1968) to conclude that the magnetic detection system of the European Robin is finely attuned to fields approximating the earth's in strength.

Next, these investigators built a large pair (diameter 2 meters) of Helmholtz coils around their octagonal test cage. By rotating these coils and adjusting the strength of the electric current passing through them, they could generate a fairly homogeneous static dc field and control both its direction and intensity. In a series of experiments encompassing both spring and fall migration seasons, they found that the orientation of caged robins could be altered quite predictably by changing the direction of the horizontal component of the artificial magnetic field (Merkel and Wiltschko, 1965; Wiltschko and Merkel, 1965; Wiltschko, 1968; Merkel, 1971). In Fig. 23 I have pooled, replotted, and recalculated all of the results from experiments performed during the autumn migration seasons on robins that were housed and tested under magnetic fields approximating the earth's in strength. Although tested in a closed room, the distribution of the nightly means shows a clear southward trend (mean bearing 187°; Rayleigh probability <0.001) (Fig. 23a). When magnetic north was rotated to coincide with geographic east-southeast (Fig. 23b), the robins kept orienting toward magnetic south although this corresponded to geographic northwest (expected direction 299°; actual mean bearing 328°; Rayleigh probability = 0.007). The robins also maintained a magnetic south-southwest preference when magnetic north was rotated to coincide with geographic west (expected direction 97°; actual mean bearing 128°; Rayleigh probability = 0.045) (Fig. 23d). In only one condition did the birds shift in an unpredictable manner. When magnetic north was rotated 45° counterclockwise (to geographic northwest) in both 1964 and 1965, the mean directions of the robins rotated clockwise, producing a deviation of 80° between predicted and observed headings (expected direction 142°; actual mean bearing 222°; Rayleigh probability <0.001) (Fig. 23c). The Wiltschkos have recently extended their investigations to include European warblers (*Sylvia communis, cantillans,* and *borin*) and report these species also show magnetically directed *Zugunruhe* (Wiltschko and Merkel, 1971; Wiltschko and Wiltschko, 1973; and unpublished observations).

In their most recent experiments, the Wiltschkos have manipulated the inclination as well as the direction of the artificial magnetic field. When the polarity of the vertical component was reversed (from −66°

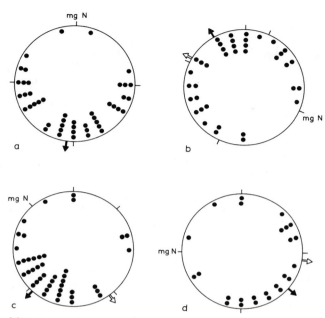

FIG. 23. Migratory orientation of caged European Robins (*Erithacus rubecula*) tested during the autumn migration seasons while in the "absence" of meaningful visual cues. (a) In the presence of a magnetic field of normal direction and strength (pooled from Wiltschko, 1968, Figure 4b and 6b). (b) When the horizontal component of the magnetic field was altered by Helmholtz coils such that magnetic north coincided with geographic east–southeast. (From Wiltschko, 1972, Figure 2a). (c) When the horizontal component of the artificial magnetic field was situated in the northwest. (From Wiltschko, 1968, Figure 9b and Merkel, 1971, Figure 1b.) (d) When magnetic north was aligned with geographic west (From Wiltschko, 1968, Figure 14d.) These diagrams include all of the results obtained to date from experiments performed during the autumn migration season on robins that were housed and tested under magnetic fields approximating the earth's in strength. Each dot represents the mean direction of one bird on one night of testing (here grouped into 15° sectors). Black arrows show overall mean direction. Open arrows give the "expected" direction (186°, see a) measured relative to magnetic north. See text for full discussion.

to +66°), robins tended to reverse their orientation even though the horizontal direction of magnetic north remained unchanged (Fig. 24A and B). Orientation also changed if the horizontal component of the field was reversed, while the inclination was held constant at −66°. But if the horizontal direction of the magnetic field was reversed 180° *and* the inclination was changed (from −66° to +66°), the birds did not alter their directional behavior (Wiltschko, 1972; Wiltschko and Wiltschko, 1972). In an artificial field with a strong horizontal but a zero vertical component, activity became random (Fig. 24c).

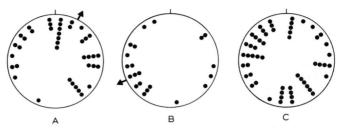

Fig. 24. Migratory orientation of captive European Robins tested in the spring of 1971 under the following artificial magnetic fields: (A) Normal magnetic information available; magnetic north = 360°, inclination = +66°, total intensity = 0.46 G. (B) Horizontal component normal (360°) but polarity of the vertical component reversed (−65°). Total intensity = 0.46 G. (C) Horizontal component much stronger but still oriented to 360°. Vertical component eliminated (inclination = 0°). Total intensity = 0.46 G. Symbols as in Fig. 23. (From Wiltschko and Wiltschko, 1972).

On the basis of these surprising results, Wiltschko and Wiltschko (1972, p. 62) hypothesize that European Robins "do not use the polarity of the magnetic field for detecting the north direction. The birds derive their north direction from interpreting the inclination of the axial direction of the magnetic field lines in space, and they take the direction on the magnetic north–south axis for 'north' where field lines and gravity vector form the smaller angle."

Although most workers in the orientation field have been stimulated by the findings of the Frankfurt group, many have questioned the biological significance of the data. In part, this is because the degree of orientation found by Merkel and the Wiltschkos is relatively weak, frequently demonstrable only on second-order statistical analysis. The activity of an individual robin is recorded by microswitches on eight directional perches in an octagonal cage and a mean direction is calculated. All such means representing one experimental situation then are pooled together. Differing numbers of birds and different numbers of replicates per bird are included in these diagrams. These means are the raw data from which a grand mean bearing is calculated. Frequently this resultant vector attains only borderline significance, even though considerable sample sizes are employed. Merkel (1971) and Wiltschko (1972; W. Wiltschko, personal communication) state that few of their birds display a clearcut orientation and that 40% of the individual bird-nights do not deviate significantly from uniform distributions. The fact that statistical significance is reached by pooling the means from such bird-nights suggests that the robins are displaying extremely weak but consistent directional preferences under the test conditions. Although one may wish to debate the biological meaning of such weak tendencies, there is a definite shift in these preferences when the direction of the magnetic field is altered.

In 1970, I published "negative" data from an attempt to observe orientation under visually cueless conditions (Emlen, 1970b). Ironically, these data actually provide confirmation of some of Merkel and Wiltschko's findings. They also point up the difficulties involved in attaching biological significance to statistically significant data. Twelve Indigo Buntings were placed in "cueless" cages for from 5 to 13 nights each during the autumn migration season. The pooled total activity for each individual bird is shown in Fig. 25. Statistical treatment of these data (Rayleigh test) showed no case where the activity distribution deviated significantly from a random or uniform distribution.

I also calculated the mean direction for each bird on each night by vector analysis. These means are shown in Fig. 26. A second-order analysis, calculating the mean of the means for each individual bird, revealed only two cases in which the distribution of means differed from random (birds P35 and R45). It was on the basis of these analyses that I concluded "birds tested repeatedly in the presence of a normal geomagnetic field but in a 'visually cueless' chamber failed to develop any significant directional preferences" (Emlen, 1970b, p. 223).

Since publishing that statement, I have performed two additional analyses of these results. First, all of the activity from all of the birds on all of the nights was pooled (Fig. 27A). Although the sample size became enormous (N of independent activity units = 7531), these data still failed to deviate from a uniform distribution (probability = 0.63 by Rayleigh test; probability = 0.17 by "v" test with 180° as the predicted direction). Finally, I pooled all of the nightly means from all of the birds and calculated the grand mean (Fig. 27B). Recall that this is the test statistic employed by Merkel and the Wiltschkos. Analyzed in this way, the Indigo Bunting data showed a highly significant trend in the appropriate migration direction, to the south-southeast (mean direction 173°; probability <0.001 by Rayleigh test; probability <0.001 by "v" test). Thus, although the concentration of orientation for individual birds or bird-nights was extremely low, the pooled results show a definite nonrandom clustering in the migration direction.

Another source of confusion arises from attempts to replicate "cueless" experiments. This concerns the relative advantages and disadvantages of different cage designs. Merkel and the Wiltschkos employed an eight-sided cage equipped with radially aligned perches. In such a cage, a bird is prone to circle, and a great deal of extraneous "noise" is introduced into the system (Merkel, 1971; H. C. Howland, 1973). Further,

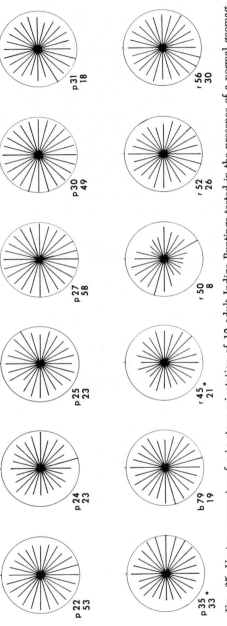

FIG. 25. Vector summaries of migratory orientation of 12 adult Indigo Buntings tested in the presence of a normal geomagnetic field but in the "absence" of visual cues. Diagrams plotted as in Fig. 11. (From Emlen, 1970b.)

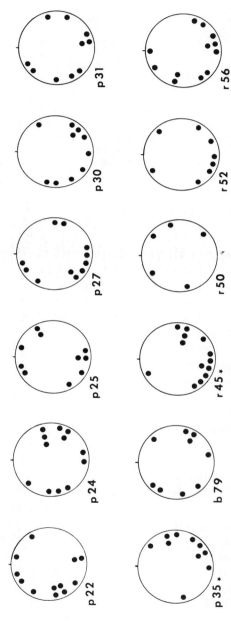

FIG. 26. Nightly mean headings of individual buntings tested in the presence of a normal geomagnetic field in the "absence" of visual cues. (From Emlen, 1970b.)

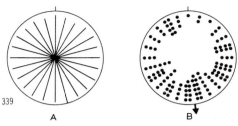

FIG. 27. Pooled results of migratory orientation of Indigo Buntings tested in the "absence" of meaningful visual cues. (A) Vector summary of *all* of the activity from *all* of the birds (total sample size = 7531 activity units). (B) Summary of all nightly means from all of the birds (N = 122). Arrow denotes mean direction (173°). See text for full discussion. (Recalculated and replotted from data of Emlen, 1970b.)

when a bird depresses a particular radial perch (e.g., northeast) its body axis typically is aligned perpendicular to this direction (the bird is facing northwest or southeast).

Most attempts to obtain oriented *Zugunruhe* in the absence of visual cues have been performed in cages with tangentially arranged perches. A perch registration to the northeast in such a cage should indicate that the bird is facing either northeast or southwest. Interestingly, all attempts to repeat Merkel and Wiltschko's "cueless" results performed in such cages have yielded negative results (Perdeck, 1963; Shumakov, 1965, 1967b; Wallraff, 1965b; Liepa, 1970; Shumakov and Vinogradova, 1970). Recently, Merkel (1971) reported that he, too, was unable to obtain orientation in a closed room when he switched his birds to a cage with tangential perches. To add to the confusion, Wallraff (1972) borrowed one of Merkel's original radial-perch cages and *was* able to repeat their observations. His robins oriented in the migratory direction, although they had failed to do so in previous tests with the other cage design.

In an attempt to better comprehend the controversial data from magnetic orientation experiments, we invited the Wiltschkos to come to Cornell and join us for a year of collaborative studies. Using Indigo Buntings, we set about to repeat "visually cueless" and magnetic deflection experiments, using both funnel (Emlen and Emlen, 1966) and radial-perch cage designs. Extreme care was taken to minimize potential artifact cues and to rigidly control for possible sources of unconscious bias (Howland, 1973). Preliminary results (S. T. Emlen, N. J. Demong, W. Wiltschko, R. Wiltschko, and S. Bergman, unpublished observations) indicate that buntings in early spring selectively oriented to the north under control conditions, but shifted to the east-southeast when the horizontal component of the magnetic field was deflected clockwise 120°. The concentration of orientation for most birds was low, but their

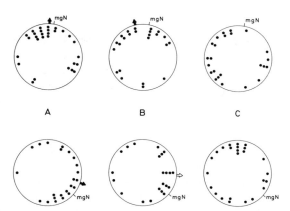

FIG. 28. Pooled mean directions of Indigo Buntings tested in a "visually cueless" environment. Top: With a normal magnetic field (mN = 10°). (A) Results from early in the spring migration season, April 24 to May 15. Mean direction = 3°; Rayleigh probability = 0.002. (B) Results from mid-season, May 16–31. Mean direction = 347°; Rayleigh probability = 0.019. (C) Results late in the migration season, June 1–15. Mean direction = 261°; Rayleigh probability = 0.265. Bottom: With the magnetic field shifted 120° clockwise (mN = 130°). (A) Early season. Mean direction = 110°; Rayleigh probability = 0.011. (B) Mid-season. Mean direction = 95°; Rayleigh probability = 0.089. (C) Late in the season. Mean direction = 1°; Rayleigh probability = 0.49. The data distributions under the two magnetic conditions are significantly different throughout the migratory period (using the Watson and Williams F test, $p < 0.01$ in A and B, $p < 0.05$ in C).

consistency was great enough to give a clear statistical separation between the pooled bearings under the two experimental conditions (magnetic north equals 10°; magnetic north equals 130°) (Fig. 28A and B). Later in the season, possibly when migratory motivation was waning, the orientation of birds under these "visually cueless" conditions deteriorated considerably (Fig. 28C).

In summary, second-order poolings of Indigo Bunting data reveal a significant tendency for orientation to the south in autumn and north in spring under visually cueless conditions, and this orientation shifts predictably in accordance with an artificial deflection of the magnetic field. Interestingly, the orientation is more clear-cut and the deflection more pronounced in the radial-perch cage design.

The biological explanation for these discrepancies remains unclear. But, at the very least, they should serve to emphasize the importance of a little-studied aspect of navigation experiments: the effect of different cage designs.

4. Homing Experiments

The older literature is full of accounts of homing experiments performed on birds to which small magnets had been attached to the head, dorsum, or wings (Wodzick: *et al.*, 1939; Yeagley, 1947, 1951; Gordon, 1948; Matthews, 1951b, 1952; Bochenski *et al.*, 1960). The species used included gulls, ducks, and storks, as well as homing pigeons. With the exception of Yeagley's original experiment, all studies yielded negative results; homing performance was unaffected by the magnetic disturbances. However, most of these experiments were performed under sunny conditions, close to the home location, or both. Consequently, alternative celestial and topographic cues might have been available.

Keeton's (1969) finding that pigeons can take up an initial bearing toward home when released under total overcast made it possible, for the first time, to study the importance of magnetic information in the absence of celestial input. He glued miniature magnets (weighing approximately 3 gm) to the backs of experienced homing pigeons. These magnets were somewhat variable in strength but were designed to produce a field of roughly 0.5 G in the vicinity of the birds' heads. Control birds wore a nonmagnetic brass bar of similar size and weight. All birds were housed together and had identical exercise and training schedules. For the critical tests, Keeton transported the birds to release locations 27–50 km distant and released them under total overcast. The bars were attached to the birds just prior to release; "brasses" and "magnets" were released alternately.

During preliminary experiments, performed when the sun was visible, both groups of birds oriented toward the home loft (Keeton, 1971, 1972). This confirmed the findings of previous investigators; magnets did not disrupt homing. But when releases were conducted under total overcast, the two groups of birds showed significant differences in their initial departure behavior. In five of seven tests, the birds wearing brass bars oriented toward home, while those with magnetic bars scattered at random (Keeton, 1971, 1972). In one additional release, both groups of birds departed toward home, while in the other, neither group showed any preferred orientation. Figure 29 shows the pooled bearings from these series of experiments. These results provide some of the clearest evidence to date of a magnetic effect upon bird orientation. They also serve to emphasize the importance of designing tests that take into account the redundancy of orientation systems.

The behavior of young, inexperienced pigeons was quite different from that of the experienced adults. Keeton and Gobert (1970) previously had reported that first-flight birds were unable to orient homeward when the sun was not visible. It appeared that homing experience was necessary for the maturation of the "overcast system" or, perhaps, for its coupling with some directional reference provided by other cues. Attaching a magnet to a young bird produces yet another decrement

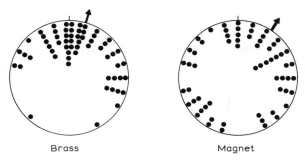

Brass   Magnet

FIG. 29. Pooled departure bearings of experienced homing pigeons released under conditions of total overcast at distances 27 to 50 km from the home loft. Home direction = 360°. Left: Results from birds with brass bars glued to their backs. Mean direction = 15°; Rayleigh probability <0.001. Right: Results from birds with a magnetic bar attached. Mean direction = 31°; Rayleigh probability = 0.04. (Data recalculated from Experimental Series III in Keeton, 1971, 1972.)

in its orientation capabilities. Such birds departed randomly even when released on clear, sunny days (Keeton, 1971, 1972). This suggests that during their first flights, young birds cannot tolerate the elimination or disruption of either solar or magnetic information. As the birds become older and more experienced, less information is required and disorientation results only when both inputs are disrupted simultaneously.

Interestingly, the magnets used in Keeton's tests did not seriously disrupt the pigeons' ability to reach the home loft. Homing speeds frequently were somewhat slower than for control birds, but only occasionally was this difference statistically significant. Further, homing success, as measured by the number of birds returning to the loft, was comparable in the two groups. This suggests that whatever disruptive effect the magnets produce, it is short-lived and the pigeons are able to compensate for the disturbance and reorient toward the home loft soon after being released.

Walcott and Green (1974) have designed a pair of miniature Helmholtz coils that fit around the head of a pigeon. The larger of the two coils slips over the head and fits like a collar around the neck while the smaller is glued directly to the top of the head. In experimental birds the coils are connected to a battery, thereby producing a magnetic field of approximately 0.6 G. Control birds are similarly equipped, but the batteries are not connected to the coils. When pigeons were released at localities 60–90 km west of the home loft under sunny conditions, no effects of the coils were detected. When released under overcast, however, the behavior of the pigeons differed depending on the polarity of the magnetic field produced by the coils. If the current produced a field where magnetic north pointed toward the larger coil around the neck of the pigeon (Sup) the birds tended to depart in a normal homeward direction. But if the current in the coils was reversed (Nup), the birds tended to reverse their direction and departed nonrandomly in the direction away from home (Fig. 30).

These results must be considered preliminary, awaiting the outcome of additional, replicate releases. But they raise the exciting possibility of a common thread running between migratory and homing studies of magnetic orientation. The difference between Nups and Sups may be interpretable in terms of the Wiltschkos' model. They suggested that birds use the inclination of the magnetic field, taking north to be the direction in which the magnetic field vector and gravity vector make the smallest angle. Reversing the current flow through the coils around a pigeon's head would reverse the magnetic field vector which, in turn, would reverse the direction interpreted as north (Walcott and Green, 1974).

The lure of finding a new sensory modality is attracting an increasing number of laboratories into studies of the biological effects of magnetism. No longer can magnetism be brushed aside on theoretical grounds. The recent surge of positive results demands an explanation. Southern's work with gulls and that of Keeton with pigeons shows a deterioration of direction-finding capability when magnetic information is disturbed. It is tempting to conclude from this that magnetic cues are directly involved in the orientation process. But an equally plausible alternative is that a magnetic storm or the presence of a strong bar magnet disrupts the normal sensory input to some other component of the orientation system. At present, the work of Merkel and Wiltschko and Walcott and Green stands as the most tantalizing evidence that magnetic information is actively used by birds in the selection of flight directions.

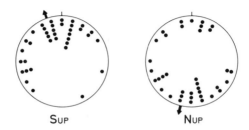

SUP   NUP

FIG. 30. Vanishing bearings of pigeons equipped with miniature Helmholtz coils. Birds were released 59 to 92 km west of the home loft under total overcast. Diagrams ploted with 360° = home direction. SUP and NUP are defined in the text. Data are pooled from Walcott and Green (1974).

V. Conclusions

One of my purposes in writing this review was to point out the major areas in migratory orientation research where active work either is being or should be performed. In so doing I have stressed that a migrant has many potential sources of directional information at its disposal. These range from topographic, through meteorological and inertial, to magnetic and celestial cues.

But the subdivision of this chapter into discussions of different sensory cues contradicts my second principal aim. That was to stress the interrelationship of different sources of directional information—to emphasize the redundant nature of orientation systems. The realization that different cues function as components in a complex navigational repertoire opens an entirely new forefront in orientation research. What are the relative advantages of one type of cue over another? Does the simultaneous use of multiple components result in increased accuracy of orientation? Is there a hierarchy in the relative importances of different types of information?

We have just reached the stage in which these interactions between components are being studied. The principal technique is to manipulate two or more cues simultaneously in such a way as to provide the animal with conflicting information. Preliminary findings suggest that a differential weighting of different cues does exist. In the case of experienced homing pigeons, clock-shifted birds make predictable errors when released under sunny conditions but orient homeward under overcast. This implies that the sun compass takes precedence over the overcast system when the two are in conflict. Clock-shifted birds also vanish in the predicted incorrect direction when released at familiar sites, including locations within a few miles of the home loft. Consequently, the sun compass must also predominate over familiar landmarks in relative importance to the pigeon. W. T. Keeton (personal communication) plans to conduct experiments in which experienced birds wearing magnets are released under total overcast from familiar sites. The results should enable him to rank the relative importance of landmarks with the overcast system.

Similar speculations can be made for migrants. Not all species would be expected to have similar weighting systems, however, and one must be careful not to overgeneralize. In the case of nocturnal songbird migrants, stellar cues seem to predominate over any possible geophysical (including magnetic) effects in at least four species. When the starry skies of a planetarium are adjusted so that "stellar north" does not coincide with either geographic or magnetic north, Indigo Buntings (Emlen,

1967a), European warblers (Sauer, 1957), and White-throated Sparrows (S. A. Gauthreaux, Jr., personal communication) shift their orientation in keeping with the new position of the artificial sky. Wiltschko *et al.* (1971, in preparation) believe that an alternative weighting system, in which magnetic information dominates over celestial cues, functions in the European Robin and in some *Sylvia* warblers. They report being able to deflect the *Zugunruhe* orientation by means of Helmholtz coils, even when the caged bird is provided with a view of a 95° sector of the natural night sky. The findings are difficult to interpret, however, since the zone of the sky that may be of most importance for northern hemisphere migrants (the northern circumpolar region) was blocked from the bird's view during the conflict experiments.

Although experimental evidence is not yet available, I expect that the position of sunset will be found to play a vital role in the initial selection of direction by the great majority of nocturnal fliers. Sunset seems a logical cue to use in the transfer between cues and reference systems employed during daylight and those operative after dark. Several fragments of information support this idea. Kramer (1949, 1951), in his pioneering studies of caged *Zugunruhe*, found that songbirds frequently oriented directly toward the horizon glows produced by city lights. But if birds were placed in their outdoor cages during the daytime and exposed to the normal sunset transition into night, they appeared less confused and normally took up directions appropriate for the migration season, ignoring the phototactic attraction of the same city lights. Hebrard's (1972) observation that nocturnal songbird migrants oriented appropriately under total overcast if the previous sunset had been visible, yet became disoriented when a view of the sun, including sunset, was obscured for long periods of time, is also of interest here. Finally, the fact that the position of sunset can be used directly as an orientational cue is born out by some of my own work (S. T. Emlen and N. J. Demong, unpublished radar observations) in which White-throated Sparrows consistently selected the appropriate spring migration direction when released and tracked during the transition period after the sun's disk had dropped below the horizon yet before the stars were visible to the human eye.

Winds aloft probably play a secondary role in migratory orientation. Although a direction may be maintained by flying at an angle relative to the wind, the selection of that initial direction presumably is based on other cues. When S. T. Emlen and N. J. Demong (unpublished observations) radar tracked White-throated Sparrows that were released artificially in the air space, the birds' behavior differed depending on the meteorological situation. Under clear skies in the spring, the birds consistently departed to the northeast; but they reverted to a downwind tendency under total overcast. This suggests that celestial predominated over downwind orientation in this species.

Finally, the indirect evidence from numerous radar studies tends to place topographic cues near the bottom of the hierarchy for passerine night migrants. However, it is possible that the weighting scheme changes at some point along the migratory route and that topographic features may play a more central role in the final phases of locating a specific breeding or wintering location.

The relative importance of different sources of information also changes with age and experience. Young birds frequently require several types of input in order to select an appropriate direction. These different sources of information appear to be integrated with each other and into a general orientational framework during early development. For example, Indigo Buntings learn star patterns and come to relate them to a directional reference axis provided by the rotational motion of the night sky (Emlen, 1969b, 1970a, 1972). Birds deprived of viewing the night sky in early ontogeny are unable to select the migratory direction. Of equal interest, birds given exposure as young are later able to select their direction under stationary planetarium skies. The implication is that celestial rotation is essential in providing a reference framework in which star patterns take on meaning. Once this framework is learned, however, rotation per se is not necessary and star patterns alone are sufficient for direction determination. A parallel story concerning the possible integration of magnetic and celestial reference systems is being proposed by the Wiltschkos (in preparation).

Young, inexperienced birds are generally much more plastic than adults in their orientation behavior (Emlen, 1972; S. T. Emlen, N. J. Demong and E. Gwinner, unpublished observations; Keeton and Gobert, 1970; Keeton, 1972). Not only are they receptive to a wide variety of cues, but the hierarchy of relative importance of different inputs is less fixed. To understand the adaptiveness of this, consider a young migrant, familiar only with cues available at its site of hatching. When it departs on its first migratory journey, it may travel 2000–3000 km, bringing it to southerly latitudes where the sky overhead is totally unfamiliar and where magnetic cues and prevailing wind patterns are vastly different from those to which it previously has been exposed. Selection and imprinting to a specific overwintering location has yet to occur. Obviously, behavioral plasticity and a receptivity to a wide range of orientational cues would be of crucial importance for such young birds.

As the bird matures and gains experience, this plasticity decreases. The navigational capabilities may change (e.g., Perdeck, 1958, 1967), accuracy improves (Emlen, 1969b, 1972), the amount of information needed for direction determination decreases (Emlen, 1969b, 1970a,

1972; Keeton and Gobert, 1970; Keeton, 1972, 1974), and the relative weighting of cue systems becomes more rigid. These trends seem to hold for both migratory birds and for homing pigeons.

As stated earlier, "the problem for the ornithologist interested in navigation is no longer searching for *the* mechanism of orientation. It is studying any and all components of the orientation systems and trying to decipher how these are integrated into a navigational complex that determines the directional behavior of the bird at any place in time or space."

The field of bird orientation has come a long way since Kramer's (1961) review in the predecessor to this volume. We have developed new techniques for studying the behavior of free-flying migrants. We have proposed various models to explain the functioning of different components in the orientation system. We have enumerated the redundant ways a bird could maintain its migratory direction. But we have a long way to go. We are no closer to an understanding of the "map" than Kramer was two decades ago. We have little knowledge of the types or accuracy of the navigational capabilities of most migrants. In short, the sum total of our knowledge today is still insufficient to explain how an individual migrant finds its way over the thousands of kilometers between its breeding territory and its overwintering destination.

ACKNOWLEDGMENTS

I thank members of the Cornell Orientation Group, particularly William T. Keeton, C. John Ralph, and W. John Richardson, for commenting on various portions of this manuscript. My own recent work has been supported by grants from the National Science Foundation (GB-13046X and GB-35199X) and the National Aeronautics and Space Administration (Project Nos. KG-5597, KL-5737, KL-5919).

REFERENCES

Able, K. P. (1972). Fall migration in coastal Louisiana and the evolution of migration patterns in the Gulf region. *Wilson Bull.* 84, 231–242.

Able, K. P. (1973). The role of weather variables and flight direction in determining the magnitude of nocturnal bird migration. *Ecology* 54, 1031–1041.

Able, K. P. (1974a). Environmental influences on the orientation of free-flying nocturnal bird migrants. *Animal Behaviour* 22, 224–238.

Able, K. P. (1974b). Wind, track, heading and the flight orientation of migrating songbirds. *In* "The biological aspects of the bird/aircraft collision problem" (S. Gauthreaux, ed.), pp. 331–358. Air Force Office of Scientific Research, Clemsen, North Carolina.

Adler, H. E. (1963). Psychophysical limits of celestial navigation hypotheses. *Ergeb. Biol.* 26, 235–252.

Baird, J., and Nisbet, I. C. T. (1960). Northward fall migration on the Atlantic coast and its relation to offshore drift. *Auk* 77, 119–149.

Baldaccini, N. E., Fiaschi, V., Fiore, L., and Papi, F. (1971). Initial orientation of directionally trained pigeons under overcast and sunny conditions. *Monit. Zool. Ital.* 5, 53–63.

Barlow, J. S. (1964). Inertial navigation as a basis for animal navigation. *J. Theor. Biol.* **6**, 76–117.

Barlow, J. S. (1966). Inertial navigation in relation to animal navigation. *J. Inst. Navig.* **19**, 302–316.

Bellrose, F. C. (1958). Celestial orientation in wild Mallards. *Bird-Banding* **29**, 75–90.

Bellrose, F. C. (1963). Orientation behavior of four species of waterfowl. *Auk* **80**, 257–289.

Bellrose, F. C. (1964). Radar studies of waterfowl migration. *Trans. N. Amer. Wildl. Natur. Resour. Conf.* **29**, 128–143.

Bellrose, F. C. (1967a). Radar in orientation research. *Proc. Int. Ornithol. Congr., 14th, 1966* pp. 281–309.

Bellrose, F. C. (1967b). Orientation in waterfowl migration. *Proc. Annu. Biol. Colloq.* [*Oreg. State Univ.*] **27**, 73–99.

Bellrose, F. C. (1971). The distribution of nocturnal migrants in the air space. *Auk* **88**, 397–424.

Bellrose, F. C., and Graber, R. R. (1963). A radar study of the flight directions of nocturnal migrants. *Proc. Int. Ornithol. Congr., 13th, 1962* pp. 362–389.

Berger, M., and Hart, J. S. (1974). Physiology and energetics of flight *In* "Avian Biology," Vol. IV (D. S. Farner and J. R. King, eds.), pp. 415–477. Academic Press, New York.

Bergman, G., and Donner, K. O. (1964). An analysis of the spring migration of the Common Scoter and the Long-tailed Duck in southern Finland. *Acta Zool. Fenn.* **105**, 3–59.

Billings, S. M. (1968). Homing in Leach's Petrel. *Auk* **85**, 36–43.

Blokpoel, H. (1970). A preliminary study on height and density of nocturnal fall migration. *Proc. World Conf. Bird Hazards Aircraft, 1969* pp. 335–348.

Bochenski, Z., Dylewska, M., Gieseczykiewicz, J., and Sych, L. (1960). [Homing experiments on birds. XI. Experiments with swallows *Hirundo rustica* L. concerning the influence of earth magnetism and partial eclipse of the sun on their orientation.] *Zesz. Nauk. W. J. Zool.* **5**, 125–130 (in Polish).

Bruderer, B. (1971). Radarbeobachtungen über den Frühlingszug im Schweizerischen Mittelland. *Ornithol. Beob.* **68**, 89–158.

Bruderer, B., and Steidinger, P. (1972). Methods of quantitative and qualitative analysis of bird migration with a tracking radar. *NASA Spec. Publ.* **NASA SP-262**, 151–167.

Cold Spring Harbor Symposia on Quantitative Biology. (1960). "Biological Clocks," Vol. XXV. Biological Laboratory, Cold Spring Harbor, Long Island, New York.

D'Arms, E., and Griffin, D. R. (1972). Balloonists' reports of sounds audible to migrating birds. *Auk* **89**, 269–279.

Davis, L. (1948). Remarks on "The physical basis of bird navigation." *J. Appl. Phys.* **19**, 307–308.

deVries, H. (1948). Die Reizschwelle der Sinnesorgane als physikalisches Problem. *Experientia* **4**, 205–213.

Dolnik, V. R., and Shumakov, M. E. (1967). Testing the navigational abilities of birds. *In* "Bionica," pp. 500–507. Moscow.

Dorst, J. (1962). "The Migrations of Birds." Houghton, Boston, Massachusetts.

Drury, W. H., and Keith, J. A. (1962). Radar studies of songbird migration in coastal New England. *Ibis* **104**, 449–489.

Drury, W. H., and Nisbet, I. C. T. (1964). Radar studies of orientation of songbird migrants in southeastern New England. *Bird-Banding* **35**, 69–119.

Eastwood, E. (1967). "Radar Ornithology." Methuen, London.

El'Darov, A. L., and Kholodov, Yu. (1964). The effect of a constant magnetic field on the motor activity of birds. *Zh. Obshch. Biol.* **25,** 224–229.

Emlen, S. T., and Emlen, J. T., Jr. (1966). A technique for recording migratory orientation of captive birds. *Auk* **83,** 361–367.

Emlen, S. T. (1967a). Migratory orientation in the Indigo Bunting, *Passerina cyanea.* Part I. Evidence for use of celestial cues. *Auk* **84,** 309–342.

Emlen, S. T. (1967b). Migratory orientation in the Indigo Bunting, *Passerina cyanea.* Part II. Mechanism of celestial orientation. *Auk* **84,** 463–489.

Emlen, S. T. (1967c). Orientation of Zugunruhe in the Rose-breasted Grosbeak, *Pheucticus ludovicianus. Condor* **69,** 203–205.

Emlen, S. T. (1969a). Bird migration: Influence of physiological state upon celestial orientation. *Science* **165,** 716–718.

Emlen, S. T. (1969b). The development of migratory orientation in young Indigo Buntings. *Living Bird* **8,** 113–126.

Emlen, S. T. (1970a). Celestial rotation: Its importance in the development of migratory orientation. *Science* **170,** 1198–1201.

Emlen, S. T. (1970b). The influence of magnetic information on the orientation of the Indigo Bunting, *Passerina cyanea. Anim. Behav.* **18,** 215–224.

Emlen, S. T. (1971). Celestial rotation and stellar orientation in migratory warblers. *Science* **173,** 460–461.

Emlen, S. T. (1972). The ontogenetic development of orientation capabilities. *NASA Spec. Publ.* NASA SP-262, 191–210.

Emlen, S. T. (1974). Problems in identifying bird species by radar signature analyses: intra-specific variability. *In* "The biological aspects of the bird/aircraft collision problem" (S. Gauthreaux, ed.), pp. 509–524. Air Force Office of Scientific Research, Clemsen, North Carolina.

Evans, P. R. (1966). Migration and orientation of passerine night migrants in northeast England. *J. Zool.* **150,** 319–369.

Evans, P. R. (1968). Reorientation of passerine night migrants after displacement by the wind. *Brit. Birds* **61,** 281–303.

Evans, P. R. (1970). Nocturnal songbird migration. *Nature (London)* **228,** 1121.

Evans, P. R. (1972). Information on bird navigation obtained by British long-range radars. *NASA Spec. Publ.* NASA SP-262, 139–149.

Fromme, H. G. (1961). Untersuchungen über das Orientierungsvermögen nächtliche ziehender Kleinvögel (*Erithacus rubecula, Sylvia communis*). *Z. Tierpsychol.* **18,** 205–220.

Gauthreaux, S. A., Jr. (1969). Stellar orientation of migratory restlessness in the White-throated Sparrow, *Zonotrichia albicollis. Amer. Zool.* **9,** 25 (abstr.).

Gauthreaux, S. A., Jr. (1971). A radar and direct visual study of passerine spring migration in southern Louisiana. *Auk* **88,** 343–365.

Gauthreaux, S. A., Jr. (1972a). Behavioral responses of migrating birds to daylight and darkness; a radar and direct visual study. *Wilson Bull.* **84,** 136–143.

Gauthreaux, S. A., Jr. (1972b). Flight directions of passerine migrants in daylight and darkness: A radar and direct visual study. *NASA Spec. Publ.* NASA SP-262, 129–137.

Gauthreaux, S. A., Jr., and Able, K. P. (1970). Wind and the direction of nocturnal songbird migration. *Nature (London)* **228,** 476–477.

Gehring, W. (1963). Radar- und Feldbeobachtungen über den Verlauf des Vogelzuges im Schweizerischen Mittelland: der Tagzug im Herbst (1957–1961). *Ornithol. Beob.* **60,** 35–68.

Gordon, D. A. (1948). Sensitivity of the homing pigeon to the magnetic field of the earth. *Science* **108,** 710–711.

Graue, L. C. (1963). The effect of phase shifts in the day-night cycle on pigeon homing at distances of less than one mile. *Ohio J. Sci.* **63**, 214–217.

Griffin, D. R. (1940). Homing experiments with Leach's Petrels. *Auk* **57**, 61–74.

Griffin, D. R. (1952). Bird navigation. *Biol. Rev. Cambridge Phil. Soc.* **27**, 359–400.

Griffin, D. R. (1955). Bird navigation. In "Recent Studies in Avian Biology" (A. Wolfson, ed.), pp. 154–197. Univ. of Illinois Press, Urbana.

Griffin, D. R. (1969). The physiology and geophysics of bird navigation. *Quart. Rev. Biol.* **44**, 255–276.

Griffin, D. R. (1972). Nocturnal bird migration in opaque clouds. *NASA Spec. Publ.* **NASA SP-262**, 169–188.

Griffin, D. R. (1973). Oriented bird migration in or between opaque cloud layers. *Proc. Amer. Phil. Soc.* **117**, 117–141.

Hamilton, W. J., III. (1962a). Bobolink migratory pathways and their experimental analysis under night skies. *Auk* **79**, 208–233.

Hamilton, W. J., III. (1962b). Celestial orientation in juvenile waterfowl. *Condor* **64**, 19–33.

Hamilton, W. J., III. (1962c). Does the Bobolink navigate? *Wilson Bull.* **74**, 357–366.

Hamilton, W. J., III. (1966). Analysis of bird navigation experiments. In "Systems Analysis in Ecology" (K. E. F. Watt, ed.), pp. 147–178. Academic Press, New York.

Hassler, S. S., Graber, R. R., and Bellrose, F. C. (1963). Fall migration and weather, a radar study. *Wilson Bull.* **75**, 56–77.

Hebrard, J. J. (1972). Fall nocturnal migration during two successive overcast days. *Condor* **74**, 106–107.

Hilditch, C. D. M., Williams, T. C., and Nisbet, I. C. T., (1973). Autumnal bird migration over Antigua, W. I. *Bird-Banding* **44**, 171–179.

Hochbaum, H. A. (1955). "Travels and Traditions of Waterfowl." Branford, Newton, Massachusetts.

Hoffmann, K. (1954). Versuche zu der in Richtungsfinden der Vögel enthaltenen Zeitschätzung. *Z. Tierpsychol.* **11**, 453–475.

Hoffmann, K. (1958). Repetition of an experiment on bird orientation. *Nature (London)* **181**, 1435–1437.

Howland, H. C. (1973). Orientation of European Robins to Kramer cages. *Zeit. Tierpsychol.* **33**, 295–312.

Hunzinger, E. (1935). Durchschneidung aller Bogengänge bei der Taube. *Pfluegers Arch. Gesamte Physiol. Menschen Tiere* **236**, 52–58.

Keeton, W. T. (1969). Orientation by pigeons; is the sun necessary? *Science* **165**, 922–928.

Keeton, W. T. (1970). Do pigeons determine latitudinal displacement from the sun's altitude? *Nature (London)* **227**, 626–627.

Keeton, W. T. (1971). Magnets interfere with pigeon homing. *Proc. Nat. Acad. Sci. U.S.* **68**, 102–106.

Keeton, W. T. (1972). Effects of magnets on pigeon homing. *NASA Spec. Publ.* **NASA SP-262**, 579–594.

Keeton, W. T. (1974). The orientational and navigational basis of homing in birds. In "Recent Advances in the Study of Behavior" Academic Press, New York.

Keeton, W. T., and Gobert, A. (1970). Orientation by untrained pigeons requires the sun. *Proc. Nat. Acad. Sci. U.S.* **65**, 853–856.

Kenyon, K. W., and Rice, D. W. (1958). Homing of Laysan Albatrosses. *Condor* **60**, 3–6.

Kramer, G. (1949). Über Richtungstendenzen bei der nächtlichen Zugunruhe gekäfigter Vögel. *In* "Ornithologie als biologische Wissenschaft" (E. Mayr and E. Shüz, eds.), pp. 269–283. Carl Winter, Heidelberg.

Kramer, G. (1950). Orientierte Zugaktivität gekäfigter Singvögel. *Naturwissenschaften* **37**, 188.

Kramer, G. (1951). Eine neue Methode zur Erforschung der Zugorientierung und die bisher damit erzielten Ergebnisse. *Proc. Int. Ornithol. Congr., 10th, 1950* pp. 269–280.

Kramer, G. (1952). Experiments on bird orientation. *Ibis* **94**, 265–285.

Kramer, G. (1953). Die Sonnenorientierung der Vögel. *Zool. Anz. 17 Suppl. Verh. Deut. Zool. Ges. Freiburg, 1952* pp. 72–84.

Kramer, G. (1955). Ein weiterer Versuch, die Orientierung von Brieftauben durch jahreszeitliche Änderung der Sonnenhöhe zu beeinflussen. *J. Ornithol.* **96**, 173–185.

Kramer, G. (1957). Experiments in bird orientation and their interpretation. *Ibis* **99**, 196–227.

Kramer, G. (1961). Long distance orientation. *In* "Biology and Comparative Physiology of Birds" (A. J. Marshall, ed.), Vol. 2, pp. 341–371. Academic Press, New York.

Kramer, G., and von St. Paul, U. (1950). Stare, *Sturnus vulgaris,* lassen sich auf Himmelsrichtung dressieren. *Naturwissenschaften* **37**, 526–527.

Kreithen, M., and W. T. Keeton (1974). Attempts to condition homing pigeons to magnetic stimuli. *J. Comp. Physiol.* **91**, 355–362.

Lack, D. (1962). Radar evidence on migratory orientation. *Brit. Birds* **55**, 139–158.

Lack, D. (1969). Drift migration: A correction. *Ibis* **111**, 253–255.

Lack, D., and Eastwood, E. (1962). Radar films of migration over eastern England. *Brit. Birds* **55**, 388–414.

Lack, D., and Lockley, R. M. (1938). Skokholm bird observatory homing experiments. I. 1936–37. Puffins, Storm Petrels, and Manx Shearwaters. *Brit. Birds* **31**, 242–248.

Liepa, V. K. (1970). [Migratory restlessness of Robins in Kramer cages under uniform artificial light.] *In* "Ornithology in the USSR," *Proc. Baltic Ornithol. Conf., 7th, 1970,* Vol. 2, pp. 123–127. Ashkhabad (in Russian).

Löhrl, H. (1959). Zur Frage des Zeitpunktes einer Prägung auf die Heimatregion beim Halbandschnapper (*Ficedula albicollis*). *J. Ornithol.* **100**, 132–140.

Martin, D. D., and A. H. Meier (1973). Temporal synergism of corticosterone and prolactin in regulating orientation in the migratory White-throated Sparrow (*Zonotrichia albicollis*). *Condor* **75**, 369–374.

Matthews, G. V. T. (1951a). The sensory basis of bird navigation. *J. Inst. Navig.* **4**, 260–275.

Matthews, G. V. T. (1951b). The experimental investigation of navigation in homing pigeons. *J. Exp. Biol.* **28**, 508–536.

Matthews, G. V. T. (1952). An investigation of homing ability in two species of gulls. *Ibis* **94**, 243–264.

Matthews, G. V. T. (1953a). Sun navigation in homing pigeons. *J. Exp. Biol.* **30**, 243–267.

Matthews, G. V. T. (1953b). Navigation in the Manx Shearwater. *J. Exp. Biol.* **30**, 370–396.

Matthews, G. V. T. (1955). "Bird Navigation," 1st ed. Cambridge Univ. Press, London and New York.

Matthews, G. V. T. (1961). 'Nonsense' orientation in Mallard, *Anas platyrhynchos,* and its relation to experiments on bird navigation. *Ibis* **103a**, 211–230.

Matthews, G. V. T. (1963). The astronomical bases of 'nonsense' orientation. *Proc. Int. Ornithol. Congr., 13th, 1962* pp. 415–429.

Matthews, G. V. T. (1964). Individual experience as a factor in the navigation of Manx Shearwaters. *Auk* **81**, 132–146.

Matthews, G. V. T. (1968). "Bird Navigation," 2nd ed. Cambridge Univ. Press, London and New York.

Merkel, F. W. (1938). Zur Physiologie der Zugunruhe bei Vögeln. *Ber. Ver. Schles. Ornithol.* **23**, Sonderheft, 1938, 1–72.

Merkel, F. W. (1956). Untersuchungen über tages- und jahresperiodische Aktivitäts-änderungen bei gekäfigten Zugvögelin. *Z. Tierpsychol.* **13**, 278–301.

Merkel, F. W. (1971). Orientation behavior of birds in Kramer cages under different physical cues. *Ann. N.Y. Acad. Sci.* **188**, 283–294.

Merkel, F. W., and Fromme, H. (1958). Untersuchungen über das Orientierungs-vermögen nächtlich ziehender Rotkehlchen (*Erithacus rubecula*). *Naturwissen-schaften* **45**, 499–500.

Merkel, F. W., and Wiltschko, W. (1965). Magnetismus und Richtungsfinden zugun-ruhiger Rotkehlchen (*Erithacus rubecula*). *Vogelwarte* **23**, 71–77.

Merkel, F. W., Fromme, H. G., and Wiltschko, W. (1964). Nichtvisuelles Orien-tierungsvermögen bei nächtlich zugunruhigen Rotkehlchen! *Vogelwarte* **22**, 168–173.

Mewaldt, L. R., and Rose, R. G. (1960). Orientation of migratory restlessness in the White-crowned Sparrow. *Science* **131**, 105–106.

Mewaldt, L. R., Morton, M. L., and Brown, I. L. (1964). Orientation of migratory restlessness in *Zonotrichia. Condor* **66**, 377–417.

Meyer, M. E. (1964). Discriminative basis for astronavigation in birds. *J. Comp. Physiol. Psychol.* **58**, 403–406.

Meyer, M. E., and Lambe, D. R. (1966). Sensitivity of the pigeon to changes in the magnetic field. *Psychon. Sci.* **5**, 349–350.

Michener, M. C., and Walcott, C. (1966). Navigation of single homing pigeons: Airplane observations by radio tracking. *Science* **154**, 410–413.

Michener, M. C., and Walcott, C. (1967). Homing of single pigeons—analysis of tracks. *J. Exp. Biol.* **47**, 99–131.

Miselis, R., and Walcott, C. (1970). Locomotor activity rhythms in homing pigeons (*Columba livia*). *Anim. Behav.* **18**, 544–551.

Myres, M. T. (1964). Dawn ascent and re-orientation of Scandinavian thrushes (*Turdus spp.*) migrating at night over the northeastern Atlantic Ocean in autumn. *Ibis* **106**, 7–51.

Nisbet, I. C. T. (1955). Atmospheric turbulence and bird flight. *Brit. Birds* **48**, 557–559.

Nisbet, I. C. T., and Drury, W. H. (1967). Orientation of spring migrants studied by radar. *Bird-Banding* **38**, 173–186.

Orgel, A. R., and Smith, J. C. (1954). Test of the magnetic theory of homing. *Science* **120**, 891–892.

Palmgren, P. (1937). Auslösung der Frühlingszugunruhe durch Wärme bei gekäfig-ten Rotkehlchen, *Erithacus rubecula* (L.). *Ornis Fenn.* **14**, 71–73.

Palmgren, P. (1938). Studien über den zeitlichen Ablauf der Zuggerregung bei gekäfigten Kleinvögeln. I. *Ornis Fenn.* **15**, 1–16.

Palmgren, P. (1949a). On the diurnal rhythm of activity and rest in birds. *Ibis* **91**, 561–576.

Palmgren, P. (1949b). Studien über die Tagesrhythmik gekäfigter Zugvögel. *Z. Tierpsychol.* **6**, 44–86.

Pennycuick, C. J. (1960). The physical basis of astronavigation in birds: Theoretical considerations. *J. Exp. Biol.* **37**, 573–593.

Perdeck, A. C. (1957). Stichting Vogeltrekstation Texel Jaarverslag over 1956. *Limosa* **30**, 62–75.

Perdeck, A. C. (1958). Two types of orientation in migrating Starlings *Sturnus vulgaris* L., and Chaffinches *Fringilla coelebs* L., as revealed by displacement experiments. *Ardea* **46**, 1–37.

Perdeck, A. C. (1963). Does navigation without visual cues exist in Robins? *Ardea* **51**, 91–104.

Perdeck, A. C. (1967). Orientation of Starlings after displacement to Spain. *Ardea* **55**, 194–202.

Potapov, R. L. (1966). Navigation ability of certain passerines. *Dokl. Akad. Nauk SSSR* **171**, 226–228.

Rabøl, J. (1969). Orientation of autumn migrating garden warblers (*Sylvia borin*) after displacement from western Denmark (Blavand) to eastern Sweden (Ottenby). A preliminary experiment. *Dan. Ornithol. Foren. Tidsskr.* **63**, 93–104.

Rabøl, J. (1970). Displacement and phaseshift experiments with night-migrating passerines. *Ornis Scand.* **1**, 27–43.

Rabøl, J. (1972). Displacement experiments with night-migrating passerines (1970). *Z. Tierpsychol.* **30**, 14–25.

Ralph, C. J., and Mewaldt, L. R. (1975). Timing of site fixation upon the wintering grounds in sparrows. *Auk* (in press).

Rawson, K. S., and Rawson, A. M. (1955). The orientation of homing pigeons in relation to change in sun declination. *J. Ornithol.* **96**, 168–172.

Reille, A. (1968). Essai de mise en évidence d'une sensibilité du pigeon au champ magnétique à l'aide d'un conditionnement nociceptif. *J. Physiol.* (*Paris*) **60**, 85–92.

Richardson, W. J. (1971). Spring migration and weather in eastern Canada: A radar study. *Amer. Birds* **25**, 684–690.

Richardson, W. J. (1972). Autumn migration and weather in eastern Canda: A radar study. *Amer. Birds* **26**, 10–17.

Richardson, W. J. (1974). Bird migration over southeastern Canada, the western Atlantic, and Puerto Rico: a radar study. Unpublished Ph.D. Thesis, Cornell University.

Richardson, W. J. (1975). Autumn migration over Puerto Rico and the western Atlantic, a radar study. *Ibis* (in press).

Rogers, D. T., Jr., and Odum, E. P. (1966). A study of autumnal post-migrant weights and vernal fattening of North American migrants in the tropics. *Wilson Bull.* **78**, 415–433.

Rüppell, W. (1934). Heimfinde-Versuche mit Rauchschwalben (*Hirundo rustica*) und Mehlschwalben (*Delichon urbica*) von H. Warnat (Berlin-Charlottenburg). *Vogelzug* **5**, 161–166.

Rüppell, W. (1936). Heimfindeversuche mit Staren und Schwalben 1935. *J. Ornithol.* **84**, 180–198.

Rüppell, W. (1937). Heimfindeversuche mit Staren, Rauchschwalben, Wendhälsen, Rotruckwürgen und Habichten 1936. *J. Ornithol.* **85**, 120–135.

Rüppell, W. (1944). Versuche über Heimfinden ziehender Nebelkrähen nach Verfrachtung. *J. Ornithol.* **92**, 106–133.

Rüppell, W., and Schüz, E. (1948). Ergebnis der Verfrachtung von Nebelkrähen (*Corvus corone cornix*) wahrend des Wegzuges. *Vogelwarte* **15**, 30–36.

Sauer, E. G. F. (1957). Die Sternenorientierung nächlich ziehender Grasmücken (*Sylvia atricapilla, borin* und *curruca*). *Z. Tierpsychol.* **14**, 29–70.

Sauer, E. G. F. (1958). Celestial navigation by birds. *Sci. Amer.* **199**, 44–49.

Sauer, E. G. F. (1961). Further studies on the stellar orientation of nocturnally migrating birds. *Psychol. Forsch.* **26**, 224–244.

Sauer, E. G. F. (1963). Migration habits of Golden Plovers. *Proc. Int. Ornithol. Congr., 13th, 1962* pp. 454–467.

Sauer, E. G. F., and Sauer, E. M. (1960). Star navigation of nocturnal migrating birds. The 1958 planetarium experiments. *Cold Spring Harbor Symp. Quant. Biol.* **25**, 463–473.

Schlichte, H. J. (1973). Untersuchungen über die Bedeutung optischer Parameter für das Heimkehrverhalten der Brieftaube. *Z. Tierpsychol.* **32**, 257–280.

Schlichte, H. J., and Schmidt-Koenig, K. (1971). Zum Heimfindevermögen der Brieftaube bei erschwerter optischer Wahrnehmung. *Naturwissenschaften* **58**, 329–330.

Schmidt-Koenig, K. (1958). Experimentelle Einflussnahme auf die 24-Stunden-Periodik bei Brieftauben und deren Auswirkungen unter besonderer Berücksichtigung des Heimfindevermögens. *Z. Tierpsychol.* **15**, 301–331.

Schmidt-Koenig, K. (1960). Internal clocks and homing. *Cold Spring Harbor Symp. Quant. Biol.* **25**, 389–393.

Schmidt-Koenig, K. (1961). Die Sonne als Kompass im Heim-Orientierungssystem der Brieftauben. *Z. Tierpsychol.* **18**, 221–244.

Schmidt-Koenig, K. (1965). Current problems in bird orientation. *In* "Advances in the Study of Behavior" (D. S. Lehrman, R. A. Hinde, and E. Shaw, eds.), Vol. 1, pp. 217–278. Academic Press, New York.

Schmidt-Koenig, K. (1972). New experiments on the effect of clock shifts on homing in pigeons. *NASA Spec. Publ.* **NASA SP-262**, 275–282.

Schmidt-Koenig, K., and Schlichte, H. J. (1972). Homing in pigeons with reduced vision. *Proc. Nat. Acad. Sci. U.S.* **69**, 2446–2447.

Schmidt-Koenig, K., and Walcott, C. (1973). Flugwege und Verbleib von Brieftauben mit getrübten Haftschalen. *Naturwissenschaften* **60**, 108–109.

Schneider, G. H. (1906). Die Orientierung der Brieftauben. *Z. Psychol. Physiol. Sinnesorg.* **40**, 252–279.

Schüz, E. (1949). Die Spät-Auflassung ostpreussischer Jungstörche in West-Deutschland durch die Vogelwarte Rossitten 1933. *Vogelwarte* **15**, 63–78.

Shumakov, M. E. (1965). [Preliminary results of the investigation of migrational orientation of passerine birds by the round-cage method.] *In* "Bionica," pp. 371–378. Moscow (in Russian).

Shumakov, M. E. (1967a). [An investigation of the migratory orientation of passerine birds.] *Vestn. Leningrad. Uni., Biol. Ser.* **1967**, (3), 106–118.

Shumakov, M. E. (1967b). [Experiments to determine the possibility of magnetic orientation in birds.] *In* "Questions on Bionics," pp. 519–523. Voprosy (in Russian).

Shumakov, M. E., and Vinogradova, N. (1970). [Orientation of birds prevented from using visual cues.] *In* "Ornithology in the USSR," *Proc. Baltic Ornithol. Conf., 7th, 1970*, Vol. 2, pp. 135–143. Ashkhabad (in Russian).

Slepian, J. (1948). Physical basis of bird navigation. *J. Appl. Phys.* **19**, 306.

Sobol, E. D. (1930). Orienting ability of carrier pigeons with injured labyrinths. *Mil.-Med. Z.* **1**, 75; *Biol. Abstr.* **8**, 15425.

Sokolov, L. V. (1970). [Experimental study, using Kramer cages, of the orientational abilities of birds with various types of movement.] *In* "Ornithology in the USSR,"

Proc. Baltic Ornithol. Conf., 7th, 1970, Vol. 2, pp. 129–133. Ashkhabad (in Russian).

Southern, W. E. (1959). Homing of Purple Martins. *Wilson Bull.* **71**, 254–261.

Southern, W. E. (1968). Experiments on the homing behavior of Purple Martins. *Living Bird* **7**, 71–84.

Southern, W. E. (1969). Orientation behavior of Ring-billed Gull chicks and fledglings. *Condor* **71**, 418–425.

Southern, W. E. (1971). Gull orientation by magnetic cues: A hypothesis revisited. *Ann. N.Y. Acad. Sci.* **188**, 295–311.

Southern, W. E. (1972). Influence of disturbances in the earth's magnetic field on Ring-billed Gull orientation. *Condor* **74**, 102–105.

Steidinger, P. (1968). Radarbeobachtungen über die Richtung und deren Streuung beim nächtlichen Vogelzug im Schweizerischen Mittelland. *Ornithol. Beob.* **65**, 197–226.

Steidinger, P. (1972). Der Einfluss des Windes auf die Richtung des nächtlichen Vogelzuges. *Ornithol. Beob.* **69**, 20–39.

Stewart, R. W. (1956). A new look at the Reynolds Stresses. *Can. J. Phys.* **34**, 722–725.

Tucker, V. A. (1971). Flight energetics in birds. *Amer. Zool.* **11**, 115–124.

Varian, R. H. (1948). Remarks on 'A preliminary study of a physical basis of bird navigation.' *J. Appl. Phys.* **19**, 306–307.

Viguier, C. (1882). Le sens d'orientation et ses organes chez les animaux et chez l'homme. *Rev. Phil.* **14**, 1–36. [Cited in Matthews (1968).]

Vleugel, D. A. (1954). Waarnemingen over de nachttrek van lijsters (*Turdus*) en hun waarschijnlijke oriëntering. *Limosa* **27**, 1–19.

Vleugel, D. A. (1959). Über de wahrscheinlichste Methode der Wind-Orientierung ziehender Buchfinken (*Fringilla coelebs*). *Ornis Fenn.* **36**, 78–88.

Vleugel, D. A. (1962). Über nächtlichen Zug von Drosselen und ihre Orientierung. *Vogelwarte* **21**, 307–313.

von Riper, W., and Kalmbach, E. R. (1952). Homing not hindered by wing magnets. *Science* **115**, 577–578.

von St. Paul, U. (1953). Nachweis der Sonnenorientierung bei nächtlich ziehenden Vögeln. *Behaviour* **6**, 1–7.

von St. Paul, U. (1956). Compass directional training of Western Meadowlarks (*Sturnella neglecta*). *Auk* **73**, 203–210.

Wagner, G. (1972). Topography and pigeon orientation. *NASA Spec. Publ.* **NASA SP-262**, 259–273.

Walcott, C. (1972). The navigation of homing pigeons: Do they use sun navigation? *NASA Spec. Publ.* **NASA SP-262**, 283–292.

Walcott, C., and Green, R. P. (1974). Orientation of homing pigeons altered by a change in the direction of an applied magnetic field. *Science* **184**, 180–182.

Walcott, C., and Michener, M. C. (1967). Analysis of tracks of single homing pigeons. *Proc. Int. Ornithol. Congr., 14th, 1966* pp. 311–329.

Walcott, C., and Michener, M. (1971). Sun navigation in homing pigeons: Attempts to shift sun coordinates. *J. Exp. Biol.* **54**, 291–316.

Walcott, C., and Schmidt-Koenig, K. (1973). The effect on pigeon homing of anesthesia during displacement. *Auk* **90**, 281–286.

Wallraff, H. G. (1960a). Können Grasmücken mit Hilfe des Sternenhimmels navigieren? *Z. Tierpsychol.* **17**, 165–177.

Wallraff, H. G. (1960b). Does celestial navigation exist in animals? *Cold Spring Harbor Symp. Quant. Biol.* **25**, 451–461.

Wallraff, H. G. (1965a). Über das Heimfindevermögen von Brieftauben mit durchtrennten Bogengängen. Z. Vergl. Physiol. 50, 313–330.

Wallraff, H. G. (1965b). Versuche zur Frage der gerichteten Nachtzug-Aktivität von gekäfigten Singvögeln. Verh. Deut. Zool. Ges. Jena pp. 339–356.

Wallraff, H. G. (1967). The present status of our knowledge about pigeon homing. Proc. Int. Ornithol. Congr., 14th, 1966 pp. 331–358.

Wallraff, H. G. (1970). Über die Flugrichtungen verfrachteter Brieftauben in Ahhängigkeit vom Heimatort und vom Ort der Freilassung. Z. Tierpsychol. 27, 303–351.

Wallraff, H. G. (1971). Homing of pigeons after extirpation of their cochlea and lagenae. Nature New Biol. 326, 223–224.

Wallraff, H. G. (1972). An approach toward an analysis of the pattern recognition involved in the stellar orientation of birds. NASA Spec. Publ. NASA SP-262, 211–222.

Wallraff, H. G. (1972). Nicht-visuelle Orientierung zugunruhiger Rotkehlchen (Erithacus rubecula). Z. Tierpsychol. 30, 374–382.

Wilkinson, D. H. (1949). Some physical principles of bird orientation. Proc. Linn. Soc. London 160, 94–99.

Williams, T. C., Williams, J., Teal, J., and Kanwisher, J. (1971). Tracking radar studies of bird migration in or near cloud layers. Nat. Res. Counc. Can., Ass. Comm. Bird Hazards Aircraft Field Note No. 58, pp. 1–13.

Williams, T. C., Williams, J., Teal, J., and Kanwisher, J. (1972). Tracking radar studies of bird migration. NASA Spec. Publ. NASA SP-262, 115–128.

Williams, J. M., Williams, T. C., and Ireland, L. C. (1974). Bird migration over the North Atlantic. In "The Biological aspects of the bird/aircraft collision problem" (S. Gauthreaux, ed.), pp. 359–382. Naval Office of Scientific Research, Clemsen, North Carolina.

Wiltschko, W. (1968). Über den Einfluss statischer Magnetfelder auf die Zugorientierung der Rotkchlchen (Erithacus rubecula). Z. Tierpsychol. 25, 537–558.

Wiltschko, W. (1972). The influence of magnetic total intensity and inclination on directions preferred by migrating European Robins (Erithacus rubecula). NASA Spec. Publ. NASA SP-262, 569–578.

Wiltschko, W., and Merkel, F. W. (1965). Orientierung zugunruhiger Rotkehlchen im statischen Magnetfeld. Verh. Deut. Zool. Ges. Jena 1965 pp. 362–367.

Wiltschko, W., and Merkel, F. W. (1971). Zugorientierung von Dorngrasmücken (Sylvia communis) im Erdmagnetfeld. Vogelwarte 26, 245–249.

Wiltschko, W., and Wiltschko, R. (1972). The magnetic compass of European Robins, Erithacus rubecula. Science 176, 62–64.

Wiltschko, W., and Wiltschko, R. (1973). Grasmücken benutzen den Magnetkompass auch bei Sternsicht. Naturwissenschaften 60, 553–555.

Wiltschko, W., Hock, H., and Merkel, F. W. (1971). Outdoor experiments with migrating European Robins in artificial magnetic fields. Z. Tierpsychol. 29, 409–415.

Wodzicki, K., Puchalski, W., and Liche, H. (1939). Untersuchungen über die Orientation und Geschwindigkeit des Fluges bei Vögeln. V: Weitere Versuche an Störchen. J. Ornithol. 87, 99–114.

Yeagley, H. L. (1947). A preliminary study of a physical basis of bird navigation. J. Appl. Phys. 18, 1035–1063.

Yeagley, H. L. (1951). A preliminary study of a physical basis of bird navigation. II. J. Appl. Phys. 22, 746–760.

Chapter 4

CIRCADIAN AND CIRCANNUAL
RHYTHMS IN BIRDS

Eberhard Gwinner

I. Introduction

Adaptation to an environment implies adaptation to its temporal variations. The most conspicuous temporal changes in environmental conditions are periodic. The rotation of the earth about its axis, its annual
rotation around the sun, and the monthly rotation of the moon around
the earth bring about tidal, daily, lunar, and annual periodicities in

the environment that have provided challenges and opportunities for evolutionary adaptation. As a result, many biological activities are organized with regard to these four dominating environmental rhythms.

Examples of diurnal and annual rhythms in birds are countless. Many biological processes vary with the time of day and thus exhibit diel rhythms. Similarly, numerous physiological and behavioral functions are organized on an annual basis, particularly in temperate-zone birds. Tidal and lunar rhythms, on the other hand, are of minor significance in birds. Tidal rhythms have been described for the activity of birds that take food from the intertidal zone (e.g., Boecker, 1967). Lunar rhythms, related to the variations in nocturnal light intensity, have been found in the times of onset and end of activity (Wynne-Edwards, 1930, in *Caprimulgus*; Hjorth, 1968, in *Lyrurus*) and in the intensity of nocturnal restlessness expressed by caged nocturnal migrants during the migratory seasons (Brehm, 1828, Gwinner, 1967b; Brown and Mewaldt, 1968; Smith *et al.*, 1969). The breeding cycle of the Sooty Tern (*Sterna fuscata*) on Ascension Island corresponds to the period of 10 lunar months (Chapin and Wing, 1959). But since nothing is known about the control of lunar and tidal rhythms in birds, no further reference will be made to them in this chapter.

The rigid temporal relationship between diurnal and annual biological rhythms on the one hand and the environmental cycles that they match on the other has been taken as evidence that the latter are the direct cause of the former. Numerous observations seem to support such a conclusion. The fact that the activity times of day-active birds are strongly correlated with seasonal and latitudinal changes in day length (for reviews, see Aschoff and Wever, 1962b; Aschoff, 1969) has been taken as evidence for a direct coupling of diurnal activity with light intensity. Such a conjecture is supported by field observations that show that on cloudy days the onset of activity is often delayed and the termination advanced (e.g., Scheer, 1952; Leopold and Eynon, 1961; Hjorth, 1968). Moreover, many day-active species of birds have been observed to sleep and show roosting behavior in the day when light intensity was decreased, for example, during solar eclipses (e.g., Ottow, 1912; Ehrström, 1954; Kullenberg, 1955; Schildmacher, 1955) or in experimental situations.

Comparable observations on the dependence of annual periodicities on seasonal changes in day length have led to the still prevailing concept that photoperiod exerts only a direct, causal influence on the physiological systems controlling overt annual periodicity. Observations frequently cited in support of such a view include, among others, the following: (1) The annual rhythms of birds, displaced across the equator from one to the other hemisphere usually get in phase with the new conditions within the first year (see Aschoff, 1955, for review). (2) Annual cycles of birds have been compressed or extended over a wide range of periods by simple manipulation of photoperiodic conditions (Rowan, 1929; Miyazaki, 1934; Burger, 1947; Damste, 1947); up to five biological cycles can be induced within one calendar year (Wolfson, 1954; for reviews, see Farner, 1961; Farner and Lewis, 1971).

Such findings demonstrate beyond doubt the participation of environmental stimuli such as light intensity or photoperiod in the control of both diel and annual rhythms. It has become clear, however, in the last decades, that the mode of action of light is often indirect. This is indicated by observations in numerous species of animals that diurnal rhythms can occur even under constant conditions; and annual periodicities have recently been shown to persist in some species in the absence of annual variations in photoperiod and other environmental variables.

This chapter reviews some aspects of these periodicities. The restriction to birds prohibits the discussion of several general topics such as the physiological mechanisms of endodiurnal rhythms. For the same reason, current models—physiological and mathematical—can at best only be touched upon. Several recent books and symposia can be consulted for a discussion of these and other aspects (Withrow, 1959; Chovnick, 1960; Cloudsley-Thompson, 1961; Wolf, 1962; Goodwin, 1963; Hague, 1964; Harker, 1964; Aschoff, 1965a; Remmert, 1965; Richter, 1965; Rohles, 1969; Sollberger, 1965; Kalmus, 1966; Beck, 1968; Sweeney, 1969; Benoit and Assenmacher, 1970; Brown et al., 1970; Conroy and Mills, 1970; Lofts, 1970; Menaker, 1971a; Bünning, 1973).

II. Properties of Circadian and Circannual Rhythms under Constant Conditions

A. THE DEMONSTRATION OF CIRCADIAN AND CIRCANNUAL RHYTHMS

Examples demonstrating the endogenous nature of daily (or diel) and annual rhythms in birds are shown in Figs. 1 and 2. Figure 1 shows daily activity rhythms of three European Starlings (*Sturnus vulgaris*) obtained from continuous recordings of perch contacts. The Starling on the upper left is kept in an artificial light–dark cycle with 12 hours of light and 12 hours of darkness (LD 12:12) for the first 20 days of the experiment. As expected from a day-active bird, the Starling's activity is essentially restricted to the photofraction of the day. On day 21, lights are turned off and the bird is then kept in constant darkness for 3 months. Two important results are obvious: (1) The bird's activity remains organized on a periodic basis; times of activity alternate with times of rest. (2) The period of the rhythm deviates from 24 hours;

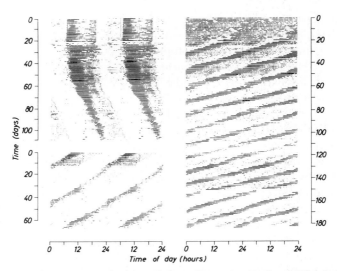

FIG. 1. Locomotor activity rhythms of three European Starlings. The daily activity recordings from the event recorder have been pasted underneath each other on charts, and each chart has then been duplicated. *Upper left:* The bird was kept for the first 20 days under a 24-hour light–dark cycle with lights from 7 AM to 7 PM. At day 21, lights were turned off and the bird's activity rhythm was allowed to free-run in constant dark. *Lower left and right:* Activity recordings of Starlings in constant light (light intensity about 0.7 lux). (E. Gwinner, unpublished data.)

successive onsets and ends of the main activity time occur about 0.2 hours later each day, indicating an average period of about 24.2 hours. The periods of the activity rhythms of the birds shown on the lower left and on the right side of Fig. 1 are, in contrast, shorter than 24 hours. Their average values are 23.3 and 21.5 hours, respectively.

It is this deviation of the period of the activity–rest cycle from 24 hours that shows its true endogenous nature. Depending on the individual bird and on various environmental conditions (Section II,D), the subjective day is either slightly longer or slightly shorter than the natural day; this excludes diurnal variables in the environment as possible causes of the observed periodicities. To emphasize the significance of such free-running rhythms, Halberg (1959) introduced the term circadian (from Latin: *circa* "about" + *dies* "day") for this class of biological periodicities.

Circadian rhythms were first recognized in plants (de Mairan, 1729; Sachs, 1857; Semon, 1905; Stoppel, 1910; Stoppel and Kniep, 1911; Pfeffer, 1911). Subsequently, it has become clear that they are ubiquitous among eukariotic organisms (see reviews above). In birds, besides loco-motor activity (see Aschoff, 1962, for review; see particularly Figures

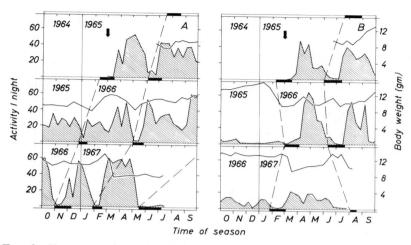

Fig. 2. Variations of *Zugunruhe* (hatched curve), body weight (unhatched curve), and molt (solid bar) in two Willow Warblers held for 28 months in a continuous LD 12:12 cycle (200:0.2 lux). Successive years are displayed one beneath the other. The dashed lines connect the onsets of corresponding molts in successive years. Activity/night = number of 10-minute intervals during which a bird is active each night; mean values for successive thirds of a month are plotted. The arrow indicates the start of the experiment. (From Gwinner, 1968a.)

4, 9, 14, 15, 18, such functions as body temperature (e.g., Winget *et al.,* 1965; von St. Paul and Aschoff, 1968; Aschoff, 1970b; Aschoff and Pohl, 1970a), energy metabolism (e.g., Aschoff and Pohl, 1970a,b; Pohl, 1970), food uptake (e.g., Pohl, 1968a; Aschoff and Pohl, 1970a), and the preferred intensity of illumination (e.g., Wahlström, 1960, 1965a,b; Gwinner, 1966b; Aschoff *et al.,* 1968; Heppner and Farner, 1971) have been shown to be controlled by circadian rhythmicity. Moreover, circadian rhythms provide essential mechanisms for sun-compass orientation (Chapter 3) and photoperiodic time measurement (Section IV,B).

Figure 2 illustrates that endogenous circannual periodicities exist in birds as well. It depicts variations in nocturnal activity and molt of two Willow Warblers (*Phylloscopus trochilus*) kept for 28 months under a continuous 12-hour daily photoperiodic regime. Nocturnal restlessness is typically shown by caged individuals of this nocturnal migrant species during the spring and fall migratory seasons (*Zugunruhe*). Two complete molts are carried out, one in winter (prenuptial molt) and one in summer (postnuptial molt). It is clear from Fig. 2 that this general pattern persists over the whole period of observation. Intense *Zugunruhe* and molt continue to recur approximately twice a year at regular intervals. Of particular interest is the behavior of bird A. The phases of its molt and *Zugunruhe* cycle shift progressively forward; both pre- and postnuptial molt, for example, occur on the average 3 months earlier each year than the year before; the period of the molt cycle is shorter than 1 year. In other words, the subjective calendar of this bird, deprived of external seasonal information, deviates from a true solar calendar. Its subjective years last only about 9 months. In bird B, the period is, in contrast, slightly longer than 1 year.

As in the case of circadian rhythms it is again the deviation of the period from the natural year that allows the conclusion that the biological rhythm must be endogenous. It can be referred to as circannual.

Annual periodicities persisting with a period different from 12 months under annually constant conditions have first been described by Pengelley and Fisher (1963) for Golden-Mantled Ground Squirrels (*Citellus lateralis*). Since that time, circannual rhythms have been convincingly demonstrated for various functions in mollusks, arthropods, fish, reptiles, birds, and mammals (for recent reviews, see Pengelley and Asmundson, 1971; Gwinner, 1971). The evidence available for birds is summarized in Table I. Distinctions have been made among three categories of results. Category 1 comprises cases in which at least two cycles with a period clearly deviating from 12 months have been measured. Such results are considered convincing evidence for the existence of circannual rhythms. Category 2 summarizes results that suggest the existence of circannual rhythms. They are not considered convincing evidence, however, because the period is indistinguishable from 12 months, and therefore the action of uncontrolled environmental variables cannot be rigorously excluded. In category 3, finally, are summarized results of experiments that revealed rhythms with periods deviating to such an extent from 12 months that it is not clear whether they should be called circannual.

In view of the obvious scarcity of critical experiments, it is difficult, at present, to evaluate the ubiquity of circannual rhythms among birds. Various experiments, in which birds have been kept for more than 1 year under constant conditions, have failed to reveal endogenous annual rhythms (e.g., Dancker, 1964; Farner and Follet, 1966; Haase, 1973). Such negative findings should be interpreted with great care, however, since the range of permissive conditions for the expression of circannual rhythms may be held within very narrow limits. Thus, European Starlings show a well-defined annual rhythm of testicular size if kept in a continuous 12-hour photoperiod, but no periodicity is observed in birds kept in daily photofractions of 11 hours or less and 13 hours or more (Schwab, 1971). Similar photoperiodic effects on circannual rhythmicity have been found in the Willow Warbler (Gwinner, 1971), in the Red-billed Quelea (*Quelea quelea*) (Lofts, 1962, 1964), and in the Whitethroat (*Sylvia*

TABLE I

EVIDENCE FOR CIRCANNUAL RHYTHMS IN BIRDS

Species	Condition	Time in constant conditions (months)	Function(s) showing periodicity	Maximal number of complete cycles[a]	Estimated period (months)[a]	References
			Category 1			
Phylloscopus trochilus	LD 12:12	28	Molt, Zugunruhe	3	9; 11.8[b]	Gwinner, 1967a, 1968a, 1971
Sylvia borin	LD 10:14	34	Body weight, molt, Zugunruhe, gonadal size	3	10.6[c]	Berthold *et al.*, 1971a, 1972b,c
Sylvia atricapilla	LD 12:12				10.5[c]	
Sylvia cantillans	LD 16:8	19	Body weight, molt, Zugunruhe	2	11.3[d]	
Sylvia undata	LD 10:14	19	Molt, Zugunruhe	2	10.8[d]	Berthold, 1974
Sylvia melanocephala				2	10.5[d]	
Sylvia sarda				2	10.4[d]	
Sturnus vulgaris	LD 12:12	28	Testicular size	2	(10)	Schwab, 1971
			Category 2			
Sylvia communis	LD 12:12	21	Body weight	1	(12)	Merkel, 1963
Quelea quelea	LD 12:12	29	Testicular size	2	(12)	Lofts, 1964
Spiza americana	LD 12:12	21	Body weight, molt, Zugunruhe	1	(12)	Zimmerman, 1966
Zonotrichia leucophrys	LD 8:16	13	Body weight, molt	1	(12)	King, 1968
	LD 20:4				(12)	
Phylloscopus collybita	LD 12:12	15	Molt, Zugunruhe	1	(12)	Gwinner, 1971
Sturnus vulgaris	LL	15	Molt, testicular size	1	(13)	Gwinner, 1973
			Category 3			
Pekin duck	LL	54	Testicular size	(5)	(4)	Benoit *et al.*, 1955, 1956, 1959; Benoit, 1970
	DD	70		(7)	(8)	
Erithacus rubecula	LD 8:16	30	Body weight	(2)	(15)	Merkel, 1963
	LD 18:6	16	Body weight	(2)	(7)	

[a] Numbers in parentheses are rough estimates.

[b] Values for individual birds.

[c] Mean values based on data of body weight, molt, and Zugunruhe of 25 *Sylvia borin* and 22 *S. atricapilla*.

[d] Mean values based on data of summer molt of at least 6 individuals of each species.

communis) (Merkel, 1963). Comparative studies of closely related species kept in identical conditions indicate, on the other hand, that there may be differences among species in the participation of circannual rhythms (*Phylloscopus* warblers: Gwinner, 1971; *Zonotrichia:* King, 1968; several examples from mammals: Pengelley and Kelly, 1966; Heller and Poulson, 1969).

B. INHERITANCE OF CIRCADIAN AND CIRCANNUALS RHYTHMS

The statement that a diurnal or an annual rhythm is endogenous does not imply that it is also inherited. It may, for instance, have become imprinted in early life. There is ample evidence, however, in various organisms, that both circadian and circannual rhythms are innate (Bünning, 1973, for review; Pengelley and Asmundson, 1971). In birds, direct demonstrations of genetic control are lacking, but results of various experiments make it virtually certain that these rhythms are inherited. Chickens raised from the egg in constant darkness developed circadian rhythms in locomotor activity (Aschoff and Meyer-Lohmann, 1954). Furthermore, three species of Old World warblers (*Phylloscopus trochilus, Sylvia borin, S. atricapilla*) showed free-running annual periodicities of gonadal size, molt, body weight, and *Zugunruhe,* even if transferred to annually constant conditions at an age of 10 weeks or less (Gwinner, 1967a, 1968a, 1972b; Berthold *et al.,* 1972 b,c), i.e., before they had had an opportunity to learn the duration of a year.

C. CIRCADIAN AND CIRCANNUAL RHYTHMS AS
OSCILLATORS—TERMINOLOGY

1. A General Oscillator Model

The continuation of daily and annual rhythms in animals isolated from periodic changes in the environment brings to mind the behavior of physical oscillators in the absence of periodic driving agents. Both circadian and circannual rhythms, therefore, have been described as autonomous oscillating systems (Pittendrigh, 1960; Aschoff, 1960; cf. Aschoff *et al.,* 1965, for terminology).

Some of the concepts and the terminology that are now generally used to describe these periodicities can easily be understood by discussing the behavior of the autonomous oscillator diagrammed in Fig. 3. The oscillator consists of a nonlinear spring with one end fixed to a rigid surface and a mass attached to the free end. Once set in motion, the system oscillates as indicated in Fig. 3A, where the position of the mass at successive times is shown. The mass will move up and down and reach corresponding positions (e.g., its lowest point) at regular

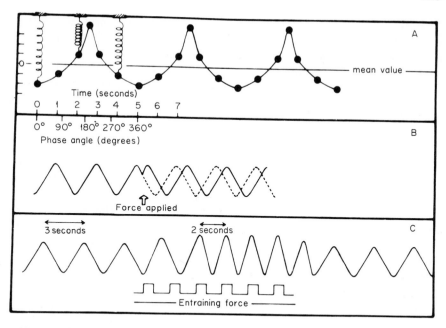

FIG. 3. Theoretical behavior of an oscillating nonlinear spring: (A) free running, (B) phase shifted by a single stimulus, (C) entrained by a periodic force. In A, the position of the mass on the end of the spring is plotted as a function of time. The period of the oscillation for this system is 5 seconds. In B, a curve is drawn representing the variation of position as a function of time for another oscillating system. An instantaneous force, applied during the third cycle, causes a phase delay of 90°. This can be seen as the difference between the dashed and the solid line that represents the system after application of the force. In C, the curve represents an oscillator with a free-running period of 3 seconds. After the first three cycles, a force with a period of 2 seconds is imposed on the system as represented by the square wave in the lower part of the figure. After two transient cycles, the oscillator becomes entrained, and its period is now the same as that of the force. After the periodic force is discontinued, the oscillator begins to free-run once again with its initial free-running period. (After Menaker, 1968c).

intervals. The time required to complete one such cycle is called the period τ ($\tau = 5$ seconds in the example). The reciprocal of τ is referred to as frequency ($1/\tau$). The instantaneous state of the oscillation at any particular time of the period is referred to as its phase. The value on the abscissa corresponding to each phase is called the phase angle φ. The arithmetic mean of all phases is the mean value, or the level, of the oscillation. The difference between the maximum and the minimum value of the oscillation is called the range of the oscillation, and the difference between the maximum (or the minimum) value and the mean value is called the amplitude.

In the absence of external forces, the system will oscillate with a frequency typical for the system, i.e., with its natural frequency. The oscillator is then said to be free-running. It can be reset (displaced in time) by external forces, i.e., it can be phase shifted by $\Delta\varphi$ (Fig. 3B). During a phase shift, it usually takes the oscillation several cycles to regain a new steady state. The cycles between two steady states are called transient cycles.

The period of the oscillator can be modified by appropriate periodic forces in such a way that it assumes the period of the driver (Fig. 3C). The forced, or driven, oscillator is then said to be synchronized with the forcing, or driving, oscillation. If both driving and driven system are self-sustaining (i.e., if they persist indefinitely without periodic input), the term "synchronization" can be replaced by the term "entrainment." The driving oscillation may then be called an entraining agent or, if it entrains a biological system, a *Zeitgeber* (Aschoff, 1954, 1958). Its period is T. Entrainment is realized if $\tau = T$.

In the entrained state, the driving and the driven systems assume a characteristic phase relationship ψ. The phase angle of the entraining agent is designated Φ and that of the entrained oscillation φ, so the phase-angle difference between them is $\psi = \Phi - \varphi$ ($+$ if the phase angle of the driven oscillation leads that of the driver).

2. Circadian and Circannual Rhythms: Damped or Self-Sustained Oscillations?

Physicists and engineers distinguish between damped oscillations, which decrease in amplitude and eventually cease if the energy input from outside the system is constant, and self-sustained oscillations, which continue to oscillate indefinitely under such conditions (e.g., Aschoff, 1963b). Both are designated as endogenous (Klotter, 1960).

Circadian rhythms seem to belong to the second category. Even though overt periodicities may cease under certain conditions, for instance under high initensities of illumination (Section II,E,1), circadian ryhthms persist, as a rule, undamped for many periods. Circadian activity rhythms in birds have been measured for up to 450 cycles (Eskin, 1971; Gwinner, 1973).

The properties of circannual rhythms are still much too obscure to allow any definite conclusions. In ground squirrels the annual rhythm of hibernation has been shown to persist undamped for seven cycles, i.e., for the whole life of the animal (Pengelley and Asmundson, 1969). In birds, the maximum number of unambiguous circannual cycles persisting without decay is three (Berthold *et al.*, 1972b), or seven, if one considers the 8-months testicular cycle of domestic Pekin ducks in constant light as circannual (see Table I). Other circannual rhythms have shown tendencies to disappear with time (Gwinner, 1971; Zimmermann, 1966, for birds; also Pengelley and Kelly, 1966; Heller and Poulson, 1969, for mammals), but whether this indicates the damping of the underlying oscillator(s), the desynchronization of a population of oscillators that share the control of the overt rhythm(s) (Section E,1), or simply the uncoupling of the overt function from the driving system cannot be ascertained in any of these cases.

D. VARIATIONS OF THE PERIOD

1. Extreme Values of the Period

The period τ of autonomous oscillators is a property of the system. It can be modified by external conditions. However, in both technical and biological oscillations, the range of possible periods is limited. Free-running circadian periods usually vary between $\tau \approx 20$ hours and $\tau \approx 28$ hours (Bruce, 1960). In birds, the extremes measured so far are approximately 22 hours and 26 hours (Lohmann, 1967). Circannual rhythms, in contrast, appear to be more variable, especially if one includes the rhythms in category 3 of Table I. The wider range of free-running periods of circannual rhythms corresponds with their larger range of entrainment (Section III,B).

2. "Spontaneous" Variations of the Period

Under the same experimental conditions, the period of circadian rhythms is typically different in different individuals. Moreover, τ may change with time in an individual (Aschoff et al., 1962; Palmer, 1964; Eskin, 1971; Heppner and Farner, 1971). Not much attention has been devoted to this phenomenon and nothing seems to be known about its meaning. Some authors have speculated that the variations in τ might be the result of changes in the physiological state of the animal (e.g., due to its changes from summer to winter conditions or vice versa) (Rawson, 1960). This possibility seems reasonable, since there are data showing that τ of bats tested in summer is different from that of conspecifics tested in winter (Menaker, 1961). On the other hand, experiments with European Starlings and Common Redpolls (Acanthis flammea) have failed to reveal such seasonal variations of τ (Gwinner and Turek, 1971; Pohl, 1972a). Castration and injection of testosterone have no demonstrable effect on the period of locomotor activity rhythms of the European Starling (Gwinner, 1974a).

Variations in τ may often result from the circadian system not yet being in a steady state. Eskin (1971), in an extensive study of the locomotor activity rhythm of the House Sparrow (Passer domesticus), has shown that it takes at least 8 weeks in constant darkness before a relatively stable period is achieved. Before this steady state is reached, the rhythm goes through a characteristic pattern of transients. τ usually increases for the first 30 to 50 days. Then for about a week there is a temporary decrease of τ, after which it again increases. After about another 30 days, τ reaches its largest value and then decreases for a few weeks. Only then does it remain relatively constant. The general pattern of these transients seems to be independent of the time of year at which the experiment begins, and individual deviations from the general pattern seem to be persistent; i.e., they can be observed in successive experiments. The findings of Eskin are of considerable importance since they suggest that many of the experiments discussed below may have been carried out at a time when the circadian system investigated had not yet reached its final steady state. The implications of such a possibility for current models are not known.

3. Effects of Current Conditions on the Period

a. *Circadian Rhythms.* Of the external variables that are known to affect the period of circadian rhythms, the action of only two has been studied in some detail—light intensity and temperature. Other factors, such as mechanical noise (Lohmann and Enright, 1967) or certain chemicals (e.g., heavy water; Palmer and Dowse, 1969; Snyder, 1969), may influence the period of circadian rhythms as well, but in birds only the effects of light and temperature have been investigated well enough to justify a detailed discussion.

i. *Light intensity.* It has been known for a long time, that the period of circadian rhythms in a variety of organisms depends on light intensity (early review, Aschoff, 1960). An example from birds is shown in Fig. 4. It depicts the rhythm of locomotor activity of a Chaffinch (*Fringilla coelebs*) in constant light of varying intensity. It is clear that the period length of the activity rhythm changes with the intensity of illumination; the higher the light intensity, the shorter the circadian period. Such a dependence of τ on light intensity is typical for the majority of day-active birds studied so far, as shown in Fig. 5. It holds even for blinded House Sparrows (Menaker, 1968b). Preliminary data on night-active birds indicate the reverse relationship between light intensity and circadian frequency (Fig. 5).

Similar qualitative differences in the dependence of τ on light intensity between other day- and night-active organisms have led Aschoff (1952, 1958, 1959b) to formulate what is now known as Aschoff's rule (Pittendrigh, 1960; Hoffmann, 1965). It states that in constant light ". . . with increasing intensity of illumination light active animals increase their

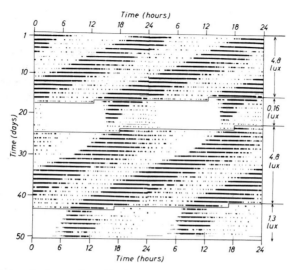

FIG. 4. Activity rhythms of a Chaffinch under varying intensities of illumination (right ordinate). For further explanations see Fig. 1. (From Aschoff, 1967a.)

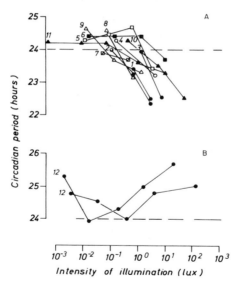

FIG. 5. Free-running circadian period as a function of light intensity in day- and night-active birds. (A) ● = *Fringilla coelebs*, ○ = *Carduelis chloris*, □ = *Carduelis spinus*, ■ = *Acanthis flammea flammea*, □ = *A. f. cabaret*, △ = *Carpodacus mexicanus*, ▲ = *Sturnus vulgaris*. (B) *Tyto alba*. References: (1) Aschoff, 1966b; (2) Aschoff *et al.*, 1962; (3) Pohl, 1968a; (4) Aschoff *et al.*, 1968; (5) H. Pohl, unpublished; (6) Pohl, 1972a; (7) H. Pohl, unpublished; (8) Enright, 1966a; (9) E. Gwinner, unpublished; (10) Hoffmann, 1960; (11) E. Gwinner, unpublished; (12) Erkert, 1969.

spontaneous frequency, while dark active animals decrease it" (Aschoff, 1960).

Recently, exceptions to this generalization have been discovered, particularly in arthropods and mammals (summaries in Rensing and Brunken, 1967; Hoffmann, 1965, 1967). In birds, the validity of the rule was based on the comparison of a number of day-active species with only one night-active species, the Barn Owl (*Tyto alba*). Recently, Erkert (1969), in an extensive study, has found that the dependence on light intensity of the circadian period in this very species is more complex than the early data had suggested; τ increases with increasing light intensity only in high intensities of illumination, while it decreases with increasing light intensity in low intensities of illumination. The reverse holds for one day-active species, the European Siskin (*Carduelis spinus*) (see Fig. 5). The existence of such optimum curves was early suspected by Aschoff (1960). Other examples are known from insects (Rensing and Brunken, 1967).

A change in sign in the dependence of the period of circadian activity rhythms on light intensity is also indicated by the results summarized in the lower graph of Fig. 6. Migratory birds of two species tested during the migratory seasons (i.e., when they typically show both diurnal and nocturnal activity) tend to increase the period of their activity rhythm

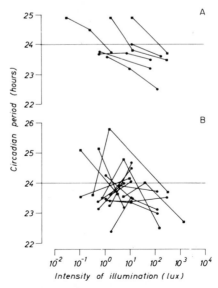

Fig. 6. Free-running circadian period as a function of light intensity in two species of migratory passerine birds—*Erithacus rubecula* (●) and *Phoenicurus phoenicurus* (■). (A) Birds tested during periods of molt; (B) birds tested during one of the migratory seasons. (E. Gwinner, unpublished data.)

with increasing light intensity in the range between 10^{-1} and 10^1 lux, while they tend to decrease the period with increasing light intensity under higher intensities of illumination. The same species tested during their postnuptial molt (i.e., at a time of the year at which the birds are typically exclusively diurnal) show the steady decrease of τ with increasing light intensity predicted by Aschoff's rule for a diurnal species (Fig. 6, upper graph). These seasonal differences may be related to the presence or absence of nocturnal activity.

Figure 4 suggests that, correlated with the light-intensity-dependent change in τ, two other parameters of the circadian activity rhythm become modified: (1) under the lower light intensities the main activity time α seems to be shorter than under the higher intensities. (2) The total amount of activity seems to be smaller under lower intensities than under higher intensities. The same has been found in a variety of other diurnal organisms, while the reverse seems to be common in a number of nocturnal species. Such observations have been the basis for the formulation of the circadian rule (Aschoff, 1960), which states, "Spontaneous frequency $(1/\tau)$, ratio of activity time to rest time (α/ρ), and amount of activity should increase with increasing intensity of constant illumination in light-active animals, and decrease in dark-active animals. Or expressed in a more general form, frequency, α/ρ-ratio and amount of activity should be positively correlated" (Hoffmann, 1965).

Aschoff (1960, 1964b), Aschoff and Wever (1962d), and Wever (1964) have used the generalization summarized in the circadian rule as the basis for a model. It assumes, that locomotor activity is controlled by a single circadian oscillator that crosses a hypothetical threshold twice each period. When the oscillation is above the threshold, the animal is active, when it is below it, the animal rests (Fig. 7). In addition, it is assumed that the level of the oscillation is a function of light intensity; it rises with light intensity relative to the threshold in day-active organisms, while it falls with increasing light intensity in night-active organisms. As a result, both the α/ρ ratio and the amount of activity will increase as a function of light intensity in diurnal organisms but decrease in nocturnal organisms.

To explain the whole circadian rule, one can assume that the intensity of illumination independently affects both level and frequency of the circadian oscillation. An alternative has been proposed by Wever (1964a, 1965). In self-sustaining oscillations, level and frequency may be internally coupled, the frequency rising with increasing level. Light intensity thus may affect only the level of the oscillation directly and may thereby indirectly cause a change in frequency. The most general form of the circadian rule therefore reads, "For all birds (possibly for all animals) there is one circadian oscillator, the level of which is correlated to light

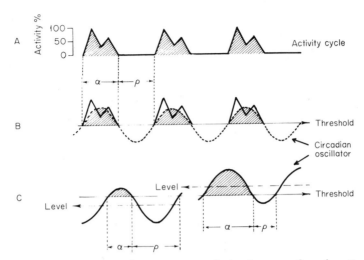

FIG. 7. A model to explain the positive correlation between the α/ρ ratio and the amount of activity. (A) Schematic representation of an activity rhythm with activity time α and rest time ρ. (B) Its description by an oscillation that passes through a threshold twice per cycle. (C) Changes in the α/ρ ratio and in the amount of activity described as the result of changes in the position of the level relative to the threshold. Shaded area indicates activity. (From Aschoff, 1967a.)

intensity, with a positive sign in diurnal species, and with a negative sign in nocturnal species" (Aschoff, 1967a).

Today, the major theoretical difficulty of this appealing model is the fact that it is based on the assumption that locomotor activity is controlled by a single oscillator. This assumption is contradicted by an increasing body of empirical data suggesting that more than one oscillator controls locomotor activity (Section II,E). Moreover, many exceptions to the circadian rule have recently been found (Rensing and Brunken, 1967; Hoffmann, 1965, 1967, for summaries; see also Gwinner and Turek, 1971; Pohl, 1972b), which cast doubt on the general validity of the circadian rule. On the other hand, various specific predictions of the model have been verified in experiments with the Chaffinch and man (Aschoff et al., 1971). A further evaluation of the model and of the rules on which it is based will clearly require more empirical data. In birds, data on night-active species and on nonpasserine day-active species are needed.

ii. *Temperature.* One of the most intriguing properties of circadian rhythms is their remarkable insensitivity to temperature (review, Sweeney and Hastings, 1960; Wilkins, 1965; Bünning, 1973). Q_{10} values are usually close to 1. This phenomenon, observed in organisms at all levels of organization, has led to various speculations about the nature of the basic rhythmic system that cannot be discussed here (e.g., Pittendrigh, 1954, 1960; Pittendrigh and Bruce, 1959; Bünning, 1973).

Instead of stressing the gross temperature independence of circadian rhythms, one can emphasize the slight but systematic temperature effects that do occur in most organisms studied, including birds. The period of locomotor activity rhythms decreases slightly with increasing temperature in both the House Finch (*Carpodacus mexicanus*) (Enright 1966a) and the Chaffinch (Pohl, 1968a). The same is true in a variety of other organisms, but the reverse effect also occurs (Aschoff, 1960; Bünning, 1973; Pohl, 1968a).

In a number of plants and animals, τ changes with increasing temperature in the same direction as it does with increasing light intensity. This is true, for instance, in the two species of finches mentioned above (cf. Fig. 5). Hence, "light-active" organisms may also be "warm-active" (i.e., τ may decrease with an increase in both light intensity and temperature), while "dark-active" organisms may be "cold-active" (Aschoff, 1960). Although there are exceptions to this rule (Aschoff, 1960; Rensing and Brunken, 1967), it seems to hold for all vertebrates in which the effects of both light and temperature have been studied. Unfortunately, data for night-active birds are again completely lacking.

b. Circannual Rhythms. In birds, only the effects of photoperiod on circannual rhythms have been studied (cf. Table I), and so far, no definite conclusions can be drawn. In White-crowned Sparrows (*Zonotrichia leucophrys gambelii*) (King, 1968) and in two species of *Sylvia* (Berthold *et al.*, 1972b,c) the circannual period in molt and body weight appears to be independent of photoperiod. In contrast, data from two other species suggest that the circannual period decreases with increasing photoperiod (European Robin, *Erithacus rubecula*, Merkel, 1963; Pekin duck, Benoit *et al.*, 1955, 1956, 1959; see also Goss, 1969b, for a mammal). One is tempted to draw an analogy between this relationship and that described by Aschoff's rule for the dependence of the circadian period on light intensity, but clearly more data are required to substantiate such a conjecture.

4. Effects of Past Conditions on the Period—Aftereffects

Pittendrigh (1960) first drew attention to the fact that the period of free-running circadian rhythms is often influenced by the conditions to which the organism was previously exposed. Thus, the period T of the entraining light–dark cycle, the photoperiod, and the light intensity of the previous conditions may cause aftereffects (review, Pittendrigh, 1960; Eskin, 1971). In birds, aftereffects of T and photoperiod have been studied. Eskin (1971) placed House Sparrows in DD after entrainment to 20-, 22-, 24-, and 28-hour light–dark cycles. After 22-hour entrainment, the birds tended to have shorter τ values than after 24-hour entrainment, while after 28-hour entrainment, τ values were longer. If the birds were previously entrained to light–dark cycles of 12–15 hours of light per 24 hours, the free-running τ in DD was significantly shorter than if they had been entrained to a light–dark cycle of 2–6 hours of light per

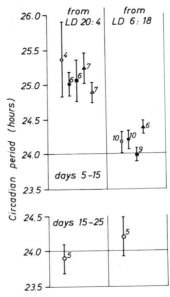

Fig. 8. Demonstration of aftereffects of photoperiod on free-running circadian activity rhythms of male White-crowned Sparrows ($\bigcirc$, $\bullet$, $\blacksquare$) and male Golden-crowned Sparrows ($\triangle$, $\blacktriangle$). The birds were kept for about 2 weeks under a 20-hour (*left*) or a 6-hour (*right*) photoperiod and were then released into constant dim light of about 0.02 lux (White-crowned Sparrows) or into constant darkness (Golden-crowned Sparrows). The figure shows the mean circadian period (with standard error) of several groups of birds, calculated from estimates of the period between day 5 and 15 (*top*) or day 15 to 25 (*bottom*) after the transfer to constant conditions. Identical symbols are from birds of the same experiment. Birds pretreated with long photoperiods had significantly longer free-running periods between days 5 and 15 in constant conditions than birds pretreated with short photoperiods. However, these differences wane with time, as indicated by those two groups of White-crowned Sparrows that were kept for another 10 days in constant conditions (open circles; compare upper with lower graph). All birds pretreated with LD 20:4 initiated testicular growth under these conditions, while the birds pretreated with LD 6:18 did not. However, the differences in the circadian period between long- and short-day birds are not related to differences in testicular size, since Golden-crowned Sparrows, castrated prior to the long-day pretreatment (open triangle) had similar long periods as conspecifics that were not castrated (closed triangles). Numbers at the symbols refer to group size. (E. Gwinner, unpublished data.)

24 hours (Eskin, 1971). White-crowned Sparrows and Golden-crowned Sparrows (*Zonotrichia atricapilla*) had longer τ values after entrainment to a long photoperiod than to a short one (Fig. 8). The longer periods of the birds pretreated with long photoperiods are not due to testicular growth, since castrated birds show the same effect.

In all cases in which free-running rhythms were studied long enough, aftereffects finally decayed (Fig. 8), though sometimes not until after 3–4 months. True steady-state aftereffects are unknown.

E. THE MULTIOSCILLATOR SYSTEM

1. The Circadian System

Only recently has it become clear that different biological functions within a single organism (possibly even within a single cell) may be controlled by different circadian oscillations; the organism can be considered a circadian system (Pittendrigh, 1967a, 1974; Menaker, 1974). The most compelling evidence comes from the observation in several species that under constant conditions different functions may free-run with different circadian frequencies. Such internal desynchronization has been studied most thoroughly in man (e.g., Aschoff, 1967b, 1970a; Aschoff *et al.*, 1967), but there is also evidence from some other mammals (Pittendrigh, 1960, 1961, 1967b, 1974; Hoffmann, 1969c, 1971; Pohl, 1972b). The free-running activity rhythms of hamsters, ground squirrels, palm squirrels, and tree shrews have been observed to split into two or more components, which free-run with different frequencies for a time but then often regain synchrony with a new relationship.

Similar phenomena have recently also been described for birds (Pohl, 1971; Gwinner, 1974a). In Starlings kept under conditions of constant dim light, the free-running circadian rhythm of locomotor activity tends to split into two components under the influence of testosterone; the splitting occurred in some intact birds whose testes grew and in several castrated birds after injection of testosterone (Fig. 9). These results suggest that testosterone affects (directly or indirectly) the mutual coupling of two or two groups of circadian oscillators controlling locomotor activity in this species (Gwinner, 1974a).

Even in the synchronized state, the separation of portions of the activity can sometimes be observed. Figure 10 shows how dusk activity of a European Robin "breaks loose" to join with the nocturnal activity (*Zugunruhe*). There is evidence that *Zugunruhe* itself is governed by (an) independent oscillation(s) (Aschoff, 1967a; McMillan *et al.*, 1970).

Such observations support the view that the bimodal pattern of locomotor activity frequently observed in birds and other organisms may reflect the action of two coupled oscillators, one controlling "morning"

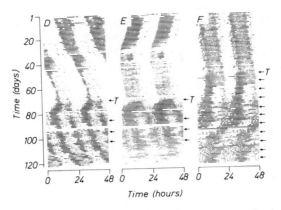

FIG. 9. The effect of testosterone on the locomotor activity rhythm of three male Starlings kept in conditions of continuous illumination (0.2 lux). All birds were castrated prior to the beginning of the experiment and injected with 2.5 mg testosterone (T) on the days indicated by arrows at the right-hand margin. Following the first testosterone injection, activity splits into two components in all three birds. In birds E and F, the second component finally merges with the first component in such a way that its beginning temporarily represents the onset of the single activity time. For further explanations see Fig. 1. (After Gwinner, 1974a. Copyright 1974 by the American Association for the Advancement of Science.)

activity, the other one "evening" activity. Moreover, they suggest that arrhythmicity of the activity pattern, often observed in high light intensities, may be the result of internal desynchronization of two or more oscillations, each controlling a particular portion of activity. This view is supported by the fact that testosterone-induced splitting of the circadian locomotor rhythm in the European Starling is often followed by continuous and apparently arrhythmic activity (Gwinner, 1974a).

The occasional observation of internal desynchronization gives rise to the question of how synchrony is normally maintained between the

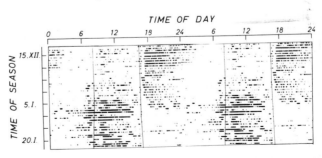

FIG. 10. Dissociation of components of locomotor activity of a European Robin under natural conditions of illumination. On December 25, dusk activity splits off from sunset to join with nocturnal activity. Vertical lines indicate sunrise and sunset. For further explanations see Fig. 1. (E. Gwinner, unpublished data.)

different circadian oscillators. Recent findings suggest that in birds the pineal gland may be involved in some way in the coupling of various circadian rhythms. Whereas pinealectomy has only slight effects on the circadian rhythm of locomotor activity in House Sparrows (Gaston and Menaker, 1968), White-crowned sparrows (Gaston, 1971) and White-throated Sparrows (*Zonotrichia albicollis*) (McMillan, 1972) kept under a 24 hour light–dark cycle, the same operation completely abolishes this rhythm as well as the circadian rhythms of body temperature (Binkley *et al.*, 1971) and uric acid excretion (E. Mackey and M. Menaker, unpublished data) in House Sparrows kept in constant dark (Fig. 11). Implantation of pineals into arrhythmic pinealectomized House Sparrows resulted in some cases in the resumption of a circadian rhythmicity, suggesting that the pineal exerts its effect on the circadian system hormonally (Gaston, 1971). Even though the interpretation of these results is by no means clear as yet, they could be interpreted on the assumption that the pineal is involved in the coupling of the various circadian oscillators that jointly control a particular function (e.g., locomotor activity) (see Menaker, 1971b, 1974, for a detailed discussion).

2. The Circannual System

In *Sylvia* warblers kept for 2 years or more under seasonally constant conditions, the various circannual rhythms may drastically change their phase relationship with each other. This holds, for instance, for the rhythms of testicular size and molt (Fig. 12) or for the rhythms of body weight and migratory restlessness. The result of such "internal dissociation" is that, for instance, molt or migratory fattening may coincide with any phase of the circannual testicular cycle. Such findings suggest that, as in circadian rhythms, different functions may be controlled by different endogenous oscillations (Berthold *et al.*, 1972b,c; Gwinner, 1974b).

Recently, Enright (1970) and Mrosovsky (1970) have proposed that circannual rhythms might result from ". . . a sequence of linked stages, each taking a given amount of time to complete and then leading into the next with the last stage linked back into the first again" (Mrosovsky, 1970). The data presented in this section clearly argue against such a mechanism.

3. Interactions of Circadian and Circannual Rhythms with Other Periodicities

In higher organisms, circadian and circannual rhythms represent only two of a large spectrum of endogenous oscillations whose periods range from less than 1 second (e.g., EEG rhythms) to 1 year. "Several of these rhythms interact with each other in a particular manner and recent findings suggest that we have to consider the spectrum, at least in parts,

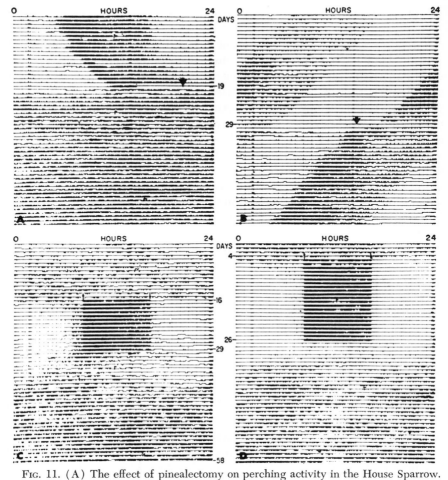

Fɪɢ. 11. (A) The effect of pinealectomy on perching activity in the House Sparrow. The bird is in constant darkness and constant temperature (about 23°C) throughout the period recorded. On day 19, the pineal organ was removed (indicated by the large arrow). Within 2 days after the operation, no circadian rhythm is discernible. (B) The effect of a sham pinealectomy on the rhythmic activity of a bird in constant dark. The operation was performed on day 29 (large arrow). (C) and (D) show entrainment patterns of two pinealectomized sparrows. The beginning and the end of the daily light period are marked with arrows. The dense black bars during the light fraction indicate intense perching activity. In (C), days 1 to 15 demonstrate arrhythmic activity in constant dark. On days 16 to 29, the birds received 8 hours of light followed by 16 hours of darkness per 24-hour period, and from day 30 to 58, the bird was once again in constant dark. After the light cycle was discontinued, about 8 days were required for this bird to reestablish an arrhythmic pattern. In (D), days 1–4 show arrhythmic activity in constant dark, on days 5–26, the bird is on LD 8:16; and on days 27–58, the bird was in constant dark. Notice the decay of rhythmicity on days 27–33, illustrating the transition to arrhythmicity more clearly than does (C). The pattern of this decay, with activity onsets occurring earlier and activity terminating later each day, is characteristic of pinealectomized birds released in constant darkness from LD entrainment. For further explanations see Fig. 1. (From Gaston and Menaker, 1968. Copyright 1968 by the American Association for the Advancement of Science.)

not as an assembly of unrelated processes, but as a hierarchy of more or less coupled oscillations which provide temporal order within the organism" (Aschoff, 1967b). It also seems possible that low-frequency rhythms emerge from high-frequency rhythms—for instance, by frequency demultiplication or as a beat oscillation. Such possibilities have been considered for lunar rhythms (Bünning and Müller, 1961), and they have been used in models for circadian rhythms (Pavlidis, 1969a).

In this context, the question has been asked whether circannual rhythms might result from frequency demultiplication of circadian rhythms (Hamner, 1971; Gwinner, 1973). In a number of cases, such a possibility seems unlikely, since the period of circadian rhythms of animals kept in a 24-hour light–dark cycle appears to depend on photoperiod (Merkel, 1963; Goss, 1969a,b). Moreover, in conditions of constant light, the light-intensity-dependent variability of circannual periods may be much larger than the range of periods that can be expected from circadian rhythms (Benoit et al., 1955, 1956, 1959; Benoit, 1970). On the other hand, such a mechanism seems plausible in animals in which the circannual period is rather invariable and independent of photoperiodic conditions (King, 1968; Berthold et al., 1972b,c).

III. Entrainment of Circadian and Circannual Rhythms

A. Zeitgebers

In the normal environment, circadian and circannual rhythms usually have exactly the period of the natural day or year. They are synchronized (entrained) by periodic factors in the environment that are referred to as entraining agents, synchronizers, or Zeitgebers (Section II,C).

1. How to Demonstrate a Zeitgeber

Three types of experiments are commonly used to test whether an environmental cycle is effective as a Zeitgeber (Aschoff, 1960) (cf. Fig. 13).

a. Catching the free-running rhythm. The free-running rhythm is exposed for several periods to the cycle to be tested. If the cycle is effective as a Zeitgeber, the endogenous rhythm should assume the period of the environmental rhythm. After the cycle has been removed, the endogenous rhythm should resume its free-running period. An example is shown in Fig. 15.

b. Driving the endogenous rhythm with different frequencies. An effective Zeitgeber should entrain the endogenous rhythm to various periods within certain limits. A change of the period of the cycle to be tested, therefore, should result in a corresponding

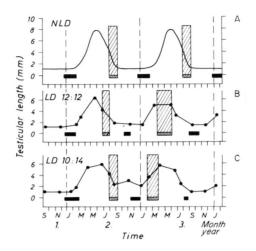

FIG. 12. Internal dissociation of the circannual rhythms of testicular size and molt in two Garden Warblers (*Sylvia borin*) kept for 29 months under a constant 12- and 10-hour photoperiod, respectively (B and C). Top (A) shows schematically the normal temporal relationship between the testis and molt cycles in free-living individuals. Solid bars indicate prenuptial molt; shaded bars with shaded columns on top indicate postnuptial molt. In the two birds kept under constant photoperiodic conditions, the cycles of testicular size and postnuptial molt have lost their normal temporal relationship relative to each other. (From Gwinner 1974b, after Berthold *et al.*, 1972c.)

change of the endogenous rhythm. An example is provided in Fig. 17.

c. *Phase shifting the endogenous rhythm.* A phase shift of the cycle to be tested should result in a corresponding phase shift of the endogenous rhythm. An example is shown in Fig. 15.

Hoffmann (1969b) has emphasized that in all three types of experiments it is often necessary to establish the free-running period of the endogenous rhythm before and after exposing it to the prospective *Zeitgeber.* The reason for this precaution is the observation in many organisms that environmental periodicities may exert direct influences on the overt function rather than entraining the underlying oscillation. Thus, low temperature may inhibit locomotor activity almost completely in mice, while high temperature may facilitate it. As a result, a 24-hour temperature cycle may produce a 24-hour activity pattern simulating entrainment by direct inhibition or stimulation even though the rhythm actually free runs. The decision whether the cycle to be tested exerts such masking effects or whether it actually entrains the oscillation can

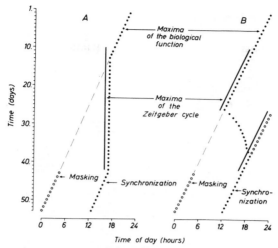

FIG. 13. Schematic representation of an experiment to test whether an environmental cycle acts as a *Zeitgeber* or as a masking agent. The position at successive days of the phase-angle reference point of a circadian rhythm (dashed line) and that of an environmental cycle (solid line) are plotted. *Case A:* The period of the circadian oscillation is different from the period of the environmental cycle. Exposed to the environmental cycle, it follows its period. Whether this indicates synchronization or masking can only be decided after the environmental cycle has been removed. In the case of synchronization, the phase-angle reference point of the circadian rhythm should start off from approximately the position to which it had been moved by the action of the environmental cycle (solid circles). In the case of masking it should reappear on the extension of the line that connects the successive positions of the circadian phase-angle reference point of the previous free runs (open circles). *Case B:* The period of the circadian rhythm is identical with that of the environmental cycle. It follows a phase shift of the environmental cycle. The decision whether this is due to synchronization or to masking depends, again, on the behavior of the circadian rhythm after removal of the environmental cycle. The behavior of the circadian rhythm during the days following the phase shift gives additional information: synchronization is suggested if the circadian rhythm follows the phase shift only slowly (through transients), while masking is suggested if the phase shift is followed instantaneously. (From Hoffmann, 1969b.)

then be made only by comparing the free runs from before and after exposure to the cycle, as illustrated schematically in Fig. 13.

2. *Cycles Effective as Zeitgebers*

Only a few of the many rhythmic variables in the environment are known to be effective as *Zeitgebers* (review in Hoffmann, 1969b; Bruce, 1960; Aschoff, 1963b). In birds, as in other organisms, the most powerful *Zeitgeber* of circadian rhythms is the daily variation of light intensity,

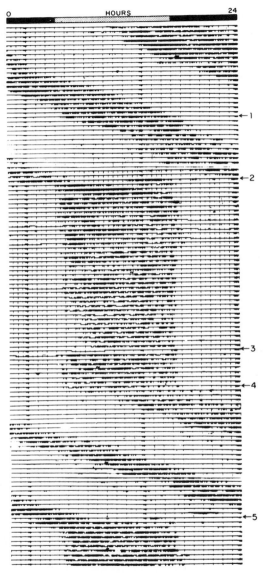

FIG. 14. Experiment demonstrating the participation of encephalic photoreceptors in the entrainment of the circadian locomotor activity rhythm of the House Sparrow by light. The bilaterally enucleated bird was exposed throughout the experiment to the light–dark cycle indicated at the top of the figure (black bars = darkness; stippled bar = dim green light of approximately 0.02 lux). The arrows at the right indicate the days on which various experimental treatments were performed. The light–dark cycle is of subthreshold level to which the activity rhythm does not entrain. Plucking of feathers from the back of the bird has no effect (1). However, plucking

i.e., the alternation between day and night (see Figs. 1 and 17 for examples). Menaker and his co-workers (Menaker, 1968a,b; Gaston and Menaker, 1968), in a series of elegant studies, have recently convincingly demonstrated that the entrainment response to light in the House Sparrow is mediated in part by extraretinal photoreceptors that are probably located in the brain (Fig. 14).

Acoustic stimuli may also entrain circadian rhythms. Figure 15 shows as an example the entrainment of the locomotor activity rhythm of a Serin (*Serinus serinus*) with a 24-hour cycle of species specific song (12 hours of song each day) (Gwinner, 1966a). Similarly, activity rhythms of House Sparrows could be entrained by $4\frac{1}{2}$ hours of tape-recorded birdsong played each day (Menaker and Eskin, 1966). Even unspecific mechanical noise, presented daily for 12 hours has been shown to be a sufficient though weak *Zeitgeber* for circadian activity rhythms of Chaffinches and Greenfinches (*Carduelis chloris*) (Lohmann and Enright, 1967).

Temperature cycles, though important *Zeitgebers* in many poikilotherm organisms, seem to be of little importance in homoiotherms, including birds (Hoffmann, 1969b; Bruce, 1960). Diurnal temperature cycles with a range of 17°–20°C failed to entrain the circadian activity rhythm of the House Finch (Enright, 1966a) and of the House Sparrow. Only cycles with a range of more than 30°C provided an effective *Zeitgeber* in the latter species (Eskin, 1971).

While the ecological significance of light cycles as the primary *Zeitgeber* is obvious, the significance of temperature and sound cycles remains to be demonstrated. It seems likely that these secondary *Zeitgebers* are important in environments in which there are no clear diurnal variations in light intensity (e.g., in the Arctic in summer and winter). The extent to which they participate in determining the phase relationship between circadian rhythms and the natural day (see below) is unknown.

Zeitgebers of circannual rhythms in birds have not yet been demonstrated. It seems likely, however, that the annual variations of photoperiod provide the most important entraining agent of circannual rhythms in the temperate zones. Indirect evidence supporting this conjecture has been summarized by Aschoff (1955) and Gwinner (1971, 1973).

---

of feathers from the top of the head (which increases the amount of light reaching the brain by a factor of at least 10^2) results in the entrainment of the rhythm (2). At (3), feathers, which had by now regrown, were again plucked from the head. Injection of india ink under the skin of the head (which reduces the amount of light reaching the brain by a factor of about 10) at (4) abolishes the entrainment response. The rhythm becomes reentrained after some of the head skin was removed and the ink deposit was scraped from the skull. For further explanations see Fig. 1. (From Menaker, 1968a.)

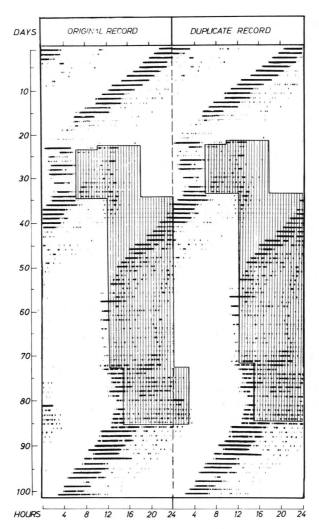

Fig. 15. Entrainment of the circadian activity rhythm of a Serin by cycles of species-specific song. The bird was kept in a soundproof chamber at constant dim illumination of about 5 lux. From the days 22 to 85, the song of the species was transmitted to the chamber for 12 hours per day (shaded area). For further explanations see Fig. 1. (From Gwinner, 1966a.)

B. VARIABLES DETERMINING THE EFFECTIVENESS OF A *Zeitgeber*— RANGE OF ENTRAINMENT—RELATIVE COORDINATION

1. Circadian Rhythms

An environmental cycle effective as a *Zeitgeber* under one set of conditions may be ineffective under other conditions. Theoretically, the success of entraining a self-sustaining system should depend on at least two factors: (1) the strength of the *Zeitgeber*, (2) the difference between the period of the *Zeitgeber* and that of the oscillation it entrains. Systematic investigations into the factors determining the strength of a *Zeitgeber* are scarce in birds. Temperature cycles were effective as *Zeitgebers* in House Sparrows only if the range of the cycle was large (Eskin, 1971). Some data suggest that the strength of a light *Zeitgeber* decreases as the intensity ratio between the light and the dark fraction of the cycle decreases (West and Pohl, 1973) and as the time ratio between the light and the dark fraction of the cycle approaches very large or very small values (Enright, 1965). Moreover, rectangular light–dark cycles appear to be weaker *Zeitgebers* than light–dark cycles with interposed twilights (Wever, 1967b; see also Hoffmann, 1969b, for a review).

Even a very strong *Zeitgeber* can be expected to entrain a self-sustaining oscillation only within certain limits. The situation is illustrated schematically in Fig. 16. Here, the period of a self-sustaining circadian oscillation is shown as a function of the period of a driving (*Zeitgeber*) oscillation. The period of the circadian oscillation assumes the period of the driving oscillation only if the two are close. Outside this range of entrainment the circadian oscillation free runs. The range of entrainment is wider with a stronger *Zeitgeber* than with a weaker one.

An example of the limited range of entrainment is shown in Fig. 17. The circadian activity rhythm of a Greenfinch exposed to light–dark

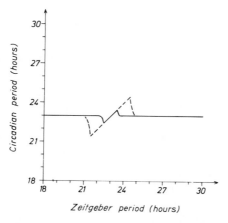

Fig 16. The behavior of a circadian rhythm (natural period: 23.0 hours) under the influence of a *Zeitgeber* with varying period (solid line = weak, broken line = strong). The circadian rhythm assumes the period of the *Zeitgeber* rhythm only within a limited range of periods. The width of this range of entrainment is larger with a stronger ·*Zeitgeber*. (From Hoffmann, 1969b, after Wever, 1965.)

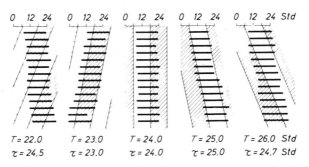

FIG. 17. Circadian activity rhythms of a Greenfinch exposed to light–dark cycles of varying period. Hatched areas = low light intensity (1 lux); white areas = high light intensity (5 lux); black bars = activity times; T = period of the light–dark cycle; τ = period of the activity rhythm. (From Hoffmann, 1970, after R. Wever, unpublished data.)

cycles (5:1 lux) of various frequencies becomes entrained only if the *Zeitgeber* period is between $T = 23.0$ and $T = 25.0$ hours. If the *Zeitgeber* period is $T = 22.0$ hours or $T = 26.0$ hours, the activity rhythm free runs with an average period of about 24.6 hours (R. Wever, unpublished data, 1970). In an extensive study, Eskin (1971) has determined the range of entrainment of locomotor activity rhythms of the House Sparrow exposed to light–dark cycles (250:0 lux; 6 hours of light per cycle). The upper limit of entrainment was found to be between $T = 28.0$ hours and $T = 28.7$ hours, while the lower limit of entrainment was found to be between $T = 17.8$ hours and $T = 15.8$ hours. The large difference in the range of entrainment of the Greenfinch and the House Sparrow study may be explained by the large difference in the range of the *Zeitgeber* oscillations.

If the potentially synchronizing cycle is slightly beyond or below the range of period values to which entrainment is possible, the driven system will free run with a frequency that changes periodically. This expectation from general oscillator theory has been verified in a variety of experiments. The phenomenon is usually referred to as relative coordination (von Holst, 1939; Aschoff, 1965c; Enright, 1965) or oscillatory free run (Swade and Pittendrigh, 1967; Pavlidis, 1969b). Figure 18 shows as an example the behavior of the activity rhythm of a Chaffinch under the influence of a daily light signal (15 minutes). This light–dark cycle is not sufficient to entrain the activity rhythm, but it clearly affects its period in a manner depending on the phase hit by the light pulse. The period is shortest when the signal occurs shortly before activity onset and longest when it occurs soon after the termination of activity. The phenomenon of relative coordination is of some significance for the understanding of entrainment mechanisms, since it demonstrates an essential

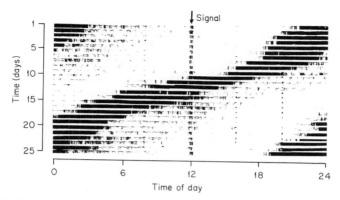

FIG. 18. Circadian activity rhythm of a Chaffinch in constant dim light (0.5 lux) which was interrupted daily at 12 o'clock by a 15-minute light signal (50 lux). For further explanations see Fig. 1. (From Aschoff, 1965c.)

prerequisite for entrainment: a periodic variation in the responsiveness of the circadian system to the effective stimuli of the *Zeitgeber*. Data demonstrating this property of circadian rhythms are the basis of phase–response curves, which are discussed in more detail in the next paragraph.

2. Circannual Rhythms

The extensive study by Goss (1969a,b,) on the Sika Deer (*Cervus nippon*) revealed that the range to which the circannual rhythm controlling antler growth and shedding can be entrained by photoperiodic cycles is extremely wide. Goss (1969b) altered the naturally occuring changes in photoperiod so that the experimental animals experienced $\frac{1}{2}$, 2, 3, 4, or 6 photoperiodic cycles within one calendar year. The circannual antler cycle could be entrained to $\frac{1}{2}$, 2, and even 3 cycles per year; only the 4 and 6 cycles per year failed to entrain the rhythm. Comparable investigations in birds are lacking, but Goss' findings suggest that the range of entrainment of circannual rhythms may be wide in this group as well. Results of experiments in which the period of annual rhythms in birds could be considerably altered by photoperiod manipulations (see Section I) are therefore compatible with the circannual rhythm concept.

C. PHASE–RESPONSE CURVES AND A MODEL OF ENTRAINMENT OF CIRCADIAN RHYTHMS

According to Aschoff's (1965c) classification, there are at least six different ways of obtaining a response curve. The classic procedure is to apply a single light pulse of a given duration and intensity at various phases of the circadian rhythm and to determine the steady-state phase shift. Figure 19 shows responses of the locomotor rhythm of sparrows in DD to 6-hour light signals. An advancing phase shift is obtained if the signal is applied shortly before the onset of activity (Fig. 19A), while a delaying phase shift of different magnitude is obtained from a signal given during the latter half of the activity time (Fig. 19B).

Using this technique, Eskin (1969, 1971) has obtained the only complete response curve available for birds. Figure 20 describes the direction and the magnitude of the resulting phase shifts as a function of the phase angle φ at which the light pulse began. Zero phase angle is taken as the onset of activity. It is clear that advancing phase shifts occur when the light signal starts before or shortly after activity onset, while delaying phase shifts result from light signals applied later in the cycle. The maximal possible phase shift is about 8 hours.

Pittendrigh (1965, 1966b, 1967a; Pittendrigh and Minis, 1964) has used the information contained in phase–response curves as the basis for a model of entrainment of the eclosion rhythm in *Drosophila*. Its starting point is the empirical fact that in each steady-state cycle of entrainment the period τ of the circadian rhythm is corrected by the action of the *Zeitgeber* so that it assumes its period T. The model specifies that this correction is achieved by a discrete instantaneous phase shift $\Delta\varphi = \tau - T$ caused by the transition from one light intensity to another. In addition, the assumption is made that the phase shift caused by

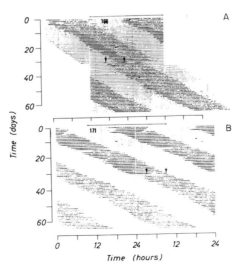

Fig. 19. Circadian activity rhythms of two House Sparrows in constant dark. Each bird received a single 6-hour pulse of light between the arrows in the middle of the records. For further explanations see Fig. 1. (After Eskin, 1969.)

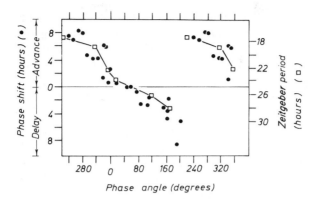

FIG. 20. The phase-response curve of the House Sparrow's circadian activity rhythm to 6-hour light signals (solid circles). On the abscissa is plotted the phase angle of the bird's activity rhythm at which the light pulse was presented ($360° = 1\ \tau$). Zero degrees is the onset of activity. A pulse at 80° is a pulse that began about 5.3 hours after the onset of activity. The magnitude and direction of the phase shift are plotted on the left ordinate. The pulse at 80° thus causes a phase delay of about 0.5 hours. The open squares represent the phase shifts corresponding to the phase angles measured in experiments with varying Zeitgeber periods. On the right ordinate are plotted the periods of the light–dark cycle that correspond to the phase-shift values of the left ordinate for $\tau = 25$ hours (= the mean τ value of Sparrows in DD). (After Eskin, 1969.)

light is that described by the phase–response curve for 15-minute light pulses.

The model has been shown to be valid in a variety of critical tests with *Drosophila pseudoobscura* and some other insects. Recently, Eskin (1969, 1971) has tested it in the House Sparrow in two ways, taking the phase–response curve for 6-hour light pulses as a basis.

1. If the assumptions made by the model are correct, one should be able to predict the phase-angle difference between the loco-motor activity rhythm and the Zeitgeber in entrainment experiments with Zeitgebers of varying periods, because a particular correction can only be obtained if the light signal falls on a particular phase angle of the cycle. If, for instance, T is 28 hours and τ is 25 hours, the circadian rhythm must phase-delay 3 hours each day. From the response curve (Fig. 20), it can be seen that such a phase shift is achieved only if the signal falls on phase angle 175°. Since the onset of the light cycle and the onset of activity are the phase reference points, the onset of activity should precede the onset of lights by 175°.

   Conversely, one can relate the phase shifts achieved by the entrainment to different Zeitgeber periods to the phase-angle differ-

ences measured under the respective *Zeitgeber* cycles and then
compare these phase shifts with those described by the response
curve to 6-hour light pulses. Figure 20 shows that the phase shifts
corresponding to the phase-angle differences of sparrows entrained to *Zeitgeber* periods ranging from 18 to 28 hours (light
time 6 hours) are in good agreement with the phase shifts measured from free-running birds.

2. If the assumptions made by the model are correct, one should
   be able to derive the limits of the range of entrainment from
   the response curve. Since the largest advance and delay phase
   shifts were about 8 hours (Fig. 20), it should not be possible
   to change the period of the circadian rhythm in experiments
   with varying T values by more than 8 hours in either direction;
   i.e., if τ is assumed to be 25.0 hours, the limits of entrainment
   should be close to $T = 33$ hours and $T = 17$ hours. However,
   one of the properties of the model is that steady-state entrainment
   is only possible when the light signal falls at points on the response curve where the slope is less than -2 (Pittendrigh, 1966b).
   This property of the model eliminates the 8.5-hour delay phase
   shift and the 5.0-hour delay phase shift at 189° and at 204°,
   respectively, from considerations in predicting the range of entrainment. The largest delay phase shift obtained within the
   part of the response curve where the slope is less than -2 was
   4.7, which leads to a predicted upper limit of entrainment of
   $T = 25.0 + 4.7 = 29.7$ hours. Correspondingly, the largest advance phase shift was 8.2 hours, from which the lower limit
   of entrainment can be calculated as $T = 25.0 - 8.2 = 16.8$ hours.
   The range of entrainment actually measured was from about
   $T = 28.2$ hours to $T = 15.8$–17.8 hours.

In spite of the remarkably good agreement between predictions and
results in these two tests, it should be emphasized that a general application of this model for predicting features of entrainment in vertebrates,
particularly in birds, faces various problems. It is known from several
organisms that response curves obtained with light pulses are different
from response curves obtained with light steps, and response curves
for "light on" steps may be different from those for "light off" steps
(Aschoff, 1965c). In birds, sudden transitions from dark to light or vice
versa produce response curves different from those obtained from slow
transitions (J. Aschoff, unpublished results). Moreover, the response
curve for one kind of signal depends on the state of the circadian system;
thus, the response curves of birds for light pulses of a given duration
and intensity change conspicuously with changes in the free-running

period (J. Aschoff and U. von St. Paul, unpublished results). It is clear that all these variations cannot be neglected in many instances if quantitative predictions are to be derived from response curves (see Aschoff, 1963b, 1965c, 1969, for discussion).

D. PHASE RELATIONSHIP BETWEEN CIRCADIAN RHYTHMS AND THEIR *Zeitgebers*

The discussion of response curves has made it clear that the function of synchronization is not simply period control (i.e., equal periods of entraining and entrained rhythm), but phase control, "a clearly defined and stable phase angle difference between the biological oscillation and the *Zeitgeber*" (Aschoff, 1963b). The result of this latter function is well known. It finds its expression, for instance, in the fact that there are diurnal and nocturnal species and that the members of either group can be further separated according to the time of day at which they begin activity.

Despite this gross species specificity of the phase relationship between circadian rhythms and their *Zeitgebers,* a slight variability in the phase-angle difference can be observed. The factors, both external and internal, responsible for such variations have been analyzed in part. They are essentially identical with those on which the success of entrainment depends (Section III,B). The following paragraphs summarize the results of pertinent experimental studies. The extent to which the rules derived from these laboratory investigations can be used to explain the field situation will be subsequently considered.

1. How to Measure the Phase-Angle Difference

The measurement of the phase-angle difference requires the definition of phase-angle reference points on both oscillations. If, as in most of the cases to be discussed, the phase angle difference between a locomotor activity rhythm and a light–dark cycle is to be established, the onset or end of activity and the onset or end of the light cycle suggest themselves as phase-angle reference points. There is no obvious reason why the comparison of the midpoint of activity with the midpoint of the light period should be less justified. This latter approach has been proposed by Aschoff (1964b, 1965d, 1967a, 1969; Aschoff and Wever, 1962b) for the following reasons: (1) Empirically, the results are usually most uniform when computed from the midpoint. (2) If the level-threshold model discussed in Section II,D is valid, both onset and end of activity may vary as a result of changes in the level-threshold relation, even if the actual phase relationship remains constant (Fig. 7). In this case, the use of the midpoint would give rise to a smaller error than the use of onset or end (for a detailed discussion, see Wever, 1965; Aschoff, 1967a).

2. Factors Determining the Phase-Angle Difference— Laboratory Investigations

a. *The Period Relationship between the Circadian Rhythm and the Zeitgeber.* It can be expected from general oscillator theory that within the range of entrainment the phase-angle difference ψ depends on the ratio between the natural period τ_n and the period T of the *Zeitgeber.* ψ should increase or become less negative as T/τ_n increases (e.g., Aschoff, 1960, 1965d; Wever, 1965). This prediction has been tested in the following ways:

1. Varying the *Zeitgeber* period T. ψ should increase or become more positive with increasing T. This prediction has been verified in experiments with several organisms (Aschoff, 1960; Enright, 1965). An example for birds is shown in Fig. 17.

2. Entraining organisms with different free-running periods to the same *Zeitgeber.* If a group of organisms of one species is kept in constant conditions, a considerable variability of τ values can typically be observed. If subsequently entrained to the same *Zeitgeber,* a corresponding variation of ψ values is to be expected. The ψ values of the different organisms should be negatively correlated with their previous or subsequent τ. This hypothesis has been shown to hold in both reptiles (Hoffmann, 1963, 1969b) and birds (Aschoff and Wever, 1962a, 1966; Eskin, 1971) (see Fig. 21).

3. Varying the natural period τ_n. ψ should increase or become less negative with a decrease of the natural period τ_n. Assuming that the factors known to affect τ_n in the free-running state (Section II,D) will affect τ_n also in the entrained state, one should expect ψ to increase or become less negative with an increase of the intensity of those factors that cause a decrease of τ in the free-running state. Among the factors known to affect τ in birds, temperature and light-intensity are the most obvious.

The effects of temperature on ψ have been tested in the House Finch and in the Chaffinch entrained to light–dark cycles (Enright, 1966b; Pohl, 1968b). Since in these species τ decreases with increasing temperature, one should expect ψ to increase if the environmental temperature is increased. A positive correlation between ψ and temperature could, in fact, be observed. Experiments with a mammal were also in general agreement with the expectation (Pohl, 1968b).

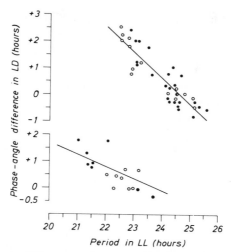

FIG. 21. Phase-angle difference ψ between the onset of activity and lights on as a function of the free-running circadian period τ measured under constant conditions before and after entrainment. Results from experiments with six male (solid circles) and four female (open circles) Chaffinches. *Top:* ψ in LD (200:0.5 lux), τ in DD (0.5 lux); *bottom:* ψ in LD (30:0.5 lux), ψ in LL (30 lux). (After Aschoff and Wever, 1966.)

An obvious way to test the effects of light intensity on ψ would be to vary the background intensity in organisms entrained to a *Zeitgeber* other than a light–dark cycle. Results of current experiments with finches entrained to an acoustical *Zeitgeber* are still ambiguous (J. Aschoff and U. von St. Paul, unpublished).

Predictions about ψ can also be made for organisms entrained to a light–dark cycle if one makes the assumption that light intensity during entrainment is effectively averaged over the light and the dark fraction of the cycle. If this assumption is valid, ψ can be expected to increase or become less negative with an increase of the average light intensity in such organisms in which τ decreases with light intensity under constant conditions (light-active organisms) (see Section II,D), and vice versa for organisms in which τ increases with light intensity (dark-active organisms).

The average light intensity of a light–dark *Zeitgeber* with a defined period can be altered in two ways: (1) by altering the light intensity of the light half of the cycle, of the dark half of the cycle, or of both, (2) by altering the light–dark ratio.

The first type of experiment has been carried out with the feral pigeon (Schmidt-Koenig, 1958) and with various finches (Aschoff and Wever, 1962a; Pohl, 1970). The results were in agreement with the prediction;

increasing the light intensity at day and/or night resulted in an increase of ψ as expected for these day-active birds.

The second type of experiment has been carried out with various species of finches (Aschoff and Wever, 1965; West and Pohl, 1973) and with the House Sparrow (Eskin, 1969). The results were by and large also in agreement with the expectation. An example from an experiment with the European Bullfinch (*Pyrrhula pyrrhula*) is shown in Fig. 22. The comparison of the midpoint of activity with the midpoint of the light period reveals a positive correlation between ψ and photoperiod as expected for these day-active birds. Data from similar experiments with other organisms have been summarized by Aschoff (1960, 1965d).

b. The Strength of the Zeitgeber. It has to be postulated on theoretical grounds that the strength of a *Zeitgeber* will affect not only the limits of entrainment (Section III,B), but also the phase-angle difference. Unfortunately, experimental investigations of this aspect are rare (Hoffmann, 1969a,b) and apparently lacking in birds. The neglect of this subject is certainly due to the fact that the parameters that determine

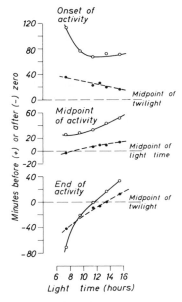

FIG. 22. Time of onset, end, and of midpoint of activity in Bullfinches as a function of photoperiod in an artificial light–dark cycle with interposed twilights of either 160 minute (solid circles) or 480 minute (open circles) duration. Onset and end of activity are related to the midpoint of the twilight; midpoint of activity is related to the midpoint of the light time; the duration of the light time is measured from the midpoint of dawn twilight to the midpoint of dusk twilight. (From Aschoff, 1969.)

the strength of a *Zeitgeber* are largely unknown (Section III,B). Some of the inconsistencies mentioned below may eventually become reconciled when more is known about this subject.

c. The Duration of Twilight. The action of most of the variables affecting the phase-angle difference discussed above were to be expected on the basis of the well-substantiated assumption that entrained circadian rhythms behave like self-sustaining oscillations under the influence of a driving oscillation. The discovery in some birds and mammals that the duration of twilight also affects the phase-angle difference was, in contrast, predicted by a special model of entrainment (Wever, 1964, 1965, 1967a,b). According to this model, the phase-angle difference should become more positive as the duration of twilight increases in both light- and dark-active organisms. Confirmative data are available from four species of day-active finches and one night-active mammal (Aschoff and Wever, 1965; Wever, 1967b). Figure 22 provides an example from one of the birds.

3. The Phase-Angle Difference under Natural Conditions— The Rules for Season and Latitude

Under natural conditions, most organisms are exposed to variations of several of the factors known to affect the phase relationship between circadian rhythms and their *Zeitgebers*. Most conspicuous are seasonal changes in photoperiod and in the duration of twilight, variations that can be expected to cause predictable seasonal variations in the phase-angle difference (Section III,D). The upper part of Fig. 23 illustrates the seasonal changes of photoperiod and twilight duration for a location on latitude 45°N. The lower part of the figure describes schematically the expected changes in the phase-angle difference between corresponding phases of a circadian rhythm and the natural day for both light- and dark-active organisms. If ψ depended on photoperiod alone, one should expect it to be most positive or least negative in midsummer and least positive or most negative in midwinter in light-active organisms. The reverse should hold for dark-active organisms. If, on the other hand, ψ depended on the duration of twilight alone, the phase angles of both light- and dark-active organisms should be most advanced in midsummer, most delayed at the times of the equinoxes, and they should show a second peak in midwinter. If, as can be expected, both variables are effective simultaneously, ψ should show the seasonal changes described by the solid curve. Since both the seasonal fluctuations of photoperiod and duration of twilight increase with latitude, these annual variations of the phase-angle difference should become more conspicuous the closer one approaches the poles.

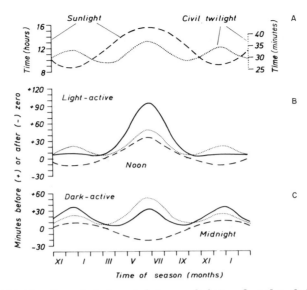

Fig. 23. (A) Duration of photoperiod (= sunlight) and civil twilight at latitude 45°N as a function of season. (B) and (C) Examples for predicted variations of the phase-angle difference between midpoint of activity and midpoint of photoperiod of a light-active and of a dark-active animal—assuming that only the photoperiod is effective (dashed line); assuming that only the twilight is effective (dotted line); assuming that both photoperiod and twilight are effective (solid line). (After Wever, 1967b.)

These predictions have been tested in a series of studies by Aschoff and his co-workers (Aschoff and Wever, 1962b; Aschoff, 1964b, 1967a, 1969; Aschoff *et al.*, 1970; see also Blume, 1963, 1964, 1965), exploiting both literature data and results of their own experiments. In view of the fact that only two out of many possible phase-determining variables have been used as the basis for the predictions, the results obtained from studies of free-living birds were in remarkably good agreement with the hypothesis. At least nine species, seven of which are diurnal birds, have been shown to behave as predicted by the solid curves in Fig. 23. Figure 24 summarizes some of these results. All six avian species show the expected maximum phase lead in midsummer. In spring and fall, ψ shows the smallest values. There are indications of a second, smaller peak of phase lead in midwinter. The range of the seasonal changes of ψ increases with latitude as predicted by the model. While data on nocturnal birds are lacking, a nocturnal mammal has been shown to follow the predictions as well.

While most species studied in the field follow the predicted course of phase-angle difference between the activity rhythm and the light–dark

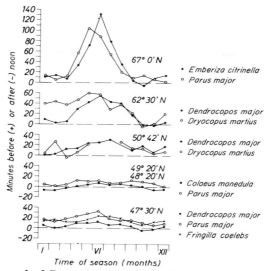

FIG. 24. Phase-angle difference between the midpoint of activity and the midpoint of the light time in six species of day-active birds as a function of season and latitude. *Coloeus monedula = Corvus monedula.* (From Aschoff, 1969.)

cycle, most species studied in captivity did not (Aschoff, 1969; Daan, 1972; Daan and Aschoff, 1975). Thus, four species of finches studied at the Arctic Circle had largest values in spring and smallest in summer (Daan, 1972). The asymmetry around the solstices suggest that other variables than those related to the light–dark cycle play an important role in determining the phase relationship. The maximum phase lead in spring observed in many birds is possibly related to the breeding condition (Aschoff and Wever, 1962b; Aschoff, 1969) and may be under hormonal control (Gwinner and Turek, 1971).

4. Conclusions

The results discussed in the preceding paragraphs have to some extent been predicted by theoretical considerations; many of the experiments have, in fact, been designed to test specific predictions. Some of the results provide compelling cases for the fruitfulness of the oscillator analogy. Many of the findings may be of considerable significance for the proper understanding of mechanisms in which circadian clocks are involved. The necessity of knowing the exact phase relationship between the circadian oscillation measuring day length and the natural day will be discussed in Section IV,B. In sun-compass orientation, the directional choice will depend on the phase relationship between the circadian clock used for compensating the sun's movement and the natural day (Chapter 3). It is obvious that changes in ψ with season and latitude imply that seasonal and latitudinal corrections have to be made in order for the sun-compass mechanism to give the correct directions. These and other problems have hardly been recognized so far.

IV. Adaptive Functions

The ultimate causes that have led to the near universality of circadian rhythms among eukaryotic organisms are still obscure. Possible advantages of endogenous self-sustained rhythmicity have been discussed with regard to economy (Aschoff and Wever, 1962c), environmental relationships (Pittendrigh, 1958; Aschoff, 1959a, 1964a, 1965b, 1967b), and internal temporal order (Pittendrigh, 1960, 1961, 1966a), but Enright's statement (1970) holds: "to date, no convincing evolutionary basis has been proposed to account for one of the most remarkable characteristics of circadian rhythms: their long-term, self-sustained persistence under constant conditions." The selection pressures that have given rise to circannual rhythms are even more obscure.

Despite the uncertainty about the original selecting forces, it is clear that, once evolved, circadian rhythms have become exploited for many purposes. Most striking is the utilization of circadian rhythms as clocks in sun-compass orientation (Chapter 3). They are advantageous as mechanisms enabling "time memory" (von Frisch, 1965) and guaranteeing synchrony between conspecifics, which is necessary, for instance, for the success of reproduction (Volume I, Chapter 8). The following discussion will focus on two other adaptive functions of circadian and circannual rhythms—their function as "programming agents" and the role of circadian clocks in photoperiodic time measurement.

A.   Temporal Adjustment to Periodic Environments

It is well known that numerous biological functions are restricted to the times of day or year at which they are most advantageously carried out. In birds, this is most obvious in the annual timing of such functions as reproduction and migration that take place at optimal seasons. The same activities may at the same time show diurnal periodicities. Thus, nest-building behavior of *Phylloscopus* warblers is restricted to the early morning hours when the dry grass used as construction material is still wet and flexible from the dew (Geissbühler, 1954). And many normally day-active birds migrate at night, which among other things, may be of thermoregulatory advantage (Dorka, 1966).

Such examples of doing the right thing at the right time (Aschoff, 1964a) illustrate the phenomenon of temporal adaptation to periodic environments. It is achieved in many instances through the action of entrained circadian and circannual rhythms. However, the significance of these endogenous periodicities as mechanisms allowing temporal adaptation is more than just crude timing. They are utilized by many organisms for advance preparation to future conditions. This function will be exemplified in the following paragraphs.

1. Circannual Rhythms

Circannual rhythms may enable young migratory birds to prepare for fall migration by completing plumage development in time and by storing fat before migration starts. In warblers of the genera *Phylloscopus* and *Sylvia*, this rather complex seasonal sequence of events is preprogrammed in the circannual organization in a species-specific manner (Gwinner, 1969, 1971; Berthold *et al.*, 1970). This is illustrated in Figs. 25 and 26. Figure 25 shows the successive stages of plumage development and the variations in body weight and *Zugunruhe* of hand-raised Willow Warblers (*Phylloscopus trochilus*) and Chiffchaffs (*Phylloscopus collybita*) under natural lighting conditions. The Chiffchaff is a short-distance migrant wintering in the Mediterranean area; it starts migration in late fall. The Willow Warbler, in contrast, is a long-distance migrant wintering in central and southern Africa; it starts migration in late summer. These differences in migratory behavior are correlated with differences in the speed of premigratory development. In the early-migrating species (Willow Warbler), feather development begins and ends earlier and

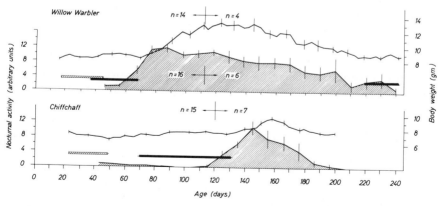

FIG. 25. Body weight (unhatched curves), *Zugunruhe* (hatched curves), plumage development (hatched bars), and molt (solid bars) in Willow Warblers and Chiffchaffs raised under natural photoperiodic conditions of their breeding grounds. Horizontal lines at the ends of the bars and vertical lines at the curve points indicate standard errors. (After Gwinner *et al.*, 1971.)

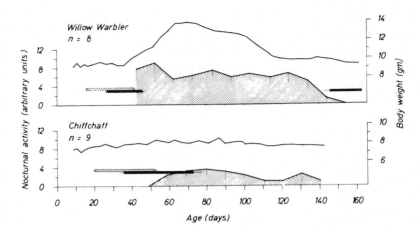

FIG. 26. Body weight (unhatched curves), *Zugunruhe* (hatched curves), plumage development (hatched bars), and molt (solid bars) in Willow Warblers and Chiffchaffs raised from an age of 9 days in a constant 12-hour photoperiod (LD 12:12, 200:0.02 lux; 20° ± 1°C). For further explanations see Fig. 25. (After Gwinner *et al.*, 1971.)

proceeds faster than in the late-migrating species. As a result, *Zugunruhe* and the accompanying increase in weight can begin earlier in the early migrant. Moreover, the long-distance migrant becomes fatter and reaches higher intensities of *Zugunruhe* than the short-distance migrant.

Figure 26 shows that most of these differences observed under natural lighting conditions can also be seen in birds raised under a constant photoperiod. There are, it is true, differences between the behavior of these birds and that of their conspecifics kept under natural photoperiodic conditions, indicating effects of external variables. On the other hand, it is clear that those species differences that persist under identical and constant conditions must be due to species differences in the endogenous temporal organization of the birds. Similar conclusions have been drawn for other species and racial differences observed in constant conditions for up to 3 years (Gwinner, 1971, 1972b; Berthold *et al.*, 1970).

In the example discussed above, the significance of the endogenous control mechanisms is obviously the proper timing of an adaptive sequence of events that has to proceed in the species-specific manner so that migration can commence at the proper time of the year. However, there is evidence that circannual rhythmicity may have adaptive value quite apart from crude timing and preparation. A striking example is provided by the probable significance of endogenously controlled migratory activity. In two species of *Phylloscopus* and in two species of *Sylvia* kept under the natural lighting conditions crudely simulating those they would have experienced during their migration, the temporal variations of *Zugunruhe* reflect fairly accurately the actual pattern of migration (Gwinner, 1968a,b, 1972a; Berthold *et al.*, 1972a). Onset and end of *Zugunruhe* coincide approximately with the onset and the end of actual migration, and the period of the most intense *Zugunruhe* coincides approximately with the period of maximal migratory speed. Moreover, differences in the duration and total amount of *Zugunruhe* between the Willow Warbler and the Chiffchaff, on the one hand, and between the Garden Warbler (*Sylvia borin*) and the Blackcap (*S. atricapilla*), on the other, are similar to differences in the distance actually covered during migration (Fig. 27). These data suggest the hypothesis that the distance traveled by these birds during their first fall migration is determined in part by an endogenous species-specific program that produces just enough migratory activity during a migratory season for the birds to reach their wintering area. Evidence supporting this hypothesis has been published elsewhere (Gwinner, 1968a, 1972b; Berthold, 1973). Since young warblers, like other first-year migrants, are probably capable of direction orientation only (Chapter 3), such a mechanism might be of vital importance for the termination of migration in the species-specific wintering grounds.

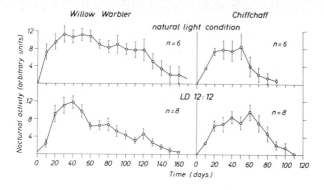

FIG. 27. Temporal variations of autumnal *Zugunruhe* in two groups of young Willow Warblers and Chiffchaffs. Natural light conditions, birds were kept in natural photoperiodic conditions of their breeding grounds; LD 12:12, birds were transferred from natural photoperiodic conditions of their breeding grounds to a 12-hour photoperiod (LD 12:12; 200:0.02 lux) between September 15 and 20 to crudely simulate the photoperiodic conditions experienced by these two species under free-living conditions. Mean values for successive thirds of a month are plotted. *Zugunruhe* curves of individual birds have been normalized with regard to onset of *Zugunruhe*. Vertical bars indicate standard errors. Under either condition, the Willow Warblers exhibit two to three times as much *Zugunruhe* as the Chiffchaffs. This corresponds with their two to three times longer migratory route. (From Gwinner, 1972a.)

2. Circadian Rhythms

Endogenous programming with the function of preparation for future conditions is certainly a major function of circadian rhythms as well. In homoiothermic organisms, including birds, body temperature shows daily fluctuations that are controlled by circadian rhythmicity. The phase relationship between the temperature cycle and the circadian activity rhythm shows clear adaptive aspects. Body temperature begins to increase at a time of day long before activity begins, so that when the animal starts to move it is fully able to cope with the demands of activity (Aschoff, 1963a).

In natural conditions, many birds show a bimodal pattern of locomotor activity. The two peaks have often been interpreted as being caused by dawn and dusk or by "lights on" and "lights off," but since it has been demonstrated (Hoffmann, 1960; Aschoff, 1966a) that both peaks may be characteristic for free-running activity rhythms under constant conditions as well, such an interpretation is no longer acceptable. The double-peaked pattern must be a property of the circadian system, a property that may well have evolved as an adaptation to (been ultimately caused by) conditions prevailing at dawn and dusk, but which is now proximately controlled by circadian rhythmicity.

Circadian programming goes even further. In European Starlings, the preference for brighter or dimmer light depends on the phase of their free-running circadian cycle (Gwinner, 1966b). If allowed to choose between two or three boxes with different constant light intensities, they prefer the darker box for sleeping and the brighter box for activity; this preference for the brighter box is, by and large, positively correlated with the intensity of their locomotor activity. However, during the evening peak of activity there is a strong tendency in some birds to prefer the dimmer intensities; this may be related to searching shelter for roosting. A preference for darkness during dusk activity is even more pronounced in European Robins when allowed to select their preferred light intensity in a circular apparatus. Outside the migratory season, a short period of evening activity is concentrated in the darkest sections of the apparatus (Fig. 28, upper recording), while the preceding bulk of "diurnal" activity is performed in the brighter section. During the migratory season, the dark-activity at the end of subjective day is supplemented by several hours of *Zugunruhe* (characterized by specific calls known from nocturnally migrating robins) in the dark section (Fig. 28, lower recording). Obviously, in both the European Starling and European Robin, the naturally occuring variations in light intensity have become incorporated as an "expectation" in their endogenous circadian organization.

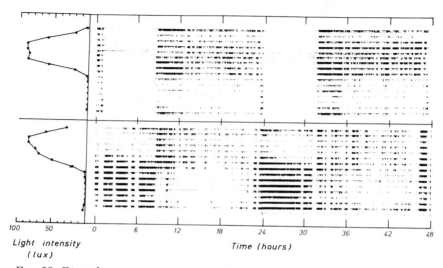

Light intensity
(lux)

Time (hours)

FIG. 28. Diurnal variations in preferred light intensity of a European Robin. The bird is kept in a circular chamber (30 cm × 34 cm; circumference, 650 cm) with sixteen perches distributed in equal intervals around the circle. The light intensity at each perch is shown on the left side of the graph. The perches are mounted on microswitches connected to an event recorder so that it can be determined at what time of the day the bird is active in which light intensity. On the upper graph, the behavior of the bird at two successive days during the time of postnuptial molt is shown. Periods of activity alternate with periods of rest on a circadian basis. Activity is concentrated in the brighter section of the apparatus, except for a brief period of "evening" activity, which is performed in the darker part. On the lower graph, the behavior of the same bird during the time of autumnal migration is shown. Activity is again organized on a circadian basis, but now the period of rest is replaced by a period of activity in the dark section and can be identified as *Zugunruhe*. Eight other robins showed the same seasonal differences. (E. Gwinner, unpublished data.)

B. CIRCADIAN MECHANISMS IN PHOTOPERIODIC TIME MEASUREMENT

1. *Bünning's Hypothesis and Evidence Supporting It in Birds*

In 1936, Bünning proposed that an endogenous daily rhythm in photosensitivity is causally involved in photoperiodic response mechanisms in plants. The first half of the cycle of this oscillation is light-requiring (photophil), the second half of the cycle is dark-requiring (scotophil). Bünning assumed that photoperiodic induction occurs when, under long day conditions, daylight extends into the scotophil half of the cycle. Since in this model the photoperiodic response depends on the coincidence of light and a sensitive phase of a circadian rhythm, it has been referred to as the "coincidence model" of photoperiodic induction (Pittendrigh and Minis, 1964).

Strong evidence supporting Bünning's general proposition has been obtained from various plants (e.g., photoperiodic induction of flowering: Bünsow, 1960; Hamner, 1960; Hamner and Takimoto, 1964; Hillman, 1964) and a few insects (photoperiodic induction or termination of diapause: Bünning and Joerrens, 1960, 1962; Saunders, 1970). Recent data strongly suggest that a circadian rhythm in photosensitivity is also involved in photoperiodic response mechanisms in birds.

W. Hamner (1963, 1966) used the method first employed by K. Hamner for investigations of photoperiodic phenomena in plants (Hamner, 1960; Hamner and Takimoto, 1964) to test whether the testicular response of birds to long days is mediated by a circadian oscillation. He exposed photoresponsive male House Finches to light–dark cycles of varying period length (T = 12, 24, 36, 48, 60, 72 hours). T was varied by varying the duration of darkness while the duration of light was held constant at 6 hours per cycle (Fig. 29). By the end of the experiments after

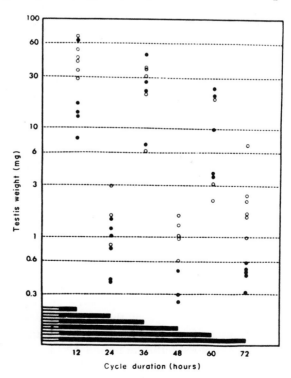

FIG. 29. Testicular response of the House Finch as a function of the duration of a dark period, separating two standard light periods of 6 hours each. Solid and open circles refer to data from two experiments. Black bars indicate duration of darkness; open bars indicate duration of light. (From Hamner, 1963. Copyright 1963 by the American Association for the Advancement of Science.)

22 and 33 days, respectively, the testes of the birds kept in the 12-, 36-, and 60-hour cycle had responded as though they had been kept in a long-day photoperiod, while the testes of the birds kept in the 24-, 48-, and 72-hour cycles had maintained their initial immature state (Fig. 29). Comparable data from the same type of experiment have recently been obtained by Follet and Sharp (1969) for the induction of testicular and ovarian growth in the Japanese Quail and by Turek (1972) for both the induction of testicular growth and the termination of photorefractoriness in the White-crowned Sparrow and Golden-crowned Sparrow.

These results are not only fully predicted by the Bünning hypothesis, but at the same time cannot be reasonably interpreted by any obvious alternative model. The fact that gonadal growth occurs (or fails to occur) during cycles of which the periods differ by 24 hours implies that a daily changing response system continues to oscillate in constant dark with a period close to 24 hours for at least 66 hours. If the recurrent 6 hour light signal occurs in the second half of the cycle (as in the case of the 12-, 36-, and 60-hour cycle), induction takes place; if it occurs in the first half (as in the 24-, 48-, and 72-hour cycle), induction fails to take place.

Equally convincing evidence in support of Bünning's hypothesis has been obtained in the House Finch with another type of experiment, also originally used in plants (Schwabe, 1955; Melchers, 1958; Bünning, 1960; Bünsow, 1960). The animals are exposed to a short, noninductive light period followed by an extended (more than 24 hours) dark period; this dark period is then interrupted by short flashes of light at varying times. Using a 6-hour main photoperiod and one-hour light flashes, Hamner (1964, 1965) obtained results supporting the interpretation of his previous experiments—induction of testicular growth in male House Finches took place when the light pulses were administered 12 or 36 hours after the onset of the main photoperiod, but not if the light pulses were administered 24 or 48 hours after the onset of the main light period. Again, these results can best be explained by assuming that there is a circadian rhythm in photoresponsiveness that continues in constant darkness. If the one-hour light pulse falls on the beginning of the second half cycle of the postulated circadian rhythm (hours 12 and 36, respectively) the light schedule is interpreted as "long-day" and testicular maturation is initiated. But if the light pulse falls on the beginning of the first half cycle (hours 24 and 48, respectively) the light schedule is interpreted as "short-day" and the testes remain in an immature state.

Since the experiments discussed so far demonstrate both the endogenous circadian nature of the underlying process and its daily variations in photosensitivity, they provide the best evidence available for birds in support of Bünning's proposition. Less striking evidence comes from such night-interruption experiments in which the dark period did not exceed 24 hours minus the main light period. Farner (1965) and Follet and Sharp (1969) obtained considerable gonadal recrudesence in the White-crowned Sparrow and in the Japanese Quail, respectively, when the night of an otherwise short day photoperiod was interrupted by light pulses of 15 minutes or 2 hours duration occuring between about 12 and 20 hours after the onset of the main photoperiod. The same light stimulus was essentially ineffective at other times of the night. Similar results have been obtained in the Greenfinch (Murton *et al.*, 1970a). While these findings are clearly in agreement with Bünning's hypothesis, they do not rigorously exclude the alternative model of an "hour-glass"-type time measuring device.

Results of other experiments indicate that a Bünning oscillator may also be involved in inducing gonadal regression. Menaker and Eskin (1967) exposed House Sparrows with mature testes to a cycle of 14 hours of dim green light and 10 hours of darkness. This cycle entrains the activity rhythm, but the intensity of the light is below that necessary for photoperiodic effectiveness. The birds thus respond with testicular regression as though exposed to constant darkness. The same is true of sparrows that experience a 75-minute pulse of bright white light early in the dim light period. However, if a similar light pulse is administered late in the dim-light period, the testes fail to regress as is the case in sparrows living in long-day conditions. Thus, depending on the phase at which it occurs, a short period of bright light may simulate a short- or a long-day photoperiod. Recent results of Murton *et al.* (1970b) from the same species are consistent with such a conclusion.

2. *Special Problems*

In the original coincidence model, Bünning implied that the rhythm which measures day lengths is rigidly locked onto the beginning of the light period and hence that its phase relationship with the light–dark cycle is independent of the photoperiodic conditions. From previous paragraphs (Section III,D) it is clear, however, that this assumption is unlikely. The phase of circadian systems can be expected to depend on the duration of light and on many other properties of the entraining light–dark cycle. Pittendrigh (Pittendrigh and Minis, 1964, 1971; Minis, 1965; Pittendrigh, 1966b) in particular has repeatedly stressed the implications of this fact for the design and interpretation of experiments carried out to test Bünning's hypothesis. Since the lighting conditions simultaneously phase the oscillation and induce the photoperiodic response, the understanding of the latter effect demands the knowledge of the former mechanism.

While such complications impose serious restraints on the interpretation of experiments as long as the phase of the postulated rhythm is unknown, they provide powerful tools for rigorously testing Bünning's hypothesis if predictions about the phase of this oscillation can be made. Striking examples are known from plants (Hillman, 1964; Bünsow, 1960) and insects (Pittendrigh, 1966b) and also from birds. Menaker (1965) worked with House Sparrows in which testicular regression can be delayed by long photoperiods. He subjected birds with mature testes to asymmetrical skeleton photoperiods consisting of a 4-hour main photoperiod followed by 2-hour light breaks occurring after varying intervals of intermittent darkness (Fig. 30A). The birds responded as shown in Fig. 30B. Assuming that the oscillation measuring day length is rigidly locked onto the main photoperiod, the results indicate two peaks of maximum sensitivity to light between hours 10 and 20. In reality, however, probably none of the groups ever experienced a skeleton photoperiod longer than 16 hours, as indicated by the activity rhythms of

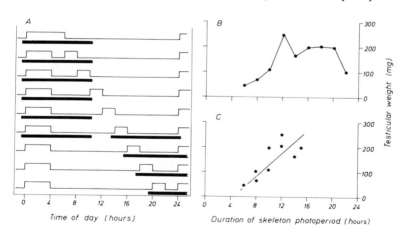

Fig. 30. Testicular response of House Sparrows to asymmetric skeleton photoperiods. In A, the skeleton photoperiods are shown to which nine pairs of birds had been exposed. The bars beneath each photoperiodic schedule indicate schematically at what time of day the birds were active. Under 4L:10D:2L, the activity of one of the two birds began relative to the onset of the 4-hour photoperiod, while that of the other bird began relative to the onset of the 2-hour photoperiod. In B and C, the testicular response of the birds is shown as a function of the duration of the skeleton photoperiod (measured from the onset of one to the end of the subsequent light pulse). In B it is assumed that the birds under all conditions interpret the onset of the 4-hour photoperiod as dawn. In C it is assumed that the birds interpret the onset of that photoperiod as dawn, relative to which they began activity. (After Menaker, 1965.)

these birds (Fig. 30A). Birds exposed to light cycles in which the onset
of the 4-hour main photoperiod and the end of the 2-hour light interruption were 14 or fewer hours apart interpreted the onset of the 4-hour
light period as dawn, while those exposed to cycles in which this interval
was longer than 16 hours, chose the onset of the 2-hour interruption
as dawn. The two birds experiencing the intermediate photoperiod
4L:10D:2L behaved nonuniformly. One bird started activity relative
to the onset of the 4-hour light period, the other one relative to the
2-hour light period. Assuming that the circadian oscillation controlling
the photoperiodic response is identical with or behaves like that controlling locomotor activity, the data on the testicular response should be
plotted as in Fig. 30C. The bimodal testicular response curve of Fig.
30B, which is inconsistent with a simple Bünning model, is then replaced
by a unimodal curve. And while the photoinducible phase of the rhythm
seems to extend over at least a 10-hour portion of the subjective night
in Fig. 30B, no conclusions can be derived about its width if the changing
phase relationships are taken into account. Data from skeleton photoperiods simulating day lengths longer than 16 hours are not available.

Changing phase relationships between the postulated photoperiodic
response rhythm and the entraining light–dark cycle can be expected
not only under the extreme situation of skeleton photoperiods, but also
under more natural lighting conditions. As discussed in Section III,D,2,
the phase relationship between activity rhythms and the light–dark cycle
is a function of the photoperiod. Results from plants and insects strongly
suggest that the same is true for the rhythm controlling their photoperiodic response. Data published by Follet and Sharp (1969) can also
be interpreted in such terms. In interrupted-night experiments of the
type discussed above, the phase of maximum photosensitivity was a
function of the duration of the main photoperiod. The maximum testicular response of Japanese Quail to 15-minute light pulses occurred later
if the main photoperiod was increased. Differences in the shape of the
testicular response curve as a possible function of the duration of the
main photoperiod are also available for the House Finch (Hamner,
1968).

It must be reemphasized that the oscillation measuring day length
has never been directly analyzed. Its existence is inferred from the properties of the photoperiodic response under various lighting regimes. As
in the experiments discussed above, its existence can only be tested
by testing predictions that can be derived from known properties of
circadian functions measured simultaneously in the experimental animals.

Rigorous predictions of the latter kind have been tested in the House
Finch by Hamner and Enright (1967). One of their experiments con-
sisted of exposing photosensitive birds to 22- and 26-hour light–dark

cycles with one short (3-hour or 6-hour) main photoperiod. The bird's activity rhythms behaved as expected from circadian systems entrained to *Zeitgeber* periods different from 24 hours. Under the 26-hour light cycle the activity rhythms assumed a more positive phase-angle difference relative to the light–dark cycle than under the 22-hour cycle. In other words, the locomotor activity of the birds experiencing the 26-hour cycle anticipated light on by as much as 10 hours, while that of the birds in the 22-hour cycle usually coincided with the light period. On the assumption that the activity rhythm reflects the behavior of the postulated photosensitivity rhythm, one therefore should expect testicular development in the birds in the 26-hour cycle, since light should fall on the second, sensitive half cycle of the circadian rhythm. On the other hand, the testes would be expected to remain undeveloped in the birds on the 22-hour cycle, since light should never hit the photosensitive phase. By and large, the predictions were confirmed by the results. There was significantly more testicular growth in the birds entrained to the 26-hour cycles than in the birds entrained to the 22-hour cycles. Moreover, there was a positive correlation consistent with Bünning's hypothesis between the magnitude of the phase lead and the final testicular size.

3. External versus Internal Coincidence

The results summarized in the previous paragraphs are, by and large consistent with the hypothesis that the photoperiodic response depends on external coincidence, i.e., the coincidence of light with a particular phase of the circadian oscillation. However, recently, Pittendrigh (1960, 1972) has pointed out that most, if not all, published results could be explained by a model of internal coincidence as well. This model assumes that changes in the photoperiodic conditions will provoke changes in the phase relationship between two or more circadian oscillations within the organism and that under some photoperiods these oscillations will be so phased that internal coincidence will occur: ". . . only under some photoperiods would critical phase-points of two (internal) oscillations coincide, thus closing a switch and effecting photoperiodic induction" (Pittendrigh, 1972).

This model of internal coincidence seems feasible, since it is now well established that more than one circadian oscillation exists in an individual organism (e.g., Pittendrigh, 1960, 1974; Aschoff, 1965b; Menaker, 1974) (see Section II,E), and that the phase relationship between different oscillations is not constant but subject to considerable variations (Pittendrigh, 1960; Aschoff, 1967c). Moreover, some recent findings in birds suggest that seasonal changes in the functional state may, in fact, be the result of changes in the phase relationship between different circadian hormonal rhythms (Meier and Martin, 1971; Meier et al., 1971; Dusseau and Meier, 1971; Martin and Meier, 1973).

The model of internal coincidence implies that a photoperiodic response is not necessarily dependent on light, but only on the proper internal state of the multioscillatory circadian system. Therefore, photoperiodic responses might be realizable by entraining the organism (i.e., effecting the phase relationship between the various oscillators) with Zeitgebers other than light. Recent experiments with an insect in which photoperiodic responses could be induced by particular temperature cycles are consistent with the model of internal coincidence (Saunders, 1973). Evidence for birds is not yet available.

ACKNOWLEDGMENTS

Previously unpublished results on Zonotrichia were supported by a grant from the National Science Foundation (GB-5969 X) to Dr. D. S. Farner. The manuscript has greatly benefited from the valuable suggestions of Professor J. Aschoff, Dr. S. Daan, Dr. J. Enright, Dr. K. Hoffmann, and Dr. H. Pohl.

REFERENCES

Aschoff, J. (1952). Frequenzänderungen der Aktivitätsperiodik bei Mäusen im Dauerlicht und Dauerdunkel. Pfluegers Arch. Gesamte Physiol. Menschen Tiere 255, 197–203.

Aschoff, J. (1954) Zeitgeber der tierischen Tagesperiodik. Naturwissenschaften 41, 49–56.

Aschoff, J. (1955). Jahresperiodik der Fortpflanzung bei Warmblütern. Stud. Gen. 8, 742–776.

Aschoff, J. (1958). Tierische Periodik unter dem Einfluss von Zeitgebern. Z. Tierpsychol. 15, 1–30.

Aschoff, J. (1959a). Zeitliche Strukturen biologischer Vorgänge. Nova Acta Leopold. 21, 147–177.

Aschoff, J. (1959b). Periodik licht- und dunkelaktiver Tiere unter konstanten Umgebungsbedingungen. Pfluegers Arch. Gesamte Physiol. Menschen Tiere 270, 9.

Aschoff, J. (1960). Exogenous and endogenous components in circadian rhythms. Cold Spring Harbor Symp. Quant. Biol. 25, 11–28.

Aschoff, J. (1962). Spontane lokomotorische Aktivität. In "Handbuch der Zoologie" (J.-G. Helmcke, H. von Lengerken, and D. Starck, eds.), Vol. 8, pp. 1–76. de Gruyter, Berlin.

Aschoff, J. (1963a). Gesetzmässigkeiten der biologischen Tagesperiodik. Deut. Med. Wochenschr. 88, 1930–1937.

Aschoff, J. (1963b). Comparative physiology: Diurnal rhythms. Annu. Rev. Physiol. 25, 581–600.

Aschoff, J. (1964a). Survival value of diurnal rhythms. Symp. Zool. Soc. (London) 13, 79–98.

Aschoff, J. (1964b). Die Tagesperiodik licht- und dunkelaktiver Tiere. Rev. Suisse Zool. 71, 528–558.

Aschoff, J., ed. (1965a). "Circadian Clocks." North-Holland Publ., Amsterdam.

Aschoff, J. (1965b). Significance of circadian rhythms for space flight. Proc. Int. Symp. Bioastronaut. Explor. Space, 3rd, 1964 pp. 262–276.

Aschoff, J. (1965c). Response curves in circadian periodicity. In "Circadian Clocks" (J. Aschoff, ed.), pp. 95–111. North-Holland Publ., Amsterdam.

Aschoff, J. (1965d). The phase-angle difference in circadian periodicity. In "Circadian Clocks" (J. Aschoff, ed.), pp. 262–276. North-Holland Publ., Amsterdam.

Aschoff, J. (1966a). Circadian activity pattern with two peaks. Ecology 47, 657–662.

Aschoff, J. (1966b). Circadian activity rhythms in Chaffinches (Fringilla coelebs) under constant conditions. Jap. J. Physiol. 16, 363–370.

Aschoff, J. (1967a). Circadian rhythms in birds. Proc. Int. Ornithol. Congr. 14th, 1966 pp. 81–105.

Aschoff, J. (1967b). Adaptive cycles: Their significance for defining environmental hazards. Int. J. Biometeorol. 11, 255–278.

Aschoff, J. (1967c). Phasenbeziehungen zwischen den circadianen Perioden der Aktivität und der Kerntemperatur beim Menschen. Pfluegers Arch. Gesamte Physiol. Menschen Tiere 295, 173–183.

Aschoff, J. (1969). Phasenlage der Tagesperiodik in Abhängigkeit von Jahreszeit und Breitengrad. Oecologia 3, 125–165.

Aschoff, J. (1970a). Desynchronization and resynchronization of human circadian rhythms. Aerosp. Med. 40, 844–849.

Aschoff, J. (1970b). Circadian rhythm of activity and of body temperature. In "Physiological and Behavioral Temperature Regulation" (J. D. Hardy, A. P. Gagge, and J. A. Stolwijk, eds.), pp. 905–919. Thomas, Springfield, Illinois.

Aschoff, J., and Meyer-Lohmann, J. (1954). Angeborene 24-Stunden-Periodik beim Kücken. Pfluegers Arch. Gesamte Physiol. Menschen Tiere 260, 170–176.

Aschoff, J., and Pohl, H. (1970a). Rhythmic variations in energy metabolism. Fed. Proc., Amer. Soc. Exp. Biol. 29, 1541–1952.

Aschoff, J., and Pohl, H. (1970b). Der Ruheumsatz von Vögeln als Funktion der Tageszeit und der Körpergrösse. J. Ornithol. 111, 38–47.

Aschoff, J., and Wever, R. (1962a). Über Phasenbeziehungen Zwischen biologischer Tagesperiodik und Zeitgeberperiodik. Z. Vergl. Physiol. 46, 115–128.

Aschoff, J., and Wever, R. (1962b). Beginn und Ende der Aktivität freilebender Vögel. J. Ornithol. 103, 1–27.

Aschoff, J., and Wever, R. (1962c). Biologische Rhythmen und Regelung. In "Bad Oeynhauser Gespräche 5," pp. 1–15. Springer-Verlag, Berlin and New York.

Aschoff, J., and Wever, R. (1962d). Atkivitätsmenge und α:ρ-Verhältnis als Messgrössen der Tagesperiodik. Z. Vergl. Physiol. 46, 88–101.

Aschoff, J., and Wever, R. (1965). Circadian rhythms of finches in light-dark cycles with interposed twilights. Comp. Biochem. Physiol. 16, 507–514.

Aschoff, J., and Wever, R. (1966). Circadian period and phase-angle difference in Chaffinches (Fringilla coelebs L.). Comp. Biochem. Physiol. 18, 397–404.

Aschoff, J., Diehl, I., Gerecke, U., and Wever, R., (1962). Aktivitätsperiodik von Buchfinken (Fringilla coelebs) unter konstanten Bedingungen. Z. Vergl. Physiol. 45, 606–617.

Aschoff, J., Klotter, K. and Wever, R., (1965). Circadian vocabulary. In "Circadian Clocks" (J. Aschoff, ed.), pp. x–xiv. North-Holland Publ., Amsterdam.

Aschoff, J., Gerecke, U., and Wever, R. (1967). Desynchronization of human circadian rhythms. Jap. J. Physiol. 17, 450–457.

Aschoff, J., von St. Paul, U., and Wever, R. (1968). Circadiane Periodik von Finkenvögeln unter dem Einfluß eines selbstgewählten Licht-Dunkel-Wechsels. Z. Vergl. Physiol. 58, 304–321.

Aschoff, J., Gwinner, E., Kureck, A., and Müller, K. (1970). Diel rhythms of Chaffinches Fringilla coelebs L., tree shrews Tupaia glis L. and hamsters Meso-

cricetus auratus L. as a function of season at the Arctic Circle. *Oikos, Suppl.* 13, 91–100.

Aschoff, J., Gerecke, U., Pohl, H., Rieger, P., von St. Paul, U., and Wever, R. (1971). Interdependent parameters of circadian activity rhythms in birds and man. In "Biochronometry" (M. Menaker, ed.), pp. 3–27. Nat. Acad. Sci., Washington, D.C.

Beck, S. D. (1968). "Insect Photoperiodism. "Academic Press, New York.

Benoit, J. (1970). Discussion to F. Halberg: Body temperature, circadian rhythms and the eye. In "La photorégulation de la reproduction chez les oiseaux et les mammifères" (J. Benoit and I. Assenmacher, eds.), pp. 520–528. Paris.

Benoit, J., and Assenmacher, I., eds. (1970). "La photorégulation de la reproduction chez les oiseaux et les mammifères." CNRS, Paris.

Benoit, J., Assenmacher, I., and Brard, E. (1955). Evolution testiculaire du Canard domestique maintenu à l'obscurité totale pendant une longue durée. *C. R. Acad. Sci.* 24, 251–253.

Benoit, J., Assenmacher, I., and Brard, E. (1956). Etude de l'évolution testiculaire du Canard domestique soumis trés jeune à un éclairement artificiel permanent pendant deux ans. *C. R. Acad. Sci.* 242, 3113–3115.

Benoit, J., Assenmacher, I., and Brard, E. (1959). Action d'un éclairement permanent prolongé sur l'évolution testiculaire du canard Pékin. *Arch. Anat. Microsc. Morphol. Exp.* 48, 5–12.

Berthold, P. (1973). Relationships between migratory restlessness and migration distances in six *Sylvia* species. *Ibis* 115, 594–599.

Berthold, P. (1974). Circannuale Periodik bei Grasmücken (*Sylvia*). III. Periodik der Mauser, der Nachtunruhe und des Körpergewichtes, bei mediterranen Arten mit unterschiedlichem Zugverhalten. *J. Ornithol.* 115, 251–272.

Berthold, P., Gwinner, E., and Klein, H. (1970). Vergleichende Untersuchung der Jugendentwicklung eines ausgeprägten Zugvogels, *Sylvia borin*, und eines weniger ausgeprägten Zugvogels, *S. atricapilla. Vogelwarte* 25, 297–331.

Berthold, P., Gwinner, E., and Klein, H. (1971a). Circannuale Periodik bei Grasmücken (*Sylvia*). *Experientia* 27, 399.

Berthold, P., Gwinner, E. Klein, H. and Westrich, P. (1972a). Beziehungen zwischen Zugunruhe und Zugablauf bei Garten-und Mönchsgrasmücke (*Sylvia borin* und *S. atricapilla*). *Z. Tierpsychol.* 30, 26–35.

Berthold, P., Gwinner, E. and Klein, H. (1972b). Circannuale Periodik bei Grasmücken. I. Periodik der Mauser, des Körpergewichtes und der Nachtunruhe bei *Sylvia atricapilla* und *S. borin* unter verschiedenen konstanten Bedingungen. *J. Ornithol.* 113, 170–190.

Berthold, P., Gwinner, E., and Klein, H. (1972c). Circannuale Periodik bei Grasmücken. II. Periodik der Gonadengrösse bei *Sylvia atricapilla* und *S. borin* unter verschiedenen konstanten Bedingungen. *J. Ornithol.* 113, 407–417.

Binkley, S., Kluth, E., and Menaker, M. (1971). Pineal function in Sparrows: Circadian rhythms and body temperature. *Science* 174, 311–314.

Blume, D. (1963). Die Jahresperiodik von Aktivitätsbeginn und -ende bei einigen Spechtarten. I. Teil. *Vogelwelt* 84, 161–184.

Blume, D. (1964). Die Jahresperiodik von Aktivitätsbeginn und -ende bei einigen Spechtarten. II. Teil. *Vogelwelt* 85, 11–18.

Blume, D. (1965). Ergänzende Mitteilungen zu Aktivitätsbeginn und -ende bei einigen Spechtarten unter besonderer Berücksichtigung des Grauspechtes (*Picus canus*). *Vogelwelt* 86, 33–42.

Boecker, M. (1967). Vergleichende Untersuchungen zur Nahrung- und Nistökologie der Flußseeschwalbe (*Sterna hirundo* L.) und der Küstenseeschwalbe (*Sterna paradisaea* Pont.). *Bonn. Zool. Beitr.* 18, 15–126.

Brehm, C. L. (1828). Der Zug der Vögel. *Isis* 21, 912–922.

Brown, F. A., Jr., Hastings, J. W., and Palmer, J. D. (1970). "The Biological Clock: Two Views." Academic Press, New York.

Brown, J. L., and Mewaldt, L. R. (1968). Behavior of sparrows of the genus *Zonotrichia* in orientation cages during the lunar cycle. *Z. Tierpsychol.* 25, 668–700.

Bruce, V. G. (1960). Environmental entrainment of circadian rhythms. *Cold Spring Harbor Symp. Quant. Biol.* 25, 29–48.

Bünning, E. (1936). Die endonome Tagesperiodik als Grundlage der photoperiodischen Reaktion. *Ber. Deut. Bt. Ges.* 54, 590–608.

Bünning, E. (1960). Circadian rhythms and the time measurement in photoperiodism. *Cold Spring Harbor Symp. Quant. Biol.* 25, 249–256.

Bünning, E. (1973). "The Physiological Clock." 3rd ed. Springer-Verlag, Berlin and New York.

Bünning, E., and Joerrens, G. (1960). Tagesperiodische antagonistische Schwankungen der Blauviolett- und Gelbrot-Empfindlichkeit als Grundlage der photperiodischen Diapause-Induktion bei *Pieris brassicae*. *Z. Naturforsch. B* 15, 205–213.

Bünning, E., and Joerrens, G. (1962). Versuche über den Zeitmessvorgang bei der photoperiodischen Diapause-Induktion von *Pieris brassicae*. *Z. Naturforsch. B* 17, 57–61.

Bünning, E., and Müller, D. (1961). Wie messen Organismen lunare Zyklen? *Z. Naturforsch. B* 16, 391–395.

Bünsow, R. C. (1960). The circadian rhythm of photoperiodic responsiveness in *Kalanchoe*. *Cold Spring Harbor Symp. Quant. Biol.* 25, 257–260.

Burger, J. W. (1947). On the relation of day length to the phases of testicular involution and inactivity of the spermatogenetic cycle of the Starling. *J. Exp. Zool.* 105, 259–268.

Chapin, J. P., and Wing, W. (1959). The wideawake calendar 1953 to 1958. *Auk* 76, 153–158.

Chovnick, A., ed. (1960). *Cold Spring Harbor Symp. Quant. Biol.* 25, 1–524.

Cloudsley-Thompson, J. L. (1961). "Rhythmic Activity in Animal Physiology and Behaviour." Academic Press, New York.

Conroy, R. T. W. L., and Mills, J. N. (1970). "Human Circadian Rhythms." Churchill, London.

Daan, S. (1972). Säsongvariationen i den circadiana aktivitatsrytmen hos fyra finkarter vid polcirkeln. *Fauna Flora* 67, 211–214.

Daan, S., and Aschoff, J. (1975). Effects of season and latitude on circadian rhythms of locomotor activity of captive birds and mammals. (In preparation.)

Damste, P. H. (1947). Experimental modification of the sexual cycle of the Greenfinch. *J. Exp. Biol.* 24, 20–35.

Dancker, P. (1964). Mehrjährige Beobachtungen zur Tages- und Jahresperiodik von Finkenvögeln. Dissertation, Nat. Freiburg.

de Mairan (1729). Observation botanique. *Hist. Acad. Roy. Sci.* (Paris) p. 35.

Dorka, V. (1966). Das jahres- und tageszeitliche Zugmuster von Kurz- und Langstreckenziehern nach Beobachtungen auf den Alpenpässen Cou/Bretolet (Wallis). *Ornithol. Beob.* 63, 165–223.

Dusseau, J. W., and Meier, A. H. (1971). Diurnal and seasonal variations of plasma adrenal steroid hormone in the White-throated Sparrow *Zonotrichia albicollis*. *Gen. Comp. Endocrinol.* 16, 399–408.

Ehrström, C. (1954). Fåglarnas uppträdande under solförmörkelsen den 30. juni 1954. *Vår Fågelvärld* 15, 1–28.

Enright, J. T. (1965). Synchronization and ranges of entrainment. In "Circadian Clocks" (J. Aschoff, ed.), pp. 112–124. North-Holland Publ., Amsterdam.

Enright, J. T. (1966a). Temperature and the free-running circadian rhythm of the House Finch. *Comp. Biochem. Physiol.* 18, 463–475.

Enright, J. T. (1966b). Influences of seasonal factors on the activity onset of the House Finch. *Ecology* 47, 662–666.

Enright, J. T. (1970). Ecological aspects of endogenous rhythmicity. *Annu. Rev. Ecol. Syst.* 1, 221–238.

Erkert, H. (1969). Die Bedeutung des Lichtsinnes für Aktivität und Raumorientierung der Schleireule (*Tyto alba guttata* Brehm). *Z. Vergl. Physiol.* 64, 37–70.

Eskin, A. (1969). The sparrow clock: Behavior of the free-running rhythm and entrainment analysis. Ph.D. Thesis, University of Texas, Austin.

Eskin, A. (1971). Some properties of the system controlling the circadian activity-rhythm of sparrows. In "Biochronometry" (M. Menaker, ed.), pp. 55–78. Nat. Acad. Sci., Washington, D.C.

Farner, D. S. (1961). Comparative physiology: Photoperiodicity. *Ann. Rev. Physiol.* 23, 71–96.

Farner, D. S. (1965). Circadian systems in the photoperiodic responses of vertebrates. In "Circadian Clocks" (J. Aschoff, ed.), pp. 355–369. North-Holland Publ., Amsterdam.

Farner, D. S. and Follett, B. K. (1966). Light and other environmental factors affecting avian reproduction. *J. Anim. Sci.* 25, Suppl., 90–118.

Farner, D. S., and Lewis, R. A. (1971). Photoperiodism and reproductive cycles in birds. *Photophysiology* 6, 325–370.

Follett, B. K., and Sharp, P. J. (1969). Circadian rhythmicity in photoperiodically induced gonadotrophin release and gonadal growth in the Quail. *Nature (London)* 223, 968–997.

Gaston, S. (1971). The influence of the pineal gland on the circadian activity rhythm in birds. In "Biochronometry" (M. Menaker, ed.), pp. 541–548. Nat. Acad. Sci., Washington, D.C.

Gaston, S., and Menaker, M. (1968). Pineal function: The biological clock in the Sparrow? *Science* 160, 1125–1127.

Geissbühler, W. (1954). Beiträge zur Biologie des Zilpzalps, *Phylloscopus collybita. Ornithol. Beob.* 51, 71–99.

Goodwin, B. (1963). "Temporal Organization in Cells." Academic Press, New York.

Goss, R. J. (1969a). Photoperiodic control of antler cycles in deer. I. Phase shift and frequency changes. *J. Exp. Zool.* 170, 311–324.

Goss, R. J. (1969b). Photoperiodic control of antler cycles in deer. II. Alterations in amplitude. *J. Exp. Zool.* 171, 223–234.

Gwinner, E. (1966a). Entrainment of a circadian rhythm in birds by species-specific song cycles (Aves, Fringillidae: *Carduelis spinus, Serinus serinus*). *Experientia* 22, 765.

Gwinner, E. (1966b). Tagesperiodische Schwankungen der Vorzugshelligkeit bei Vögeln. *Z. Vergl. Physiol.* 52, 370–379.

Gwinner, E. (1967a). Circannuale Periodik der Mauser und der Zugunruhe bei einem Vogel. *Naturwissenschaften* 54, 447.

Gwinner, E. (1967b). Wirkung des Mondlichtes auf die Nachtaktivität von Zugvögeln.– Lotsenversuche an Rotkehlchen (*Erithacus rubecula*) und an Gartenrotschwänzen (*Phoenicurus phoenicurus*). *Experientia* 23, 227.

Gwinner, E. (1968a). Circannuale Periodik als Grundlage des jahreszeitlichen Funktionswandels bei Zugvögeln. Untersuchungen am Fitis (*Phylloscopus trochilus*) und am Waldlaubsänger (*Ph. sibilatrix*). *J. Ornithol.* 109, 70–95.

Gwinner, E. (1968b). Artspezifische Muster der Zugunruhe bei Laubsängern und ihre mögliche Bedeutung für die Beendigung des Zuges im Winterquartier. Z. Tierpsychol. 25, 843–853.

Gwinner, E. (1969). Untersuchungen zur Jahresperiodik von Laubsängern: Die Entwicklung des Gefieders, des Gewichts und der Zugunruhe bei Jungvögeln der Arten Phylloscopus bonelli, Ph. sibilatrix, Ph. trochilus und Ph. collybita. J. Ornithol. 110, 1–21.

Gwinner, E. (1971). A comparative study of circannual rhythms in Warblers. In "Biochronometry" (M. Menaker, ed.), pp. 405–427. Nat. Acad. Sci., Washington, D.C.

Gwinner, E. (1972a). Endogenous timing factors in bird migration. In "Animal Orientation and Navigation." (S. R. Gallet et al., eds.). pp. 321–338. NASA, Washington, D.C.

Gwinner, E. (1972b). Adaptive functions of circannual rhythms in Warblers. Proc. Int. Ornithol. Congr., 15th, 1970, pp. 218–236.

Gwinner, E. (1973). Circannual rhythms in birds: Their interaction with circadian rhythms and environmental photoperiod. J. Reprod. Fert. 19, Suppl., 51–65.

Gwinner, E. (1974a). Testosterone induces, "splitting" of circadian locomotor activity rhythms in birds. Science 185, 72–74.

Gwinner, E. (1974b). Adaptive significance of circannual rhythms in birds. In "Physiological Adaptation to the Environment" (F. Vernberg, ed.). Intext, New York.

Gwinner, E., and Turek, F. (1971). Effects of season on circadian activity rhythms of the Starling. Naturwissenschaften 58, 627–628.

Gwinner, E., Berthold, P. and Klein, H. (1971). Untersuchungen zur Jahresperiodik von Laubsängern. II. Einfluss der Tageslichtdauer auf die Entwicklung des Gefieders, des Gewichts und der Zugunruhe bei Phylloscopus trochilus und Ph. collybita. J. Ornithol. 112, 253–265.

Haase, E. (1973). Zur Kontrolle von Fortpflanzungszyklen bei Vögeln. Untersuchungen am Bergfinken. J. Comp. Physiol. 84, 375–431.

Hague, E. B., ed. (1964). Photo-neuro-endocrine effects in circadian systems, with particular reference to the eye. Ann. N.Y. Acad. Sci. 117, Art. 1.

Halberg, F. (1959). Physiological 24-hour periodicity: General and procedural considerations with reference to the adrenal cycle. Z. Vitam.-, Hormon- Fermentforsch. 10, 225–296.

Hamner, K. C. (1960). Photoperiodism and circadian rhythms. Cold Spring Harbor Symp. Quant. Biol. 25, 269–277.

Hamner, K. C., and Takimoto, A. (1964). Circadian rhythms and plant photoperiodism. Amer. Natur. 48, 295–322.

Hamner, W. M. (1963). Diurnal rhythm and photoperiodism in testicular recrudescence of the House Finch. Science 142, 1294–1295.

Hamner, W. M. (1964). Circadian control of photoperiodism in the House Finch demonstrated by interrupted-night experiments. Nature (London) 203, 1400–1401.

Hamner, W. M. (1965). Avian photoperiodic response-rhythms: evidence and inference. In "Circadian Clocks" (J. Aschoff, ed.), pp. 379–388. North-Holland Publ., Amsterdam.

Hamner, W. M. (1966). Photoperiodic control of the annual testicular cycle in the House Finch, Carpodacus mexicanus. Gen. Comp. Endocrinol. 7, 224–233.

Hamner, W. M. (1968). The photorefractory period of the House Finch. Ecology 49, 211–227.

Hamner, W. M. (1971). On seeking an alternative to the endogenous reproductive rhythm hypothesis in birds. *In* "Biochronometry" (M. Menaker, ed.), pp. 448–462. Nat. Acad. Sci., Washington, D.C.

Hamner, W. M., and Enright, J. T. (1967). Relationship between photoperiodism and circadian rhythms of activity in the House Finch. *J. Exp. Biol.* **46**, 43–61.

Harker, J. E. (1964). "The Physiology of Diurnal Rhythms." Cambridge Univ. Press, London and New York.

Heller, H. C., and Poulson, T. L. (1969). Circannian rhythms. II. Endogenous and exogenous factors controlling reproduction and hibernation in Chipmunks (*Eutamias*) and Ground Squirrels (*Spermophilus*). *Comp. Biochem. Physiol.* **33**, 357–383.

Heppner, F., and Farner, D. S. (1971). Training White-crowned Sparrows, *Zonotrichia leucophrys gambelii*, in self-selection of photoperiod. *Z. Tierpsychol.* **28**, 62–68.

Hillman, W. S. (1964). Endogenous circadian rhythms and the response of *Lemna perpusilla* to skeleton photoperiods. *Amer. Natur.* **48**, 323–328.

Hjorth, J. (1968). Significance of light in the initiation of morning display of the Black Grouse (*Lyrurus tetrix* L.). *Vitrevy* **5**, 39–94.

Hoffmann, K. (1960). Versuche zur Analyse der Tagesperiodik. I. Der Einfluss der Lichtintensität. *Z. Vergl. Physiol.* **43**, 544–566.

Hoffmann, K. (1963). Zur Beziehung zwischen Phasenlage und Spontanfrequenz bei der engogenen Tagesperiodik. *Z. Naturforsch.* **186**, 154–157.

Hoffmann, K. (1965). Overt circadian frequencies and circadian rule. *In* "Circadian Clocks" (J. Aschoff, ed.), pp. 87–94. North-Holland Publ., Amsterdam.

Hoffmann, K. (1967). Kritik des Erlinger Modells. *Nachr. Akad. Wiss. Goettingen, Math. Physi. Kla.*, **10**, 132–133.

Hoffmann, K. (1969a). Zum Einfluss der Zeitgeberstärke auf die Phasenlage der synchronisierten circadianen Periodik. *Z. Vergl. Physiol.* **62**, 93–110.

Hoffmann, K. (1969b). Die relative Wirksamkeit von Zeitgebern. *Oecologia* **3**, 184–206.

Hoffmann, K. (1969c). Circadiane Periodik bei Tupajas (*Tupaia glis*) in konstanten Bedingungen. *Verh. Deut. Zool. Ges.* **63**, 265–274.

Hoffmann, K. (1970). Zur Synchronisation biologischer Rhythmen. *Verh. Deut. Zool. Ges.* **64**, 266–273.

Hoffmann, K. (1971). Splitting of the circadian rhythm as a function of light intensity. *In* "Biochronometry" (M. Menaker, ed.), pp. 134–146. Nat. Acad. Sci., Washington, D.C.

Kalmus, H., ed. (1966)." Regulation and Control in Living Systems." Wiley, New York.

King, J. (1968). Cycles of fat deposition and molt in White-crowned Sparrows in constant environmental conditions. *Comp. Biochem. Physiol.* **24**, 827–837.

Klotter, K. (1960). General properties of oscillating systems. *Cold Spring Harbor Symp. Quant. Biol.* **25**, 185–187.

Kullenberg, B. (1955). Biological observations during the solar eclipse in southern Sweden (Province Öland). *Oikos* **6**, 51–60.

Leopold, A., and Eynon, A. E.(1961). Avian daybreak and evening song in relation to time and light intensity. *Condor* **63**, 269–293.

Lofts, B. (1962). Photoperiod and the refractory period of reproduction in an equatorial bird, *Quelea quelea*. *Ibis* **104**, 407–414.

Lofts, B. (1964). Evidence of an autonomous reproductive rhythm in an equatorial bird (*Quelea quelea*). *Nature* (*London*) **201**, 523–524.

Lofts, B. (1970). "Animal Photoperiodism," Stud. Biol. No. 25. Arnold, London.

Lohmann, M. (1967). Ranges of circadian period length. *Experientia* **23**, 788.

Lohmann, M., and Enright, J. T. (1967). The influence of mechanical noise on the activity rhythm of Finches. *Comp. Biochem. Physiol.* **22**, 289–296.

McMillan, J. P. (1972). Pinealectomy abolishes the circadian rhythm of migratory restlessness. *J. Comp. Physiol.* **79**, 105–112.

McMillan, J. P., Gauthreaux, S. A., and Helms, C. W. (1970). Spring migratory restlessness in caged birds. A circadian rhythm. *BioScience* **20**, 1259–1260.

Martin, D. D., and Meier, A. H. (1973). Temporal synergism of corticosterone and prolactin in regulating orientation in the migratory White-throated Sparrow (*Zonotrichia albicollis*). *Condor* **75**, 369–375.

Meier, A. H., and Martin, D. D. (1971). Temporal synergism of corticosterone and prolactin controlling fat storage in the White-throated Sparrow (*Zonotrichia albicollis*). *Gen. Comp. Endocrinol.* **17**, 311–318.

Meier, A. H., Martin, D. D., and MacGregor, R. (1971). Temporal synergism of corticosterone and prolactin controlling gonadal growth in Sparrows. *Science* **173**, 1240–1242.

Melchers, G. (1958). Die Beteiligung der endonomen Tagesrhythmik am Zustandekommen der photoperiodischen Reaktion der Kurztagpflanze *Kalanchoe blossfeldiana*. *Z. Naturforsch. B* **11**, 544–548.

Menaker, M. (1961). The freerunning period of the bat clock: Seasonal variations at low body temperatures. *J. Cell. Comp. Physiol.* **57**, 81–86.

Menaker, M. (1965). Circadian rhythms and photoperiodism in *Passer domesticus*. *In* "Circadian Clocks" (J. Aschoff, ed.), pp. 385–395. North-Holland Publ., Amsterdam.

Menaker, M. (1968a). Light perception by extra-retinal receptors in the brain of the sparrow. *Proc., 76th Annu. Conv. APA, 1968* pp. 299–300.

Menaker, M. (1968b). Extraretinal light perception in the sparrow. I. Entrainment of the biological clock. *Proc. Nat. Acad. Sci. U.S.* **59**, 414–421.

Menaker, M. (1968c). Biological clocks. *In* "Adaptation of Domestic Animals" (E. S. E. Hafez, ed.), pp. 141–152. Lea and Febinger, Philadelphia.

Menaker, M., ed. (1971a). "Biochronometry." Nat. Acad. Sci. Washington, D.C.

Menaker, M. (1971b). Synchronization with the photic environment via extraretinal receptors in the avian brain. *In* "Biochronometry" (M. Menaker, ed.), pp. 315–332. Nat. Acad. Sci., Washington, D.C.

Menaker, M. (1974). Aspects of the physiology of circadian rhythmicity in the vertebrate central nervous system. *In* "The Neurosciences Third Study Program" (F. O. Schmitt, ed.). MIT Press, Cambridge, Massachusetts.

Menaker, M., and Eskin, A. (1966). Entrainment of circadian rhythms by sound in *Passer domesticus*. *Science* **154**, 1579–1481.

Menaker, M., and Eskin, A. (1967). Circadian clock in photoperiodic time measurement. A test of the Bünning hypothesis. *Science* **157**, 1182–1185.

Merkel, F. W. (1963). Long-term effects of constant photoperiods on European Robins and Whitethroats. *Proc. Int. Ornithol. Congr., 13th, 1962* pp. 950–959.

Minis, D. H. (1965). Parallel peculiarities in the entrainment of a circadian rhythm and photoperiodic induction in the pink boll worm (*Pectinophora gossypiella*). *In* "Circadian Clocks" (J. Aschoff, ed.), pp. 333–343. North-Holland Publ., Amsterdam.

Miyazaki, H. (1934). On the relation of the daily period to the sexual maturity and to the moulting of *Zosterops palpebrosa japonica*. *Sci. Rep. Tohoku Imp. Univ. Ser. 4* **9**, 183–203.

Mrosovsky, N. (1970). Mechanisms of hibernation cycles in ground squirrels: Circannian rhythm or sequence of stages. *Pa. Acad. Sci.* 44, 172–175.

Murton, R. K., Lofts, B., and Westwood, N. (1970a). The circadian basis of photoperiodically controlled spermatogenesis in the Greenfinch *Carduelis chloris*. *J. Zool.* 161, 125–136.

Murton, R. K., Lofts, B. and Westwood, N. J. (1970b). Manipulation of photorefractoriness in House Sparrows by circadian light regimes. *J. Gen. Comp. Endocrinol.* 14, 107–113.

Ottow, B. (1912). Das Verhalten der Vögel während der Sonnenfinsternis vom 17. April 1912. *Ornithol. Monatsber.* 20, 97.

Palmer, J. D. (1964). Comparative studies in avian persistent rhythms: Spontaneous change in period length. *Comp. Biochem. Physiol.* 12, 273–282.

Palmer, J. D., and Dowse, H. B. (1969). Preliminary findings on the effect of D₂O on the period of circadian activity rhythms. *Biol. Bull.* 137, 388.

Pavlidis, T. (1969a). Populations of interacting oscillators and circadian rhythms. *J. Theor. Biol.* 22, 418–436.

Pavlidis, T. (1969b). An explanation for the oscillatory free-runs in circadian rhythms. *Amer. Natur.* 103, 31–42.

Pengelley, E. T., and Asmundson, S. M. (1969). Free-running periods of endogenous circannian rhythms in the Golden-mantled Ground Squirrel, *Citellus lateralis*. *Comp. Biochem. Physiol.* 30, 177–183.

Pengelley, E. T., and Asmundson, S. J. (1971). Annual biological clocks. *Sci. Amer.* 224, 72–79.

Pengelley, E. T., and Fisher, K. C. (1963). The effect of temperature and photoperiod on the yearly hibernating behavior of captive Golden-mantled Ground Squirrels (*Citellus lateralis tescorum*). *Can. J. Zool.* 41, 1103–1120.

Pengelley, E. T., and Kelly, K. H. (1966). A "circannian" rhythm in hibernating species of the genus *Citellus* with observations on their physiological evolution. *Comp. Biochem. Physiol.* 19, 603–617.

Pfeffer, W. (1911). Der Einfluss von mechanischer Hemmung und von Belastung auf die Schlafbewegungen. *Abh. Math.-Phys. Kl. Saechs. Ges. Wiss.* 32, 163–295.

Pittendrigh, C. S. (1954). On temperature independence in the clock system controlling emergence time in *Drosophila*. *Proc. Nat. Acad. Sci. U.S.* 40, 1018–1029.

Pittendrigh, C. S. (1958). Adaptation, natural selection and behavior. *In* "Behavior and Evolution" (A. Rose and G. G. Simpson, eds.), pp. 390–416. Yale Univ. Press, New Haven, Connecticut.

Pittendrigh, C. S. (1960). Circadian rhythms and the circadian organization of living systems. *Cold Spring Harbor Symp. Quant. Biol.* 25, 159–184.

Pittendrigh, C. S. (1961). On temporal organization in living systems. *Harvey Lect.* 56, 93–125.

Pittendrigh, C. S. (1965). On the mechanism of the entrainment of a circadian rhythm by light cycles. *In* "Circadian Clocks" (J. Aschoff, ed.), pp. 277–297. North-Holland Publ., Amsterdam.

Pittendrigh, C. S. (1966a). Biological clocks: The functions, ancient and modern of circadian oscillations. *In* "Science and the Sixties," Cloudcroft Symp., pp. 96–111. U.S. Air Force Office of Scientific Research.

Pittendrigh, C. S. (1966b). The circadian oscillation in *Drosophila pseudoobscura* pupae: A model for the photoperiodic clock. *Z. Pflanzenphysiol.* 54, 275–307.

Pittendrigh, C. S. (1967a). Circadian systems I. The driving oscillation and its assay in *Drosophila pseudoobscura*. *Proc. Nat. Acad. Sci. U.S.* 58, 1762–1767.

Pittendrigh, C. S. (1967b). Circadian rhythms, space research, and manned space flight. In "Life Sciences and Space Research," pp. 122–134. North-Holland Publ., Amsterdam.

Pittendrigh, C. S. (1972). Circadian surfaces and the diversity of possible roles of circadian organization in photoperiodic induction. Proc. Nat. Acad. Sci. U.S. 69, 2734–2737.

Pittendrigh, C. S. (1974). Circadian oscillations in cells and the circadian organization of multicellular systems. In "The Neurosciences Third Study Program" (F. O. Schmitt, ed.). MIT Press, Cambridge, Massachusetts.

Pittendrigh, C. S., and Bruce, V. G. (1959). Daily rhythms as coupled oscillator systems and their relation to thermoperiodism and photoperiodism. In "Photoperiodism and Related Phenomena in Plants and Animals" (R. B. Withrow, ed.), Publ. No. 55, pp. 475–505. Amer. Ass. Advan. Sci., Washington, D.C.

Pittendrigh, C. S., and Minis, D. H. (1964). The entrainment of circadian oscillations by light and their role as photoperiodic clocks. Amer. Natur. 43, 261–294.

Pittendrigh, C. S., and Minis, D. H. (1971). The photoperiodic time measurement in Pectinophora gossypiella and its relation to the circadian system in that species. In "Biochronometry" (M. Menaker, ed.), pp. 212–247. Nat. Acad. Sci., Washington, D.C.

Pohl, H. (1968a). Einfluss der Temperatur auf die freilaufende circadiane Aktivitätsperiodik bei Warmblütern. Z. Vergl. Physiol. 58, 364–380.

Pohl, H. (1968b). Wirkung der Temperatur auf die mit Licht synchronisierte Aktivitätsperiodik bei Warmblütern. Z. Vergl. Physiol. 58, 381–394.

Pohl, H. (1970). Zur Wirkung des Lichtes auf die circadiane Periodik des Stoffwechsels und der Aktivität beim Buchfinken. Z. Vergl. Physiol. 66, 141–163.

Pohl, H. (1971). Über Beziehungen zwischen circadianen Rhythmen bei Vögeln. J. Ornithol. 112, 266–278.

Pohl, H. (1972a). Seasonal change in light sensitivity in Carduelis flammea. Naturwissenschaften 59, 518.

Pohl, H. (1972b). Die Aktivitätsperiodik von zwei tagaktiven Nagern, Funambulus palmarum und Eutamias sibiricus unter Dauerlichtbedingungen. J. Comp. Physiol. 78, 60–74.

Rawson, K. S. (1960). Effects of tissue temperature on mammalian activity rhythms. Cold Spring Harbor Symp. Quant. Biol. 25, 105–113.

Remmert, H. (1965). Biologische Periodik. In "Handbuch der Biologie," (F. Gessner, ed.), pp. 335–411. Akad. Verlagsges. Frankfurt am Main.

Rensing, L., and Brunken, W., (1967). Zur Frage der Allgemeingültigkeit circadianer Gesetzmässigkeit. Biol. Zentralbl. 86, 545–565.

Richter, C. P. (1965). "Biological Clocks in Medicine and Psychiatry." Thomas, Springfield, Illinois.

Rohles, F. H., ed. (1969). "Circadian Rhythms in Nonhuman Primates." Karger, Basel.

Rowan, W. (1929). Experiments in bird migration. I. Manipulation of the reproductive cycle: Seasonal histological changes in the gonads. Proc. Boston Soc. Natur. Hist. 39, 151–208.

Sachs, J. (1857). Über das Bewegungsorgan und die periodischen Bewegungen der Blätter von Phaseolus und Oxalis. Bot. Z. 15, 814.

Saunders, D. S. (1970). Circadian clock in insect photoperiodism. Science 168, 601–603.

Saunders, D. S. (1973). Thermoperiodic control of diapause in an insect: Theory of internal coincidence. Science 181, 358–360.

Scheer, G. (1952). Beobachtungen und Untersuchungen über die Abhängigkeit des Frühgesanges der Vögel von inneren und äusseren Faktoren. *Biol. Abh.* **3,** 1–68.

Schildmacher, H. (1955). Ornithologische Beobachtungen auf Hiddensee während der Sonnenfinsternis am 30. Juni 1954. *Beitr. Vogelkunde,* **4,** 59–64.

Schmidt-Koenig, K. (1958). Experimentelle Einflussnahme auf die 24-Stunden Periodik bei Brieftauben und deren Auswirkungen unter besonderer Berücksichtigung des Heimfindevermögens. *Z. Tierpsychol.* **15,** 301–331.

Schwab, R. G. (1971). Circannian testicular periodicity in the European Starling in the absence of photoperiodic change. *In* "Biochronometry" (M. Menaker, ed.), pp. 428–445. Acad. Nat. Sci., Washington, D.C.

Schwabe, W. W. (1955). Photoperiodic cycles of length differing from 24 hours in relation to endogenous rhythms. *Physiol. Plant.* **8,** 263–278.

Semon, R. (1905). Über die Erblichkeit der Tagesperiode. *Biol. Zentralbl.* **25,** 241–252.

Smith, R. W., Brown, J. L., and Mewaldt, L. R. (1969). Annual activity patterns of caged non-migratory White-crowned Sparrows. *Wilson Bull.* **81,** 419–440.

Snyder, L. R. (1969). Circadian rhythms and sun navigation in homing pigeons. Undergraduate Thesis, Dept. of Biology, Harvard University, Cambridge, Massachusetts.

Sollberger, A. (1965). "Biological Rhythm Research." Elsevier, Amsterdam.

Stoppel, R. (1910). Über den Einfluss des Lichtes auf das Öffnen und Schliessen einiger Blüten. *Z. Bot.* **2,** 369–453.

Stoppel, R., and Kniep, H. (1911). Weitere Untersuchungen über das Öffnen und Schliessen der Blüten. *Z. Bot.* **31,** 382.

Swade, R. H., and Pittendrigh, C. S. (1967). Circadian locomotor rhythms in the Arctic. *Amer. Natur.* **101,** 431–466.

Sweeney, B. M. (1969). "Rhythmic Phenomena in Plants." Academic Press, New York.

Sweeney, B. M., Hastings, J. W. (1960). Effects of temperature upon diurnal rhythms. *Cold Spring Harbor Symp. Quant. Biol.* **25,** 87–104.

Turek, F. W. (1972). Circadian involvement in termination of the refractory period in two sparrows. *Science* **178,** 1112–1113.

von Frisch, K. (1965). "Tanzsprache und Orientierung der Bienen." Springer-Verlag, Berlin and New York.

von Holst, E. (1939). Die relative Koordination als Phänomen und als Methode zentralnervöser Funktionsanalyse. *Ergeb. Physiol., Biol. Chem. Exp. Pharmakol.* **42,** 228–306.

von St. Paul, U. and Aschoff, J. (1968). Gehirntemperaturen bei Hühnern. *Pfluegers Arch. Gesamte Physiol. Menschen Tiere* **301,** 109–123.

Wahlström, G. (1960). Self-selection of light and darkness in the canary and the effects of some drugs. *Acta Physiol. Scand.* **50,** 150–151.

Wahlström, G. (1965a). Experimental modifications of the internal clock in the canary, studied by self-selection of light and darkness. *In* "Circadian Clocks" (J. Aschoff, ed.), pp. 324–328. North-Holland Publ., Amsterdam.

Wahlström, G. (1965b). The circadian rhythm of self-selected rest and activity in the canary and the effects of barbiturates, reserpine, monoamine oxidase inhibitors and enforced dark periods. *Acta Physiol. Scand.* **65,** Suppl. 250. 1–67.

West, G. C., and Pohl, H. (1973). Effect of light-dark cycles with different LD time-ratios and different LD intensity-ratios on the activity rhythm of Chaffinches, *J. Comp. Physiol.* **83,** 289–302.

Wever, R. (1964). Zum Mechanismus der biologischen 24-Std.-Periodik. III. Mitteilung. Anwendung der Modell-Gleichung. *Kybernetik* **2**, 127–144.

Wever, R. (1965). A mathematical model for circadian rhythms. *In* "Circadian Clocks" (J. Aschoff, ed.), pp. 47–63. North-Holland Publ., Amsterdam.

Wever, R. (1967a). Ein mathematisches Modell für die circadiane Periodik. Z. *Angew. Math. Mech.* **46**, 148–157.

Wever, R. (1967b). Zum Einfluss der Dämmerung auf die circadiane Periodik. Z. *Vergl. Physiol.* **55**, 255–277.

Wilkins, M. B. (1965). The influence of temperature and temperature changes on biological clocks. *In* "Circadian Clocks" (J. Aschoff, ed.), pp. 146–163. North-Holland Publ., Amsterdam.

Winget, C. M., Averkin, E. G., and Fryer, T. B. (1965). Quantitative measurement by telemetry of ovulation and oviposition in the fowl. *Amer. J. Physiol.* **209**, 853–858.

Withrow, R. B., ed. (1959). "Photoperiodism and Related Phenomena in Plants and Animals," Publ. No. 55. Amer. Ass. Advan. Sci., Washington, D.C.

Wolf, W., ed. (1962). Rhythmic functions in the living system. *Anna. N. Y. Acad. Sci.* **98**, Art. 4.

Wolfson, A. (1954). Production of repeated gonadal, fat, and molt cycles within one year in the Junco and White-crowned Sparrow by manipulation of day length. *J. Exp. Zool.* **125**, 353–376.

Wynne-Edwards, V. C. (1930). On the waking-time of the Nightjar (*Caprimulgus e. europaeus*). *Brit. J. Exp. Biol.* **7**, 241–247.

Zimmerman, J. L. (1966). Effects of extended tropical photoperiod and temperature on the Dickcissel. *Condor* **68**, 377–387.

Chapter 5

VOCAL BEHAVIOR IN BIRDS

Fernando Nottebohm

This chapter includes a selection of topics representing basic approaches to vocal behavior. The emphasis will be on the vocalizations themselves. Full attention to the complex behavioral context in which they occur would require more space and will not be attempted here.

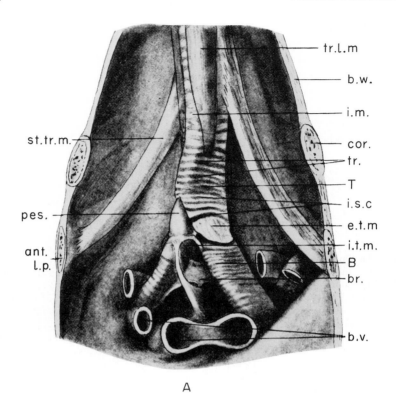

A

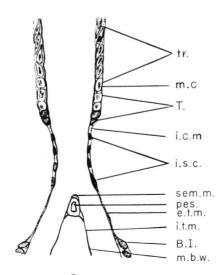

B

The text covers three relatively self-contained themes: (1) anatomy and physiology of voice production; (2) size, function, and structure of vocal repertoires; (3) vocal ontogeny in the individual and as an evolutionary problem.

I. How Vocalizations are Produced

The vocal organ of birds is the syrinx, located at the junction of trachea and bronchi. The larynx is not known to generate vocal sounds, though it may influence their amplitude and harmonic spectrum (Harris et al., 1968). Also, because of its role in regulating air flow, the larynx may affect the onset and termination of sounds.

The anatomy of the syrinx varies between orders and families of birds (e.g., Rüppell, 1933; Ames, 1971). In domestic fowl, Japanese Quail, and Mallards (Anas platyrhynchos), there is an internal tympaniform membrane on the medial wall and an external tympaniform membrane on the lateral wall of each bronchus (Fig. 1). As air is forced out of the air-sac system and into the bronchi, the internal and external tympaniform membranes are drawn into the bronchial lumen and begin to oscillate.

The relation between membrane oscillation and sound production can be demonstrated with an isolated syrinx preparation of the kind devised by Rüppell (1933). The syrinx of a domestic cock is mounted as shown in Fig. 2. The chamber containing the syrinx replicates the role of the interclavicular air sac, which surrounds the syrinx in the intact bird. Sound is produced by forcing air through the syrinx in a cephalad direction, while the air space surrounding the syrinx is maintained at a pressure above 1 atmosphere. By illuminating the tympaniform membranes with a stroboscopic flash, it can be shown that visible displacement of these membranes stops at a stroboscopic frequency identical to the fundamental frequency of the sound produced (F. Nottebohm, unpublished data). Hersch (1966) confirmed this observation with Japanese

---

FIG. 1. Syrinx of domestic fowl. (A) Ventral body wall and heart have been removed to show position of syrinx in an adult cock, ×2½. (B) Midcoronal section of the syrinx of an adult male, ×6½. *Abbreviations:* ant. l.p., anterolateral process of sternum; B, bronchial half-rings; B.I., first bronchial half-ring; br., bronchidesmus; b.v., blood vessels; b.w., body wall; cor., coracoid; e.t.m., external tympaniform membrane; i.c.m., interannular and intercartilaginous membranes; i.m., interannular membrane; i.s.c., intermediate syringeal cartilages; i.t.m., internal tympaniform membrane; m.b.w., medial bronchial wall; m.c., marrow cavity; pes., pessulus; sem.m., semilunar membrane; st. tr. m., sternotrachealis muscle; T., tympanum; tr., tracheal rings; tr. l. m., tracheolateralis muscle. (Modified from Myers, 1917.)

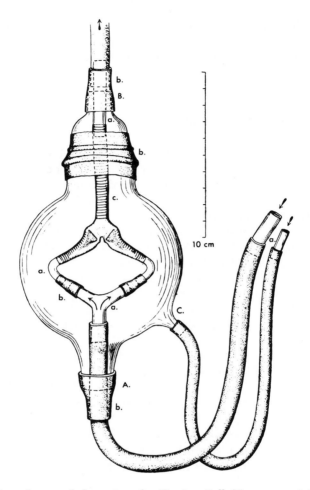

FIG. 2. Ventral view of the syrinx of a Herring Gull (*Larus argentatus*) mounted in a pressure chamber (from Rüppell, 1933). The internal tympaniform membranes are indicated by stipling on the bronchial walls. *Abbreviations:* a., glass tubing; b., rubber cuffs; c., trachea. Airflow through syrinx and trachea enters glass chamber at A., exits at B.; pressure of air volume surrounding syrinx is controlled by air inflow at C.

Quail and Mallards. He attached a blunt probe to a phonograph cartridge, then applied the probe to the internal and external tympaniform membranes. Hersch noted that all four tympaniform membranes oscillated at the same frequency, which corresponded to the fundamental frequency of the sound produced.

Gross (1964b) devoiced chickens by attaching a small piece of stainless steel mesh to each external tympaniform membrane. Gottlieb and Vandenbergh (1968) achieved the same result by covering the internal tympaniform membranes of ducklings with a coat of collodion. Membranes stiffened in this manner cannot bulge into the bronchial lumen. This restriction on bore reduction may prevent the other pair of unrestrained membranes from oscillating in a sound-generating fashion.

In many species, particularly songbirds, the external tympaniform membrane is much reduced or absent, and the internal tympaniform membrane may be the sole sound source (Setterwall, 1901; Greenewalt, 1968). The semilunar membrane, which projects into the tracheal lumen at the intersection of the two bronchi (Fig. 1), was also thought to contribute to vocal sounds in birds. However, destruction of this membrane does not result in vocal changes (Miskimen, 1951).

In most male ducks, the base of the trachea is modified into a large bony bulla, typically absent in females. This anatomical difference is correlated with sexual differences in the quality of adult sounds (Johnsgard, 1968). However, the voice of ducklings shows no sexual differences, although syringeal dimorphism is already well marked (Gottlieb and Vandenbergh, 1968). Thus, vocal differences in adult ducks of both sexes may reflect differences in central control functionally unrelated to syringeal dimorphism. Ducks produce many of their calls as they float and swim about, when their interclavicular air sac is below the water line. The syringeal bullas of males may well be an adaptation for sound propagation under water, though this possibility has not been tested. Marked sexual dimorphism has also been reported for the syrinx and vocalizations of European Coots (*Fulica atra*) (Rüppell, 1933).

Syringeal sexual dimorphism has been noticed in some songbirds. The skeletal elements and musculature of the syrinx are larger in male European Blackbirds (*Turdus merula*) and Bullfinches (*Pyrrhula pyrrhula*) than in females. The relation of such dimorphism to the nature of sounds produced remains unclear. Whereas female blackbirds are poor songsters, female bullfinches reportedly sing well in captivity (Haecker, 1900). The syringeal musculature of female Zebra Finches (*Poephila guttata*) is considerably lighter than that of males; the males of this species sing, but the females do not (Arnold, 1974).

Hérissant (1753) noted that the integrity of the interclavicular air sac is indispensable for vocalization. This was confirmed by Rüppell (1933) and his pressure chamber preparation discussed earlier. More recently, Gross (1964a) measured the air pressure inside the interclavicular air sac of vocalizing domestic fowl and found it to be slightly above 1 atmosphere. A similar increase in pressure in the interclavicular air sac, coincident with the duration of vocalization, has been described in the European Starling (*Sturnus vulgaris*) (Gaunt et al., 1973). However, if air is withdrawn at the laryngeal end of the trachea in an anesthetized House Sparrow (*Passer domesticus*) sound is produced even if the interclavicular air sac is open and the syrinx fully exposed (Miskimen, 1951). Presumably, under those conditions the pressure differential across the tympaniform membranes is sufficient to induce membrane oscillation and sound production.

Extrinsic and intrinsic syringeal muscles modulate the fundamental frequency of vocalizations. In chickens and Japanese Quail, the tracheolateralis muscle pulls the trachea cephalad, stretching the syringeal membranes, while the sternotrachealis muscle pulls the trachea caudad causing these same membranes to bulge into the bronchial lumen (Fig. 3). The higher fundamental frequencies are produced in the latter position (Hersch, 1966). The number of syringeal muscles varies among groups of birds. Oscine songbirds, with five to seven pairs of intrinsic syringeal muscles, achieve a refined control of membrane tension and bronchial bore which in turn influences the quality of their vocalizations (Miskimen, 1951).

The number of intrinsic syringeal muscles can be a misleading guide to the vocal talents of a species. As an example, the Superb Lyrebird (*Menura novaehollandiae*) is an accomplished songster and mimic, though its syrinx has only three pairs of intrinsic muscles (Thorpe, 1961).

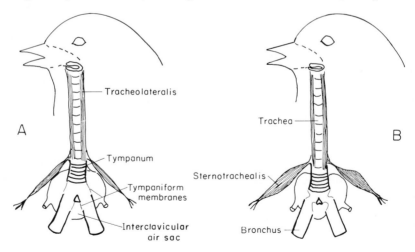

FIG. 3. Highly diagrammatic representation of syringeal action. (A) The tracheolateralis muscles are contracted, while the sternotrachealis muscles are relaxed. Under this condition, the syringeal membranes are taut, and the bore from bronchi to trachea is maximally open; air flow through the syrinx generates low-frequency sounds or no sound. (B) Tracheolateralis muscle relaxed and sternotrachealis contracted. The syringeal membranes bulge inward, and the bore leading from bronchi to trachea is maximally reduced; air flow through the syrinx generates high-frequency sounds. (Modified from Hersch, 1966.)

Syringeal muscles may also play an important role in respiration. Chaffinches (*Fringilla coelebs*) have difficulty breathing after their syringeal muscles are denervated. Presumably, under these circumstances the internal tympaniform membranes are drawn into the bronchial lumen during inspiration, obstructing air flow. Whereas a frightened Chaffinch will breathe faster, following syringeal denervation the number of inspirations per minute is reduced below the resting level. If the disturbance is maintained, the bird asphyxiates (Nottebohm, 1971). Thus, syringeal anatomy, musculature, and innervation must be viewed with regard to their dual functions, both vocal and respiratory.

The extent to which the trachea and syringeal membranes interact in sound production remains controversial. Myers (1917) and Rüppell (1933) showed that shortening the tracheal length of a domestic cock causes a rise of the fundamental frequency. Hersch (1966) has sought to clarify the physical basis of this observation by having birds vocalize in different gas mixtures. Replacement of nitrogen by helium in the air breathed by a bird has an effect equivalent to that of a uniform reduction in the size of its gas-filled resonating chambers. Yet this substitution of gases leads to no change in the fundamental frequency of utterances of various songbirds and galliform species (Hersch, 1966). Therefore, the influence of tracheal length on frequency of the fundamental may reflect changes in resistance to air flow or in turbulence patterns, rather than a rigid coupling of sound source and resonating cavity. In several genera, e.g., *Grus, Aramus, Platalea,* and *Phonygammus,* the disproportionately long trachea is arranged in complex loops and coils. The influence on vocalizations of such an elongated and contorted windpipe remains unclear (Rüppell, 1933).

II. The Dual Sound Source Theory

Several authors have been aware that the syrinx of the majority of birds comprises two separate sound sources, one in each bronchus (review in Greenewalt, 1968). The first strong indication that birds can control these two acoustical sources independently was furnished by Potter *et al.* (1947). They noted that Brown Thrashers (*Toxostoma rufum*) include in their song two simultaneous notes with unrelated frequencies. Borror and Reese (1956a) described many instances of this phenomenon in songs of the Wood Thrush (*Hylocichla mustelina*). Greenewalt (1968) notes that two unrelated frequencies occur in the vocalizations of a broad variety of songbirds, as well as in representatives of more primitive groups, such as the Eared Grebe (*Podiceps nigricollis*), the American Bittern (*Botaurus lentiginosus*), and the Greater Yellowlegs (*Tringa melanoleuca*). Greenewalt (1968) and Stein (1968) attribute the production of such unrelated frequencies to the two internal tympaniform membranes.

This interpretation has received some recent experimental support. In the Chaffinch, each bronchus performs as a separate sound source with independent muscular and neural control. Each syringeal half is innervated by the tracheosyringealis branch of the ipsilateral hypoglossus nerve. Section or denervation of the muscles on one half of the syrinx distorts or suppresses some notes while the remainder is unaffected (Fig. 4). When unrelated notes show temporal overlap in the intact bird, one of the overlapping sounds disappears after unilateral denervation of the syrinx, while the other one remains unchanged (Nottebohm, 1971).

Chaffinches with a unilaterally denervated syrinx often produce poorly modulated sounds bearing little or no relation to the notes they replace. These sounds probably arise as the expiratory flow of air passes the internal tympaniform membrane, which now oscillates without muscular control. A few days after section of one hypoglossus nerve, the operated bird is able to silence the denervated half of the syrinx. Presumably, this is done by stopping or reducing expiratory flow of air through that bronchus.

When two sustained sounds show a frequency mismatch, they interact in a predictable manner. Given two fundamentals of 750 and 800 Hz, for example, sound waves will be in phase and reinforce each other

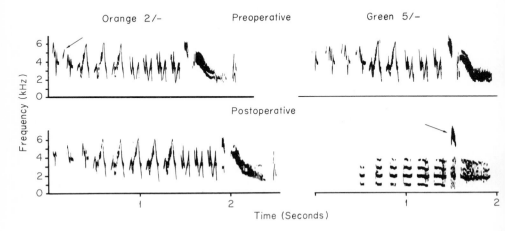

Fig. 4. Two adult male Chaffinches shared a same song theme while intact. Then Orange 2/— had its right hypoglossus cut and lost one song element, indicated by arrow. Green 5/— had its left hypoglossus cut and lost all song elements but one, also indicated by arrow; other sounds produced by this latter bird after operation can be attributed to air flow through the denervated bronchus. (Nottebohm, 1971.)

every fifteenth cycle of the longer wavelength $[750/(800 - 750) = 15]$; that is to say, with a periodicity of 20 msec $[(1 \text{ sec}/750) \times 15 = 20 \text{ msec}]$; maximum interference will also occur every 20 msec. These recurring fluctuations in sound amplitude are known as beats; in sound spectrograms, they appear as evenly spaced fluctuations in the darkness of the sound trace. As relative differences in frequency between two sounds become greater, the periodicity of beats will become shorter. When two sounds undergo unrelated changes in frequency, their interactions lack periodicity. Beats should not be confused with amplitude modulation at the sound source. When interplay of two fundamentals is not apparent in sound-spectrographic analysis, beats and source-generated amplitude modulations can be told apart by oscillographic inspection (Greenewalt, 1968). Beats are particularly well demonstrated when two birds render close copies of the same song theme simultaneously (Fig. 5.).

Most birds have one sound source in each bronchus, yet sound-spectrographic analysis of their song reveals only one sound trace at a time, with no evidence of beats. Absence of beats in this case can be explained in various ways: (1) The syrinx performs as two symmetrical halves,

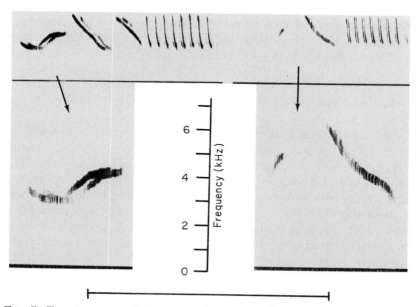

Fig. 5. Two instances of almost simultaneous rendering of a same song theme by two Rufous-collared Sparrows (*Zonotrichia capensis*) recorded in the field in Tucumán Province, Argentina. Notes in temporal overlap have been blown up to show better occurrence of beats.

so that membrane tension and expiratory air flow are identical in both bronchi. (2) Frequencies are matched as a result of muscular tuning of tympaniform membranes in both bronchi. (3) The tympaniform membranes of both bronchi are physically coupled so that they cannot drift into separate frequencies. (4) Only one bronchus at a time generates sound.

In some cases at least, syringeal symmetry is not necessary to produce fundamental frequencies free of beats (Gottlieb and Vandenbergh, 1968). Neither are beats produced by isolated syrinx preparations lacking muscular control (Hersch, 1966; Harris *et al.*, 1968). In the Chaffinch, pure tones with no beats result from the use of one sound source at a time; the other sound source is presumably kept silent by control of membrane tension or reduction of air flow through its corresponding bronchus (Nottebohm, 1971).

It has been claimed that three different and simultaneous fundamentals can be recognized in parts of the courtship song of the Gouldian Finch (*Chloebia gouldiae*) (Thorpe, 1961). Borror and Reese (1956a) have noted the temporal overlap of four unrelated notes in the song of the Wood Thrush. If confirmed, these observations reflect on yet undescribed mechanisms of sound production. The considerable anatomical and vocal differences between species of birds advise against premature attempts to produce a detailed and general theory of syringeal performance. As we shall see later, even the two halves of the syrinx may participate to a very unequal extent in vocal behavior.

III. Size and Function of Vocal Repertoire

Avian vocalizations have been traditionally divided into song and calls, though this distinction is somewhat arbitrary. The term "song" is usually reserved for loud and sustained vocalizations delivered seasonally by males in possession of a breeding or courting territory. However, many songbirds can be heard singing even while migrating toward their breeding destination. In the majority of species, females sing infrequently or not at all. There are many exceptions to this, and they will be discussed later. Though song may be triggered by a variety of social stimuli, particularly auditory ones, it is often delivered in the absence of an obvious addressee or external trigger.

In most species, song advertises the presence of a paired or unpaired male in breeding condition, as well as the existence of a defended territory. In some cases, different song functions are, to some extent, separated into different song types. The Grasshopper Sparrow (*Ammodramus savannarum*), the Black-throated Green Warbler (*Dendroica virens*), some viduine whydahs (*Vidua regia, V. fischeri,* and *V. chalybeata*), the Eurasian Pigmy-Owl (*Glaucidium passerinum*), and the Red-billed Leio-thrix (*Leiothrix lutea*) have two song types (Smith, 1959; Morse, 1967; Nicolai, 1964; König, 1968; Thielcke and Thielcke, 1970). One of these song types is given in the presence of females, or under circumstances where females seem to be the main target; the other kind of song is given by birds engaged in territorial defense. Morse (1970) studied the Black-throated Green Warbler in small islands having only one breeding pair of warblers. Under those conditions, the song thought to be given to other males is almost absent, while the one directed at females is frequent throughout the season.

Calls tend to be shorter and of simpler structure than song. Some calls are seasonal and sex typical, while others are heard round the year or are produced by both sexes. The context in which calls are delivered, the identity of the calling individual, and the response elicited from conspecifics suggest the information conveyed and the function of a call. In this manner, various calls are said to maintain contact between foraging individuals, attract individuals to food, solicit copula-tion, denote alarm, threaten, etc.

In a social context, messages refer to the probability of subsequent behavior by the signaler. This probability is influenced by the identity of the sender and receiver, as well as by other contextual circumstances. Early in the spring, the song of a male Chaffinch conveys at least two simultaneous messages—readiness to attack trespassing males and pre-disposition to mate (Marler, 1956a). Later in spring, a paired male Chaffinch will also attack trespassing females, so that the message con-veyed by a single signal may undergo seasonal changes. In this manner, the potential number of messages conveyed by a system of signals can exceed the number of the signals themselves. These considerations emphasize the need for descriptive accounts of vocal repertoires, to be followed by statistical treatment of signaling context and consequences.

Assigning communicatory significance to particular vocalizations can be difficult. Canaries paired in a breeding cage and observed until the female started laying give as many as fourteen different calls, grouped into five different categories (Mulligan and Olsen, 1969). The first three categories denote mild, medium, and high-level arousal. The re-maining two, attack and mating calls, occur in a narrower context. Thus, whereas some calls broadcast an obvious message and are good predictors of the subsequent behavior of the signaller, others are more generalized.

Table I presents the size of the adult vocal repertoire in various species. Variation in the number of vocalizations per species is not easy to interpret. Even assuming that all counts are equally complete, we have to consider the tendency of some authors to count variants of

TABLE I

SIZE OF ADULT VOCAL REPERTOIRE IN VARIOUS SPECIES

Species	Common name	Adult vocalizations[a]			Reference
		Both Sexes	Males	Females	
Fringilla coelebs	Chaffinch	12	11	5	Marler, 1956b
Parus major	Great Tit	18	16	6	Gompertz, 1961
Ploceus cucullatus	Village Weaverbird	13	12	6	Collias, 1963
Passerina cyanea	Indigo Bunting	9	8	6	Thompson and O'Hara Rice, 1970
Sylvia communis	Whitethroat	26	26	24	Sauer, 1954
Sturnella magna	Eastern Meadowlark	12	9	9	Lanyon, 1957
Agelaius phoeniceus	Red-winged Blackbird	20	13	6	Orians and Christman, 1968
Streptopelia "risoria"	Ring Dove	5	5	4	Miller and Miller, 1958
Gallus "domesticus"	Domestic Fowl	26	16	22	Baümer, 1962
Colinus virginianus	Bobwhite Quail	22	21	20	Stokes, 1967
Lophortyx californicus	California Quail	16	13	12	Williams, 1969
Tringa solitaria	Solitary Sandpiper	7	7	7	Oring, 1968
Tringa ochropus	Green Sandpiper	7	7	7	Oring, 1968

[a] Each different type of call counts as one; territorial song counts as one, too.

a call as new vocalizations, whereas other authors include such variants within the range of variability of each call type. Some vocalizations, such as those used in territorial advertisement, to solicit copulation, or to announce the presence of an aerial predator, are very stereotyped, so that on successive repetitions and in different contexts these vocalizations show little or no grading in amplitude, length, or structure. Other calls, used in agonistic interactions with conspecifics or denoting alarm or arousal, often occur as graded series and thus are presumably capable of reflecting a broad range of moods. Examples of graded call systems can be found in the Great Tit, the Whitethroat, and domestic fowl (Table I).

Table I suggests that communicatory needs differ between species and between sexes within species. Female passerines are markedly less vocal than males. This disparity is not restricted to territorial vocalizations. Males also deliver a greater variety of alarm notes. We must assume that in these species the survival of the progeny is best served by a relatively silent hen and a more vocal cock. This situation can also be described by saying that males assume a greater variety of communicatory roles. In other species, such as Bobwhite Quail and sandpipers, both male and female share in territorial defense, incubation, and parental duties. In those species, the two sexes show little or no vocal dimorphism. In domestic fowl, the greater vocal repertoire of females can be correlated with nest activities and with the complex hierarchical interactions of hens in a flock; the polygamous cocks do not partake in incubation or nest defense and under natural conditions space out territorially. As in many passerines, cocks, but not hens, give the aerial alarm call.

Communicatory needs are likely to differ between species in still other ways (Armstrong, 1963; Collias, 1963). So, for example, Village Weaverbirds (*Ploceus cucullatus*) are colonial, and males display in front of nests they build before the arrival of females. Pair formation occurs at the colony. The intense competition for females is correlated with a variety of male vocal displays. These displays attract females to a nest and then induce them to enter it. Collias (1963) notes that half of the various vocal signals used by this species are closely related to its highly specialized mating and nesting habits.

Adult Hill Mynas (*Gracula religiosa*) produce a variety of vocalizations, some typical of each individual, others similar in both sexes, and still others typical of each sex. The complexities of the vocal repertoire of this species far exceed that of others mentioned above, though its advantages remain unclear (Bertram, 1970).

A paucity of vocal signals need not necessarily mean that social interactions are rare. The Columbiformes as a group have very small vocal repertoires, yet many doves and pigeons are extremely sociable in their breeding and feeding habits (Goodwin, 1967). But even the columbiformes seem garrulous when compared to species such as the Turkey Vulture (*Cathartes aura*), in which adults emit only hoarse, hissing sounds (Miskimen, 1957). The correlations between social systems and vocal repertoires have received little attention so far, though this promises to be a most interesting area of study.

IV. The Communicatory Valence of Different Signal Characteristics

Vocalizations are complex signals, and there is no reason to suppose that each of their characteristics is equally important for the fulfillment of a communicatory function. For example, the song of male White-throated Sparrows (*Zonotrichia albicollis*) consists of pure, whistled notes of sustained frequency and no detectable harmonics. Such songs generally end with notes divided into three roughly equal parts (triplets). The frequency spectrum of song notes ranges from 2150 to 6500 Hz (Falls, 1969). Figure 6a and 6b shows two common song themes of this species. Recorded male songs elicit territorial defense from resident males in breeding condition. Reproductive females respond to male

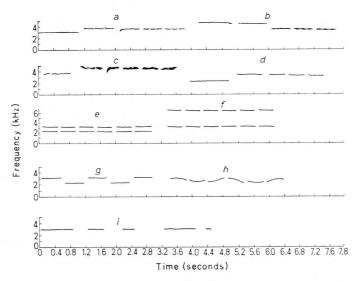

Fig. 6. Song-playback experiments with White-throated Sparrows (*Zonotrichia albicollis*). (a) and (b) Two different wild-type male White-throated Sparrow songs, (c) female song. Artificial songs = wild-type song lacking terminal triplets (d); two unrelated tones (e); with first harmonic (f); alternating pitch (g); varying pitch (h); and random timing (i) (Falls, 1969.)

song with a precopulatory trill and may even assume a posture of solicitation (Falls, 1969).

Because of its simple structure, the song of the White-throated Sparrow can be readily simulated with an audio oscillator. In this manner, Falls (1969) synthesized close replicas to natural song, while other versions departed from the wild-type pattern in various ways. The response of territorial males to playbacks was used to establish the sufficient characteristics for species recognition in this context. Normal timing, frequency range, and sustained pitch of notes were proven to be important, whereas songs lacking terminal triplets had normal effectiveness (Fig. 6d). Modifications greatly reducing the response of males were: making notes variable in pitch (Fig. 6h), adding an unrelated tone to the fundamental (Fig. 6e), adding a second harmonic (Fig. 6f), too many changes in pitch between successive notes (Fig. 6g), and overly long silent periods between notes (Fig. 6i). Territorial White-throated Sparrows respond more strongly to playbacks of alien song than to those of neighbors. This discrimination is presumably based on the fine detail of song, which therefore encodes individual identity.

In the European Robin (*Erithacus rubecula*), each song is composed of some eight different phrases. Successive phrases within a song fall above or below the average song frequency at 4 kHz. Songs composed of repetitions of a single phrase, or including only high-frequency or low-frequency phrases, elicit marginal responses from territorial males (Brémond, 1967).

The song repertoire of individual European Robins may include several hundred phrases (Brémond, 1968a). Successive songs rendered by an individual are different. Yet constant repetition of a song elicits from a male robin 88% of the responsiveness elicited by a normal sequence of different songs. Thus, the enormous repertoire of phrases available to each individual is hardly necessary for purpose of species recognition, though there is a price to be payed by habituation to the monotonous repetition of a single song (Brémond, 1967).

The European Wren (*Troglodytes troglodytes*) and the Nightingale (*Erithacus megarhynchos*) do not react to each other's song. It is possible to synthesize songs obeying the syntactic rules of wren song, yet composed of elements extracted from Nightingale song. In this case, the interspecific barrier is weakened, so that wren and Nightingale males respond aggressively to this artificial song (Brémond, 1968b). Clearly, the vocal identity of these two sympatric species is determined by the syntactic arrangement of elements within the song, and much less so by the auditory characteristics of the isolated elements.

Emlen (1972) concludes that the most important parameters eliciting species recognition by Indigo Bunting (*Passerina cyanea*) song are rhythmic cadence and the structure of individual notes, with syntax playing a minor role, if any. Thus, for the purposes of specific identification, different species attach different significance to the various parameters of natural song. In all studies reported here, males were used as a bioassay of the effectiveness of various song parameters. It remains to be seen whether species recognition by females is based on a similar weighing of song parameters.

V. The Adaptive Significance of Call and Song Structure

A. LOCATABILITY

In two classic papers, Marler (1955, 1956b) noted that whereas for some calls it is important that location or identity of the caller be conveyed, for others identity may be less important and location a disadvantage. He noted that the physical characteristics of the sound signal used could be expected to vary according to whether location was to be conveyed or not. Marler argued that, as demonstrated for humans (review in Stevens and Davis, 1938) birds locate sounds by making binaural comparisons of differences in phase, intensity, and time of arrival. The efficiency of location would vary with the frequency of the signal. Phase difference is most useful at low frequencies, because the information it provides becomes ambiguous when the wavelength is less than twice the distance between the ears. Intensity differences are most useful at high frequencies, since obstructions in the path of a sound, such as a head, only cast appreciable sound shadows when their dimensions are greater than the wavelength of the signal. Time differences should be useful throughout the auditory range, localization being enhanced by sounds with many interruptions or with marked frequency or amplitude modulations. The above principles were derived from work with pure tones and binaural hearing in humans. Natural sounds with more than one frequency can also be binaurally located by comparison of frequency spectra at both ears (review in Erulkar, 1972).

Despite this wealth of theoretical expectations (also see Chapter 7, Volume III, by J. Schwartzkopff), the ability of birds to locate sounds has received little experimental attention. Konishi (1973a,b) has recently shown that Barn Owls (*Tyto alba*) tested in total darkness can find prey by sound when one of their ears is plugged. Binaural barn owls tested under similar conditions make little use of time cues; a sustained pure tone with several interruptions is located with no greater ease than a continuous tone of the same frequency. Taking into consideration the distance between a Barn Owl's ears, Konishi (1973b) calculated that use of phase differences in the binaural arrival of sounds should improve location in the frequency range of 1–4 kHz. However, owls make large errors within this range, leading one to question their use of phase differences. Barn Owls locate sounds best within the 6–8 kHz frequency range, with improved accuracy when the cues used include a broad spectrum of frequencies rather than single pure tones. Thus, sound location in this species is aided predominantly by binaural differences in intensity, preferably measured simultaneously over a broad spectrum of frequencies. (For a discussion of the monaural ability to locate sounds, see Batteau, 1967, and Konishi, 1973b). It would be premature to conclude that Konishi's (1973a,b) findings apply to birds in general, and more information is needed on this issue. It is probably fair to argue that signals designed to facilitate location by conspecifics, such as the territorial song or calls of many birds, will cover a broad frequency spectrum; their exact acoustic characteristics will presumably be influenced by the head dimensions and frequency thresholds typical of each species and of the species that prey upon them.

Marler (1955, 1956b) indicates that there is a conflict in many animals between the need to appear conspicuous to their own species and inconspicuous to predators or prey. Small passerines have two distinct responses to a hawk or owl. If the bird of prey is perched, they make themselves conspicuous by mobbing behavior, which attracts other bird's attention to the predator; if it is in flight, they dash for cover and hide. Passerine mobbing calls are loud, have a sudden onset, and cover a broad spectrum of frequencies, and thus are easy to locate. In contrast, aerial alarm calls are usually much fainter, high-frequency pure tones with an almost imperceptible beginning and ending (Figs. 7 and 8). Species especially subject to the attack of aerial predators may have extended these ventriloquial characteristics to other social calls.

B. Sound Propagation in Different Environments

Functional constraints on structure of vocalizations do not end there. To serve its functions, territorial song must be audible at a considerable distance. This sets an upper limit to frequency, since, as it increases, the rate of attenuation with distance also rises sharply. Sound transmission is also influenced by vegetation. Short wavelengths are reflected by tree trunks, whereas longer wavelengths bend around such obstacles. In keeping with this, Ficken and Ficken (1962) report that warblers (family Parulidae) that forage in the tree tops have higher pitched songs than do related species that forage on the forest floor. Sound reflection may affect the use of vocal signals in still another way; echoes, which might confuse the reception of complex temporal patterns, would be less troublesome in the case of more sustained pure tones.

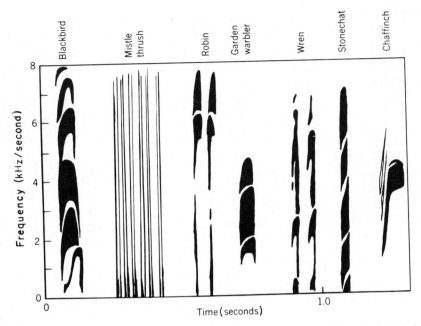

Fig. 7. Sound spectrograms of calls used by several species of British birds while mobbing an owl. They share the characteristic of being easy to locate. (After Marler, 1957.)

Morton (1970) has tested the propagation of pure tones below the canopy of a neotropical forest. He found that sound attenuation increases rapidly above 2500 Hz, while least attenuation occurs between 1500 and 2500 Hz. Long-distance communication sounds recorded from sixty-nine species of birds from the lower levels of the forest were found to be pure tonelike with an average frequency emphasis at 2200 Hz (Fig. 9). Morton also theorized that in open habitats such as grasslands, temperature- and wind-stratified environments would tend to distort frequency-coded information. The speed of sound in the warmer air near the ground is faster than in the cooler air above, causing the wavefront to deflect upward. Higher frequencies deflect upward more steeply than lower frequencies because of the shorter wavelength. Higher frequencies are scattered by turbulence more than lower frequencies, also because of the difference in wavelength. A temporal code would be unaffected by such variables. In fact, sustained pure tones are rare in the song of grassland species, while highly patterned and rhythmic frequency modulations are common. Forest canopy species and species of secondary growth and edge habitats have vocal patterns displaying characteristics of both grassland and low-level forest birds.

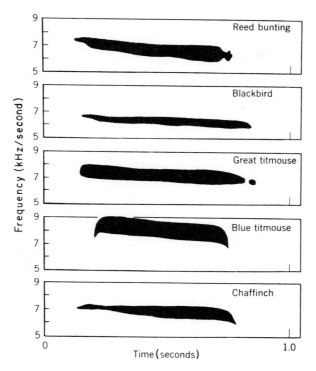

FIG. 8. Sound spectrograms of calls of five species of British birds used when a hawk flies over. They share characteristics that presumably make them difficult to locate by an aerial predator. (After Marler, 1957.)

An independent study of the same problem was conducted in tropical Africa (Chappuis, 1971). A spectral analysis of recordings showed that, as a rule, the denser the habitat, the lower the pitch. The songs of thirty-five deep-forest bird species recorded near the ground had an average frequency of 1850 Hz; those of twenty bird species typical of semiopen habitats, such as parkland savannah and the upper layers of the forest, 2230 Hz; the songs of thirty-five savannah species had an average frequency of 2720 Hz. Concomitantly with changes in average frequency, the harmonic spectra are broad in the open habitat and narrow in the closed one. Birds living in closed habitats increase the carrying power of their song by concentrating the emitted energy in the fundamental frequency (Chappuis, 1971).

C. SPECIFIC DISTINCTIVENESS

To serve its functions, song has to be easily recognized by all conspecifics, and thus the auditory environment created by all signalling

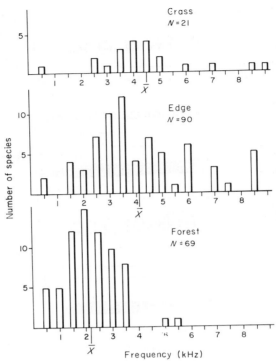

FIG. 9. Frequency distribution of the frequencies emphasized in the vocalizations (predominantly song) produced by Panamanian birds in three types of habitats. "Grass" refers to birds of open grassy savannahs; "edge" refers to birds living in secondary growth with dense underbrush and young trees; "forest" refers to birds living in the lower level of mature tropical forest. In each histogram, each species is entered only once, and N refers to the number of species sampled per habitat. The mean frequency emphasized ($\bar{X}$) for each habitat category appears below each histogram. Significant differences in emphasized frequencies occur only between forest birds and those from the other two habitats. (After Morton, 1970.)

sympatric species must exert a strong influence on song structure. Females responding to the song of other species would be at a disadvantage, and thus it is probably females that exert the greater pressure in favor of song specificity.

The breadth of a communicatory channel will be determined by the variability of the signals it includes. If it is true that each species needs a private auditory channel of minimum ambiguity, we might expect to find narrower communicatory channels in the species-rich environments of temperate and tropical avifaunas, while broader channels may occur in the poorer faunas of high latitudes, high altitudes, deserts, or oceanic islands. This expectation has not been rigorously tested, but supporting evidence has been reported. The songs of the Blue Tit (*Parus caeruleus*) on Tenerife and the Goldcrest (*Regulus regulus*) in the Azores have been noted to be more variable and include a greater diversity of sounds than do the songs of their mainland counterparts (Lack and Southern, 1949; Marler and Boatman, 1951; Marler, 1960). Both island forms occur in the absence of closely related species. On large land masses, the presence of congeners increases the risk of hybridization, and consequently there is a greater need for unambiguous species identification. Thielcke (1969) notes that these comparisons have taken British birds as the mainland form, when in fact song variability differs even among continental populations.

In complex avifaunas, species identity will be conveyed by improbable signals; frequency modulation, temporal patterning, and presence or absence of harmonics will encode this improbability. In these species, information coding will be enhanced by noise-free fundamentals, often resulting in rather musical sounds. In simpler avifaunas, species identity may be conveyed by less structured sounds. The song of the Austral Thrush (*Turdus falcklandii*) in the Falkland Islands is said to be harsher than that of the Patagonian and Tierra del Fuego populations (Reynolds, 1935). Chaffinch song in the Azores and Canary islands is said to be less elaborate than in the continental forms (Lack and Southern, 1949; Marler and Boatman, 1951). Mockingbirds of the genus *Mimus* are amongst the most talented songsters of the world. Yet the closely related Galapagos genus *Nesomimus* is said to have a hoarse, less structured song (J. Hatch, personal communication). The song of the Bananaquit (*Coereba flaveola*) is richer and more musical in Trinidad than in Aruba, where it is hoarse and whispered. The avifauna of Trinidad is many times richer than that of Aruba. Asian and African weaver birds (Ploceinae) use a variety of courtship sounds, including complex songs; yet the Seychelles Fody (*Foudia sechellarum*) uses a colorless trill in courting (Crook, 1969).

D. POPULATION CHARACTERISTICS

Variables influencing the quality of song in different avifaunas need much more study. Observations reported for island avifaunas are likely to reflect the characteristics of the few individuals that survived the hazards of distance and colonization. This founder effect is further complicated by the usually small populations of island species and the consequent possibility of genetic drift. Furthermore, there is no systematic evaluation of the extent to which breeding densities of any one species may influence its vocal patterns. Preliminary observations on the Rufous-collared Sparrow (*Zonotrichia capensis*) suggest that higher breeding densities and marked ecological gradients are correlated with reduced interindividual song variability (Nottebohm, 1969b). In this same species, King (1972) has noted that interindividual song variability at any one locality may change dramatically during different phases of the breeding season.

Marler *et al.* (1962) reared Oregon Juncos (*Junco hyemalis* subsp.) in varying degrees of sound isolation. Birds raised in the richer auditory environment had more song types and a more elaborate syllable structure, ". . . derived not from imitations but from an unspecific stimulation to improvise." Hatch (1967) has made similar observations on Common Mockingbirds (*Mimus polyglottos*). More recently, P. Marler (unpublished observations) has noted that when deaf canaries are reared in pairs, they produce songs with more note types than when reared singly. In this case, visual stimulation may elicit more singing, which in turn leads to the development of more note types. These kinds of variables must be taken into account when comparing the vocal repertoires of poor and rich avifaunas.

E. PHYSIOLOGICAL CONSTRAINTS

Vocal frequencies in birds are dependent on the size of the sound-generating membranes and on their tension, as determined by the out-flowing air, the interclavicular air sac pressure, and the action of the syringeal muscles. These factors may vary independently of each other, and thus it is not surprising that there is no strict correlation between the size of a bird and the highest frequencies it produces (Greenewalt, 1968). However, it is quite possible that the lowest vocal frequencies are correlated with the size of these membranes, because larger membranes are physically necessary for the production of low-frequency sounds. Miller (1934) observed that female Great Horned Owls (*Bubo virginianus*), whose trachea is equal to or longer than that of the male but whose syringeal membranes are smaller, produce higher-pitched notes than males. The Flammulated Owl (*Otus flammeolus*) is a small bird with unexpectedly low-pitched calls. Miller (1934) noted that the syringeal membranes of this bird are thickened and rugose and thus presumably oscillate at a lower frequency.

Although the dimensions of the syringeal membranes and the number of syringeal muscles may influence some characteristics of the sounds produced, it is unlikely that different patterns of sound modulation in closely related congeners are determined by differences in syringeal anatomy. This is clearly exemplified by the song of *Phaethornis longuemareus* and *P. guy*, two neotropical hummingbirds that are sympatric and deliver their songs at communal assemblies as they perch a few feet above the forest floor. Whereas the song of *P. longuemareus* is composed of a series of marked frequency modulations, that of *P. guy* is of a much simpler structure (Fig. 10). The syrinx is a conservative anatomical trait, and it is unlikely that peculiarities in its structure would account for such differences in the complexity of modulation. More likely, vocal differences in closely related species are attributable to differences in their neural substrate.

Vocal behavior must not interfere with respiratory needs. Vocalizations probably occur in most cases during expiration (Rüppell, 1933; Miskimen, 1951). Calder (1970) argues that in species such as canaries inspiration does occur between successive notes in a song. The physiological significance of the brief inspirations noted by Calder needs further appraisal, but the observation stresses the fact that length of song and spacing of its notes may reflect respiratory needs. The rarified atmosphere of higher altitudes may impose physiological constraints on song length and spacing of its components. Recordings of *Zonotrichia capensis* in Argentina indicate a gradual reduction of song length with increasing altitude (Nottebohm and Nottebohm, 1975). Song length may also correlate with availability of food and with temperature. At sea level, the song of Patagonian *Zonotrichia capensis* is briefer than the song of birds in the richer and warmer eastern pampas (J. R. King, personal communication). Longer songs presumably require more energy. Song length and persistence of singing may be a measure of the cost of food and homeothermy at any particular place. When the energy cost of these two items is high, singing time will be expensive and we may expect shorter and more sporadic songs. In species that learn their song by reference to auditory cues, a reduced amount of singing time and practice may lead to simpler patterns.

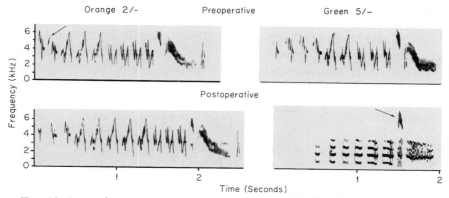

FIG. 10. Song of two sympatric neotropical hummingbirds, *Phaethornis guy* (A) and *P. longuemareus* (B), recorded in Trinidad. Both species are forest dwellers and sing a few feet above the ground in communal display arenas.

F. Interspecies Signal Sharing

It is important to note that for many vocalizations specific distinctiveness may not be necessary or may even be undesirable. Thus, all birds in an area will gain if they respond to the predator alarm calls of other species. Interspecific similarities in calls of this category are in fact found. Also, many species forage in mixed flocks during the nonbreeding season. Contact calls in these species show only moderate specific divergence. The extent to which various vocalizations show specific distinctiveness will reflect the various contexts in which such vocalizations are used (Marler, 1957).

G. Taxonomic Affinity

From the preceding considerations, it will be clear that vocalizations will often be misleading predictors of phylogenetic relationships. Similar selective pressures will determine parallel evolution of some call types, while other selective pressures will tend to maximize the differences between closely related species. Similarities and differences between the vocal repertoires of various species will be further affected by whether such species are sympatric or allopatric.

Despite these cautionary remarks, one is also struck at times by the occurrence of vocal similarities within a group of related species. As an example, Smith (1970) has noted the resemblances in the song of three congeneric flycatchers (Tyrannidae), *Sayornis phoebe*, *S. nigricans*, and *S. saya* (Fig. 11). All three species inhabit open country and enter forested regions along watercourses.

Further discussion of the taxonomic value of vocalizations will be found in Marler (1957), Andrew (1957), Löhrl (1963), Crook (1969), Lanyon (1969), Morton (1970), Smith (1970), Thielcke (1970), Güttinger (1970), and Smith and Vuilleumier (1971). It is probably safe to say that the taxonomic value of vocal signals is greater in birds that develop their species-typical vocal repertoires in the absence of environmental models. Species that learn their vocal repertoire from environmental models are likely to show greater intrafamilial, intrageneric, and intraspecific variability.

VI. The Neural Substrate of Vocal Behavior

An earlier section focused on the peripheral implementation of vocal behavior. Control of this behavior by various brain areas will be discussed in this section. Karten and Hodos (1967) were the first to publish

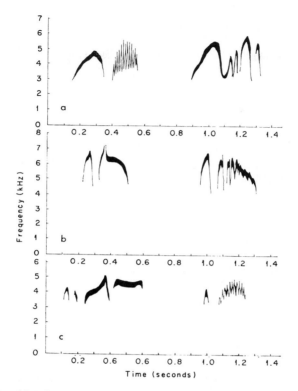

FIG. 11. Songlike displays of three New World flycatchers of the genus *Sayornis*—(a) *S. phoebe*, (b) *S. nigricans*, (c) *S. saya*. Each of these species has two different songs, as shown by the traces on the left and right halves of each sound spectrogram. Context and relative incidence of each song type varies between species. (After Smith, 1970.)

a brain atlas for any species of bird, that of the pigeon. Their designations will be followed here.

Kalischer (1905) searched for brain sites related to vocal behavior in parrots (*Amazona*) and cockatoos (*Cacatua*) and reported that alarm calls could be evoked by stimulating the archistriatum. He also was able to remove parts of the right and left hemispheres in these birds, an operation which was followed by temporary losses of vocal skill.

More recently, Brown (1965a, 1971) has undertaken a detailed search of brain loci from which vocalizations can be elicited by electrical stimulation. These studies were conducted with Red-winged Blackbirds tested under local anesthesia or with chronically implanted electrodes. Responsive areas were found (1) in and near the nucleus intercollicularis of the midbrain (2) in the septum and anteroventral area of the diencepha-

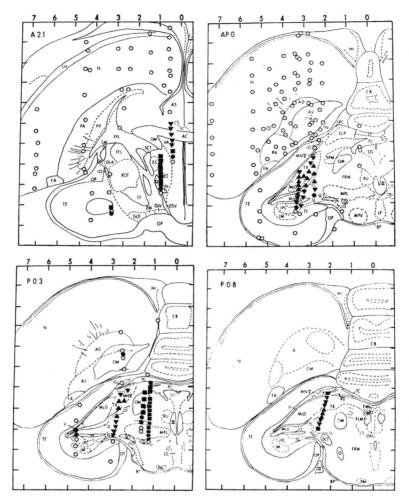

FIG. 12. Brain loci of Red-winged Blackbirds from which vocalizations can be elicited by electrical stimulation. Results of stimulating with different current intensities are indicated by five different symbols: ○, negative for vocalization at 500 μA; ●, positive for vocalization at thresholds of 310–500 μA; ■, positive for vocalization at thresholds of 160–300 μA; ▼, positive for vocalization at thresholds of 60–150 μA; ▲, positive for vocalization at thresholds of 10–50 μA. *Abbreviations:* AC, anterior commissure; AD, archistriatum, pars dorsalis; AS, area septalis; AV, archistriatum, pars ventralis; BP, brachium pontis; CB, cerebellum; CS, nucleus commisuralis septi; DLA, nucleus dorsolateralis anterior; DLP, nucleus dorsolateralis posterior; DSV, decussatio supraoptica ventralis; EI, nucleus entopeduncularis inferior; EM, nucleus entomamillaris; ES, nucleus entopeduncularis superior; EW, nucleus of Edinger-Westphal; FA, tractus frontoarchistriatalis; FLM, fasciculus longitudinalis medialis; FPL, lateral forebrain bundle; FRM, formatio reticularis medialis; GLD, nucleus geniculatus lateralis dorsalis; GLV, nucleus geniculatus later-

lon, (3) in the archistriatum, and (4) in and near the occipitomesence-phalic tract; a few high threshold points were also found in the hyperstriatum and neostriatum (Fig. 12). Thresholds for evoked vocalizations were lowest in the midbrain. Whereas some of the vocalizations elicited by electrical stimulation were indistinguishable from calls given by wild birds, others were clearly abnormal. The most frequently evoked calls belonged to the various categories of general alarm calls; song was never elicited in this manner.

Studies by other authors on brain sites responsible for vocal behavior tend to agree with Brown's findings, though there might be species and methodological differences. Electrical stimulation of the midbrain has been shown to elicit vocalizations in pigeons (Popa and Popa, 1933; Delius, 1971), in domestic fowl (Putkonen, 1967, p. 52; Murphey and Phillips, 1967), in the Mallard (Maley, 1969), in Japanese Quail (Potash, 1970a,b), and in gulls (*Larus argentatus* and *L. fuscus*) (Delius, 1971). Interestingly, Potash (1970a) reports that, "The exact physical structure of the vocalizations elicited from a given electrode depended upon the intensity–frequency parameters used. In some instances brain-stimulation elicited vocalizations resembled natural calls only when elicited by a relatively narrow range of intensity–frequency combinations." Potash also notes that electrodes implanted in close neuroanatomical proximity tend to produce similar sequences of vocalizations.

In addition to these sites, Åckerman (1966) has elicited bow-cooing in pigeons by stimulating electrically preoptic and anterior hypothalamic areas. Delius (1971) reports evoked cooing following electrical stimulation of the nucleus ovoidalis of the diencephalon in pigeons and of the posteromedial nucleus of the hypothalamus in pigeons and gulls.

Some of the brain areas reportedly involved in vocal control are part of the main auditory pathway. This is the case with the nucleus mes-

alis ventralis; HI, hippocampus; IM, nucleus isthmi magnocellularis; IN, tractus infundibularis; IO, tractus isthmo-opticus; IP, nucleus interpeduncularis; IPC, nucleus isthmi parvocellularis; LC, nucleus linearis caudalis; LH, lamina hyperstriatica; LLD, nucleus lemnisci, pars dorsalis; MLD, nucleus mesencephalicus lateralis dorsalis; MLV, nucleus mesencephalicus lateralis, pars ventralis; MNV, nucleus mesencephalicus nervi trigemini; MPL, nucleus profundus lateralis; MPV, nucleus profundus ventralis; N, neostriatum; OM, tractus occipitomesencephalicus; OMD, nuclus nervi occulomotorii, pars dorsalis; OMV, nucleus nervi occulomotorii, pars ventralis; OP, tractus opticus; PA, paleostriatum augmentatum; PC, posterior commissure; PM, nucleus pontis medialis; PP, paleostriatum primitivum; Q, tractus quintofrontalis; ROT, nucleus rotundus; RU, nucleus ruber; SCT, tractus striocerebellaris et tegmentalis; SG, substantia grisea periventricularis; SL, nucleus semilunaris; SM, tractus cortico-septomesencephalicus; SPC, nucleus superficialis parvocellularis; SPM, nucleus spiriformis medialis; TC, tectal commissure; TE, cortex of the optic lobe; TFL, tractus thalamofrontalis lateralis; TS, tractus tectospinalis; TT, tractus tectothalamicus; TX, torus externus; VA, area ventralis anterior. (After Brown, 1971.)

encephalicus lateralis dorsalis of the midbrain and the nucleus ovoidalis of the diencephalon (Karten, 1968). However, Newman (1970) and Potash (1970b), working with Red-winged Blackbirds and Japanese Quail, respectively, have found that the threshold for eliciting vocalizations is lower in the nucleus intercollicularis of the midbrain than in the adjacent nucleus mesencephalicus lateralis dorsalis. This suggests a more direct vocal involvement for the former nucleus, less so for the latter one. Brown (1971) also emphasizes that in general, stimulation of known sensory areas in the brain of Red-winged Blackbirds does not evoke vocalization, and thus the idea that vocalizations are evoked mainly by activation of sensory pathways to produce "hallucinations" is not confirmed by available data.

Delius (1971) has proposed an efferent vocal system in the bird brain composed of the archistriatum, occipitomesencephalic tract, periventricular nuclei of the diencephalon, and nucleus intercollicularis of the midbrain, which could be tentatively extended to medullar levels. In support of such a hierarchial organization, Brown (1971) notes that vocalizations evoked from midbrain sites are relatively simple and invariant when compared to those evoked from septal and diencephalic sites. Calls evoked by forebrain stimulation may be rather complex in abnormal ways and sometimes contain elements of more complex vocalizations, which are not elicited in complete form. Lesions in the area of the midbrain from which vocalizations are evoked by electrical stimulation cause long-lasting loss of alarm calls; such birds also show a loss of locomotor excitability and escape reactions. In the Red-winged Blackbirds operated in this manner, recovery of the alarm call response takes from 10 to 84 days (Brown, 1965b).

Most of the vocalizations evoked by electrical stimulation of the avian brain correspond to calls usually given in an agonistic context by frightened or aggressive animals. When chronically implanted electrodes are used with awake and unrestrained birds, the vocalizations elicited may (Åckerman, 1966; Putkonen, 1967; Delius, 1971; Potash, 1970a) or may not (Maley, 1969) be accompanied by other behavior with which they normally occur. In one case described by Potash (1970a), a posture typical of a call was elicited at a lower threshold than that triggering the simultaneous display of call plus posture. It seems possible that some of the vocal behavior elicited may be secondary to the evocation of a particular mood. One may wonder to what extent individual vocalizations have a neural substrate independent of the mood eliciting them. Reporting on the mapping of speech areas in the human cortex, Penfield (in Penfield and Roberts, 1959) states that, "Application of an electrical current to one of these areas produces local interference or aphasic arrest. . . . The man cannot for the moment use the area. There is no positive response to the stimulation. No movement is produced and no sensation. No positive psychical process is set in motion. The stimulation does not summon words to the mind of the man nor does it cause him to speak." Even granting the enormous differences in the organization of the human and avian brain, Penfield's observations should sound a cautionary note. Electrically elicited calling may not be the best way to map primary vocal areas in the bird brain.

Based on a functional asymmetry of the hypoglossal nerves innervating the syrinx of some songbirds, Nottebohm (1970, 1971) has suggested a possible laterality in the central mechanisms controlling song in these species. In no case have brain-stimulation experiments with birds provided any evidence to bolster this possibility. This apparent discrepancy may mean that dominance is restricted to the hypoglossal level or may reflect on limitations of electrically evoked vocalization as a way to map primary vocal areas.

VII. Hormonal Variables Affecting Vocal Behavior

Andrew (1969) has recently reviewed the influence of testosterone on avian vocalizations. A critical evaluation of the influence of gonadal hormones on the hypothalamic integration of courtship in Ring Doves has been presented by Hutchison (1970). Hamilton (1938) was the first to show that testosterone injections cause domestic chicks to crow. Soon thereafter, this hormone was shown to induce song in female Canaries (Baldwin et al., 1940) and in castrated male Chaffinches (Collard and Grevendal, 1946). Bow-cooing in male Ring Doves is suppressed by castration (Erickson and Lehrman, 1964). This behavior reappears upon insertion of minute amounts of testosterone propionate crystals in the anterior and preoptic areas of the hypothalamus (Hutchison, 1967, 1971). Testosterone concentrating cells occur in the nucleus intercollicularis of the Chaffinch midbrain (Zigmond et al., 1973); electrical stimulation of this nucleus elicits a variety of calls (see previous section). In nature, male birds are by far the more talented songsters. Song is usually discontinued as the testes enter into eclipse.

The importance of testosterone for the occurrence of song and other territorial calls in some birds is thus firmly established. This observation has to be reconciled with the fact that females of many avian species sing, though usually not as much as males. The hen of the European Robin sings quite freely in the autumn, when male and female robins defend individual territories (Lack, 1943). The same author also reports two instances of female robin song after pair formation. This occurred in one case when a pair moved into and claimed a new territory, and in the other case, when a pair was defending its territory against a rival pair. Female Chaffinches sing occasionally and then usually incomplete songs (Marler, 1956a). Female Red-winged Blackbirds have a song typical of their sex, which Orians and Christman (1968) call a "chatter"; the chatter is given by territorial females, though less than half as often as male song. Females of the Eastern Phoebe sing early in the nesting season, their song being very much like that of males (Smith, 1969). Female White-throated Sparrows produce songs like that of males, except that they tend to be shorter and quavering (Fig. 6c). Female Rufous-collared Sparrows give a song indistinguishable from the male counterpart; female song in this species probably disappears as the breeding season is fully under way (Nottebohm and Nottebohm, 1975). Hill Mynahs of both sexes produce the same categories of sounds and are equally vocal (Bertram, 1970). Male and female Ring Doves produce the bow-cooing call referred to earlier until they are 3–4 months old. Bow-cooing disappears from the female repertoire as the birds get into breeding condition, which occurs at 5–6 months of age (Nottebohm and Nottebohm, 1971).

Observations such as these raise the question as to whether female song is as rare and inconsequential as often represented. It seems possible that it is a deliberate feature of female behavior, perhaps influencing the bird's socialization and choice of partner and, in some cases at least, aiding in territorial defense. In some species, males and females engage in complex duetting patterns that are believed to be important in the synchronization of breeding behavior and in the reinforcement of the pair bond (e.g., Thorpe and North, 1965; Thorpe, 1966; Hooker and Hooker, 1969; Bertram, 1970; R. B. Payne, 1971). There is no information on hormonal or neural substrates of female song.

VIII. Vocal Ontogeny

A. Stages in Song Development

The egg-bound chicks of many species begin to call soon after they pierce the inner shell membrane and gain access to the egg's air space (Vince, 1969). Subsequent vocal development can follow various different strategies, so that a greater or lesser number of vocal stages precedes the onset of adult vocalizations. Vocal development is particularly complex in species that learn their songs from conspecifics, as exemplified in the Chaffinch (Poulsen, 1951; Thorpe, 1954, 1958a; Marler, 1956a,b; Nottebohm, 1968, 1970).

Soon after hatching, hungry young Chaffinches give food-begging calls. In fledglings, these calls gradually change into a more rambling type of vocalization known as "subsong." Subsong is a loose aggregation of notes, some variable and others stereotyped, given at a low volume, often while the young birds doze. It is believed to be of the nature of vocal practice, with no communicatory significance. Subsong subsides during the winter months but recurs early in the spring, growing in complexity. After 1–3 weeks, interspersed with subsong occur vague and variable imitations of adult song. These vocal patterns, known as "plastic song," become louder and more stereotyped, until a time is reached when successive repetitions of a song theme are virtually identical and remain stable on successive days. Song is said to have "crystallized" and is now known as "full song." Figure 13 summarizes these vocal stages, which in the Chaffinch cover a period of some 11 months. Vocal development follows a similar course in the Song Sparrow (*Melospiza melodia*) (Mulligan, 1966), the common Cardinal (*Cardinalis cardinalis*) (Lemon and Scott, 1966), and the Indigo Bunting (*Passerina cyanea*) (O'Hara Rice and Thompson, 1968) among others.

A subsong stage of vocal development has been noticed in many Passeriformes (Thorpe, 1958b; Armstrong, 1963) and in Orange-winged Parrots (*Amazona amazonica*) (F. Nottebohm, unpublished data). In both these groups, vocalizations are developed by reference to auditory information. In Galliformes and Columbiformes, the absence of a subsong stage is complete. As we shall see, neither group develops its vocalizations by reference to auditory information. Subsong in juvenile birds has been compared to babbling in children. In both cases this kind of behavior may constitute an important step in the acquisition of the vocal practice that precedes vocal imitation (Marler, 1970a,b; Nottebohm, 1970).

B. THE IMPORTANCE OF AUDITORY VARIABLES

The contribution of auditory feedback to vocal ontogeny can be tested by removal of both cochleas, which renders a bird chronically deaf (review in Konishi and Nottebohm, 1969). Domestic chicks hatched

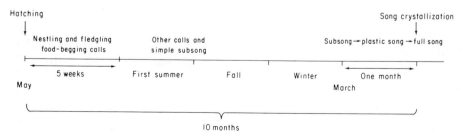

FIG. 13. Summary of vocal development, by stages, during the first year of a young male Chaffinch. (After Nottebohm, 1971.)

in an incubator and deafened on their first day after hatching develop a normal adult vocal repertoire (Konishi, 1963). Ring Doves, deafened at 5 days after hatching also develop normal vocalizations (Nottebohm and Nottebohm, 1971). When doves are foster-reared by parents of a different species, their vocalizations are not influenced by this misleading model (Whitman, 1919; Lade and Thorpe, 1964). Normal calls are also reported for turkeys deafened soon after hatching (Schleidt, 1964).

Canaries develop normal song when hand-reared from an early nestling stage (Metfessel, 1935; Poulsen, 1959). When Canaries are hatched in chambers exposed to 100 dB of white noise and then deafened at 1 month of age by removal of both cochleas, they are deprived of auditory feedback over the entire span of their vocal development. The song they produce as adults includes highly aberrant screeches and hisses as well as stretches of silent song. In the latter case, the bird holds a singing posture, with open bill and pulsating throat, chest, and abdomen, but emits no sound. The audible song of these deaf birds resembles that of intact adults in that it includes a succession of distinct phrases as well as some notes of normal structure and tonality. In this species, auditory feedback is necessary for the elimination of silent song and aberrant sounds and for the generation of large song repertoires (Marler et al., 1973, and unpublished observations).

Song Sparrows hatched and reared by Canary foster parents and prevented from hearing conspecifics develop song of fairly normal, but not perfect, wild-type quality. Song Sparrows deafened as juveniles develop a hoarse song totally lacking the structure of the wild-type pattern (Mulligan, 1966). Thus, Song Sparrows are more dependent than Canaries on auditory feedback for normal vocal development. However, both species are able to imitate the song of other conspecifics, and in this sense are facultative mimics (Poulsen, 1959; Mulligan, 1966).

In still another group of birds, development of wild-type song is fully dependent on exposure to environmental sources. Chaffinches, Eastern and Western Meadowlarks (Sturnella magna and S. neglecta), White-crowned Sparrows (Zonotrichia leucophrys), Common Cardinals, and Zebra Finches (Poephila guttata), hand-reared from an early nestling age, produce abnormal song lacking the detailed patterns of wild conspecifics (Poulsen, 1951; Thorpe, 1954, 1958a; Lanyon, 1957; Marler and Tamura, 1964; Immelmann, 1969; Lemon and Scott, 1966; Dittus and Lemon, 1969). The song of White-crowned Sparrows, Chaffinches, and Common Cardinals deafened before onset of song is even more aberrant than that of their hand-reared counterparts (Konishi, 1965; Nottebohm, 1967, 1968; Dittus and Lemon, 1970). The subsong and calls of Chaffinches deafened at an early age are also abnormal (Nottebohm, 1967,

1972a). The calls of adult Common Cardinals deafened as juveniles are no different from those of wild adults (Dittus and Lemon, 1970).

The effects of deafening on vocal development remain speculative. One obvious result of this operation is to prevent access to auditory models and auditory feedback. If each stage of vocal development depends on an integration of the motor, proprioceptive, and auditory information generated by preceding stages, then loss of hearing would disrupt this developmental interaction. But deafening may also bring about hormonal changes. Deaf Chaffinches tend to be reluctant songsters, though their rate of singing will increase if implanted with exogenous testosterone. Deafening may influence the hormonal background of the developing bird, affecting vocalizations only in an indirect manner. However, a male Chaffinch castrated before it developed its song and implanted with exogenous testosterone will then develop normal wild-type song. Other young male Chaffinches deafened at the same age and also implanted with testosterone developed highly aberrant songs. Thus, given a similar testosterone level, Chaffinches deafened before the onset of song develop abnormal song, whereas hearing birds develop normal song (Nottebohm, 1967, 1968, 1969a). It seems probable that in Chaffinches, at least, interdiction of auditory feedback has a direct effect on vocal development.

The effect of deafening on amount of singing probably varies between species. Loss of hearing has no appreciable effect upon amount of singing in European Bullfinches (*Pyrrhula pyrrhula*) and Oregon Juncos (*Junco hyemalis* ssp.) and the effect is small in Ring Doves (Schwartzkopff, 1958; Konishi, 1964; Nottebohm and Nottebohm, 1971). However, surgically devocalized Budgerigars (*Melopsittacus undulatus*) undergo gonadal regression. Brockway (1967) attributes this effect to the induced loss of self-generated auditory stimulation; these birds could hear other sham-operated male budgerigars, which retained a breeding gonadal condition, and the corresponding vocal repertoire. Though the interpretation of this latter experiment is arguable, it seems likely that the relation between auditory stimulation and vocal behavior might be a complex one and differs between species.

Deaf birds may also sing less because they are shielded from auditory stimuli likely to trigger song, and this may be reflected in the quality of the patterns developed. Deaf Canaries kept in pairs develop greater song repertoires than deaf Canaries kept singly. However, birds in both groups develop notes of comparable quality. Thus, social stimulation may influence the size of song repertoires, while presence or absence of auditory feedback influences the quality of the component notes (P. Marler, unpublished observations).

C. EFFECTS OF DEAFENING AT VARIOUS STAGES OF VOCAL ONTOGENY

The gradual development of vocalizations in the Chaffinch has already been described. The effects of deafening birds at 3 months of age, at 7–8 months, at 10 months, at 11 months, and as adults is shown in Fig. 14. Loss of hearing in adult birds does not affect any of their

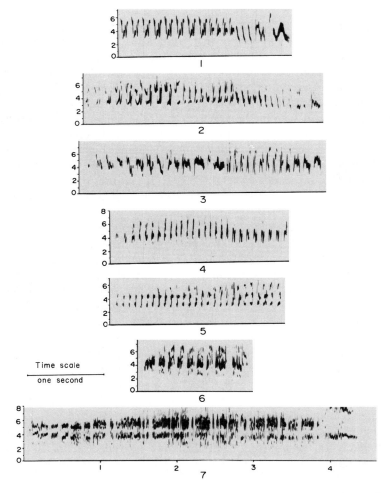

FIG. 14. Songs of male Chaffinches deafened at different stages of vocal development. (1) Bird deafened as adult, after it had developed stable song; (2, 3, and 4) birds deafened in their first spring after they had been in plastic song stage for 12, 2, and 1 days, respectively; (5) bird deafened in the middle of its first winter, when 7 months old; (6) bird deafened at 107 days of age; (7) bird deafened at 88 days. (After Nottebohm, 1970.)

vocalizations. Birds deafened while in plastic song show a deterioration of song patterns evolved up to that time (Fig. 15). The song of Chaffinches deafened early during their first spring comes closer to the wild-type pattern than that of birds deafened in the middle of their first winter. The most aberrant song patterns are those of individuals deafened during their first summer, when 3 months old. Their song as adults is a dissonant screech almost totally lacking in pattern. It seems reasonable to conclude that in this species the auditory–motor experience of each vocal stage contributes to the quality of the final song pattern (Nottebohm, 1968).

Common Cardinals establish their song repertoire during their first breeding season. If deafened after then, their song patterns undergo gradual deterioration (Dittus and Lemon, 1970). This finding differs from the stable song retained by White-crowned Sparrows and Chaffinches deafened as adults (Konishi and Nottebohm, 1969). The significance of these differences is not clear.

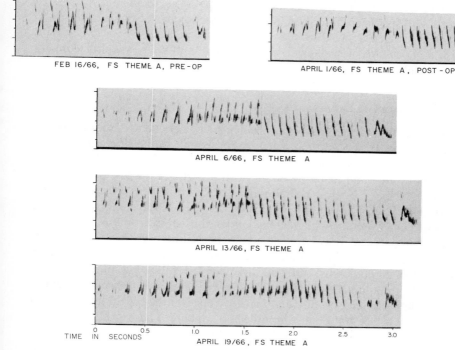

FEB 16/66, FS THEME A, PRE-OP

APRIL 1/66, FS THEME A, POST-OP

APRIL 6/66, FS THEME A

APRIL 13/66, FS THEME A

TIME IN SECONDS

APRIL 19/66, FS THEME A

Fig. 15. Deterioration of song in a 10-month-old male Chaffinch deafened after it had produced plastic song during a period of 12 days. As time goes by, similarity with the preoperative pattern decreases. (Nottebohm, 1968.)

D. The Critical Period for Song Learning

Chaffinches, Common Cardinals, White-crowned Sparrows, and Zebra Finches learn their song during their first year of life and do not change it thereafter. In the case of the latter two species, commitment to the model to be imitated occurs before this model is converted into a motor pattern (Marler and Tamura, 1964; Marler, 1970a; Immelmann, 1969). It is not clear whether the termination of the critical period for song learning comes about (1) because no new models are acquired as auditory memories, (2) because there is a loss of motor plasticity, or (3) because of motivational factors. In the Chaffinch, the critical period for song learning can be extended for an extra year and presumably for longer. A juvenile male castrated before coming into song and implanted with testosterone 1 year later will copy a model presented then. In this case, the end of song learning is not an age-dependent phenomenon, but is correlated with the first occurrence of high testosterone levels (Nottebohm, 1969a).

In other birds, learning of new vocal patterns recurs on successive years. This has been described in Cardueline Finches (Mundinger, 1970). I had a Gray Parrot (*Psittacus erithacus*) that at 20 years of age and with a repertoire of well over one hundred imitations was still copying new sounds.

E. Hypoglossal Dominance

It was noted earlier that each bronchus can behave as an independent sound source. However, their roles are not equal. In the Chaffinch cutting the left hypoglossus results in the loss of most of the components of song, subsong, and calls. Section of the right hypoglossus results in the loss of few, if any, components (Fig. 4). When either the right or the left hypoglossus is cut before the bird has started developing its song, normal song will develop under the control of the intact side. Thus, before vocal development commences, both syringeal halves and both hypoglossi are potentially equivalent, and the normal dominance of the left hypoglossus can still be reversed following injury of that nerve (Nottebohm, 1971, 1972a). The left hypoglossus also plays a dominant role in sound production in the Canary and the White-crowned Sparrow (F. Nottebohm, unpublished data). These phenomena of hypoglossal dominance in birds are reminiscent of hemispheric dominance in humans. In both cases neural dominance is related to the learning of complex behavior (Nottebohm, 1970).

Intriguingly, the left midbrain auditory projection of the Oilbird (*Steatornis caripensis*) and of the Saw-whet Owl (*Aegolius acadicus*) is larger than its corresponding counterpart on the right half of the brain (Cobb, 1964). Both these species engage in complex problems of auditory localization—of obstacles in the case of the Oilbird (Griffin, 1953), and of prey in the case of the owl (R. S. Payne, 1971). Neural asymmetries in vocal control may be causally related to the requirements of complex auditory processing, rather than to the learning of complex output patterns per se.

F. VOCAL IMITATION AND CHOICE OF MODELS

In species that develop their song by reference to auditory variables, not all auditory inputs are equally important for normal vocal development. To understand vocal development in these species, the ontogeny of auditory preferences is as important as that of the motor patterns themselves.

The juveniles of some songbird species will imitate selectively conspecific song (Thorpe, 1958a; Marler, 1970a). Konishi (1970) discovered that the upper frequency range of auditory sensitivity varies among songbirds. When species vocalizations fall within the high-frequency range (i.e., 6–10 kHz), auditory curves are displaced in the same direction. Clearly, though, this correspondence is not sufficient to dictate an auditory preference for conspecific vocalizations. White-crowned Sparrows do not imitate the song of Song Sparrows, though both species are sympatric and share the same frequency band (Marler and Tamura, 1964). Marler (1964) has suggested that a clear-cut preference for conspecific song presupposes the availability of an auditory template with which environmental sounds are matched, permitting recognition of the correct model in the absence of other conspecific cues. The developmental basis of such an auditory bias remains unclear (Nottebohm, 1972a).

In some cases, the correct model to be imitated is identified by a social bond. European Bullfinches and Zebra Finches imitate the song of their fathers or foster fathers (Nicolai, 1959; Immelmann, 1969). In some African shrikes, the pair bond determines the sharing of antiphonal duetting patterns (Thorpe and North, 1965). In some cardueline finches, males and females share a same set of flight calls. Other calls are shared by members of a nonbreeding flock (Mundinger, 1970). Parasitic viduine finches imitate the song and calls of their foster parents, though they also have a species-specific vocal repertoire (Nicolai, 1964, 1973).

There are species that imitate in the wild a broad variety of sympatric birds. This group includes the Superb Lyrebird, the scrub-birds (*Atrichornis* sp.), the Common Mockingbird, and several others (e.g., Chisholm, 1958; Borror and Reese, 1956b; Armstrong, 1963). The parameters that determine the sounds to be imitated are not known. Though these birds have established their reputation as mimics of other species, it seems probable that each individual gleans most of its imitations from neighboring conspecifics with which it countersings. This certainly seems to be the case with mockingbirds, which often can be heard ordering the delivery of their song themes so as to match theme by theme those of another mockingbird singing nearby.

IX. The Evolution of Vocal Learning

A. THE PHYLOGENY OF VOCAL LEARNING

"Vocal learning" will refer here to the influence of auditory information, including auditory feedback, on vocal development. The only documented case of vocal learning in mammals is that of man, and this makes the frequent incidence of this phenomenon in birds all the more intriguing.

Three kinds of evidence point to the occurrence of vocal learning: (1) abnormal vocal development in individuals denied access to conspecific sounds, (2) abnormal vocal development in birds deafened early in life; (3) imitation of environmental sounds. Instances (1) and (3) refer to vocal imitation. Instance (2) refers to a role played by auditory feedback during normal ontogeny. Actually, all species known to depend on auditory inputs for normal vocal development are also known to be able to imitate some environmental sounds.

In most orders of birds vocal ontogeny probably excludes auditory influences, so that genetic or other developmental influences predominate. Examples of this type of vocal ontogeny have been described for Galliformes and Columbiformes. Avian taxonomists consider these two orders closer in morphology and possibly behavior to the ancestral avian stock than are groups such as Passeriformes and Psittaciformes, in which vocal learning has been demonstrated; incomplete evidence also points to vocal learning in toucanets and hummingbirds (Wagner, 1944; Snow, 1968). Since the latter four groups are phylogenetically unrelated, vocal learning must have developed *de novo* in each of them. The alternative hypothesis, that vocal learning was discarded by the remaining twenty-five orders of birds, is far more improbable. It seems, then, that vocal development in birds shows a phylogenetic trend from auditory-independent strategies to more open and environmentally dependent vocal ontogenies.

The extent to which vocal learning is represented in each of the four avian groups where it reportedly occurs is unclear. The number of Oscine families recognized by taxonomists varies from thirty-six to fifty-four (review in Sibley, 1970). Vocal learning is known to occur in twenty-one families of the suborder Passeres and in the two families of the suborder Menurae (review in Nottebohm, 1972b). Vocal learning seems to occur very broadly among Psittaciformes, though most of the reports to date come from people that have kept them as pets. Outside of these two orders, only one species of hummingbird, *Phaethornis longuemareus*, and one toucanet, *Aulacorhynchus prasinus*, are thought to imitate vocal sounds (Wagner, 1944; Snow, 1968; Wiley, 1971).

Vocal learning occurs in groups that have evolved complex song repertoires or vocal dialects. At least three of these groups have also undergone intense speciation. An evaluation of selective pressures that may have led to vocal learning is presented in Nottebohm (1972b).

B. STAGES LEADING TO VOCAL LEARNING IN BIRDS

With the exception of Oilbirds and swiftlets (*Collocalia* sp.) (Griffin, 1953; Medway, 1959), which produce sounds to echo-locate objects in the dark, vocalizations in birds seem to have evolved for purposes of communication. Consequently, vocal learning must have evolved from a genetically prescribed motor ontogeny in such a way that all intervening stages served an adaptive communicatory function. This need not have been difficult. Even among species that do not learn their song and calls, conspecific vocalizations are recognized by auditorily naive individuals (Gottlieb, 1971). In these cases selective responsiveness presupposes the occurrence of an auditory template (Marler, 1964). This very auditory template could then have provided a reference point for song learning, so that even as the variety of the song repertoire increased, it still reflected the species-specific paramenters of the template.

As pointed out earlier, all birds that rely on vocal learning as a strategy for vocal development can also imitate at least other conspecifics. Thus we may wonder which came first, improvisation or vocal imitation. Birds that improvise their song repertoire store it as a memory that is "imitated" in future renderings until crystallization of the song repertoire is achieved. To the extent that other conspecifics develop patterns acceptable to the species' template, the imitation of self-generated themes and of those produced by other conspecifics may be an almost inevitable result of vocal learning. Thus, the primacy of vocal improvisation or of vocal imitation need not be an arguable point; both may have arisen together. If this is so, then vocal ontogenies strictly dependent on imitation of other individuals may be considered as specialized, secondary adaptations.

The significance of facultative imitation deserves one final comment. Hand-reared Oregon Juncos kept communally in a laboratory cage or in a sound-proofed booth share some of the song themes they develop

(Marler *et al.*, 1962). There is no evidence of song imitation in wild Oregon Juncos (Konishi, 1964). A singing bird provides a model he can imitate in further renderings, or which he can alter by improvisation. The self-produced song will be heard louder and nearer than that of other birds. However, other conspecifics singing at close range may provide a stimulus of comparable loudness, which may thus qualify as a model. Laboratory examples of facultative imitation may reflect on the rather artificial proximity at which birds are kept; it need not mean that the species in question imitates in the wild or that its learning abilities evolved to imitate other conspecifics. The loudness–proximity cues for conspecific imitation would predict greater homogeneity of song patterns when singing birds occur naturally at high densities. A correlation of this kind has been described for the Argentinian Rufous-collared Sparrow (Nottebohm, 1969b). Of course, a greater chance to countersing with nearby neighbors may also increase exposure to their song, which then becomes more accessible for imitation.

REFERENCES

Ackerman, B. (1966). Behavioural effects of electrical stimulation in the forebrain of the pigeon. I. Reproductive behaviour. *Behaviour* 26, 323–338.

Ames, P. L. (1971). The morphology of the syrinx in passerine birds. *Bull. Peabody Mus.* 37, 1–194.

Andrew, R. J. (1957). A comparative study of the calls of *Emberiza* spp. *Ibis* 99, 27–42.

Andrew, R. J. (1969). The effects of testosterone on avian vocalizations. *In* "Bird Vocalizations" (R. A. Hinde ed.), p. 97. Cambridge Univ. Press, London and New York.

Armstrong, E. A. (1963). "A Study of Bird Song." Oxford Univ. Press, London and New York.

Arnold, A. P. (1974). Behavioral effects of androgen in Zebra Finches (*Poephila guttata*) and a search for its site of action. Ph.D. thesis, Rockefeller University, New York.

Baldwin, F. M., Goldin, H. S., and Metfessel, M. (1940). Effects of testosterone propionate on female Roller Canaries under complete song isolation. *Proc. Soc. Exp. Biol. Med.* 44, 373–375.

Batteau, D. W. (1967). The role of the pinna in human localization. *Proc. Roy. Soc. London, Ser. B* 168, 158–180.

Baümer, E. (1962). Lebensart des Haushuhns, dritter Teil—über seine Laute und allgemeine Ergänzungen. *Z. Tierpsychol.* 19, 394–416.

Bertram, B. (1970). The vocal behavior of the Indian Hill Mynah, *Gracula religiosa*. *Anim. Behav. Monogr.* 3, 81–92.

Borror, D. J., and Reese, C. R. (1956a). Vocal gymnastics in Wood Thrush songs. *Ohio J. Sci.* 56, 177–182.

Borror, D. J., and Reese, C. R. (1956b). Mockingbird imitations of Carolina Wren. *Bull. Mass. Audubon Soc.* 40, 246–250.

Brémond, J. C. (1967). Reconnaissance de schémas reactogènes liés à l'information contenue dans le chant territorial de rouge-gorge (*Erithacus rubecula*). *Proc. Int. Ornithol. Congr., 14th, 1966* p. 217.

Brémond, J. C. (1968a). Recherches sur la semantique et les éléments vecteurs d'information dans les signaux acoustiques du rouge-gorge (*Erithacus rubecula*). *Terre Vie* **2**, 109–220.

Brémond, J. C. (1968b). Valeur spécifique de la syntaxe dans le signal de défense territoriale du troglodyte (*Troglodytes troglodytes*). *Behaviour* **30**, 66–75.

Brockway, B. F. (1967). The influence of vocal behavior on the performer's testicular activity in Budgerigars (*Melopsittacus undulatus*). *Wilson Bull.* **79**, 328–334.

Brown, J. L. (1965a). Vocalization evoked from the optic lobe of a songbird. *Science* **149**, 1002–1003.

Brown, J. L. (1965b). Loss of vocalizations caused by lesions in the nucleus mesencephalicus lateralis of the Redwinged Blackbird. *Amer. Zool.* **5**, 693.

Brown, J. L. (1971). An exploratory study of vocalization areas in the brain of the Redwinged Blackbird (*Agelaius phoeniceus*). *Behaviour* **39**, 91–127.

Calder, W. A. (1970). Respiration during song in the Canary (*Serinus canaria*). *Comp. Biochem. Physiol.* **32**, 251–258.

Chappuis, C. (1971). Un exemple de l'influence du milieu sur les emissions vocales des oiseaux: L'évolution des chants en forêt équatoriale. *Terre Vie* **2**, 183–202.

Chisholm, A. H. (1958). "Bird Wonders of Australia." Michigan State Univ. Press, Ann Arbor.

Cobb, S. (1964). A comparison of the size of an auditory nucleus (n. mesencephalicus lateralis, pars dorsalis) with the size of the optic lobe in twenty-seven species of birds. *J. Comp. Neurol.* **122**, 271–280.

Collard, J., and Grevendal, L. (1946). Etudes sur les caractères sexuels des Pinsons, *Fringilla coelebs* et *F. montifringilla*. *Gerfaut* **2**, 89–107.

Collias, N. E. (1963). A spectrographic analysis of the vocal repertoire of the African Village Weaverbird. *Condor* **65**, 517–527.

Crook, J. H. (1969). Functional and ecological aspects of vocalization in weaverbirds. *In* "Bird Vocalizations" (R. A. Hinde ed.), p. 265. Cambridge Univ. Press, London and New York.

Delius, J. D. (1971). Neural substrates of vocalization in gulls and pigeons. *Exp. Brain Res.* **12**, 64–80.

Dittus, W. P. J., and Lemon, R. E. (1969). Effects of song tutoring and acoustic isolation on the song repertoires of cardinals. *Anim. Behavior.* **17**, 523–533.

Dittus, W. P. J., and Lemon, R. E. (1970). Auditory feedback in the singing of cardinals. *Ibis* **112**, 544–548.

Emlen, S. T. (1972). An experimental analysis of the parameters of bird song eliciting species recognition. *Behaviour* **41**, 130–171.

Erickson, C. J., and Lehrman, D. S. (1964). Effect of castration of male Ring Doves upon ovarian activity of females. *J. Comp. Physiol. Psychol.* **58**, 164–166.

Erulkar, S. D. (1972). Comparative aspects of spatial localization of sounds. *Physiol. Rev.* **52**, 237–360.

Falls, J. B. (1969). Functions of territorial song in the White-throated Sparrow. *In* "Bird Vocalizations" (R. A. Hinde ed.), p. 207. Cambridge Univ. Press, London and New York.

Ficken, M. S. and Ficken, R. W. (1962). The comparative ethology of wood warblers: A review. *Living Bird* **1**, 103–122.

Gaunt, A. S., Stein, R. C., and Gaunt, S. L. L. (1973). Pressure and air flow during distress calls of the Starling, *Sturnus vulgaris* (Aves; Passeriformes), *J. Exp. Zool.* **183**, 241–261.

Gompertz, T. (1961). The vocabulary of the Great Tit. *Brit. Birds* **54**, 369–418.

Goodwin, E. (1967). "Pigeons and Doves of the World." British Museum (Natural History), London.

Gottlieb, G. (1971). "Development of Species Identification in Birds: An Inquiry into the Prenatal Determinants of Perception." Univ. of Chicago Press, Chicago, Illinois.

Gottlieb, G., and Vandenbergh, J. G. (1968). Ontogeny of vocalization in duck and chick embryos. *J. Exp. Zool.* **168**, 307–326.

Greenewalt, C. H. (1968). "Bird Song: Acoustics and Physiology." Smithson. Inst. Press, Washington, D.C.

Griffin, D. R. (1953). Acoustic orientation in the Oil Bird, *Steatornis. Proc. Nat. Acad. Sci. U.S.* **39**, 884–893.

Gross, W. B. (1964a). Voice production by the chicken. *Poultry Sci.* **43**, 1005–1008.

Gross, W. B. (1964b). Devoicing the chicken. *Poultry Sci.* **43**, 1143–1144.

Güttinger, H. R. (1970). Zur Evolution von Verhaltensweisen und Lautäusserungen bei Prachtfinken (*Estrildidae*). *Z. Tierpsychol.* **27**, 1011–1075.

Haecker, V. (1970). "Der Gesang der Vögel, seine anatomischen und biologischen Grundlagen." Fischer, Jena.

Hamilton, J. B. (1938). Precocious masculine behaviour following administration of synthetic male hormone substance. *Endocrinology* **23**, 53–57.

Harris, C. L., Gross, W. B., and Robeson, A. (1968). Vocal acoustics of the chicken. *Poultry Sci.* **47**, 107–112.

Hatch, J. G. (1967). Diversity of the song of mockingbirds (*Mimus polyglottos*) reared in different auditory environments. Ph.D. Dissertation, Duke University, Durham, North Carolina.

Hérissant, M. (1753). Recherches sur les organes de la voix des quadrupèdes et de celle des oiseaux. *Acad. Roy. Sci. Mem.* pp. 279–295.

Hersch, G. L. (1966). Bird voices and resonant tuning in helium-air mixtures. Ph.D. Dissertation, University of California, Berkeley.

Hooker, T., and Hooker, B. I. (1969). Duetting. In "Bird Vocalizations" (R. A. Hinde, ed.), p. 185. Cambridge Univ. Press, London and New York.

Hutchison, J. B. (1967). Initiation of courtship by hypothalamic implants of testosterone propionate in castrated doves (*Streptopelia risoria*). *Nature (London)* **216**, 591–592.

Hutchison, J. B. (1970). Influence of gonadal hormones on the hypothalamic integration of courtship behaviour in the Barbary Dove. *J. Reprod. Fert., Suppl.* **11**, 15–41.

Hutchison, J. B. (1971). Effect of hypothalamic implants of gonadal steroids on courtship behaviour in Barbary Doves (*Streptopelia risoria*). *J. Endocrinol.* **50**, 97–113.

Immelmann, K. (1969). Song development in the Zebra Finch and other estrildid finches. In "Bird Vocalizations" (R. A. Hinde, ed.), p. 61. Cambridge Univ. Press, London and New York.

Johnsgard, P. A. (1968). "Waterfowl: Their Biology and Natural History." Univ. of Nebraska Press, Lincoln.

Kalischer, O. (1905). Das Grosshirn der Papagien in anatomischer und physiologischer Beziehung. *Abh. Preuss. Akad. Wiss.* **4**, 1–105.

Karten, H. J. (1968). The ascending auditory pathway in the pigeon (*Columba livia*). II. Telencephalic projections of the nucleus ovoidalis thalami. *Brain Res.* **11**, 134–153.

Karten, H. J., and Hodos, W. (1967). "A Stereotaxic Atlas of the Brain of the Pigeon (*Columbia livia*)." Johns Hopkins Press, Baltimore, Maryland.

King, J. R. (1972). Variation in the song of the Rufous-collared Sparrow, *Zonotrichia capensis*, in Northwestern Argentina. Z. *Tierpsychol.* 30, 344–373.

König, C. (1968). Lautäusserungen vom Rauhfusskauz (*Aegolius funereus*) und Sperlingskauz (*Glaucidium passerinum*). *Vogelwelt* 1, 115–138.

Konishi, M. (1963). The role of auditory feedback in the vocal behavior of the domestic fowl. Z. *Tierpsychol.* 20, 349–367.

Konishi, M. (1964). Song variation in a population of Oregon Juncos. *Condor* 66, 423–436.

Konishi, M. (1965). The role of auditory feedback in the control of vocalization in the White-crowned Sparrow. Z. *Tierpsychol.* 22, 770–783.

Konishi, M. (1970). Comparative neurophysiological studies of hearing and vocalizations in songbirds. Z. *Vergl. Physiol.* 66, 247–272.

Konishi, M. (1973a). Locatable and non-locatable acoustic signals for Barn Owls. *Amer. Natur.* 107, 775–785.

Konishi, M. (1973b). How the owl tracks its prey. *Amer. Sci.* 61, 419–424.

Konishi, M., and Nottebohm, F. (1969). Experimental studies in the ontogeny of avian vocalizations. *In* "Bird Vocalizations" (R. A. Hinde, ed.), p. 29. Cambridge Univ. Press, London and New York.

Lack, D. (1943). "The Life of the Robin," p. 27. Witherby, London.

Lack, D., and Southern, H. N. (1949). Birds on Tenerife. *Ibis* 91, 607–626.

Lade, B. I., and Thorpe, W. H. (1964). Dove songs as innately coded patterns of specific behaviour. *Nature* (*London*) 202, 366–368.

Lanyon, W. E. (1957). The comparative biology of the meadowlarks (*Sturnella*) in Wisconsin. *Publ. Nuttall Ornithol. Club* No. 1.

Lanyon, W. E. (1969). Vocal characters and avian systematics. *In* "Bird Vocalizations" (R. A. Hinde, ed.), p. 291. Cambridge Univ. Press, London and New York.

Lemon, R. E., and Scott, D. M. (1966). On the development of song in young cardinals. *Can. J. Zool.* 44, 191–197.

Löhrl, H. (1963). The use of bird calls to clarify taxonomic relationships. *Proc. Int. Ornithol. Congr., 13th, 1962* pp. 544–552.

Maley, M. J. (1969). Electrical stimulation of agonistic behaviour in the Mallard. *Behaviour* 34, 138–160.

Marler, P. (1955). Characteristics of some animal calls. *Nature* (*London*) 176, 6–7.

Marler, P. (1956a). Behaviour of the Chaffinch, *Fringilla coelebs*. *Behaviour, Suppl.* 5, 1–184.

Marler, P. (1956b). The voice of the Chaffinch and its function as a language. *Ibis* 98, 231–261.

Marler, P. (1957). Specific distinctiveness in the communication signals of birds. *Behaviour* 11, 13–39.

Marler, P. (1960). Bird songs and mate selection. *In* "Animal Sounds and Communication" (W. Lanyon and W. Tavolga, eds.), AIBS Publ. No. 7, p. 348. Amer. Inst. Biol. Sci., Washington, D.C.

Marler, P. (1964). Inheritance and learning in the development of animal vocalizations. *In* "Acoustic Behaviour of Animals" (R.-G. Busnel, ed.), p. 228. Elsevier, Amsterdam.

Marler, P. (1970a). A comparative approach to vocal learning: Song development in White-crowned Sparrows. *J. Comp. Physiol. Psychol.* 71, (monogr.), 1–25.

Marler, P. (1970b). Birdsong and speech development: Could there be parallels? *Amer. Sci.* 58, 669–673.

Marler, P., and Boatman, D. (1951). Observations on the birds of Pico, Azores. *Ibis* **93**, 90–99.

Marler, P., and Tamura, M. (1964). Culturally transmitted patterns of vocal behaviour in sparrows. *Science* **146**, 1483–1486.

Marler, P., Kreith, M., and Tamura, M. (1962). Song development in hand-raised Oregon Juncos. *Auk* **79**, 12–30.

Marler, P., Konishi, M., and Waser, M. S. (1973). Effects of continuous noise on avian hearing and vocal development. *Proc. Nat. Acad. Sci. U.S.* **70**, 1393–1396.

Medway, L. (1959). Echo-location among *Collocalia*. *Nature* (*London*) **184**, 1352–1353.

Metfessel, M. (1935). Roller Canary song produced without learning from external source. *Science* **81**, 470.

Miller, A. H. (1934). The vocal apparatus of some North American owls. *Condor* **36**, 204–213.

Miller, J. W., and Miller, L. S. (1958). Synopsis of behavior traits in the Ring Dove. *Anim. Behav.* **6**, 3–8

Miskimen, M. (1951). Sound production in passerine birds. *Auk* **68**, 493–504.

Miskimen, M. (1957). Absence of syrinx in the Turkey Vulture (*Cathartes aura*). *Auk* **74**, 104–105.

Morse, D. H. (1967). The contexts of songs in Black-throated and Blackburnian Warblers. *Wilson Bull.* **79**, 64–74.

Morse, D. H. (1970). Territorial and courtship songs of birds. *Nature* (*London*) **226**, 659–661.

Morton, E. S. (1970). Ecological sources of selection on avian sounds. Ph.D. Dissertation, Yale University, New Haven, Connecticut.

Mulligan, J. A. (1966). Singing behavior and its development in the Song Sparrow *Melospiza melodia*. *Univ. Calif., Berkeley, Publ. Zool.* **81**, 1–76.

Mulligan, J. A., and Olsen, K. C. (1969). Communication in Canary courtship calls. *In* "Bird Vocalizations" (R. A. Hinde, ed.), p. 165. Cambridge Univ. Press, London and New York.

Mundinger, P. C. (1970). Vocal imitation and individual recognition of finch calls. *Science* **168**, 480–482.

Murphey, R. K., and Phillips, R. E. (1967). Central patterning of a vocalization in fowl. *Nature* (*London*) **216**, 1125–1126.

Myers, J. A. (1917). Studies on the syrinx of *Gallus domesticus*. *J. Morphol.* **29**, 165–216.

Newman, J. D. (1970). Midbrain regions relevant to auditory communication in songbirds. *Brain Res.* **22**, 259–261.

Nicolai, J. (1959). Familientradition in der Gesangsentwicklung des Gimpels (*Pyrrhula pyrrhula* L.). *J. Ornithol.* **100**, 39–46.

Nicolai, J. (1964). Der Brutparasitismus der Viduinae als ethologisches Problem: Prägungsphänomene als Faktoren der Rassen- und Artbildung. *Z. Tierpsychol.* **21**, 129–204.

Nicolai, J. (1973). Das Lernprogramm in der Gesangausbildung der Strohwitwe (*Tetraenura fischeri* Reichenow). *Z. Tierpsychol.* **32**, 113–138.

Nottebohm, F. (1967). The role of sensory feedback in the development of avian vocalizations. *Proc. Int. Ornithol. Congr., 14th, 1966* p. 265.

Nottebohm, F. (1968). Auditory experience and song development in the Chaffinch, *Fringilla coelebs*. *Ibis* **110**, 549–568.

Nottebohm, F. (1969a). The "critical period" for song learning. *Ibis* **111**, 386–387.

Nottebohm, F. (1969b). The song of the Chingolo, Zonotrichia capensis, in Argentina: Description and evaluation of a system of dialects. Condor 71, 299–315.

Nottebohm, F. (1970). Ontogeny of bird song. Science 167, 950–956.

Nottebohm, F. (1971). Neural lateralization of vocal control in a passerine bird. I. Song. J. Exp. Zool. 177, 229–262.

Nottebohm, F. (1972a). Neural lateralization of vocal control in a passerine bird. II. Subsong, calls and a theory of vocal learning. J. Exp. Zool. 179, 35–49.

Nottebohm, F. (1972b). The origins of vocal learning. Amer. Natur. 106, 116–140.

Nottebohm, F., and Nottebohm, M. E. (1971). Vocalizations and breeding behaviour of surgically deafened Ring Doves, Streptoplia risoria. Anim. Behav. 19, 313–327.

Nottebohm, F., and Nottebohm, M. E. (1975). Ecological correlates of vocal variability in Zonotrichia capensis hypoleuca. Condor (in press).

O'Hara Rice, J., and Thompson, W. L. (1968). Song development in the Indigo Bunting. Anim. Behav. 16, 462–469.

Orians, G. H., and Christman, G. M. (1968). A comparative study of the behavior of Red-winged, Tricolored, and Yellow-headed Blackbirds. Univ. Calif., Berkeley, Publ. Zool. 84, 1–81.

Oring, L. W. (1968). Vocalizations of the Green and Solitary Sandpipers. Wilson Bull. 80, 395–420.

Payne, R. B. (1971). Duetting and chorus singing in African birds. Ostrich, Suppl. 9, 125–146.

Payne, R. S. (1971). Acoustic location of prey by Barn Owls (Tyto alba). J. Exp. Biol. 54, 535–573.

Penfield, W., and Roberts, L. (1959). "Speech and Brain Mechanisms," p. 109. Princeton Univ. Press, Princeton, New Jersey.

Popa, G. T., and Popa, F. G. (1933). Certain functions of the midbrain in pigeons. Proc. Roy. Soc., Ser. B 113, 191–195.

Potash, L. M. (1970a). Vocalizations elicited by electrical brain stimulation in Coturnix coturnix japonica. Behaviour 36, 149–167.

Potash, L. M. (1970b). Neuroanatomical regions relevant to production and analysis of vocalization within the avian torus semicircularis. Experientia 26, 1104–1105.

Potter, R. K., Kopp, G. A., and Green, H. C. (1947). "Visible Speech." Van Nostrand-Reinhold, Princeton, New Jersey (Reprint: Dover, New York, 1966).

Poulsen, H. (1951). Inheritance and learning in the song of the Chaffinch (Fringilla coelebs L.). Behaviour 3, 216–227.

Poulsen, H. (1959). Song learning in the Domestic Canary. Z. Tierpsychol. 16, 173–178.

Putkonen, P. T. S. (1967). Electrical stimulation of the avian brain. Ann. Acad. Sci. Fenn., Ser A 130, 1–95.

Reynolds, P. W. (1935). Notes on the birds of Cape Horn. Ibis [13] 5, 65–101.

Rüppel, W. (1933). Physiologie und Akustic der Vogelstimme. J. Ornithol. 81, 433–542.

Sauer, F. (1954). Die Entwicklung der Lautäusserungen vom Ei ab schalldicht gehaltener Dorngrasmücken (Sylvia c. communis, Latham) im Vergleich mit später isolierten und mit wildlebenden Artgenossen. Z. Tierpsychol. 11, 10–93.

Schleidt, W. M. (1964). Über die Spontaneität von Erbkoordinationen. Z. Tierpsychol. 21, 235–256.

Schwartzkopff, J. (1958). Soziale Verhaltensweisen bei hörenden und gehörlosen Dompfaffen (Pyrrhula pyrrhula L.). Experientia 14, 106–111.

Setterwall, C. G. (1901). Studier öfver Syrinx hos Polymyoda passeres. Dissertation, University of Lund, Sweden.

Sibley, C. G. (1970). A comparative study of the egg-white proteins of passerine birds. *Bull. Peabody Mus.* **32**, 1–131.

Smith, R. L. (1959). The songs of the Grasshopper Sparrow. *Wilson Bull.* **71**, 141–152.

Smith, W. J. (1969). Displays of *Sayornis phoebe* (Aves, Tyrannidae). *Behaviour* **33**, 283–322.

Smith, W. J. (1970). Song-like displays in *Sayornis* species. *Behaviour* **37**, 64–84.

Smith, W. J., and Vuilleumier, F. (1971). Evolutionary relationships of some South American ground tyrants. *Bull. Mus. Comp. Zool., Harvard Univ.* **141**, 179–268.

Snow, D. W. (1968). The singing assemblies of Little Hermits. *Living Bird* **3**, 47–55.

Stein, R. C. (1968). Modulation in bird sounds. *Auk* **85**, 229–243.

Stevens, S. S., and Davis, H. (1938). "Hearing, its Psychology and Physiology." Wiley, New York.

Stokes, A. W. (1967). Behavior of the Bobwhite, *Colinus virginianus. Auk* **84**, 1–33.

Thielcke, G. (1969). Geographic variation in bird vocalizations. *In* "Bird Vocalizations", (R. A. Hinde, ed.), p. 311. Cambridge Univ. Press, London and New York.

Thielcke, G. (1970). Vogelstimmen. *Verstaendl. Wiss.* **104**, 1–156.

Thielcke, G., and Thielcke, H. (1970). Die sozialen Funktionen verschiedener Gesangsformen des Sonnenvogels (*Leiothrix lutea*). *Z. Tierpsychol.* **27**, 177–185.

Thompson, W. L., and O'Hara Rice, J. (1970). Calls of the Indigo Bunting, *Passerina cyanea. Z. Tierpsychol.* **27**, 35–46.

Thorpe, W. H. (1954). The analysis of bird song. *Proc. Roy. Inst. Gt. Brit.* **35**, No. 161, 1–13.

Thorpe, W. H. (1958a). The learning of song patterns by birds, with special reference to the song of the Chaffinch, *Fringilla coelebs. Ibis* **100**, 535–570.

Thorpe, W. H. (1958b). The nature and characteristics of sub-song. *Brit. Birds* **51**, 509–514.

Thorpe, W. H. (1961). "Bird-Song," Cambridge Monogr. Exp. Biol. No. 12. Cambridge Univ. Press, London and New York.

Thorpe, W. H. (1966). Vocal imitation in the tropical Bou-Bou Shrike *Laniarius aethiopicus major* as a means of establishing and maintaining social bonds. *Ibis* **108**, 432–435.

Thorpe, W. H., and North, M. E. W. (1965). Origin and significance of the power of vocal imitation: With special reference to the antiphonal singing of birds. *Nature (London)* **208**, 219–222.

Vince, M. A. (1969). Embryonic communication, respiration and the synchronization of hatching. *In* "Bird Vocalizations" (R. A. Hinde, ed.), p. 233. Cambridge Univ. Press, London and New York.

Wagner, H. O. (1944). Notes on the life history of the Emerald Toucanet. *Wilson Bull.* **56**, 65–76.

Whitman, O. C. (1919). "Posthumous Works of Charles O. Whitman" (H. A. Carr ed.), Vol. 3. Carnegie Institute, Washington, D.C.

Wiley, R. H. (1971). Song groups in a singing assembly of Little Hermits. *Condor* **72**, 28–35.

Williams, H. W. (1969). Vocal behavior of adult California Quail. *Auk* **86**, 631–659.

Zigmond, R. E., Nottebohm, F., and Pfaff, D. W. (1973). Androgen-concentrating cells in the midbrain of a songbird. *Science* **179**, 1005–1007.

Chapter 6

INCUBATION*

Rudolf Drent

I. Scope of the Chapter

For many field observers, incubation is the relatively monotonous phase of egg-tending intercalated between the more evocative period

---

* For Professor G. P. Baerends on his sixtieth birthday, in thanks for introducing me to the Herring Gull and its egg.

of courtship and sexual display on the one hand and the intensely active phase of caring for the young on the other. The very inactivity of incubation has in fact greatly promoted its study. The parents can be observed at the nest with relative ease, and their activities are sufficiently restricted that direct observation, supported by simple recording devices, can yield reasonably complete data. As a further advantage, incubation activities are directed toward a clearly defined function; the incubating bird acts as a control system, and many of the physical parameters involved are amenable to experiment even in the field. Quite apart from the field tradition, as old as any branch of ornithology [Aldrovandi in 1603 recorded the observations of Dutch whalers regarding incubation practices of Brant (*Branta bernicla*) in arctic Greenland, the collection of eggs and adults for consumption being combined with scientific inquiry], a great deal of experimental investigation directly relevant to the natural situation has been carried out with eggs of the domestic fowl, from the point of view of perfecting artificial incubation. Fortunately, these investigations have recently been extensively reviewed (Landauer, 1967; Lundy, 1969), so this chapter will emphasize the field approach.

Although many animal species exhibit forms of egg care (regulation or manipulation of physical factors influencing development), transfer of heat by bodily contact is found only in some reptiles (Benedict, 1932), a few unusual mammals (Starck, 1965), and, in its most highly developed form, in birds. As we shall see, the converse statement (that all birds transfer heat to their eggs) is not true. The vast majority of birds manage heat transfer by sitting on the eggs, pressing a bare and highly vascular region of the abdomen tightly against them, hence the word "incubate" (from Latin *incubare,* to sit upon, see Jayakar and Spurway, 1965a). In its restricted sense, by incubation we mean "the process by which the heat necessary for embryonic development is transferred to an egg after it has been laid" (Beer, 1964). It is customary to reserve the word "brooding" for the application of heat to the hatchlings.

A chapter built around a process, as is this one, must touch on many points and is offered as an inventory of problems and phenomena, rather than an exhaustive compendium of facts about incubation.

II. Length of the Incubation Period

A. Definition

By "incubation period," we mean the development time of the embryo of a freshly laid egg, given the regular attention of the parent as typical of undisturbed incubation. In practice, the period is measured as the interval between the laying of the last egg and the hatching (i.e., young free of shell) of that egg, the assumption being that the incubation behavior of the parent is established by the time of clutch completion. Preferably, one should mark the eggs as laid so as to avoid any uncertainty, as the eggs do not always hatch in the order laid. These simple rules were early recognized (see Evans, 1891), but it is astonishing how inconsistently the term has been applied in the literature; although Heinroth (1922) described the above method very fully, both Nice (1954) and Swanberg (1950) felt called upon to cover the same ground anew.

B. FACTORS DETERMINING INCUBATION PERIOD

Heinroth (1922) presented a massive compilation, drawn to a large extent from his own painstaking investigations, of female body weight, egg weight, clutch size, and incubation period for 436 species, and in many cases, also the relative proportions of yolk and albumen in the fresh egg, as well as the weight of the newly hatched young. Heinroth compared closely related forms in which the degree of development of the young at hatching is comparable so that other factors, such as egg size, could be identified. More recent compilations have not refuted any of the generalities that he deduced; in attempting to unravel the causal web, much has since been written, but current views are little more than a selection from the alternatives posed by Heinroth.

Heinroth's data clearly showed a positive correlation between incubation period and absolute egg weight, as was confirmed in subsequent analyses (Needham, 1931; Worth, 1940). In the most recent investigation, Rahn and Ar (1974) marshal data covering 475 species and find a statistical relation between incubation period and egg weight (see Fig. 1) indistinguishible from that computed from Heinroth's original data (194 species). The relation to egg size holds within the species as well [poultry data reviewed by Landauer, 1967; Snow Petrel (*Pagodroma nivea*), Isenmann, 1970; Herring Gull (*Larus argentatus*), Parsons, 1972]. A further correlation evident in Heinroth's data was that larger birds tended to lay larger eggs [Huxley (1927) clarified this relation in a reanalysis of Heinroth's original data], again a conclusion supported by modern data (summarized in Lack, 1968). That the egg of larger birds is smaller relative to the parental body weight is a point that need not concern us here.

Heinroth noticed that long incubation periods were generally associated with long fledging periods, so that the search for factors influencing incubation period became in fact a search for factors influencing overall growth rates. Lack (1967, 1968) provides extensive documentation confirming the positive correlation between incubation and fledging periods. The most sensitive test would be to compare growth rates in the two

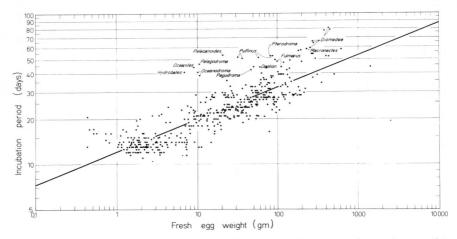

Fig. 1. Incubation period in days in relation to fresh egg weight in grams. The relation is described by a straight line when plotted on double logarithmic coordinates and is thus of exponential form [formula for the line of regression, based on data for 475 species; $\log I = \log 12.03 + 0.217 \log W \ (\pm 0.092)$]. Procellariiformes are identified for comparison with Fig. 7. (Modified from Rahn and Ar, 1974.)

phases; a step is taken in this direction in Fig. 2, where incubation periods are compared with growth rates in the posthatch period. There is evidence that embryonic growth rate is correlated with posthatching growth rate within the species as well [Axelsson (1954) reviews data on the domestic fowl].

Returning to the relation between egg weight and incubation period depicted in Fig. 1, we can now frame the question, why should growth rate decline with increasing egg size? The path of causality through the maze of correlations appears to me to be the following. We start with the generalization that growth rate is inversely related to mature body size (Rensch, 1954; documented for birds by Ricklefs, 1968a, 1973), a generalization founded on impressive empirical evidence but without apparently the benefit of an accepted physiological basis. We have seen that large birds tend to lay large eggs, and taking these two facts together, a positive correlation between egg weight and incubation period will result. This is not to say that egg size alone determines incubation period, for a number of other factors intervene in this basic relationship, resulting in a large scatter of incubation periods about any given egg weight:

1. *State of development of the bird at hatching.* As has been stressed by Portmann (1955), young birds at hatching are by no means equivalent. Woodpeckers, for instance, are hatched at a relatively early stage of development compared with most other groups,

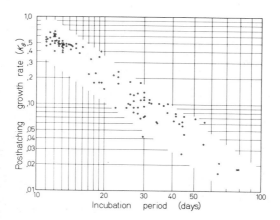

FIG. 2. Growth rate in the posthatching period, as measured by the statistic K_G, a measure of the rate of increase of body weight, in relation to incubation period, for 107 species, representing 12 orders of birds. Data on growth rate from Ricklefs (1968a, 1973), incubation periods from the original literature supplemented where necessary by various handbooks. Data for Anseriformes and Galliformes are omitted; few growth studies in wild forms are available, and aside from this, the precocial development of flight in these groups apparently affects growth so profoundly that a separate line of regression, lower than that for other birds, can be expected.

and this circumstance helps to explain the unexpectedly short incubation periods (Nice, 1954; supported by Lack, 1968).

2. *Relative proportion of reserve materials.* In altricial species, it is the rule that the yolk supplies laid down in the egg are completely exhausted by the time of hatching, whereas in precocial forms, up to half the original supply may still be present (see p. 404). The varying proportion of this reserve material in the fresh egg may reasonably be expected to obscure the correlation between development time and egg weight.

3. *Temperature.* Heinroth pointed out that the artificial incubation of the eggs of wild species, as with the use of foster parents (domestic fowl, turkey, goose), resulted in incubation periods only marginally shorter than those observed in the field, and argued that natural conditions allowed maximal embryonic growth. As will be discussed later, the rate at which definitive levels of incubation temperature are reached with respect to clutch completion shows a large margin of variability in the species investigated and might conceivably explain a part of the spread of incubation periods at a given egg weight. Temperatures during steady incubation have been measured in a large number of species (see Table I), and when viewed against the data of Barott (1937) establishing the degree of variability in incuba-

TABLE I

INTERNAL EGG TEMPERATURE AS MEASURED BY TELEMETRY DURING NATURAL
INCUBATION COMPARED TO BROOD PATCH AND BODY TEMPERATURES

Species	Temperature (°C)			Reference
	Egg	Brood patch	Body	
Struthioniformes				
Struthio camelus	36.0	—	—	Siegfried and Frost (1975a)
	—	—	38.7	Bligh and Hartley (1965)
Sphenisciformes				
Pygoscelis adeliae	33.7	—	37.9	Eklund and Charlton (1959)
Procellariiformes				
Puffinus pacificus	36–36.6	37.8	39.5	Howell and Bartholomew (1961)
Oceanodroma leucorhoa	34.8	37.4	37.9	R. H. Drent (unpublished)
Oceanites oceanicus	36	—	38.8	Beck and Brown (1972)
Pelecanoides urinatrix	35.8	—	38.0	Thoresen (1969)
Anseriformes				
Anas platyrhynchos	35–39	—	—	Caldwell and Cornwell (1975)
	—	—	41.2	Wetmore (1921)
Galliformes				
Domestic fowl	37–38	—	—	Süchting, in Groebbels (1937)
	—	40.7	—	Eycleshymer (1907)
	—	—	41.0	Yeates, in King and Farner (1961)
Gruiformes				
Gallinula chloropus	35.0	—	—	Siegfried and Frost (1975b)
Charadriiformes				
Catharacta skua	35.9	—	41.2	Eklund and Charlton (1959)
	35.6	39.3	—	Stonehouse (1956)
	36.4	39.0	41.4	Spellerberg (1969)
Larus argentatus	37.6–39.0	40.5	41.2	Drent (1970)
Sterna fuscata	38	39.6	40.5	Howell and Bartholomew (1962b)
Strigiformes				
Tyto alba	34.2	39.3	40.8	Howell (1964)
Speotyto cunicularia	35.5	39.8	40.2	Howell (1964)
Apodiformes				
Chaetura pelagica	35.2	—	—	Huggins (1941)
Stellula calliope	34.6[a]	—	—	Calder (1971)
Passeriformes				
Hirundo rustica	35.3	—	—	R. H. Drent (unpublished)
Troglodytes aedon	35.1	40.6	41.3	Kendeigh (1963); Baldwin and Kendeigh (1932)
Dumetella carolinensis	34.0	—	—	Huggins (1941)
Parus ater	35.4[a]	—	—	Haftorn (1966)
Parus caeruleus	35.5	—	—	J. A. L. Mertens (personal communication)
Sturnus vulgaris	35	—	—	G. Groot Bruinderink (unpublished)
Quelea quelea	35.2[b]	—	—	Ward (1965)
Poephila guttata	35.2	—	—	El Wailly (1966)
	—	—	40.2	Calder (1964)
Melospiza melodia	34.4	—	—	Huggins (1941)

[a] Synthetic egg.
[b] Determined after departure of parent.

tion period expected when the temperature is varied, they leave
no doubt that the spectacular variability between groups at a
given egg weight cannot be explained by differences in egg temperature alone. In particular, the contention that the very long
incubation periods of Procellariiformes are related to low incubation temperatures (Warham, 1971) is not warranted on the basis
of the measurements now available. Variability within the species,
however, is primarily due to vagaries in the attentiveness of the
individual parents (see, for example, Breckenridge, 1956; Tickell,
1962, 1968; Harris, 1969a and b; Drent, 1970; Beck and Brown,
. 1972). Another source of intraspecific variation is the effect of external weather conditions, implied in seasonal trends in incubation
period (von Haartman, 1956; Seel, 1968; MacRoberts and Mac-
Roberts, 1972; Winkel, 1970), although we must consider the
possibility that this trend is in part due to a seasonal shift in
egg size (Parsons, 1972).

4. *Ecological adjustment of growth rate.* Heinroth (1922) argued
that growth rate as a parameter exposed to natural selection
should vary in response to such factors as the intensity of predation (e.g., birds with vulnerable nesting sites showing more rapid
growth to shorten the vulnerable period when compared to related forms utilizing holes), a line of reasoning accepted by Lack
1947–1948, 1968) and Nice (1954, 1962); "The young of species
with secure, well protected nests generally remain in the nest
longer than those which are exposed to predators and inclement
weather." Lack (summarized in 1968) has argued that in species
in which the parents bring food to the young, an additional adjustment of growth rate may be made to keep the energy demands
of the brood within the foraging capabilities of the parents. In
this view, the extremely long incubation periods of the single-
brooded procellariiforms reflect a depressed overall growth rate
evolved to enable reproduction to take place in areas where the
foraging time budget of the parents leaves only a narrow margin
(Chapter 6, Volume I). This line of reasoning is particularly
compelling in species in which the brood is small (Ricklefs,
1968b), since adjustment by varying brood size either involves
relatively gigantic steps or cannot be carried further (brood of
one).

C. PROLONGED INCUBATION

In the event that the egg does not hatch at the expected time, the
parent generally continues to incubate, which is a functional response

in view of the variability in incubation period that has been observed. [Thus Beck and Brown (1972) record eggs hatching after 54 days in *Oceanites oceanicus* in which the normal period is 38–43; Tickell (1962) records an extreme of 50 days in *Pachyptila desolata,* normally 44–46; Harris (1969a) records an extreme of 65 days in *Puffinus lherminieri,* normal period averaging 49 days, and Harris (1969b) records an extreme of 51 days in *Oceanodroma castro,* normal period averaging 42 days, to mention only some examples from the Procellariiformes.] Holcomb (1970) has recently reviewed some of the records, and his data, in agreement with an earlier compilation (Skutch, 1962), indicate that this "safety margin" represents 50–100% in excess of the normal period (the mean extension in the 20 species covered by Holcomb amounts to 75%).

Eventually, incubation is terminated whether the egg has hatched or not, and, as Holcomb points out, there must be strong selective pressure against prolonging incubation beyond a reasonable limit, particularly in species in which replacement laying is possible.

III. The Brood Patch

A. STRUCTURE

Heat transfer during incubation is effected in the vast majority of birds by the close application to the eggs of a bare region of the ventral skin, known as the brood patch or incubation patch. In some species (e.g., pigeons and doves) the patch is situated in an apterium and is bare of feathers the year round; in others, it is an evanescent structure, the covering feathers being plucked from the area of the prospective patch by the parent (waterfowl, see Hanson, 1959) or the feathers falling out through hormonal action (majority of birds). Beyond defeathering (see Fig. 3), other structural changes take place (exceptionally, no histological changes are manifest see Maridon and Holmcomb, 1971). Lange (1928) enumerated the most striking differences, illustrated in Vol. III, Chapter 1, Fig. 3b, between the fully developed incubation patch and the adjacent skin as follows: (1) edema leading to a general flabbiness of the superficial skin, i.e., a folding and wrinkling, accompanied by an infiltration of leukocytes; (2) the epidermis (the cornified layer) thickens; (3) the number and dimensions of individual blood vessels increase, and strong muscle capsules develop around the arterioles. The functional interpretation given to these findings was that the looseness of the skin allowed closer contact with the eggs, and the danger of mechanical damage to the skin was lessened by the epidermal thickening; the strong hypertrophy of the vascular network was obviously related

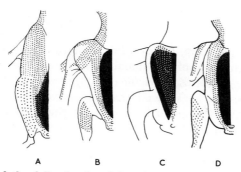

A B C D

Fɪɢ. 3. Extent of the fully developed brood patch (shown in black) in relation to pterylosis (feather tracts stippled, apteria open) in (A) a grebe, *Podiceps grisegena;* (B) a harrier, *Circus aeruginosus;* (C) a galliform, *Lophortyx californicus;* and (D) a passerine, *Corvus frugilegus.* (A, B, and D from Lange, 1928; C adapted from Jones, 1969a, where pterylosis of the appendages is not included).

to the function of the patch as a heat exchanger, with the development of the arteriole musculature seen as a shut-off mechanism for reducing heat loss when the parent was not actually incubating (Peterson, 1955). The role of the leukocytes remains conjectural. Since Lange's pioneer investigation, many aspects of incubation-patch development have been quantified, and much has been learned of the hormonal mediation of these changes. Detailed anatomic study of the vascular bed, however, has not been undertaken, nor are we well informed regarding the nature of the nerve endings in the patch.

The presence of the incubation patch is understandably closely linked with the incubation practices characteristic for the species (Jones, 1971), the most widespread pattern being that of both sexes sharing incubation and each developing a patch (Podicipediformes, Procellariiformes, Columbiformes, Piciformes, most Charadriiformes, most Gruiformes, some Falconiformes, some Passeriformes). In Galliformes, Strigiformes, most Falconiformes, Apodiformes, and some Passeriformes, only the female incubates and develops a patch, although in some species the male retains the potential (Jones, 1969a); and finally, there are some groups in which only the male incubates and develops a patch (phalaropes, jacanas, and some sandpipers in the Charadriiformes, the Turnicidae in the Gruiformes, and Tinamiformes). There are a number of exceptions to these generalizations, the most puzzling being species that incubate effectively without obvious incubation patches [late-breeding Cassin's Auklet, *Ptychoramphus aleutica,* Payne (1966) and Manuwal (1974)] or sexes without patches that incubate functionally [The male Bank Swallow, *Riparia riparia,* see Peterson (1955); further examples are given by Skutch (1962)].

B. HORMONAL CONTROL AND TIMETABLE OF DEVELOPMENT

In his pioneering study, Bailey (1952) showed that in passerine birds the seasonal formation of the incubation patch was under hormonal control. Working with *Zonotrichia leucophrys* and *Junco "oreganus"* [= *hyemalis* subsp.], he showed that (1) estrogen treatment of intact nonbreeding birds resulted in both defeathering and vascularity of the incubation patch; and (2) in hypophysectomized birds, estrogen treatment produced vascularization only, but estrogen followed by prolactin treatment led to both defeathering and vascularization. From this he concluded that estrogen controls vascularization, but that prolactin is also necessary for defeathering.

The most detailed subsequent investigation has been carried out by Hinde and his co-workers on the domestic canary (Steel and Hinde, 1963; Hinde and Steel, 1966; Hinde, 1967; Hutchison *et al.*, 1967), who besides treating intact birds with various hormone combinations, also studied ovariectomized birds. These workers conclude that defeathering in this species can be explained by the combined action of estrogen and prolactin, the effectiveness of estrogen alone in ovariectomized birds arguing against the involvement of progesterone. Changes in vascularity could be accounted for by estrogen alone. As a new element in the analysis, the mediation of the enhanced sensitivity of the skin of the incubation patch was investigated. In confirmation of the earlier studies of Hinde and Steel (1964), estrogen was found to act synergistically with another ovarian hormone (most likely progesterone) in bringing about the increased tactile sensitivity of the patch. Other studies of passerines, though less detailed, are in essential agreement, estrogen plus prolactin eliciting formation of the incubation patch in the European Starling (*Sturnus vulgaris*) (Lloyd, 1965b), the Red-winged Blackbird (*Agelaius phoeniceus*) (Selander and Kuich, 1963) and the House Sparrow (*Passer domesticus*) (Selander and Yang, 1966). In the last study, Selander and Yang were able to show that epidermal thickening was influenced by estrogen working with progesterone (the same combination effective in regulating tactile sensitivity in the canary) and, as the only discrepancy with the results of Bailey and Hinde, found that estrogen alone did not elicit full vascularity, the combination of estrogen and prolactin being essential.

More recently, several nonpasserine species have been investigated. In the California Quail (*Lophortyx californicus*), in which normally only the female incubates, Jones (1969b) found that full patch development could be elicited in nonbreeding birds of either sex by administration of the combination of estrogen plus prolactin. Unlike the case in several studies of passerines, however, treatment with estrogen alone had no effect on the ventral abdominal skin. The simplest explanation of this discrepancy is that the level of circulating prolactin differs in the various experimental subjects and that in all cases a synergism between estrogen and prolactin is necessary for vascularization as well as defeathering. As Jones points out, available data do indicate lower background levels of prolactin in quail compared with the passerines studied. Testosterone plus prolactin was also found to be effective in both sexes, and Jones surmises that this is the combination eliciting patch development in the male under field conditions (an emergency measure that occurs if the female disappears during incubation or the early brooding phase). In the domestic fowl, Jones *et al.* (1970) have shown that the classic combination of estrogen plus prolactin is capable of eliciting full patch development in immature birds.

In phalaropes, in which normally only the males form patches, Johns and Pfeiffer (1963) showed that incubation patches could be induced out of season in two species (*Phalaropus tricolor* and *P. lobatus*) in either sex by injection of testosterone plus prolactin, but not by estrogen plus prolactin. Female phalaropes are not deficient in testosterone, so the functional difference is presumably explained by the low (or perhaps even nonexistent) production of prolactin in this sex (Nicoll *et al.*, 1967), the male, on the contrary, normally secreting this hormone (Höhn and Cheng, 1967; summary in Höhn, 1967).

Although there are minor discrepancies among the studies mentioned, the broad pattern of patch induction by the action of steroid hormones combined with prolactin (defeathering, vascularization) and in combination with progesterone (epidermal thickening, sensitivity) seems a firmly established generality. Resolution of the tentative interpretation that estrogen application is in some species adequate to elicit vascularization of typical intensity must await adequate assessment of the blood levels of circulating prolactin before a major dichotomy in this respect need be concluded. It should be emphasized that in several studies the hormone-induced patches were found to be deficient in several respects compared to those formed in normal mated pairs in the field (Lloyd, 1965b; Selander and Kuich, 1963; Selander and Yang, 1966), and several lines of experimentation (Lloyd, 1965b; Hinde and Steel, 1966) tend to suggest that the manipulation of nest material, stimulation by the mate, and so forth, are intimately concerned with or mediate the release of hormones. Another general point worth comment is the finding that in most species both sexes respond to the same hormone combination, regardless of which one normally forms the patch. It can be concluded that a difference in hormone levels rather than a lack of sensitivity of the target tissue regulates the occurrence of the patch in the field, allowing the option of mobilizing the patch when it is called for. Only in the obligatory nest parasite *Molothrus ater* does the total lack of response to hormone manipulations indicate the loss of tissue sensitivity (Selander and Kuich, 1963). In other brood parasites, the hypothesis has been advanced that brood-patches fail to develop due to a lack of prolactin in the pituitary during breeding, rather than a loss in tissue sensitivity (Höhn, 1972, refers to data for *Eudynamys scolopacea*, a cuckoo).

In the House Sparrow (Selander and Yang, 1966) as in the Red-winged Blackbird (Selander and Kuich, 1963), Bank Swallow (Peterson, 1955), European Starling (Lloyd, 1965a), and domestic canary (Hinde, 1962; Hinde *et al.*, 1963; White and Hinde, 1968), thickening of the epidermis, increased vascularity, and mild edema are all apparent and defeathering is complete several days or weeks before the eggs are laid, although the most rapid increase in epidermal thickness, dermal vascularity, and edema occur shortly before the laying period (see Fig. 4). Defeathering, which usually begins several days before egg laying and is complete by the end of laying or early incubation itself requires 7–26 days in the canary, 25 days in the California Quail, "several days" in the Bank Swallow, and at least 12 days in the Red-winged Blackbird. The earlier report by Bailey (1952) of defeatherization occurring within a 24 hour period in the White-crowned Sparrow apparently involves the use of other criteria, and a closer look at this species will undoubtedly bring these results into line.

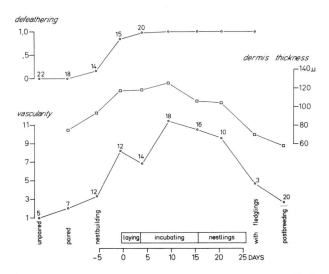

Fig. 4. Chronology of the development of the brood patch in the Red-winged Blackbird (*Agelaius phoeniceus*). Number of birds examined shown by lower graph unless otherwise indicated (by courtesy of L. C. Holcombe).

Although histological study has not yet been carried out, extensive samples of two procellariiform species have been examined in the hand. In *Oceanodroma castro*, Allan (1962) and Harris (1969b) found molt to occupy minimally 12 days, with defeathering completed no later than 20 days before laying. As in *Oceanodroma*, Beck and Brown (1972) indicate a wide variability in *Oceanites oceanicus*, with defeathering completed in some individuals as much as 50 days before laying. Most individuals had fully bare brood patches by the time of laying.

Refeathering of the patches may be delayed until the postnuptial molt (Red-winged Blackbird, Holcomb, personal communication), but in other cases, following egg loss, refeathering may begin sooner [in some individual canaries as soon as 10 days after egg removal, Hinde (1962); the same appears true of the American Sparrow Hawk (*Falco sparverius*), in which egg removal led to patch involution within the next few weeks and disappearance of the patch within approximately 1 month (Willoughby and Cade, 1964)].

In trying to forge a link between the demonstrable hormonal mediations on the one hand, and the reproductive timetable of the species on the other, we are on much less certain ground. Lehrman and Brody (1964) reported that progesterone elicited incubation behavior in the Ring Dove, and this was confirmed by the elegant work of Komisaruk (1967) involving implants of progesterone pellets in the brain. Cheng (1975) indicates that the accumulated rather than the daily dose level of progesterone is critical to the mediation of incubation behavior. Lehrman (1965) argued that incubation, once commenced, was maintained by the action of prolactin, the production of which was stimulated by the act of incubation itself or stimuli associated with it. Little direct experimental evidence is available to test how general this pattern may be, but the results of investigations of the Bengalese Finch (*Lonchura striata*) by Slater (1967) and Eisner (1969) are in line with this view. There is in any case a growing body of evidence that prolactin levels increase during the course of incubation (Breitenbach *et al.*, 1965; also the literature reviewed in Jones, 1969b). A complication in this research is that the sensitivity toward hormonal dosage is influenced by previous breeding experience (Lehrman and Wortis, 1967).

C. ALTERNATIVE SOLUTIONS

A number of groups manage incubation without the use of brood patches. The megapodes of Australasia (Frith, 1956a) lay their eggs in holes in the ground or in mounds of rotting vegetable matter and utilize natural heat for incubation. In some forms, the site is not visited after laying, but in others (see p. 372) temperature is regulated through-

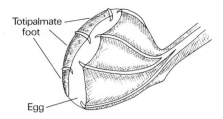

FIG. 5. Incubation in the Northern Gannet by application of the webs of the foot to the egg. (Redrawn from Nelson, 1966.)

out by elaborate mound-tending behavior. All pelecaniform birds likewise lack brood patches, and Fig. 5 shows how the Northern Gannet (*Morus bassanus*), a representative of this group, cradles the egg in the webs of the feet during incubation. To what extent the ventral body surface contributes to heating the eggs is an open question; in *Phaethon rubricauda* and *Sula sula*, Howell and Bartholomew (1962a) concluded that the feet did not provide the main source of heat. The issue is complicated by the fact that blood flow through the feet is enormously variable, an adjunct of the use of the feet as heat dissipators in thermoregulation, and extensive temperature (or, preferably, heat-flow) determinations on unrestrained birds will be needed to resolve the question.

IV. Physical Optima for Development

Due to extensive research aimed at perfecting the artificial incubation of eggs of the fowl and a few other domesticated species, optima for development are empirically well defined in three critical environmental dimensions—the gaseous environment (oxygen, carbon dioxide, and water vapor levels), the thermal environment, and the position of the eggs (the prevailing stance, as well as the frequency and type of movement to which they are subjected). This research has attempted not only to define the optima rigorously, but to gain insights into the nature of the physiological limitations of the embryo that set the optima in the first place. In the following paragraphs, we will explore the interpretation behind each of the established optima, in order to judge the expected relevance for the natural situation.

A. Gaseous Factors

Relative humidity influences the water loss from the egg, and this is in turn related to hatchability. Water loss is also affected by egg temperature and shell porosity, so it is impossible to specify a humidity optimum independent of other factors. It can be said, though, that hatchability is highest when the total water loss over the whole of incubation is between 10–12% of the initial egg weight, and this is achieved within the range 40–70% relative humidity (Barott, 1937; Robertson, 1961d; Romijn and Lokhorst, 1961, 1962; also critical discussion by Lundy, 1969). The effects of deviations in water loss are many, but little understood.

A continuous relationship exists between hatchability and oxygen concentration over a wide range. Hatchability is very sensitive to decreasing oxygen concentrations below 15%, and to rising concentrations above 40% (normal atmospheric air contains about 21% oxygen). Adequate information is not available for the range 15–40%, and further work will be needed to specify the optimum more exactly (Lundy, 1969). More is known concerning the effects of carbon dioxide concentration. Above 1%, both hatchability and growth rate of the embryo decrease; every percent rise in carbon dioxide is accompanied by a drop in hatchability of the order of 15% (Lundy, 1969).

What do these findings imply for the natural situation? In order to keep water loss within such narrow bounds, we might expect adaptations in shell porosity, in parental behavior (e.g., enhancing humidity conditions through nest structure), or both. Again, certain environments will exact adaptations for ensuring an adequate oxygen supply. At high altitudes, for instance, the partial pressure of oxygen is lowered to such an extent that eggs of the domestic fowl may require supplementary oxygen to attain normal hatching rates (Lundy, 1969). It would be of great interest to study the physiology of the egg of species normally occurring in the high mountains to see whether enhanced oxygen permeability of the shell, a reduction in metabolism of the embryo (Wangensteen et al., 1974), a greater reliance on anaerobic energy sources (Lokhorst and Romijn, 1964, 1969), or other factors are involved in adaptation to this environment. The accumulation of carbon dioxide in the air surrounding the eggs in the megapodes (see p. 375) and possibly other birds is another critical situation deserving physiological study. So far, the gaseous needs of the embryo have been discussed as though completely separate provisions could be reached for each. As has recently been stressed, however (review in Wangensteen, 1972), the three gases must be considered together because exchange between the embryo and the environment follows the same pathway.

The overall exchange between the embryo and the environment has recently been reexamined (Bartels, 1970; Wangensteen and Rahn, 1970), and the known facts can be explained on the assumption that diffusion alone is responsible. This makes it possible to treat the movements of any gas by applying the simple laws derived in other organisms. The most nearly complete data refer to the passage of water-vapor molecules, and these are believed to be representative for carbon dioxide and oxygen as well. Before summarizing this work, it will be convenient to clarify the anatomic relations.

The exchange organ of the developing embryo, the chorioallantois, is separated from the outside world by three successive barriers (see Fig. 6)—the eggshell and the two shell membranes. Numerous pores in the calcareous shell (counts range from 6000 to 17,000 pores per egg for the domestic fowl, see review by Simons, 1971) function as ventilation pores. The outer cuticle, a proteinaceous froth of air bubbles, forms a continuous but extremely thin covering—measurements varying

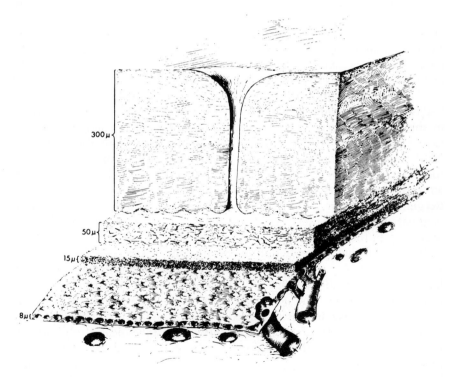

FIG. 6. The gas exchange pathway of the avian embryo, as exemplified by a scale drawing for the domestic fowl. From ambient air at the top, oxygen molecules must pass through the funnel-shaped pore in the shell, the outer shell membrane, and the thinner but more compact inner shell membrane to reach the blood in the very dense chorioallantoic capillary bed. Carbon dioxide and water travel in the reverse direction. Not shown is the hyperthin cuticle, covering the calcareous shell continuously. (From Wangensteen et al., 1970, by permission of North-Holland Publishing Company.)

from 0.5 to 12.8 mμ (Simons, 1971)—apparently functioning to protect the egg content from penetration by microorganisms without impairing gas exchange. The shell membranes are fibrous, and the resistance offered to gas transport has recently been measured. The permeability to oxygen as measured in fresh eggs is not sufficient to allow for the peak uptake observed later, and in fact, two changes take place in the course of incubation. First, the shell membranes undergo a drying process, and as the layer of adhering water molecules thins, the permeability to oxygen increases (Romijn, 1950, confirmed by Kutchai and Steen, 1971). Second, the oxygen tension in the capillary bed of the chorioallantois declines (review by Wangensteen, 1972), thus increasing the driving force for oxygen exchange. Because of the resistance of the shell, the partial pressure of carbon dioxide in the blood rises during later incubation. The eggshell membranes do not appear to form a measurable resistance to exchange of this gas. Water is lost from the egg content along the same gas exchange path, and Rahn and his associates have emphasized the argument that a compromise is reached in evolution between the opposing tendency of many pores to favor oxygen exchange and few pores in order to conserve water.

It is a well-established fact that the avian egg loses weight during incubation, and only slight errors are introduced if this weight loss is ascribed entirely to water loss (gaseous exchange with an R.Q. of 0.727 involves no weight change, and values for developing embryos are close to this in the latter half of incubation when gaseous exchange is large enough to have repercussions on weight loss) (see Kendeigh, 1940; Romijn and Lokhorst, 1960; Kashkin, 1961). The passage of water vapor through the membranes and eggshell can thus be measured quite simply by noting the weight change of the whole egg. Ar et al. (1974) measured weight loss (= water loss) of eggs belonging to 29 species under specified water-vapor pressure differences across the shell (the eggs were placed in desiccators held at a known temperature). Data from this study are plotted in Fig. 7 and allow the interpretation that there is an exponential relationship between daily water loss of the egg and the fresh-egg weight. Interestingly, the slope of the line depicting this relation is identical to the slope of the line relating basal metabolism and body weight in mature birds (Lasiewski and Dawson, 1967; Aschoff and Pohl, 1970). Bearing in mind that the relationship shown will also apply to the exchange of other gases, it is reasonable to ask whether the peak values of oxygen uptake during development in various birds might not be related to egg weight in the same fashion as metabolism is to the body weight in the hatched animal. Data on only four species are at present available [House Wren (*Troglodytes aedon*), Kendeigh, 1940; domestic duck, Kashkin, 1961; domestic fowl, Romijn and Lokhorst,

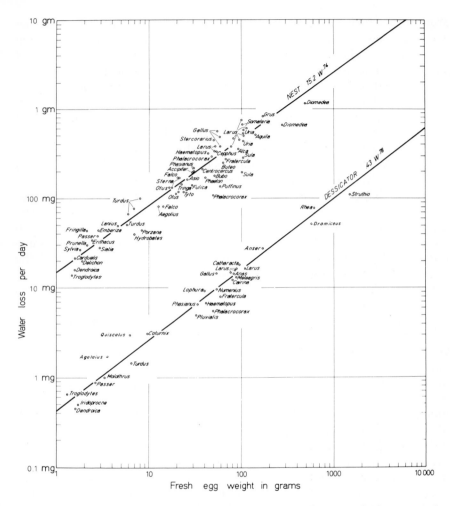

Fig. 7. Water loss of the egg during incubation in relation to fresh-egg weight. (*Top*) Total daily loss in egg weight per day under conditions of natural incubation; (*bottom*) water loss per mm Hg water-vapor pressure difference across the shell per day, as determined in nondeveloping eggs contained in desiccators. The slope of the two lines of regression is essentially equal (top, based on 57 studies on 46 species, from Drent, 1970, with additional data from Barth, 1953; Fisher, 1969; Scott, 1970; Tickell, 1968; Nelson, 1971; bottom, based on 29 species, from Ar *et al.*, 1974).

1960; Herring Gull, Drent, 1970]. In these data, the peak oxygen uptake is proportional to the 0.8 power of fresh-egg weight, but the data are too few to ascertain this slope reliably.

Returning now to the subject of water loss, field data for nearly fifty species yield a similar relation with virtually identical slope (see

Fig. 7). How does weight loss during natural incubation relate to the generalization from artificial incubation that hatch is maximal only when total water loss is kept within a certain range? Experimental data are not available, but if the daily weight losses depicted for field conditions in Fig. 7 are multiplied by the incubation period of the species in question, we arrive at the total water loss during incubation. Because of the way egg weight and incubation are related, total percentage weight loss over the entire incubation period is approximately equal for all species (averaging 16% in these data). This is in keeping with Groebbels' (1937, p. 355) earlier compilation and confirms Heinroth's (1922) supposition of equal percentage water loss, deduced from the observation that the freshly hatched young represented approximately two-thirds of the weight of the freshly laid egg, regardless of incubation period. These interrelations can readily be appreciated if we write the formula for total percent water loss during incubation (using the terminology of Ar et al., 1974):

$$F = M_{H_2O} I / W$$

where F = fractional water (weight) loss throughout incubation, M_{H_2O} = mass of water vapor lost daily, I = incubation period in days, W = fresh-egg weight. For F to be a constant fraction, the exponents relating daily water loss and incubation period to the fresh-egg weight must together yield the exponent 1.0. As we have seen (p. 336), incubation period is related to fresh-egg weight by the 0.22 exponent, so the predicted exponent relating daily water loss to fresh-egg weight would be $1.0 - 0.22 = 0.78$. As has been noted, the field data yield an exponent of 0.74, so the premise of equal fractional weight loss throughout incubation is true for the "average egg" regardless of initial weight. The reality of constraints on the tolerable fractional loss is well illustrated by taking species in which the incubation period diverges markedly from the expected or ideal value when plotted against egg weight, and examining water-loss data. The Procellariiformes, with their inordinately long incubation periods, are ideal material. Four species, covering a range in egg weight from 7 to 490 gm, are represented in the field data on weight loss (Fig. 7). Since the incubation period is longer than average (see Fig. 1), we would predict daily water loss to be lower than average if in fact a constant fractional value obtains; as can be seen, this is the case (the average percent weight loss in these four species is 16.8%, with no systematic weight influence discernible). Just how this apparent adjustability in eggshell porosity relates to other needs of the embryo (in particular in allowing entry of oxygen and egress of carbon dioxide) invites further research. As will be pointed out later, a certain amount of leeway is permitted the embryo by advancing the moment of pipping, thus allowing free access to environmental oxygen, in relation to the time of final emergence.

Another conclusion to be drawn from Fig. 7, since the lines for desiccator eggs and for eggs in the field are parallel, is that the water-vapor pressure difference across the shell during conditions of natural incubation is constant despite the range in egg weights and other variables involved. This constant difference, of about 35 torr, can be converted into a relative humidity measurement in the nest if egg temperature is accurately known (see Ar *et al.*, 1974), and the egg can in this sense be used as an indicator of humidity conditions.

B. TEMPERATURE

The pattern for the domestic fowl is the best documented (Lundy, 1969). The optimal temperature for incubation is 37°–38°C, and percentage hatch falls rapidly as temperatures deviate from this. No eggs at all survive continuous exposure to temperatures above 40.5°C or below 35°C. The significance of short-term deviations beyond these limits depends to a great extent on age, but overheating of the egg is a real danger at any time. Lethal internal temperature varies fom 42.2° to 48.3°C, depending on age. When egg temperature is lowered, difficulties also arise. If the egg is maintained for any length of time in the range between the optimum and the point of no development—the "physiological zero temperature," which has been determined as approximately 25°–27°C—various anomalies, such as disproportionate growth of the heart, may occur. Below physiological zero, there is a wide range of relative safety, however, at least for moderate periods of exposure. This region of suspended development is bordered by temperatures slightly below 0°C, which result in the formation of ice crystals in the egg that cause irreversible damage. Summing up, we can expect that, in general, the range for optimum development is narrow; death through overheating is uncomfortably close at all times; a slight fall in internal temperature, if long maintained, will lead to abnormal development, and an internal egg temperature between physiological zero (no development) and freezing can be tolerated for prolonged periods, the most important limitation being that tolerance declines with age.

Optimal holding temperatures for storing eggs prior to incubation have also been defined (Proudfoot, 1969) and can be said to fall in the same region of suspended development we have met with above (11°–13°C being variously given as the ideal temperatures, allowing the storage of eggs for several weeks without affecting hatchability, cf. Matthews, 1954). In the many species of birds in which steady incubation does not set in until completion of the clutch, a period of enforced storage inserts itself between laying and the application of temperatures sufficient to start development. Protective mechanisms designed to maintain the egg in the zone of suspended development can be expected to be involved.

C. Egg Position

Incubator research has been directed at a number of facets of egg position, including the axis of setting of the egg, the axis of rotation, the frequency of turning, the angle of turning, the number of planes of rotation, and the stage of incubation during which turning is beneficial. The conclusion appears warranted that the close imitation of the way the eggs are jostled by the sitting hen during natural incubation has in every respect proved to increase the hatching rate in incubators; this is certainly true for frequency of turning (once hourly), position of setting (blunt pole highest, angle of long axis approximately 20° from horizontal), pattern of rotation (multiplane), and is probably true for the angle of turning (optimum curve peaking at 45° alterations) but in this case field data are inadequate to specify the normal pattern.

Kaltofen (1961) and Robertson (1961b), confirming the earlier suggestion of New (1957), were able to show that the beneficial effect of turning, especially critical in the first half of incubation, was due to a reduction in the incidence of premature adhesions involving the extraembryonic membranes. Such early adhesions (the premature fusion of the chorioallantois with the inner shell membrane), mainly as a result of disruptions in the development of the amnion, lead to various difficulties later in incubation, such as interference with the uptake of albumen (see Fig. 16) or an aberrant positioning of the embryo within the egg, reducing the probability of successful hatch. Whether turning is of benefit to the embryo in other ways as well has not been critically evaluated in incubator research, but as we shall see (p. 363), field observations have led to the postulation of several supplementary functions of turning in natural incubation.

V. The Parent as an Incubator

A. Incubation Temperature and How It Is Regulated

In view of the complex temperature gradients that exist in nests during incubation, it should be clear that no single value represents the "incubation temperature." Values can, however, be specified for internal egg temperature, the upper surface of the egg (i.e., egg–brood patch inter-

TABLE II
Nest–Air Temperature[a]

Species	Nest temperature (°C)	Reference
Struthioniformes		
Struthio camelus	31.7	Siegfried and Frost (1975a)
Anseriformes		
Anas acuta	35.9	Irving and Krog (1956)
Mergus serrator	36.9	Barth (1949)
Domestic goose	33.4	Koch and Steinke (1944)
Falconiformes		
Accipiter brevipes	33.0	Ponomareva (1972)
Falco tinnunculus	35.0	Ponomareva (1972)
Falco vespertinus	34.0	Ponomareva (1972)
Galliformes		
Phasianus colchicus	35.1	Westerskov (1956)
Phasianus colchicus	35.3	Kessler (1960)
Lagopus lagopus	35.1	Barth (1949)
Charadriiformes		
Larus argentatus	35.5	Drent (1970)
Larus fuscus	34.9	Barth (1949)
Larus canus	36.2	Barth (1949)
Sterna hirundo	36.0	Barth (1949)
Columbiformes		
Streptopelia senegalensis	33.7	Ponomareva (1972)
Piciformes		
Dendrocopos leucopterus	35.5	Ponomareva (1972)
Passeriformes		
Lanius excubitor	33.0	Ponomareva (1972)
Lanius minor	32.5	Ponomareva (1972)
Troglodytes aedon	32.0	Kendeigh (1963)
Erythropygia galactotes	34.0	Ponomareva (1972)
Erithacus rubecula	34.8	Keil (1964)
Ficedula hypoleuca	33.1	Keil (1964)
Sylvia curruca	35.0	Ponomareva (1972)
Hippolais caligata	34.0	Ponomareva (1972)
Scotocerca inquieta	34.5	Ponomareva (1972)
Parus major	33.8	Keil (1964)
Parus major	35.0	Ponomareva (1972)
Parus caeruleus	37.3	Keil (1964)
Parus ater	34.3	Haftorn (1966)
Passer domesticus	33.5	Ponomareva (1972)
Passer montanus	31.5	Ponomareva (1972)
Passer ammodendri	30.5	Ponomareva (1972)
Rhodopechys obsoleta	32.5	Ponomareva (1972)

[a] Aside from Saiz and Hajek's (1968) report of nest-floor temperatures in *Macronectes giganteus*, there do not appear to be any temperature measurements of the air surrounding the egg in Procellariiformes.

face), the lower surface of the egg, the mean nest–air temperature between the eggs, and finally, the nest floor. Reliable values have been collected in Tables I and II. Internal egg temperatures typical of steady incubation range from 34° to 39°C in birds that sit on their eggs; as we shall see (p. 372) in megapodes where the eggs are incubated in mounds, egg temperature is minimally 33°–34°C. Clearly, there is no universal egg temperature during incubation, but the range of variation is small. Four orders are well represented in Table I, and it appears that Passeriformes generally experience lower developmental temperatures than many Galliformes and Charadriiformes, the Procellariiformes being intermediate. It should be noted that the present compilation does not support the tentative conclusion advanced by Warham (1971), who had fewer data to draw on, that egg temperatures in Procellariiformes were lower than in other birds, a factor contributing to the exceptionally long incubation periods observed in this group. We must now look elsewhere for an explanation of retarded growth rates in petrels and their allies (see p. 339).

Temperature is not constant through incubation, as can be seen in Fig. 8 presenting data on the Herring Gull. Temperature as measured at the top of the eggs shows a plateau value averaging 39.5°C and is virtually steady, at least from the time of clutch completion onward. Similar constancy at this site, reflecting a constant skin temperature of the sitting bird, has been reported in most other species investigated and implies that brood-patch defeathering is complete by the time the clutch is full (*Troglodytes aedon*, Baldwin and Kendeigh, 1932; *Parus major* and *P. ater*, Haftorn, 1966; *Branta canadensis*, Kossack, 1947; domestic fowl, Burke, 1925; *Phasianus colchicus*, Westerskov, 1956; *Charadrius alexandrinus* and *C. dubius*, Walters, 1958; *Larus canus*, Barth, 1949, 1955). In the Yellow-eyed Penguin (*Megadyptes antipodes*), however, Farner (1958) reported that the definitive level was not reached until the fifteenth day of incubation (average period 43 days). Penguins examined during this warm-up phase were found to exhibit less obvious vascularization of the brood patch than those beyond the second week. It is a symptom of the slow pace with which work in this field proceeds that almost twenty years after this research was done, we still do not know the time course of internal temperature during incubation for any penguin species, so we are still in the dark as to how general this slow development of a fully functional brood patch is and the role that delayed attainment of definitive temperature levels may have in explaining the longish periods of incubation typical of penguins. Baker (1968), does, however, provide corroborative evidence showing that in the Adélie Penguin (*Pygoscelis adeliae*), embryonic development is retarded during early incubation, as compared

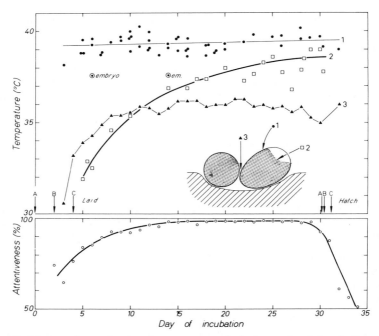

Fɪɢ. 8. Nest and egg temperatures, measured during natural incubation in the Herring Gull (*Larus argentatus*) (*top*), compared with constancy of incubation (= attentiveness) (*bottom*). Sites of temperature measurement indicated in diagram; in addition, probes were inserted adjacent to the embryo on day 6 and 14. The average time scale for laying and hatching of the three eggs is shown. (Compiled from Drent, 1970.)

with embryos of the domestic fowl; the logical experiment to sort this out would be to exchange eggs between fowl and penguin and to note embryonic development in each. The earlier suggestions of delayed attainment of plateau values in various Charadriiformes (Bergman, 1946) and several falconiform species (Holstein, 1942, 1944, 1950) are based on measurements within dummy eggs that are not reliable indicators of the temperatures applied to the egg surface. Indeed, for the Northern Goshawk (*Accipiter gentilis*), Wingstrand (1943) has shown by incubator experiments that definitive temperatures must be reached about the time of clutch completion, and not after the delay of more than a week as indicated by Holstein's data, even with the corrections given by Holstein (1944).

Air temperature between the eggs is still rising in the first week of incubation in the Herring Gull (Fig. 8), at a time when the egg–brood patch interface is being held steady (see also Fig. 9). Undoubtedly, the gradual change in parental attentiveness (as given in the bottom graph)

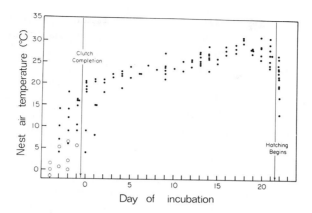

FIG. 9. Nest–air temperatures under incubating *Calidris alpina* from egg laying to nest departure (solid circles) and in unincubated nests during laying (open circles). The eggs are covered 98% of the time after clutch completion. (From Norton, 1972.)

is of influence here; the relatively large eggs of the Herring Gull are slow in warming up, and early in incubation, equilibrium values are rarely reached due to frequent interruptions. Whether what might be termed the heating power of the brood patch is still increasing in this period is a point worth study. As we have seen (p. 344), full vascularity of the brood patch is not reached until well along in incubation; conceivably, skin temperature might reflect the steady state of plateau values before the volume of blood coursing through the brood patch per time unit would reach the level of maximal heat transfer.

Internal egg temperature, as measured centrally, is lower than the surrounding air at first and rises above it from about day 14. This is the direct result of the rise of embryonic heat production, which surpasses the evaporative heat loss of the egg after approximately 10 days of development (as determined in a respirometer, Drent, 1970) and is responsible for the steadily increasing gradient thereafter. A similar influence of embryonic heat production has been reported for the domestic fowl (Romanoff, 1941; Romijn and Lokhorst, 1956, 1960) and domestic duck (Kashkin, 1961), whereas in the much smaller eggs of the House Wren, the elevation of egg temperature due to thermogenesis of the embryo is barely discernible (Kendeigh, 1940, 1963).

Since the embryo is at first a tiny structure and located on the upper surface of the yolk sphere close beneath the shell, central egg temperatures will not adequately reflect the levels actually experienced by the embryo in the early period of its life. Thermistors were inserted to lie beside the embryo during early incubation, and the embryo was found to experience a remarkably uniform thermal environment, experiencing a rise from 37.6° to 39°C during steady incubation, a record comparing favorably with that of a modern incubator. As might be surmised, measurements of temperature at various sites in the nest and within the eggs did not show a relation with external conditions at least over the range of 12°–28°C shade temperatures (Drent, 1970).

How is the relative stability of internal egg temperature achieved? Depending on the environmental circumstances, it may be surmised that compensatory behavior of the parent must be involved, and in fact a great deal of effort in field research has been devoted to this problem (see the reviews by Kendeigh, 1952; and Skutch, 1957). The basic pattern of response depends on whether the nest is exposed, with direct danger of overheating if the parent departs (in an abandoned nest of the Herring Gull, lethal internal egg temperatures were reached within 2 hours on a sunny morning at a shade air temperature of 18°C), or sheltered, in which case parental behavior can show greater variation. Constancy of incubation in the Great Tit (*Parus major*), for example—a species nesting in the shelter of a hole (nowadays often a nest box)—bears a direct relation to air temperature, as shown in Fig. 10 taken from Kluijver's (1950) study. In this case, the decline in time devoted to covering the eggs as the air temperature rises results from a simultaneous

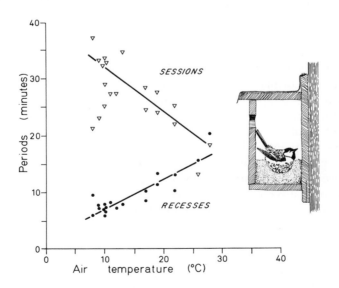

FIG. 10. Incubation rhythm of the Great Tit (*Parus major*) as registered by activity recorders mounted in nest-box entrance in relation to external air temperature. Sessions = sitting bout, recesses = absences, each point being the mean of determinations for one day. (From Kluijver, 1950.)

decrease in length of sessions, with increase in length of the intervening recesses, but all possible patterns of response occur, other species varying only the sessions, or only the recesses (see the review by von Haartman, 1956). That the variations in the parental pattern of incubation are in fact triggered by temperature stimuli was shown by von Haartman (1956) who heated the nest of the Pied Flycatcher (*Ficedula hypoleuca*), another hole-nester, experimentally for a 12 hour period and compared behavior of the incubating female with that shown during a 12 hour session of normal temperature the following day. The results were clear cut and demonstrated a shortening of sessions during the experimental period (33°C nest temperature) compared to the control period (nest temperature 16°C). On both days, the outside air temperatures were similar (16°–17°C). It would be very profitable to continue this type of work to discover whether these effects are due to air temperature in the nest directly, or whether they are in fact mediated by differences in egg temperature (for example, the egg temperature measured upon return from a recess).

In the response of the incubating bird to air temperature, there is a mechanism that apportions the time spent covering the eggs according to the needs of the embryo for heat in addition to that provided by the environment. At high air temperatures, the parent is free to devote time and energy to other activities, such as food-getting. In birds whose nests are exposed, protection of the eggs from isolation or predation may make practically continuous coverage a necessity. Even so, some degree of compensatory behavior of the parent is required, for in most cases the heat loss of the eggs, and hence the heat input by the parent needed to stabilize temperature, will be influenced by the external air temperature. In periods of continuous incubation, therefore, we should look for variations in the heat production of the sitting bird as representing what might be termed a fine adjustment in contrast to the crude adjustment afforded by variations in the time pattern of egg coverage. In fact, data on this point have recently become available.

Figure 11 shows for the Great Tit the relation between parental heat production and external air temperature at night when incubation by the female is continuous. Mertens (1967), from whose study these data are taken, measured the metabolism of the sitting bird in the nest box by means of indirect calorimetry. Since nest-box temperatures remained within the thermoneutral zone of the parent tit (throughout the range of external air temperatures depicted), the linear relation that heat production shows in relation to external air temperature must be viewed as an adjustment to compensate for varying amounts of heat loss from the clutch, determined primarily by the gradient between the nest chamber and the outside world. That parental compensation was not

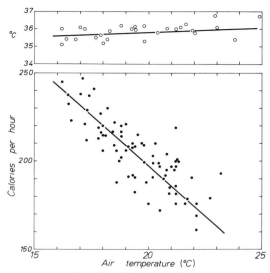

FIG. 11. Egg temperature (*top*) and heat production of the incubating female (*bottom*) of the Great Tit (*Parus major*) in relation to external air temperature. Each point is based on an hourly determination during the night when the bird sat continuously. (After data in Mertens, 1967, from Drent, 1972.)

complete is revealed by the slight decline in egg temperature with a fall in external air temperature (Fig. 11, top). This same relation has been shown in field data for other passerine species (Huggins, 1941; Kendeigh, 1963) and is similar to the registration of rectal temperature with respect to external air temperature in a homoiothermal animal (e.g., Aschoff, 1970). In fact, the heat-transfer relation between the parent and its eggs as revealed by this experiment can be thought of as though the parent treats the eggs as an extension of its own body core.

Direct experimental alteration of egg temperature is also revealing. Franks (1967) replaced the clutch of the Ring Dove with copper eggs through which water of carefully regulated temperature could be pumped. The frequency of various behavior patterns as determined during short-term exposures to egg temperatures altered through the range —4° to 62°C were compared to the control situation of normal egg temperature (38–39°C), and selected examples are plotted in Fig. 12. The immediate reaction to an abnormal egg temperature was often the adjustment of the contact between the eggs and the ventral body skin by means of trampling, resettling, and egg-shifting behavior. After several minutes, these preliminary adjustments would be followed by changes in some activities, unmistakably aimed at regulating the body temperature of the sitting parent—(see Fig. 12, top) gular flutter, shivering,

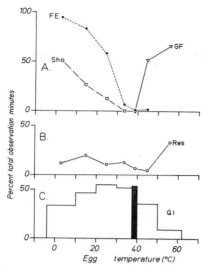

FIG. 12. Activities of the Ring Dove while incubating copper eggs whose internal temperature was varied by the experimenter. Ordinates are percent of total observation minutes in which the activity occurred. FE = feathers elevated or fluffed, SH = shivering, GF = gular flutter, Res = resettling on the eggs, QI = quiet incubation, eyes open but no other activity. Normal egg temperature, 38°–39°C, is shown by the stippled bloc. (From Franks, 1967.)

and elevation of the feathers. These doves, as confirmed by the work of McFarland and Baher (1968), do not appear to ruffle the plumage at high temperatures as do many other birds. A generally similar pattern of response has been found in experimentation with above-normal and below-normal egg temperature tests in the Herring Gull (Drent et al., 1970). Of the functionally relevant changes in behavior, only the adjustment of the egg–brood patch contact can be considered as aimed at regulating the egg temperature directly. Other actions are aimed at safeguarding the body temperature of the sitting bird, and as a corollary thereof, the stabilization of the egg temperature. As stressed earlier, the parent and its eggs act as a unit.

Continuous application of the brood patch is the only way of correcting a subnormal egg temperature, and similarly, overheated eggs in extreme cases where the surrounding air temperature exceeds the optimum for incubation, can be returned to normal only by continuous contact with the brood patches, the blood of the parent in this case carrying off the excess heat. In considering the ability of the parent to deal with extremes, therefore, it is pertinent to examine how much time the parent is able to devote to incubation in the strict sense, i.e., uninterrupted egg contact. The lower part of Fig. 12 is meant to convey an impression of this. The sharp decrease in time available for egg contact as egg temperature rises above the normal level emphasizes how precarious correction of overheating really is. The strategy of incubation adaptations must be to prevent overheating from occurring in the first place, rather than correcting aberrations afterwards. We will return to this point later.

Several recent field studies describe the reactions of incubating parents to extreme heat at exposed nests, and these support the interpretation of the experimental studies as given above. Maclean (1967) observed the Double-banded Courser (*Rhinoptilus africanus*) in the Kalahari and found that the parent frequently shaded but did not incubate the egg at air temperatures between 30° and 36°C, whereas at temperatures in excess of this the egg was always incubated. At these temperatures, convective heat loss from the egg will be inadequate to prevent a rise in temperature, so contact with the brood patch (which now functions to withdraw heat) becomes necessary. Russell (1969), who has studied the White-winged Dove (*Zenaida asiatica*) in the deserts of Arizona, points out that on hot days when ambient air temperatures reach 45°C in the shade, over-heating of the eggs can be prevented only by continuous contact with the brood patch, which is maintained at approximately 38.5°–39.5°C. The reaction of these birds is to settle more closely on the eggs at high air temperatures, and their effectiveness in coping with these conditions is illustrated in Fig. 13. How the parent is able to maintain thermal balance in these extreme conditions without even showing gular flutter is an unsolved problem deserving further study.

In summary, egg temperature in natural incubation is regulated by adjusting the time pattern of egg–brood patch contact, and adjusting the heat transfer during contact as though the eggs were part of the

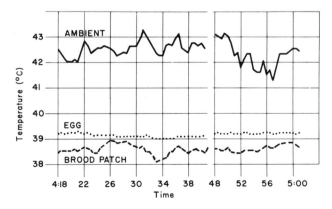

FIG. 13. Temperatures of ambient air, egg center, and brood patch as measured at the fully exposed nest of a White-winged Dove (*Zenaida asiatica*). (From Russell, 1969, by courtesy University of New Mexico Press.)

core of the parent. The main problem faced by the parent is the maintenance of its own body temperature in the face of environmental stresses
impinging on it while incubating. Finally, the behavior of the parent
must be anticipatory, in the sense of taking action before internal egg
temperature has reached dangerous extremes, since the ability of the
parent to bring aberrant temperatures back to normal is limited. The
question of what clues the parent might make use of in adjusting its
behavior to the expected needs of the embryo will be touched on later
(p. 378).

B. Egg Turning

Anyone who has watched an incubating bird for any length of time
will have noted periodic interruptions during which the parent rises,
peers down at the eggs, and makes a sweeping movement with the
bill, arching the neck and drawing the bill back among the eggs towards
the belly (a movement reminiscent of the way many birds retrieve eggs
that have rolled out of the nest). One result of such egg-poking behavior,
particularly noticeable in large clutches, is that the eggs are shifted
relative to one another in the nest bowl (e.g., Payne, 1921; Spingarn,
1934; Koskimies and Routamo, 1953; Kessler, 1960). More than two hundred years ago, de Reaumur (1751) suggested that the function of this
behavior in domestic fowl was to promote an even distribution of warmth
among the eggs of the clutch, a perfectly valid suggestion when we bear
in mind the large differences in internal temperature that have been measured between central and peripheral eggs; Huggins (1941) reported
an average difference of 5.6°C in a clutch of the Mallard (*Anas platyrhynchos*), and Mertens (1970) found a similar gradient in a clutch
of the Blue Tit (*Parus caeruleus*).

Redistribution of eggs to counteract the temperature gradients existing
in the nest is not the only function of this behavior, however, since
as we have seen (p. 353) even in nearly isothermal modern incubators,
a certain amount of tilting or turning improves hatchability. Egg turning
in the wild is doubtless significant in this way also, but an additional
and unexpected effect of egg-turning behavior was discovered by Lind
(1961) in the first large-scale attempt to measure variations in egg position in the field. By marking each of forty-four eggs with a series of
small spots spaced at 45° intervals around the blunt pole, and recording
which of the spots was uppermost during subsequent nest inspections,
Lind was able to show in the Black-tailed Godwit (*Limosa limosa*)
that there was a tendency for each egg to assume a stable position
as incubation progressed, i.e., the same mark tended to be uppermost.
Lind correlated this with the progressive development of weight asym-

metry of the egg as the embryo develops, such that the egg when placed in a dish of water will settle in a definite stance. Once the extraembryonic membranes have fused with the inner egg-shell membranes and egg content thus becomes fixed with respect to the shell, this asymmetry will be reflected in a restriction in the degree of movement of the egg about its long axis. By imitating egg-turning behavior himself using a thin stick, Lind was able to show that the eggs did not necessarily revolve about their long axes as might intuitively be supposed, but on the contrary, egg-turning behavior released the individual egg momentarily from the constraints posed by friction from the surrounding eggs, so that it was free to adopt the position of equilibrium dictated by its weight asymmetry. It is of course essential in studying this phenomenon to consider each egg individually, and it is because the data were pooled indiscriminately that the basic asymmetry of the frequency distribution of the individual eggs escaped Tinbergen (1948, 1953) in his earlier investigation of this problem. When new measurements were collected for the Herring Gull (Drent, 1970), the situation was found to be altogether comparable to that in the godwit, as will be clear from examination of Fig. 14.

Holcomb (1969), in his study of egg turning in twelve passerine species, used such conspicuous markings on the eggs that his work, as he realized himself, in fact constitutes an experiment on the influence of foreign color patches (red and orange fingernail polish) on turning behavior. In the nine species for which enough observations were collected to allow statistical analysis, it appeared that the bird adjusted egg position more often when the color blotches were visible than when the blotches were not visible, i.e., turned down in the nest. Prompted by this report, Gibb (1970) recorded similar findings in the Great Tit where eggs had been marked with a number in black ink: the numbered

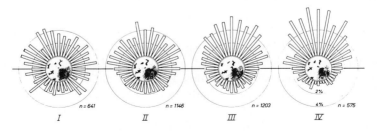

FIG. 14. Position of the Herring Gull egg in the nest during incubation. The egg is viewed from the blunt pole, and the heavy line represents the horizon. Shown in the form of a percentage frequency diagram is the position of that 10° sector of each egg that lies uppermost when the egg is in its mean position. Incubation has been subdivided into four periods. The figure is based on 3565 observations on 51 eggs. (From Drent, 1970.)

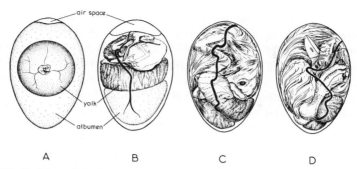

A   B   C   D

FIG. 15. Position of the Herring Gull embryo within the egg, as seen when that part of the shell lying uppermost in the nest is chipped away. Approximate ages: (A) 5 days, (B) 16–17 days, (C) 21–22 days, (D) 26 days (the egg hatches on day 27). In B, the head of the embryo is visualized as being transparent in order to show the position of the limbs. (From Drent, 1970.)

side was more often found down than up. Visual stimulation when the bird rises at the nest or returns from an absence thus appears to be one of the factors eliciting egg-turning behavior, but we cannot conclude from these observations anything about the influence of the parent on the stance of the eggs in the course of undisturbed incubation.

What significance might the progressive restriction in egg position we have seen in both the godwit and the gull have for the developing embryo? If the shell is peeled away from that part of the egg lying, on average, uppermost in the nest (or floating uppermost in the water) the pattern shown in Fig. 15 is revealed. At the earliest stage the yolk mass is still free to revolve within the shell, so the embryo is always uppermost [as pointed out by Ragosina (1963), the lighter portion of the

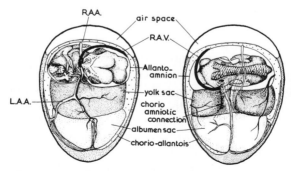

FIG. 16. *Passer domesticus* embryo of 8½ days exposed within the egg by removal of the chorioallantois, showing the chorioamniotic connection through which the albumen will be ingested during the ninth through eleventh day of incubation. RAA = right allantoic artery, LAA = left allantoic artery, RAV = right allantoic vein. (Redrawn from Witschi, 1949.)

yolk bears the embryo] and oriented as shown—the head faces away from the observer if the blunt pole is held to the left [known as "von Baer's rule," reviewed by Fargeix (1963)]. Eventually, the extraembryonic membranes and shell membranes fuse, and approximately midway in incubation, embryo position is fixed in relation to the shell and the egg becomes asymmetrical. During the whole of incubation, then, there is a predictable relation between the orientation of the embryo and the horizontal plane, at first because the contents revolve to bring the embryo uppermost, later because the frequent egg-turning behavior of the parent allows the egg to assume its equilibrium position. This fixed orientation has immediate relevance to the problems faced in escaping from the shell during hatching, and perhaps also for the attainment of the prehatching position (see p. 386).

C. Adaptiveness of the Nest

Nests function to enhance incubation effectiveness in at least three ways—by providing thermal isolation, rendering protection from predators, and in helping to maintain optimal positioning of the egg. That nests often provide crucial protection from climatic extremes has perhaps seemed so obvious that little critical research has been done in this area. Palmgren and Palmgren (1939) emphasized the significance that insulative value of the nest might have for the distributional ecology of the species and collected comparative data on a number of passerines by measuring the rate of cooling of a flask of water encased within two (or more) nests of the species in question. A number of suggestive differences were found among the fifteen species investigated. For instance, the difference in insulative value of the nests of the Chaffinch (*Fringilla coelebs*) and Brambling (*F. montifringilla*) was in the direction to be expected on the basis of their distribution, the more southerly form, (*F. coelebs*), possessing a less well insulated nest. Most nests were found to compare roughly with a layer of cotton of equal thickness as far as insulative ability is concerned. In her study of the survival value of the nest of the Long-tailed Tit (*Aegithalos caudatus*), Riehm (1970) also paid attention to this aspect and provides data showing the insulative function of the nest, due primarily to the astonishingly high number of feathers that are worked into its structure (ordinarily 1500–2000). This feature must make an important contribution toward reducing energy expenditure during incubation (see p. 395), as in this species the female incubates alone and must maintain the temperature of an egg mass equal to her own body weight of 8–9 gm [computed from data given by Heinroth and Heinroth (1926) using mean clutch size as observed by Riehm].

Particularly striking variations in nest construction are found within the hummingbirds, and Wagner (1955) has called attention to the critical role nest insulation must play in enabling the sitting female to forgo torpor during the often very cold nights. [Calder (1971) confirmed that the female *Stellula calliope* remained homoiothermic throughout the night and points out that early morning minima in the Wyoming mountains where he did his work were often close to freezing. More recently, Calder and Booser (1973) studied *Selasphorus platycercus* in the Colorado mountains, and detected hypothermia during two of the 161 incubation nights recorded.] A suggestive difference in the depth of the cup as well as the thickness of the walls between montane as compared to lowland species is illustrated in Fig. 17; the use of the fine mosses or plant cottons as a substantial nest lining necessitates the use of outer layers with water-shedding capacity (Wagner calls attention to the use of grass stems).

Within as well as among species, the nest structure may be varied in accord with local circumstances. Krüger (1965) observed the nesting of Gray Partridge (*Perdix perdix*) in large aviaries and found that during the laying phase the female altered features of the nest depending on the conditions prevailing at the time of her inspection visits (adding straw or an earth layer, opening the nest to cool at night), such alterations functioning to avoid excessive heating or cooling of the eggs before steady incubation sets in. This behavior calls to mind the intricate nest tending of the megapodes to be considered in the next section.

Varying the orientation of the nest can also serve temperature regulation; the heat of the sun may be utilized (Riehm, 1970; Dorst, 1962b) or avoided (Maclean, 1970), or the effect of cooling winds may be exploited (Ricklefs and Hainsworth, 1969). Many features of nest-site selection may have evolved to enhance the microclimate in which the eggs develop; Dorst (1962a,b) interprets a number of peculiarities of nest-site location in birds of the high Andes in this way, and Horváth (1964) discovered that seasonal shift in nest height and vegetation choice

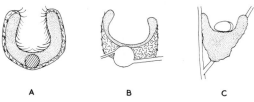

A   B   C

Fig. 17. Nest structure in hummingbirds in relation to cold stress: (A) the cross section of the deep-cupped nest of *Selasphorus platycercus* of the high mountains, the core of moss (stippled) clothed externally with lichens, internally with feathers; (B) and (C) shallower cups of warmer regions, both without feather linings, of two forms of *Amazilia* (from Wagner, 1955).

in the Rufous Hummingbird (*Selasphorus rufus*) results in the most equable microclimate being utilized at each period (cf. Calder, 1973).

Hole-nesting species are particularly interesting subjects, since choice experiments can be managed in the field without difficulty (Löhrl, 1970). It would be worthwhile to investigate the thermal advantage conferred by utilizing holes in living trees. Measurements made within living Saguaro cacti, for example, indicate how advantageous the buffering effect of living tissue may be to hole users such as the Elf Owl (*Micrathene whitneyi*) in the extreme desert environment in which these trees occur (Ligon, 1968).

Many features of nest construction by weavers can be interpreted as adaptations for protection from the rain, strong wind, or sun (Crook, 1963; Collias and Collias, 1964), and in addition, many features have been interpreted as conferring protection from predators, the second major function deserving comment. Selection of nest site is certainly involved here; Crook (1963) has pointed out that nests in homogeneous vegetation are situated far apart and are generally cryptic, while nests in colonies occur in sites protected by proximity to wasps' nests, ants, nests of raptorial birds, human habitation, or by suspension over water, features on which many naturalists in the tropics have commented [see, besides the literature reviewed by Crook, Hoesch (1940, 1956), Haverschmidt (1957), and MacLaren (1950), who established the association between *Lonchura cucullata* and the vicious red ant *Oecophylla smaragdina* by presenting data from Nigeria on the incidence of ants and bird nests in ninety-five suitable trees]. The ultimate in this type of nest-site preference is found in species that build a pendent nest at the tip of a fine supporting twig from which all projecting twiglets and leaves have been removed by the builder, making approach extremely difficult (Crook, 1963) (see Fig. 18). The long entrance tubes that many weavers build are also seen as antipredator devices, mainly aimed at tree-climbing snakes and hawks. As Collias and Collias (1971) were able to verify with first-hand observation, the weavers have the opportunity to attack the snake (which they do *en masse*) while it inserts its head into the entrance tube and before it is able to reach the nest chamber, if in fact it is able to do so without falling from its precarious hold. Protection offered against nest-robbers such as the Gabar Goshawk (*Melierax gabar*) was less effective, as some individuals at least simply tear a hole in the roof of the nest.

Hoesch (1933) described the nest of the African Penduline-tit (*Anthoscopus caroli*) and speculated on the possible functions of a false entrance leading into a blind pocket, the true entrance tube being collapsible (see Fig. 19). In the editorial comment appended to Hoesch's paper, field observations on the related *A. minutus*, a bird with a similar nest,

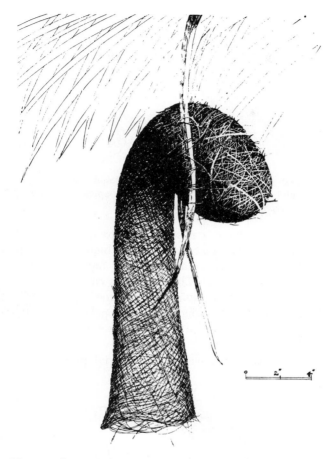

FIG. 18. The pensile nest of *Malimbus scutatus,* the long entrance tube being interpreted as an antipredator device. (From Crook, 1963.)

are quoted showing that the sitting parent often closed the entrance when leaving the nest and sometimes also when entering it to resume incubation. This combination of trap door and door mimic are seen as adaptations against egg-eating lizards and snakes.

Other comprehensive analyses of antipredator features of nest construction and situation have come from the tropics of the New World. Snow (1970) argues that much of the nest diversity typical of the tropics can be explained by selection in response to predation and indicates that the three main strategies followed are (1) to have a nest that is defendable [cf. Skutch's 1958 observations on woodpecker–toucan interactions at the nest hole; an extreme is shown by the hornbills, in which the female is sealed in the nest hole when incubation commences (see

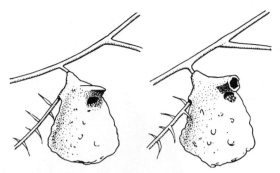

FIG. 19. The nest of *Anthoscopus caroli* in the closed state showing the false entrance pocket (*left*) and with true entrance tube opened (*right*). (Drawn from a photograph in Hoesch, 1933, from Grassé, 1950.)

Kemp 1971)], (2) to have a nest difficult to get at (such as pendent nests, or nests built on rock faces, an extreme being various swifts nesting behind waterfalls), (3) to have an extremely inconspicuous nest (either small or of light construction). These features can also operate in combination; Snow found many cotingids to reinforce (1) with (2). Examples of all of these features were amassed by Koepcke (1972) in her studies in interior Peru, and some of the delightful sketches she made are reproduced here (Figs. 20 and 21).

The covering of the eggs with nest material preceding departure of

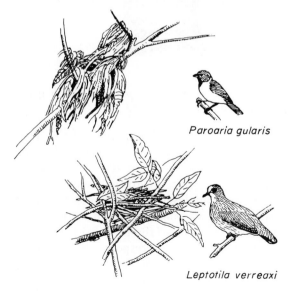

Paroaria gularis

Leptotila verreaxi

FIG. 20. Nests of *Paroaria gularis* and *Leptotila verreauxi* exemplifying protection from predators by inconspicuousness (i.e., similarity to nonnests). (From Koepcke, 1972.)

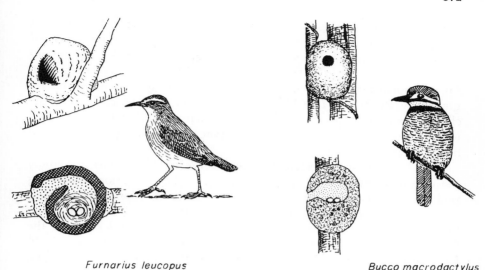

Furnarius leucopus   *Bucco macrodactylus*

FIG. 21. Nests of *Furnarius leucopus* and *Bucco macrodactylus*, defensible armored types providing protection from predators. (From Koepcke, 1972.)

the parent has been observed in many groundnesters with conspicuous eggs. One function of such behavior is undoubtedly to reduce the risk of discovery of the eggs by predators; data that support this assertion can be found in a study of *Podiceps nigricollis* in South Africa (Broek-huysen and Frost, 1968a,b). Sixty-odd nests in a colony of this grebe were inspected periodically, the condition of the eggs being recorded as covered, partially covered, or uncovered. When losses by flooding are discounted, a comparison can be drawn between nests where eggs hatched and where nests were terminated by predation [the species of predator(s) was not identified, gulls being suspected] as shown in Table III, constructed from the nest protocols presented in the papers cited. Clearly, birds that habitually covered their clutches were less prone to predation than those that failed to do so (χ^2, $P < 0.05$). Why

TABLE III

INFLUENCE OF EGG COVERING ON RISK OF PREDATION IN *Podiceps nigricollis*[a]

Fate of nests	State of eggs during nest inspections		
	Completely covered	Partially covered	Uncovered
Nests eventually hatching	36	33	10
Nests preyed upon	10	36	48

[a] Compiled from Broekhuysen and Frost (1968a,b).

the members of this colony were so dilatory in covering their eggs is unexplained, but it is often only in such extreme situations that the functions of behavior become demonstrable. The frequent disruptions of the incubating birds by the observers constituted a sort of experiment. An additional function of egg covering, as most recently discussed for grebes by Bochenski (1961), is to reduce the rate of cooling of the clutch during parental absence, a function that would repay experimental investigation in waterfowl (see the observations of cooling rates collected by Breckenridge, 1956).

Even the merest of nest scrapes may serve vital functions in helping the egg to maintain an optimal stance (see the discussion under egg position) or in helping to keep the clutch in the configuration best suited to coverage by the brood patches of the parent, a configuration that may function importantly in reducing overall heat loss, as in the four-egg clutches of waders (see p. 403). Birds without a nest of any kind, such as murres (*Uria*), who lay their eggs on bare rocky ledges, face special problems. As documented by Tuck (1961), the greatest mortality in a murre colony is due to eggs falling off the ledges or rolling away into crevices from which they cannot be retrieved. Early naturalists suggested that the pear shape of murre's eggs conferred some protection by causing the eggs to roll in a narrow arc, and measurements carried out by Uspenski (1958) showed how short the radius of rolling actually was, particularly as incubation advanced and the weight distribution within the egg changed. Recent experiments by Tschanz *et al.* (1969) have clarified this feature in a comparative analysis of eggs of *Uria aalge,* which always nest on ledges, and *Alca torda,* another alcid that usually nests in crevices or semicaves, but rarely utilizes ledges. As might be expected from their shape, *Uria* eggs roll in a much narrower arc than do *Alca* eggs, and when eggs placed on a nesting ledge were given a gentle push by the experimenter a significantly higher proportion of *Alca* eggs than *Uria* eggs dropped over the edge, and within the latter species, three shape variants also demonstrated the expected trend when arranged in a series from "typical *Uria*" to "*Alca*-like" (see Fig. 22). The applicability of these tests to the natural situation was verified by allowing parent *Uria* to incubate dummy eggs of plaster on the ledges, and observing the decrement with time. *Uria* egg types survived better than did *Alca* types, in the most extensive observation series, significantly so (42/50 compared to 35/50).

D. Nest Tending in the Megapodes

Due to the intensive research of Frith (1956b, 1957, 1959, 1962), we are well informed concerning the functional aspects of incubation

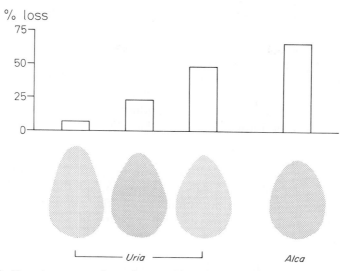

% loss

FIG. 22. Experiments on the influence of egg shape (three naturally occurring variants in *Uria aalge*, as compared with *Alca torda* of typical shape) on the risk that the egg will roll off the ledge when the sitting bird is disturbed. On each of 400 trials the egg was placed in the normal position on a narrow nesting ledge and given a gentle push by the experimentor. Shown is the percentage of trials in which the egg rolled off the ledge. (Combined from Tschanz *et al.*, 1969.)

in the megapode *Leipoa ocellata*, the Mallee Fowl. This dry-country inhabitant of Australia has an extended breeding season, dictated (so argues Frith) by the period during which operation of the mound is climatically feasible. The male tends the mound some 11 months of the year, and in this span has never been found more than 100 m from it. Preparations for breeding involve the excavation of the mound, a pit of some 1 × 3 m (where possible old mounds are used again, though not necessarily by the same individual) and the addition of a layer of leaf litter, an impressive heap filling the pit and reaching a crown some 60 cm above ground level, finally topped over by a thin layer of soil. When the heat of fermentation of the vegetable matter causes the mound temperature to approach 29°C (up to this point, the male's activities at the mound span approximately 5 months), egg-laying commences, and eggs will be laid at intervals of 6–7 days until the clutch, commonly 15–24 eggs, is completed. For somewhat longer than 2 months, the heat of fermentation of the decaying vegetable matter allows the mound to be regulated at about 33°C (there are some idiosyncrasies among individuals, the range being 32°–36°C), the problem being the dissipation of excess heat. The male tests mound temperature in the early morning by sampling sand in his mouth, and "airs" the mound when necessary by scraping away the soil covering, digging down to egg level before sunrise, then returning some sand (some 25–30 cm) for the remainder of the day, heaping up the soil covering again in the late afternoon as a protective blanket for the coming night. Excessively hot or very cool weather at this time of year will repress this airing activity; exceptionally cold weather may lead to an increase in the protective blanket (one bird building up a meter-high mound). After the burst of fermentation heat has died away, a period of less strenuous activity intervenes; the male visits the mound every 6–7 days and opens it before the sun is high in the sky, presumably to allow heat to escape, and his main concern is to avoid overheating; in hot weather, the soil blanket is increased, preventing the sun's heat from entering the mound. After mid-summer (i.e., the end of February) cooling of the mound threatens, and the amount of solar heat reaching the eggs is increased by raking away the sand cover and leaving the mound open during the heat of the day, then raking back the warmed sand at the end of the day. In this fashion, the male extends the useful period of the mound (15% compared to the 85% during which overheating is the problem).

Frith experimentally manipulated temperature at two mounds (one in which a heating coil was inserted, another in which the vegetable layer was removed) and found the Mallee Fowl to alter its behavior in a way functionally relevant to the natural situation, its actions depending on the season during which the alterations were carried out (heating of the mound in the spring leading to airing, but in the summer resulting in an increase in the soil blanket). Another experiment involved the comparison of the course of temperature in a natural mound tended by the parent with temperatures at two artificial nontended mounds, one complete with the vegetable layer, the other containing only sand. The results (Fig. 23) demonstrate that (1) heat from fermentation of the vegetable mass allows incubation temperature to be attained earlier than by use of solar heat alone; (2) the mound tending of the parent delays the decline in temperature consequent upon the end of fermentation and (3) enables a higher temperature to be maintained during the final phase when only solar heat is available. The overall success of the method compares favorably with conventional brood-patch incubation, 79% of the eggs not removed by predators hatching.

Only one other megapode, *Alectura lathami*, has been carefully studied. This species, a jungle dweller of Australia studied at the zoological gardens at Frankfurt am Main by Baltin (1969), utilizes only the heat of fermentation to achieve the incubation temperature in its mound, which consists of impressive heaps of leaf litter (field observations quoted by Baltin give 90 cm × 4 m diameter). The eggs are planted

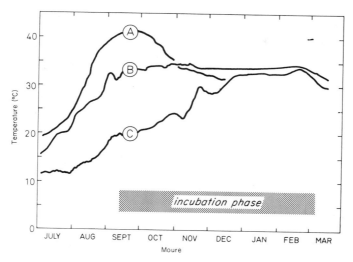

Fig. 23. Frith's experiment revealing the effectiveness of the parent Mallee-Fowl (*Leipoa ocellata*) in regulating mound temperature. Contrasted are the temperature registrations in (A) a mound of natural composition (sand + vegetable matter) but without the parent (unregulated heat of fermentation and solar heat); (B) a mound of natural composition tended by the parent (control): and (C) a mound of sand only without the parent (solar heat only). (From Frith, (1956b).

vertically in the mound and are not moved again, the clutch consisting of 22–23 eggs laid at intervals of 2–3 days. The temperature in the egg chamber as measured beside developing eggs averaged 33.7° and 33.2°C in two continuous traces during the 4 month period in which the mound contained developing eggs. Baltin noted that the heat of fermentation was enormous in the beginning, and without the periodic cooling caused by the sinking of shafts by the male and the excavation of the egg chamber for laying by the female, rapid overheating would result. However, he doubts that this behavior is primarily directed at temperature regulation, but believes instead that the main reason for the continual sorting and digging in the mound is to ensure adequate supplies of oxygen both for the developing egg and to maintain aerobic fermentation. As a concomitant, the carbon dioxide levels are kept down. In fact, measurements in the active mound often showed levels of oxygen below 10% and carbon dioxide up to 10–12% (Fig. 24). No experiments on artificial alterations of these factors were carried out, however, and Baltin does not suggest by what means the birds would measure these variables if indeed they are to be regulated. It would be worthwhile to carry out experiments like those done in the field by Frith, and at the same time we should extend the measurement of gaseous levels to other megapodes.

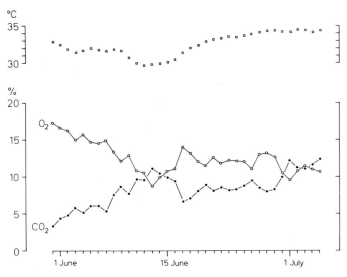

FIG. 24. Daily means for oxygen and carbon-dioxide concentrations in the mound of *Alectura lathami* during the 1965 season in the zoological gardens, Frankfurt. Shown on top is the temperature in the egg chamber. (From Baltin, 1969.)

E. ANTIPREDATOR BEHAVIOR

Several features of incubation behavior have already been interpreted as preventive measures to reduce predation (features of nest structure, egg covering, aspects of nest visitations such as the use of devious approach paths, a low frequency of nest visitation, exchange under cover of darkness), but in addition there are several direct mechanisms employed. Ground-nesting birds with inconspicuous yet vulnerable nests often show distraction displays, i.e., "the elaborate stereotyped activities performed by a *parent* bird that tend to concentrate the attention of potential predators on it and away from the nest or young" (Armstrong, 1964), when the nest is approached by predators including man. As reviewed by Simmons (1952, 1955), two principal types of distraction behavior are shown, for which Brown (1962) has coined the evocative terms "injury flight" and "rodent run" (Fig. 25). Brown argues that in the Western Sandpiper (*Calidris mauri*), injury flight, most often observed when first put off the nest, serves to deflect the predator from the nest, and that the rodent run, then brought into play, serves to draw the predator off by presenting stimuli releasing prey-seizing behavior. No information is available showing how effective this might be in coping with the arctic fox, the predator the sandpiper must most often face, but Brown quotes observations showing how various

FIG. 25. Distraction postures in the Western Sandpiper (*Calidris mauri*): (A) injury flight; (B) rodent run. (From Brown, 1962.)

waders successfully drew off dogs and a stoat by showing this behavior. He doubts that such distraction display can be functional where man is concerned as an egg gatherer. Ridpath (1972) illustrates the rodent run in the Tasmanian Native Hen (*Tribonyx mortierii*) (Fig. 26) and notes how the alarmed parent, running conspicuously in a zigzag course in the open, gives the fleeting impression of a small mammal.

Actual attack of potential predators is also widespread and has been studied particularly in terns and gulls. The extensive research of Tinbergen and his group on the Black-headed Gull (*Larus ridibundus*) (Tinbergen *et al.*, 1962; Kruuk, 1964; summarized in Tinbergen, 1967) has clarified a number of interrelated antipredator features. As an adjunct to habitat selection and the degree of dispersion of the well-camouflaged eggs, attacks *en masse* are carried out by the incubating birds when approached by predators; in the case of Herring Gulls and Carrion Crows (*Corvus corone*) these attacks are highly effective, as evidenced in the rate of egg disappearance when experimental plots of eggs were laid out both within and beyond the colony. Lemmetyinen (1971) extended this approach to terns, measuring the attack rate when a stuffed predator (*Corvus corone, C. corax, and Larus marinus*) was placed within or outside a ternery and also measuring the rate of disappearance of hen's eggs laid out in three situations. The loss of eggs was inversely correlated with the rate of attack as measured with the dummy predators.

FIG. 26. Distraction run in the Tasmanian Native Hen (*Tribonyx mortierii*) exhibited when a predator approaches the nest or young chicks (B) contrasted with the normal walking posture (A). (From Ridpath, 1972.)

In the course of incubation, the attack rate rises steadily in both species of terns (*Sterna hirundo* and *S. paradisaea*) as was earlier found in gulls (*Larus ridibundus*, Kruuk, 1964; *L. argentatus*, Drent, 1970) and appears to be a general phenomenon [Ring-necked Pheasant (*Phasianus colchicus*) Breitenbach *et al.* (1965); Common Nighthawk (*Chordeiles minor*), Gramza (1967); Red-backed Shrike (*Lanius collurio*) Gotzman (1967)] that may well be mediated by hormonal changes occurring in this period (see discussion in Breitenbach *et al.*, 1965). The functional basis for this increasing attack propensity must be that the increased risk to the parent is offset by the steadily rising value of the clutch. With every day that passes, the probability of starting a replacement clutch that will result in fledging young diminishes rapidly.

The Common Eider (*Somateria mollissima*), in which only the cryptically colored female incubates, lacks an attack response toward predators. Instead, the female sits tight, and in many cases the nest will presumably be overlooked by predators. If approach is very close, the female flushes at the last moment without covering the eggs. Discovery of the conspicuous eggs will now be inevitable, and a second line of defense is brought into action. The female defecates over the eggs as she flies off. The fluid ejected at this time is particularly noxious, quite different from normal fecal material, and Swennen (1968) has experimentally demonstrated the repellent effect this has on several potential predator species (ferret, rat).

The use of devious approach paths when returning to the nest (Ridpath, 1972) or even the simulation of food-hunting behavior when flying to the nest (Riehm, 1970) are further antipredator measures, and no doubt many more could be listed.

F.   The Organization of Incubation Behavior

The preceding pages have given an impression of how the parent actively influences the physical environment of incubation. As has been pointed out, needs such as protection of the clutch from predation must also be met, and at the same time some compromises must be reached between duties to the eggs and the nutritional demands of the parent(s). An overall view of the parent's role in incubation will now be developed, based mainly on work with the Herring Gull.

We have seen that data on parental behavior allow the definition of an optimum for egg temperature for incubation, and in a similar way, one can speak of an optimum for egg number. The data assembled in Fig. 27 are taken from a study of the Herring Gull by Baerends and his associates (1970) and illustrate clearly that when the egg number deviates from three, the normal clutch size in this species, the amount

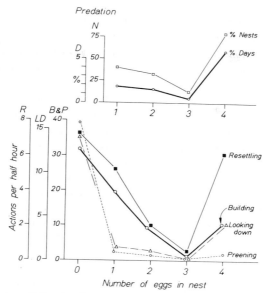

FIG. 27. Behavior of the incubating Herring Gull, as influenced by experimental alteration of clutch (*bottom:* mean frequencies of interruptive activities) compared to predation rate (*top*) as measured by percentage of days on which predation occurred, as well as total percentage of nests preyed upon. The clutches of 1, 2, and 4 are experimental, 3 being the normal number. (From Baerends *et al.,* 1970.)

of interruption in incubation increases dramatically. That such behavioral measures are significant for the eventual outcome of incubation is shown in the top section of the figure, where information on predation rate is presented (in these experiments the number of eggs was manipulated, i.e., all birds originally had a typical three-egg clutch). In some way, therefore, the Herring Gull is geared to a three-egg clutch, in correspondence to its three discrete brood patches. Can we in fact speak of a template, or inner expectation value, independent of individual experience? The work of Beer (1965) on the Black-billed Gull, *Larus bulleri,* is of great interest in this connection. This New Zealand species sometimes lays a clutch of three, but usually only two, although it still possesses the three brood patches typical of *Larus* gulls. Curiously, these birds, too, were found to incubate with least interruption on clutches of three when the range from one to four was examined by experiment, indicating that we can indeed speak of this "optimum number" as an abstraction.

Similarly, there is evidence (Baerends, 1974) for an optimum in egg size. Incubation shows the least interruption and predation is minimized with eggs of approximately normal size, but superimposed on this there is a tendency for large eggs to elicit more effective incubation behavior than small ones.

We will now assemble this kind of information in a diagram intending to show the functional structure of incubation behavior (Fig. 28). We can start by theorizing that there is an inner template or expectation for a variety of parameters and that information on the current status can be obtained either just previous to taking place at the nest or during the act of sitting itself. Subsequent actions of the bird will depend on the outcome of the comparison between the moment-to-moment measurement of the variable, and the inner expectation for that variable. What variables are in fact involved? Experiments (Baerends *et al.*, 1970) with a variety of egg models, some of which were transparent and possessed no visual value (or perhaps a slightly negative one), have demonstrated the importance of tactile feedback from the brood patches in the regulation of incubation behavior, and that temperature also plays a role has already been mentioned. The brood patches together with the legs, and probably the visual impression the bird receives before

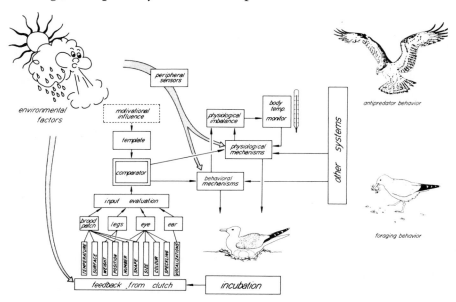

FIG. 28. Functional control of incubation in the Herring Gull, based mainly on Baerends (1959, 1970). Commencing at lower left, sensory input during incubation constitutes feedback on at least ten characteristics. The measured state of these variables is compared to the template (expectation or "optimum"), and mismatch will result in commands activating either physiological mechanisms (primarily those concerned in body temperature control) or behavioral mechanisms or both. Events are not completely predictable; hence, the template is here placed under the influence of motivational factors. Other needs of the animal may modify or interrupt incubation ("other systems"). Environmental events influence incubation directly eliciting anticipatory action ("feed forward") or via a change in feedback state of the clutch.

settling, are involved in measuring the position of the eggs in the nest bowl. The incubating birds strive to combine a perfect match of the eggs with the brood patches, with the direction of sitting preferred because of wind and other factors, such as the location of the main route of access. As shown by extensive experimentation with painted wooden egg models (not yet published in detail), information on a large number of visual attributes of the egg also influences incubation behavior. Goethe (1953, 1955) has commented on the probable significance of the vocalizations of the embryo during approximately the last 2 days of incubation, and this view is confirmed by Tschanz's (1968) demonstration of the influence of vocalizations from the egg on the incubation behavior of the parent *Uria aalge*. Furthermore, Impekoven (1973b) has shown experimentally in *Larus atricilla* that a number of behavior changes associated with incubation of pipping eggs can be elicited by calls of the embryo alone. This completes the list for the time being, but it is likely that other variables will be identified as work progresses.

There is no more than an analogy to a physical control system, since even with a completely specified sensory input, the reaction of the incubating bird is predictable only in a statistical sense. In order to incorporate this element of uncertainty in the model, the template, or inner expectation, is allowed to vary under influence of motivational factors, as suggested by Baerends (1970). Other solutions are feasible, however. For instance, in another connection, Jander (1968) introduces variability by interposing a source of "noise" between the outcome from the comparator and the effector organs, and there are doubtless other ways to incorporate this feature. Proceeding now to the ways in which the output from the comparator is translated into action, following McFarland (1971), a distinction has been made between physiological or behavioral mechanisms, but these are interwoven as shown in the diagram. Physiological mechanisms refer primarily to adjustments made to serve temperature regulation. For example, the signal "internal egg temperature below normal" might be expected to lead to increased blood flow through the brood patches and to increased heat production by the parent (shivering) in order to maintain a stable blood temperature despite the decrease experienced as the blood courses through the patches transferring heat to the eggs. A behavioral mechanism that might be elicited by the same signal would be an adjustment of the egg–brood patch contact, which might be effected by several behavior patterns (quivering, resettling, trampling, and so forth). Physiological mechanisms can also be brought into play as a by-product of behavioral actions, in addition to direct elicitation by a discrepancy in the incubation comparator as just discussed. The main reason for making this somewhat artificial separation of mechanisms in the diagram is to emphasize the importance of the body temperature control loop in the proper functioning of incubation behavior.

Finally, conflicting demands made on the bird are represented in the category "other systems." Prime among these is the need of the sitting bird to obtain the energy for its own metabolism during incubation. Some species can subsist on body fat during incubation or are fed by the mate and thus normally do not leave the nest; but in all cases, the nature of the adjustment is delicate. In birds of prey, for instance, where the male habitually forages for both himself and his partner, the eggs are abandoned when food-getting is difficult [Cavé (1968); and Southern (1970), who observed parental behavior and measured food supply at the same time, both reported these breakdowns as happening not uncommonly in *Falco tinnunculus* and *Strix aluco*]. Desertion of the nest in the face of impending starvation has also been observed in species in which the partners alternate at the nest. When survival of the parent is pitted against survival of the progeny, the former prevails. Nelson (1969) has provided a beautiful example from his work on the Red-footed Booby (*Sula sula*) on the Galapagos, where almost 70% of the eggs laid were abandoned before hatching. This species is a highly specialized feeder that travels long distances to and from the foraging areas, and Nelson argues that most desertion occurs because the off-duty partner is unable to return on schedule. Only a precarious margin of time uncommitted to feeding, in comparison to that available to other members of the group, appears to be free (Nelson, 1970). Fisher (1967) documents similar breakdowns in the Laysan Albatross (*Diomedea immutabilis*), and through regular weighings of the sitting birds was able to determine the degree of weight loss tolerated before the egg will be deserted whether the partner has returned or not. Rice and Kenyon (1962) indicate the tremendous potential range of the off-duty partner as computed from flight speeds and ringing returns, underlining the potential hazards in maintaining a regular schedule of alternation at the nest. In a totally different situation, Murton (1965) has evidence that in times of food shortage, incubation may be abandoned in the Wood Pigeon (*Columba palumbus*). Such drastic effects have not been observed in the Herring Gull, but the rhythm of foraging on the intertidal flats at low water has been shown to influence the pattern of exchange at the nest, and delayed return of the off-duty partner leads to a deterioration of incubation constancy (see Fig. 28, other systems).

We have been considering how feeding demands may influence incubation, but it is also pertinent to ask in what ways the demands of incubation may modify the feeding rhythm. For example, if a given species requires, say, one hour per day to satisfy its demands for food and water, and only one sex incubates, then the question becomes, what strategy of nest visitation will evolve? Will the bird subdivide its feeding time and make many brief excursions from the nest, or will it tend to take one long recess? Obviously, a great many factors will enter into the picture, but it may be worth considering the theoretically most efficient way of maintaining the egg at the temperature requisite to development. It is a matter of observation that when eggs are exposed upon departure of the parent, their rate of cooling exceeds the rate of heating when the parent returns. This relation is diagrammed in Fig. 29 (top). In other words, the time required to return the eggs to normal temperature will exceed the time of absence. In terms of energy, the parent is penalized for every absence, since returning the egg to normal temperature costs energy, and until the embryo is again metabolizing normally, there is no return on this energy. Now as the egg cools and the temperature difference between the egg and its surroundings become progressively smaller, the rate of cooling declines until it becomes hardly perceptible. This means that there is an asymptotic relation between the time that an egg has been exposed to cooling and the time required to return the egg to normal incubation temperatures afterward; empirical data from two species are plotted in Fig. 29.

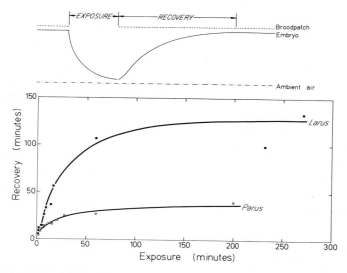

Fig. 29. Relation between cooling and heating rate of an egg in the nest of the Herring Gull (*Larus*) (based on 83 determinations) (from Drent, 1973) and the Blue Tit (*Parus*) (based on 116 determinations compiled from registrations supplied by J. A. L. Mertens). A generalized curve of internal egg temperature is shown at top to clarify the terminology. Parental absence is termed "exposure," and the time required to return the egg to the initial equilibrium temperature "recovery."

The data for the Herring Gull indicate that for an egg in this size range (100 gm) either absences above 1 minute should be avoided altogether, in view of the very unfavorable relation between exposure time and time required for recovery, or, if absences cannot be avoided, then they should be concentrated into one or two long excursions rather than many short ones. For example, if a feeding time of 1 hour is required the least delay in development, and the lowest energy expenditure, would be incurred in a recess of 60 minutes (total lost time $60 + 110 = 170$ minutes) compared to multiple recesses (thus ten excurions of 6 minutes each would total $60 + 10 \times 35 = 410$ minutes lost time). For the Blue Tit, with its much smaller eggs (1 gm) and well-insulated nest, penalties on excursions of moderate duration are much less severe (Fig. 29). Observed rhythms in these species are in accord, 75% of all interruptions of incubation in the Herring Gull being less than 1 minute and 93% 2 minutes or less, whereas in the Blue Tit 50% fall in the range of 5–8 minutes, and 75% are in the range of 3–10 minutes. Certainly, cooling and warming rates of the eggs are not the sole factors underlying the incubation rhythm, but it would nevertheless be of interest to extend this line of reasoning to other species.

Various antipredator measures may also interfere with incubation— both those aimed at protecting the eggs and those aimed at protecting the parent. The behavior involved has already been described (p. 376), and in the context of the Herring Gull, it is sufficient to note that the approach of a certain category of animals, either seen at first hand by the sitting gull or signalled by the alarm cries of its neighbors, will bring about the temporary abandonment of the nest. Removal of the eggshells, designed to remove a clue that might alert predators to the presence of newly hatched young (Tinbergen et al., 1962), is another behavior pattern that will interfere with incubation (more than 24 hours elapse between the hatching of the first and third chick).

On the question of the protection of the parent, the appearance of various birds of prey in the gullery leads to widespread panic flights in the colony [the Osprey (Pandion haliaetus) shown in Fig. 28, which takes gulls occasionally, is such an example]. Of special relevance to incubation is the nighttime vulnerability of the diurnally active Herring Gull. Outside the breeding season, gulls are extremely wary at night and prefer to sleep on isolated stretches of flat beach or out on some protected body of water where they roost in dense rafts. Commencement of incubation brings the adult into an area of high risk. Because of the demands of camouflage of the eggs, and later of the chicks, a complex terrain is chosen for the nest, where the parent is at a disadvantage and can be surprised by nightly forays of foxes, for example. What in fact happens is that nighttime incubation sets in only gradually and, interestingly enough, in relation to the density of incubating neighbors (and hence in relation to the number of potential warning devices, so to speak). Birds isolated from their fellows have been found to abandon the clutch regularly at night, up to 16 days after clutch completion.

In more general terms, the time of nest exchange has undoubtedly evolved in response to predator pressure in many species; the nocturnal exchange of Lesser Flamingos (*Phoeniconaias minor*) has been interpreted in this way (Brown and Root, 1971), as can the nocturnal exchange in many burrow-nesting sea birds; in some cases, however, the advantage can be questioned, as one set of predators is merely exchanged for another (Harris, 1969b). Skutch (1962) and Snow (1970) have drawn attention to the advantage of reducing the rate of nest visitation in cases in which the predator may be guided to the nest by observation of the parent, and this is another factor to be considered.

Turning finally to the influence of external weather conditions, we can distinguish two modes of action (see Fig. 28). On the one hand, external conditions can influence the parent through a change in the feedback from the clutch. In the case of the female Great Tit incubating at night (Fig. 11), for example, the reaction to lowered environmental temperature was probably triggered by a decline in egg temperature (leading to an increased output of heat from the brood patches) rather than by the fall in external air temperature per se. On the other hand, anticipatory action of the parent is also possible. A clear example is provided by the reaction of many open-nesting birds to the advent of a heavy shower. Skutch (1962) and Rittinghaus (1956, 1961) report how birds that were foraging away from the nest rushed back to resume incubation when driving rain commenced. Doubtless many other "predictive clues" are used by the parents in adjusting their behavior; Lind (1961) has shown that *Limosa limosa* covers the eggs at night in the laying phase particularly at low air temperatures, when there is danger that the embryos might freeze. Since the parents do not otherwise visit the nest at this time, this must be a reaction to the decrease of air temperature itself. Earlier, in the discussion on the regulation of egg temperature, it was argued that anticipatory action is vital to successful incubation. Prompt action may be needed to prevent stresses from having an effect on the egg. This we can call "feedforward," using McFarland's (1971) terminology, in distinction to the "feedback" from the clutch, and far more attention ought to be directed to finding out just what clues are utilized. We do not know, for example, exactly how a rise in nest temperature leads to a change in the pattern of attentiveness. Is this a reaction to air temperature itself, or to a change in cooling rate of the eggs? Further experimentation along the lines pioneered by von Haartman (1956) would certainly be profitable.

VI. Hatching

A. DESCRIPTION

Hatching itself is a relatively quick operation, but successful emergence depends in large measure on attainment of the correct starting position [see the discussion of malpositions in Landauer (1967, p. 205); even minor aberrations can greatly impair the probability of successful hatch] and this is the end result of a train of events commencing many days previous. The prehatching position illustrated for the Herring Gull in Fig. 15D, where the tarsal joints are located in the pointed pole of the egg, with the toes resting against the head, the head being withdrawn and tucked beneath the right wing, with the bill directed towards the air space in the blunt pole, is typical of birds in general—the essential features are identical in *Fulica atra, Limosa limosa, Larus ridibundus, Ardea cinerea, Somateria mollissima, Aptenodytes forsteri* (for references, see Drent, 1970) domestic fowl, Hamburger and Oppenheim, 1967, domestic duck, Oppenheim, 1970, and see Fig. 30, *Coturnix coturnix,* Vince, 1969. The single exception to date, the megapode *Alecturus lathami,* shows a peculiar mode of hatching (Baltin, 1969), as will be commented on later. This position is attained from the previous head-between-legs-resting-on-yolk-mass (Fig. 15C) by a gradual withdrawal

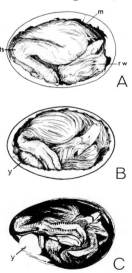

FIG. 30. Prehatching position in (A) domestic fowl, (B) *Anas platyrhynchos,* and (C) *Fulica atra.* (Combined from Hamburger and Oppenheim, 1967; Oppenheim, 1970; and Steinmetz, 1930.) m = membrane, rw = right wing, ts = tarsal joint, y = yolk sac.

of the head, the bill tip being passed up between the body and right wing. It is during this movement (termed "tucking" by Hamburger and Oppenheim, 1967) that a new type of coordinated muscular activity commences, as Oppenheim (1972) has demonstrated by sophisticated observational techniques in his comparative study of hatching behavior in ten bird species. Tucking can justifiably be reckoned as preparatory to escape from the shell. Penetration of the membranes separating the bill from the air chamber is the next significant event, and, in general, this coincides with the commencement of lung ventilation, which will in the remaining interval gradually usurp the respiratory role of the chorioallantois. The available data that show when lung ventilation starts in relation to the total incubation period are collected in Table IV. In some cases, the first respiratory movements may precede membrane penetration. Pipping, the formation of the first cracks in the shell, generally follows, but in a few species, and the Herring Gull is one of these, the first breaks on the shell surface may occur before the bill has worn its way through the membranes [Vince (1969) has designated this as "false pip"]. Observations on the external surface of the egg, made during periodic inspection visits to the nest, showed (Fig. 31) that a series of bumps started to appear about 64 hours before hatch, preceding the formation of the first hole by some 35 hours. The observations of Oppenheim (1972) on specially prepared eggs of the Laughing Gull (*Larus atricilla*) placed in a viewing incubator make clear that this track of bumps in fact reveals the path of the bill tip as the head is withdrawn during tucking (compare Fig. 15C and Fig. 31A). Oppenheim relates this to the relatively long bill and fragile eggshell and has noted similar scoring of the shell during tucking in the pigeon, and from Lind's (1961) description, this preliminary breakage also occurs

TABLE IV

CHRONOLOGY OF HATCHING IN SELECTED SPECIES

Species	Incubation period (days)	Initiation before emergence (hours)			Reference
		Breathing	Pipping	Rotation	
Domestic fowl	20–21	28–35	12.5	0.8	Vince (1970); Oppenheim (1972)
Colinus virginianus	23	55	44	0.6	Vince (1970); Oppenheim (1972)
Coturnix coturnix	16–17	40	10	—	Vince (1970)
Uria aalge	32	35	22	4.8	Tschanz (1968)

 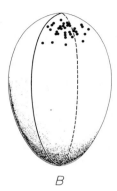

A   B

Fig. 31. Shell surface views of Herring Gull egg during hatching. (A) shows the progress of cracking in relation to the first crack made, plotted along the solid line. Each dot represents an observation on the position of the bill-tip. The broken line shows that portion of the egg which, on average, lies uppermost in the nest. (B) shows the site of the first hole, each dot representing the measurement for one egg. (From Drent, 1970.)

in the Black-tailed Godwit. In the majority of species, tucking leaves no trace on the shell, and the bill tip punctures the membranes and gains access to the air cell with the egg still intact [see, for example, the very complete description given for the fowl by Robertson (1961d) who traced head movements on the basis of X-ray photography].

Hatching itself is usually inaugurated quite suddenly when vigorous thrusts of the beak against the shell are accompanied by alternating treading movements of the legs that impart a rotatory component to the movements of the embryo. As a result, the hole broken in the shell is gradually extended in a circular fashion, the embryo rotating counterclockwise as viewed from the blunt pole. Several evanescent morphological features are associated with the hatching process. Shell breakage is facilitated by a knoblike protuberance on the upper mandible, known as the egg tooth, a structure that disappears in later life [detailed descriptions of the appearance of the egg tooth and indications of the age at which it disappears have been assembled by Clark (1961), Parkes and Clark (1964), Jehl (1968), and Sealy (1970), for a variety of species]. It is now agreed that the power source for the backward thrusts of the head bringing the egg tooth to bear against the inner egg surface resides in the Musculus complexus, called the hatching muscle, and extending from the vertebral column to the dorsal surface of the head (Fig. 32). Fisher retrieved this muscle from the oblivion of the early literature in a series of morphological studies tracing its size changes during embryonic development and in the early posthatch period in

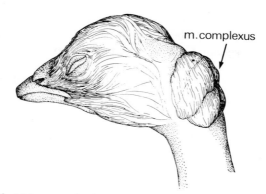

m.complexus

FIG. 32. The hatching muscle (*m. complexus*) in the domestic fowl on the day of hatching. This muscle is responsible for the vigorous back thrusts of the head during pipping and hatching, and later regresses. (From Bock and Hikida, 1968.)

the fowl (Fisher, 1958), grebes (Fisher, 1961), various gulls and terns (H. I. Fisher, 1962), and waterfowl (Fisher, 1966). As is also true for the American Coot (*Fulica americana*) (J. R. Fisher, 1962) the m. complexus attains maximal relative size at approximately the time of pipping and regresses sharply following hatching. Coincident with the pipping process, George and Iype (1963) were able to demonstrate a rapid depletion of glycogen stores in the hatching muscle, indicating active contraction in this period. Apparently only the twitch fibers play an important role in the process of shell breakage, the tonus components presumably functioning only in the adult (Bock and Hikida, 1968).

The degree of rotation varies greatly from species to species, depending in large measure on the strength of the shell and shell membranes. Tschanz (1968) noted that if the embryonic *Uria aalge* had not managed to sever the cap neatly in its first bout, it proceeded to rotate through a second revolution, and even a third, and Oppenheim (1972) experimentally induced additional rotation in *Larus atricilla* by taping shut the portion of the shell already cut. Eventually, the cap is lifted off lidlike by vigorous heaving movements of the shoulders coupled with extension of the legs, and the embryo emerges. In the Herring Gull, more than 24 hours elapse between the making of the first hole and emergence but the time devoted to the climax movements (to employ Oppenheim's term) described here is only a fraction of that time. How short the actual hatching process is, in relation to the long period elapsing while the embryo is in the prehatching position, is shown in Table IV.

Although Wetherbee and Bartlett (1962) indicate that hatching in the waders *Philohela minor* and *Catotrophorus semipalmatus* was rather unusual in that the blunt pole of the egg appeared to be torn off by convulsing movements of the embryo after only limited slit formation by the working of the bill, it seems from the brief description given that the general pattern shows great similarity to other species that tend to rotate little before emergence, namely, gulls and *Limosa limosa.* Undoubtedly highly aberrant, however, is the mode of hatching in the megapodes. In all species, the egg is buried and develops due to heat supplied by the sun, fermentation, volcanic heat, or a combination of these (review in Frith, 1956a). Hatching has been described in two species, *Leipoa ocellata* (Frith, 1962) and *Alectura lathami* (Baltin, 1969). Prior to hatching, the embryo remains in the head-between-legs position, and the shell is burst by the combined action of legs and wings, without the help of the bill, which lacks an egg tooth at this stage (in earlier embryonic life one is present). It is possible that this unique method of shattering the egg is related to the circumstance that the egg stands vertically in a relatively densely packed mound. As a further peculiarity, the egg virtually lacks an air space.

B. CRITICAL REQUIREMENTS

Considering first the relevance of egg-turning behavior for the hatching process, it is a striking fact that in all cases examined so far, the position that the embryo assumes in the egg, together with the weight asymmetry of the egg, result in the first hole formed being uppermost. It is tempting to speculate that this has functional significance. Tschanz (1968), for example, who describes this relation in *Uria aalge,* suggests that this might be important in assuring the fetus an adequate supply of oxygen, since the ledges on which the murres nest are often extremely slimy at hatching time, and upside-down eggs might be in danger of suffocating. It is also possible that there is less danger that the eggshell will be crushed when the hole is uppermost and obvious to the parent. Birds incubating hatching eggs settle in an entirely different stance and appear to apply less pressure to the eggs. In any case, a certain amount of mortality occurs at this stage by eggs flattened by the parent (personal observations on gulls). Again, it might be important for the embryo to maintain a certain position with respect to gravity, on account of requirements for adequate lung breathing, for example. Clearly, it would be rewarding to direct research to the ways in which the position of the egg influences the assumption of the correct prehatching position of the embryo, and how it influences the process of hatching itself. It is suggestive that, when air breathing has begun and the embryo is able to vocalize, one of the stimuli eliciting a type of cheeping known as the distress call is to turn the egg from the resting position. Würdinger (1970, 1975) experimented with nests of *Anser indicus.* When the pipped egg was turned upside down, the fetus started to call, and the effect on the parent was an immediate return to the nest followed by egg-turning behavior, after which the parent settled to incubate. Recordings of this distress call, when played back to parents sitting quietly, caused them to rise at the nest and poke the eggs.

Influence of the gaseous environment on hatching has received considerable attention. In a series of elegant experiments, Visschedijk (1962, 1968a,b) was able to show that the moment of pipping, but not the time of emergence, could be accelerated by sealing the pores in the shell over the air cell (by dipping the egg in paraffin) and could be retarded by making an artificial hole in this region as compared to untampered controls. Visschedijk identified the stimulus for pipping as increased carbon dioxide or reduced oxygen concentration in the air space or both. A large increase in the rate of oxygen uptake by the embryo follows pipping, suggesting that the preceding plateau in uptake may correspond with the maximal exchange permitted by the permeability of the shell. Part of the observed variability in the moment of pipping with regard to emergence (see Vince's 1970 compilation) may be explained by relative differences in oxygen permeabilities of the shell. In artificial incubation, it is customary to move the eggs upon pipping to hatching incubators held at higher humidities in order to avoid excessive drying of the membranes with consequent difficulties in hatching. As emphasized by Romijn (1950), high humidity levels preceding pipping depress the rate of oxygen diffusion through the shell and may be disastrous. Whether the incubating parent regulates humidity levels in this or indeed any phase of incubation is unknown. The stimuli that initiate hatching itself are still far from clear (see Vince's 1970 review). Although a mediation by hormonal action (thyroxine) is implicated, we still need to know the exact circumstances leading to its release, and whether other hormones may be involved (Oppenheim, 1973).

C. Synchronization of Hatch within the Clutch

So far, the hatching process has been discussed as an interaction between the behavior of the parent and the needs of the embryo. Recently, however, it has become clear that there is in addition a communication between the embryos of a clutch, effecting a synchronization of hatch in certain species. Effective incubation is established gradually during the laying period in the majority of birds [Haftorn's 1966 record for a clutch of the Coal Tit (*Parus ater*) is shown in Fig. 33], and this results in a wide range of development within the clutch when the last egg is laid. This type of observation implies that a considerable spread in time of hatching of the clutch will occur, but surprisingly this is not always the case.

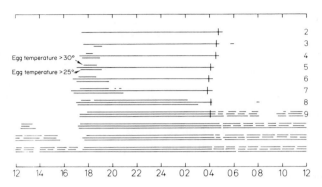

FIG. 33. Gradual establishment of steady incubation in the Coal Tit (*Parus ater*). The graph is to be read from top to bottom, each track giving the record of a 24 hour period (note the time scale in hours of the day along the bottom). The central line denotes presence of the female in the nest, the line below it periods in which the egg temperature exceeded 25°C, and the line above it periods during which the egg temperature exceeded 30°C. The vertical bars mark the time of egg laying in the early morning of successive days; shown are the 24 hour periods during which the second to ninth eggs were laid and the first three periods of incubation of the full clutch. (From Haftorn, 1966.)

Many years ago, Heinroth (1922, 1938) pointed out that in natural circumstances the entire clutch (11–13 eggs) of the Mallard tends to hatch within 2 hours [more recently, Bjärvall (1967) has indicated a range of 3–8 hours], an astonishing degree of synchronization when one considers the many factors contributing to variations in length of incubation. Of course, this synchronization would have adaptive significance in species such as the Mallard in which the mother leads the young from the nest together, not to return, for late-hatchers will be left behind (personal observations on Mallard, Common Shelduck (*Tadorna tadorna*), and Common Eider; see also Bjärvall, 1968; Ryder, 1972). Curiously, Prince *et al.* (1970) found that the range of emergence times of the eggs of a clutch of the Mallard, when incubated artificially, increased from 6 to 16 hours going from a semiwild starting population to the incubator-raised f_2 generation, a finding consistent with the hypothesis that in nature some selection pressure tends to stabilize incubation period. Beyond this, however, an active synchronizing process is involved. Faust (1960) observed the incubation habits of the Greater Rhea (*Rhea americana*) in captivity, a species in which only the male incubates, the female continuing to add eggs to the nest after incubation has started. Remarkably, despite differences in the age of the eggs of up to 9 (commonly) and 12 (exceptionally) days, hatching is highly synchronous, the entire clutch emerging within 2–3 hours. During this study, incubation period (time of laying to time of hatch) was found to vary through extremes of 27–41 days. This was strong evidence for an active synchronization process, and Faust suggested that acoustic signals exchanged between the members of the clutch prior to the start of hatching might by the synchronizing mechanism, but he was not able to follow up this lead. A very complete picture on the nature of the synchronizing influence was constructed by Vince on the basis of a series of elegant experiments with a variety of Galliformes. The summary that follows is based largely on her recent review (Vince, 1969).

Some time after respiratory movements first begin, there is a period during which a peculiar sound, termed clicking, accompanies breathing in all species investigated so far. This sound is audible when the egg is held close to the ear, and presumably is audible from one egg to another when they are in contact. Since synchronization does not occur when the eggs do not touch, and acceleration of hatching can be obtained when eggs are subjected to stimulation by sound or vibration at about the rate of naturally occurring clicks, it seems clear that clicking plays an important part in this phenomenon. Figure 34 shows the time course of events in the synchronized hatch of two clutches of the Bobwhite Quail (*Colinus virginianus*). It will be noted that the pip–click interval is the variable part of the scheme, which suggests that loud clicking

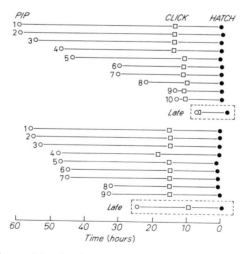

Fig. 34. Chronology of hatch of individual eggs (numbered in order of pipping) in two clutches of the Bobwhite Quail (*Colinus virginianus*) plotted in relation to mean time of hatch of the entire clutch (= 0 hours). Open circles show time of first pip, squares time of first click, and solid circles time of emergence. The eggs of each clutch were incubated in contact with one another. In each clutch, one egg was placed in the incubator 24 hours later than the rest (marked "late"), but nevertheless hatched at about the same time. Note the shortening of the pip–click interval in late-pipping eggs. (After Vince, 1969.)

in one egg might trigger clicking in another, i.e., act as an accelerator (note that even a 24 hour gap is bridged in this way), but once clicking is under way, a chain of events is activated that culminates in hatching after a given period of time. A closer look at the adjustments in timing that are being brought about is afforded in Fig. 35, where close scrutiny will reveal evidence of acceleration as well as retardation effects of the proximity of other embryos. Provisionally, a low-frequency sound produced in the "silent" phase of respiration before clicking commences has been ascribed a retarding function; i.e., the commencement of clicking in advanced embryos will be somewhat retarded as long as other embryos are producing the low-frequency sound. Quantitatively acceleration of the late embryos is the more important phenomenon, and provides a very neat system for achieving a close synchrony in time of hatch within the clutch, a fine adjustment as it were, superimposed on other factors (such as the total amount of heat received by each

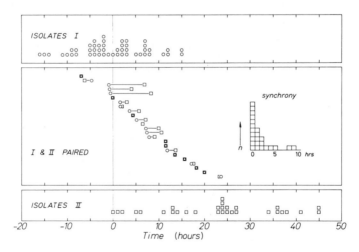

FIG. 35. Effect of contact between eggs of the Bobwhite Quail on hatching time. Circles show hatching times of individual eggs where incubation started at time I, squares give hatching times of eggs placed in the incubator 24 hours later (time II). The point of the experiment is to show an effect on hatching time when I eggs and II eggs are incubated in pairs, the eggs touching (starting 48 hours before the expected hatching time of the group I eggs) as compared to I eggs and II eggs incubated in isolation (as expected, II isolates hatch about 24 hours later than I isolates). All times are plotted with respect to mean hatching time of the I isolates (= 0 hours). For paired eggs, hatching times reflect a compromise; group I eggs in the pair hatch on average 7 hours later than their controls, and group II eggs in the pair 14 hours earlier than their controls. Acceleration of late eggs is thus quantitatively stronger than retardation of advanced eggs. Inset graph shows how closely members of a pair synchronized their hatch. (Adapted from Vince, 1968, by courtesy of Ballière, Tindall & Cassell.)

of the eggs, differences in genetic constitution, length of time elapsing between laying and commencement of incubation). The possible role of parental vocalization in bringing about synchronization is a point under active investigation (Hess, 1972; Impekoven, 1973a).

Summing up, synchronization of hatch involving an active adjustment in the timing of individual eggs as a result of mutual vocal stimulation in the hours preceding has been demonstrated by incubator trials in waterfowl (*Anser indicus*) (Würdinger, 1970, 1975), one ratite (*Rhea americana*) (Faust, 1960), and several gallinaceous birds (*Colinus virginianus*, Johnson, 1969; *Coturnix coturnix*, and possibly domestic fowl) (see Vince, 1970, 1973, for references). In all of these species, the brood leaves the nest simultaneously shortly after hatching. In certain waders (*Limosa limosa*, Lind, 1961; *Vanellus malarbaricus*, Jayakar and Spurway, 1965b), there are indications that a similar process is at work. The last eggs to pip appear to be accelerated by stimulation from the earlier eggs, in that the time they required to the moment of emergence is shorter. This might be designated incipient synchronization; that this speeding up effect has survival value has been stressed by Lind. The suggestion that the Skylark (*Alauda arvensis*) also shows synchronization of this active type (Vince, 1969) requires confirmation by experiment, since consultation of the original source (Delius, 1963) indicates that the range of hatching times reported (4–15 hours) is consistent with the observation that incubation is steady only after completion of the clutch.

VII. Energetics of Incubation

A. Time Budget Studies

Recently, a number of attempts have been made to follow individual parents during the incubation phase and keep records of the time devoted to every activity during the day. Eventually it may be possible to extrapolate from these records a quantitative impression of how energy is expended during incubation, and in any case such studies form the raw material from which to develop hypotheses about the adaptive significance of the time pattern of egg care shown by the species.

By direct observation of color-marked individuals, Verbeek (1972) managed to gather detailed records on the activities of Yellow-billed Magpies (*Pica nuttalli*) in the field. Only the female incubates, an average of 92.3% of her daylight activity (and 100% of her nocturnal activity) being spent at the nest, leaving very little time for anything else (1.7% foraging, which is a severe reduction from the 40% expended during laying; 0.6% flight; 5.4% for such other items as resting and territorial

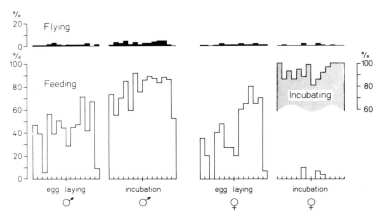

Fɪɢ. 36. Time allotted to flying (solid), feeding (open), and incubating (stippled) in *Pica nuttalli* during the daylight hours, presented in hourly histograms (double mark on time scale = 12 noon). Foraging claims approximately 42% (male) and 40% (female) of available time during egg laying, and increases to 78% in the male during incubation, as he then feeds the female (her foraging time drops to 2%). (Modified from Verbeek, 1972.)

defence). The male provides food for the sitting partner, and the portion of his day devoted to foraging doubles in this period, reaching 78% of his daylight time (Fig. 36). This example is presented here to show how sensitive the method is. When coupled with experiments such as the removal of the male (von Haartman, 1958), much can be revealed about the degree of leeway the incubation system adopted allows the parents. The use of radioactive tags to distinguish the parents (Coulson, 1972; Storteir and Palmgren, 1971; Ward, 1969) coupled with the refinement of radiotelemetry (Schladweiler, 1968; Robel, 1969) and the development of devices yielding time records of activities such as flight [such as the flight-time integrator designed for albatrosses by LeFebvre *et al.* (1967)] will make it possible to study virtually any species in this fashion.

B. ESTIMATING ENERGY EXPENDITURES DURING INCUBATION

Since in natural incubation the temperature of the air surrounding the eggs is lower than the temperature of the eggs themselves, the eggs will be subjected to a continuous heat loss and can be maintained at the requisite temperature only by heat input from the parent (contact heating from above), aided, toward the close of incubation, by the heat liberated by the developing embryo. Several avenues have been explored to arrive at the energy cost of the act of incubation, and we now have figures of considerable precision for several species. The goal of this work is to define the total energy input required for incubation and to partition this cost between the parent or parents, on the one hand, and the developing embryos on the other.

1. Monitoring Parental Heat Production

The most direct method, the measurement of heat production of the sitting parent with and without a clutch of eggs, is technically demanding but has successfully been applied in the Great Tit by Mertens (1967, 1974). By continuous sampling of air drawn from behind the female incubating in a nestbox located in a large room, heat production of the sitting parent could be computed from the gaseous exchange (indirect calorimetry). By night, the birds sits quietly, so these periods can be used as a baseline. On several occasions, heat production of single birds roosting in nestboxes before incubation commenced was measured, and these values could be compared with the heat production measured at the same air temperature during incubation, the increment representing the cost of incubation (in this particular situation, the contribution of the developing embryos was negligible, only 3 of the 8 eggs developing).

2. Measuring Food Consumption

In his study of the Zebra Finch (*Poephila guttata castanotis*) El Wailly (1966) kept pairs in captivity at ten levels of temperature and compared food intake in the incubation phase with food intake in a control period preceding the introduction of nesting material (Fig. 37), a method obviating corrections for embryonic heat production. The increment in food intake was taken to represent the added energy cost of incubation, and as expected, was found to increase as the air temperature declined, reaching 18% of the available productive energy at an air temperature of 21°C. Incubation was inadequate at the next step down (14.5°C), none of the eggs laid being hatched. In supplementary experiments, when food was offered in another form, Burley (1968) observed successful incubation at this temperature. This must be considered exceptional, as the parents sat side by side on the nest (one being in contact with the eggs) instead of alternating duties, and it may be presumed that under field conditions the constraint of foraging time would not permit this behavioral adaptation.

Although Brisbin (1969) provides detailed information on food consumption in pairs of Ring Doves incubating in captivity, a suitable control period is not available for comparison. Incubation involves a marked reduction in motor activity, and the most reliable control values would be obtained by allowing the birds to incubate a clutch of copper eggs such as those employed by Franks (1967), thermostatically maintained

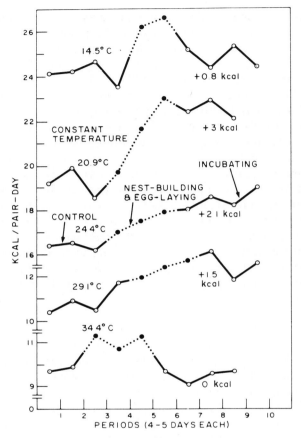

Fɪɢ. 37. Metabolized energy of *Poephila guttata castanotis* at various stages of the nesting cycle in relation to ambient temperature, based on analysis of food consumption and excreta. (From El Wailly, 1966.)

at the typical incubation temperature. Until such figures become available, estimation of the energy cost of incubation cannot reliably be given for this species (see King, 1973).

3. Parental Weight Loss

In species in which parental shifts are of long duration, estimation of energy expenditure by observing the decrement in weight loss is a distinct possibility. In petrels, for instance, a good deal of information on daily weight loss of the sitting parents is now available (Harris, 1966, 1969a,b; Beck and Brown, 1971, 1972; Fisher, 1967; Rice and Kenyon, 1962) and, providing figures could be obtained for birds inactive in the burrow without eggs to tend, the energy cost of incubation for the parent could be estimated. Waterfowl, among which the female may sit for protracted periods without feeding, also show spectacular declines in body weight during incubation shifts (Gorman and Milne, 1971; Barry, 1962; Ryder, 1967), and it might be feasible to compare weight loss in individuals incubating clutches of one, two, three, or more eggs in order to arrive at the parental cost of incubating a single egg. This would require close attention to heat loss characteristics of clutches of various sizes and could best be combined with method 4.

Harvey (1971) has taken a first step in this direction by measuring the heat loss of a fiberglass goose torso clothed in the skin and feathers of the Blue Goose (*Anser caerulescens*) when held in a wind tunnel simulating environmental conditions on the arctic breeding grounds. This work has not yet progressed to the stage that the heat expended to maintain the clutch at the requisite temperature can be specified, but the calculated weight losses during incubation were at least of the order of magnitude of the field data of other observers, so this lead seems worth following up.

4. Measuring Heat Loss of the Clutch

Kendeigh (1963) reasoned that since heat input must balance heat loss, the specification of the heat loss of the clutch during conditions of steady incubation by the parent would be equivalent to the total energy cost of incubation. Since eggs are near-spherical objects, measurement of their rate of cooling in still air is not difficult (see Kendeigh, 1973, for procedures). Kendeigh advanced the following overall equation

$$\text{Continuous incubation (cal/hour)} = nwhb(t_e - t_{na})i(1 - c)$$

where n = number of eggs in clutch, w = mean weight of the egg in grams, h = specific heat of the egg (0.80 cal/°C), b = cooling constant of the egg, t_e = egg temperature as maintained in the nest, t_{na} = nest-air temperature of the air surrounding the eggs, c = percentage of the total surface of the clutch covered by the incubating bird and hence not exposed to cooling. As pointed out by Kendeigh (1973), the cooling constant of the egg (expressed as degrees centigrade fall per centigrade degree gradient per hour) is proportional to the cube root of egg weight. The available measurements of b are displayed in Fig. 38, and it should be possible to compute the expected rate of cooling for an egg of known weight with a fair degree of accuracy on the basis of this information.

The main difficulties in applying this equation arise in specifying the temperature of the air surrounding the eggs and in measuring the portion

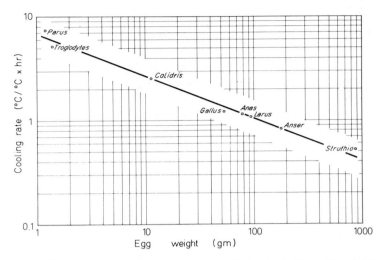

Fig. 38. Cooling constant of the avian egg in relation to fresh weight. [Data from Kendeigh (1973) supplemented by Siegfried and Frost (1975a) for *Struthio* (0.53 for an egg of 881 gm).]

of the egg surface covered by the brood patch and hence not exposed to cooling. In relatively small clutches, these problems are not insurmountable, and the equation has been applied to field data for the House Wren (Kendeigh, 1963) and the Herring Gull (Drent, 1970). In the latter study, experiments with an artificial brood patch whose heat output was known showed that, providing the air temperature to which the eggs were exposed could be measured, the equation gave results within 1% of the measured heat input by the patch (and eggs so treated hatched successfully).

By incubating eggs in individual respirometers and estimating heat output by measurements of gaseous exchange, the contribution of embryonic heat production to the total can be specified in these two species (Kendeigh, 1940; Drent, 1970), working out to one-tenth (House Wren and one-fifth (Herring Gull) of the total energy cost of incubation from clutch completion through to hatching.

The available measurements showing the parental cost of incubation are collected in Table V. Values representative for field conditions have been chosen (the data for House Wren and Herring Gull are taken directly from field measurements; for the Zebra Finch the observations at 21°C are taken to be representative on the basis of Immelmann's (1965) distributional analysis of this species; for the Great Tit the computations are based on the mean ambient air temperature in Merten's

TABLE V

ENERGY COST OF INCUBATION ESTIMATED FOR FIELD CONDITIONS

| Species | Weight (gm) | | Parental energy cost | | | | Authority |
	Parent	Clutch (fresh)	kcal/clutch–day	cal/gm egg–day	Productive energy devoted to incubation (%)	Method	
Troglodytes aedon	10	8.2	1.6 (♀ only)	195	19	4	Kendeigh (1963)
Poephila guttata	12.1	3.8	1.5 ♀ + 1.5 ♂	789	18	2	El Wailly (1966)
Parus major	18.5	18.3	2.4 (♀ only)	131	16	1	Mertens (1967)
Larus argentatus	1000	284	11.2 ♂ + 11.2 ♀	79	25	4	Drent (1970)

study area). In addition to the data from the food consumption trials used for the table, El Wailly computed energy costs according to Kendeigh's equation, but since nest-air temperatures were not measured, these computations are of doubtful validity, aside from the error made when extrapolating the cooling constant from the data on other species corrected by Kendeigh, 1973). Values reported by West (1960) and Brenner (1967) have not been included, since there are too many estimated quantities in their calculations.

A great variability in the caloric cost of incubation per gram egg is apparent in the table. This variability is related to the variability in the gradient between the eggs and the surrounding air, determined by the overall insulation. In the Herring Gull, in which incubation is for all practical purposes continuous, the nest and contents are so completely sealed off by the sitting bird that temperatures are independent of environmental levels throughout the normal range. Performance in the tit and wren, both of which are hole-nesters constructing well-insulated nests, is comparable, and very much better than in the Zebra Finch with its comparatively shoddy nest. One problem with the interpretation of the Zebra Finch data is the lack of detailed field observations; we do not know, for example, if energy costs in the field can be cut back by exploitation of solar heat (see Ward's 1965 measurements on *Quelea quelea*) or by the construction of better insulated nests. A second, more tentative, comparative measure is the maximal proportion of the productive energy available to each species that, according to these data, is devoted to incubation. In all species, the energy transferred to the eggs represents about one-fifth to one-quarter of the energy leeway at command. In the tit and Zebra Finch, the computation, based as it is on an observed increment in energy uptake (*Poephila*) or loss (*Parus*) is on reasonably safe ground, but in the other two species, as has been pointed out by King (1973) in a thoughtful review, we cannot assume without direct assessment that the caloric cost for incubation is added in its entirety to the other energy expenditures of the parent. In particular, it is possible that a portion of the required heat output can be balanced against the heat loss normally taking place from the bird in a resting state. Van Kampen and Romijn (1970) indicate that in the fowl at rest in the thermoneutral zone, as much as 60% of the total heat production may be lost from the body surface, and the possibility that a portion of the heat so dissipated might be channeled via the brood patches is worth investigation. Without adjustment of the conductance of the shell, such a savings would be negligible, since the brood patches form but a small proportion of the total surface area (in the Herring Gull, where both have been measured, the brood patches amount to 5% of the total body surface; discounting legs and bill). Clearly, further work is needed, but it is tempting at this stage to hypothesize that incubation adaptations (share of the sexes, insulative quality of the nest, rhythm of egg coverage) have evolved to keep the parental cost within certain margins (upper limits tentatively placed at about 20–25% of productive energy). This hypothesis, committed to paper, provides future workers with at least something to disprove.

One of the factors influencing energy cost, the cooling rate of the egg, is easily measured. Comparison of the cooling rate determined for an isolated egg in standardized conditions [Kendeigh (1973)—still air, suspended or otherwise held so as to avoid conductive heat losses] when compared with the rate observed in the field can give insight into the insulative quality of the nest, and into the influence of the configuration of eggs in the nest bowl. In the case of the Herring Gull, laboratory and field cooling rates did not differ appreciably (Drent, 1970), confirming that the insulative value of the nest material was mainly a prevention of conductive heat loss to the cool sand underlying the nest. In the Dunlin (*Calidris alpina*), on the other hand, Norton (1970) discovered an impressive difference between performance in the laboratory and in the field. He was able to show that the cooling constant of the egg when positioned in the typical packed mass of four eggs (the clutch size typical of most waders, in fact) in the nest was 1.95 (°C/°C hour) as compared to the 2.6 determined for the isolated egg; as Norton points out, the clutch can be seen as a compound egg (total weight 45 gm theoretical cooling constant for a single egg of this weight = 1.6) (see Fig. 38). Removing one egg destroyed this advantage, the cooling rate reverting to the level of the isolate, and Norton speculates that here may be one of the selective pressures that have led to the evolution of four-egg clutches in Scolopacidae, many of whom are high-latitude breeders, and all of whom lay relatively heavy eggs (e.g., have heavy eggs masses to incubate), as documented by Lack (1968).

A final point deserving attention when considering the energetics of incubation is the energy expenditure of the embryo. This must be financed from the yolk reserves laid down in the fresh egg, and Table VI shows the energy depletion of the avian egg during incubation, as determined from the respiratory exchange of the embryo. The figures given here for the domestic fowl are substantiated by the work of Tangl and von Mituch (1908), who measured the energy consumed during incubation by an entirely different method. On the basis of chemical analyses and determinations of the caloric value of eggs incubated for various lengths of time, these authors reported 22.94 kcal to be dissipated per egg over the whole incubation period, almost entirely at the expense

TABLE VI

TOTAL ENERGY EXPENDITURE OF EMBRYO DURING INCUBATION

Species	Energy (kcal)		Expended by embryo	Fat expended by embryo (% original)	Reference[a]
	Fresh egg				
	Total	Fat			
Domestic duck	138.7	95.1	39.7	42	Kashkin (1961)
Fowl	90.0	59.9	25.3	42	Romijn and Lokhorst (1960)
Herring Gull	144.4	96.4	57.9	60	Drent (1970)
House Wren	1.572	0.879	0.866	99	Kendeigh (1963)

[a] Caloric value of fresh egg: duck and fowl from Plimmer's value after substraction of the shell according to Romanoff and Romanoff (1949, p. 114). House Wren as given by Kendeigh et al. (1956). Fat content computed for duck and fowl from constituent proportions reported by Romanoff and Romanoff (1949, p. 314).

of the fat content, of which 40.5% was consumed. The very close agreement of these figures with those of Romijn and Lokhorst (1960) on the respiratory exchange of the domestic fowl is striking.

In the two precocial, nidifugous species (domestic duck and fowl), less than half of the original fat store is consumed during the development of the embryo, so that upon hatching the young bird can survive for a considerable period without food—Mallard, longer than 48 hours (Kear, 1965); $7\frac{1}{2}$ days (Marcström, 1966); Capercaillie (*Tetrao urogallus*), 5 days (Marcström, 1960). Not only does this give the young bird a certain amount of leeway in learning to find its food, but the considerable energy store available to it at hatching also permits long journeys between the nest and the first feeding area to be undertaken. Lack (1968) sees this as an important adaptation allowing dispersed nesting in waterfowl. In the precocial but nidicolous Herring Gull, in which the young, although hatched in an advanced state, are fed at the nest by their parents, an intermediate position is taken, with only 40% of the original fat store still available to the hatchling.

In the Adélie Penguin, like the Herring Gull precocial but nidicolous, approximately half of the original yolk store (by weight) is still present in the hatchling (Reid and Bailey, 1966), and this store enables the chick to survive if the off-duty partner, feeding at sea, is late in returning to the colony. Finally, in the House Wren, the only altricial bird in the list, practically the entire fat content of the egg is exhausted during incubation, so that the young bird is entirely dependent on the parents from the moment of hatching. Put in another way, the eggs of altricial birds contain no reserves, whereas those of precocial birds contain a considerable surplus of fat, which can be drawn upon by the chick in its first posthatching days.

VIII. Prospect

Incubation is clearly an inexhaustible subject, but there are certain "blind spots" where the need for further research is particularly pressing. Despite several bursts of effort, our knowledge of physical conditions during natural incubation is incomplete. In particular, oxygen and carbon dioxide levels have not been measured in the field, and we have no more than rough approximations concerning humidity conditions. A lack of suitable equipment has no doubt hampered work in this direction, but in the case of humidity, a lack of interest must be admitted. The weight loss of various substances—as so ingeniously applied fifty years ago by Chattock (1925)—and indeed of the eggs themselves can be used as a sensitive indicator of the humidity conditions prevailing in the nest, so that much can be done without recourse to the more elaborate methods currently available. Regarding temperature, we are better informed, but most data are restricted to measurements of air temperature between the eggs or to internal registrations from inert egg models (copper, plaster, and so forth). Particularly where the eggs are large, the temperatures actually experienced by the embryo during development may differ markedly from such records; there is thus no substitute for determinations on intact eggs. The collection of observations on these topics should be followed as soon as possible by experiments in which the various parameters are altered and the actions of the parent birds noted. The work should be so designed that besides information about which of the physical characteristics of the clutch and the nest that the parent actively regulates, insights are also obtained about the specific stimuli the bird relies upon.

The orientation of the egg in the nest, and the way the egg's position changes in the course of incubation, both with regard to the other eggs of the clutch (i.e., locations in the nest bowl), and with regard to its individual long and short axis, are further facets of the physical environment of incubation that are simple to measure, have a direct bearing on hatching success, yet have received little attention. In the three species investigated to date, position plots of individual eggs are skewed (i.e., the eggs are oriented nonrandomly during the greater part of incubation), and the generality of these findings should be established by extending the observations to other species. In evaluating the functional significance of egg-turning behavior, incubator trials will continue to play a central role, but some field experiments do suggest themselves, such as determining the hatching success of eggs consistently turned upside down during the phase in which weight asymmetry exerts its influence, or fixed in this inverted position.

Apart from examining its dependence on external air temperature and other meteorological factors, little attention has yet been directed toward the interpretation of incubation rhythm in a broader context. Evaluating the adaptive significance of the pattern of sharing of parental duties or of the finer patterning (duration of uninterrupted spells of sitting, duration of the intervening breaks) can be approached through comparison of related species, in addition to systematic analysis of variability within the species (extremes of season, habitat, and range might be especially rewarding). Such study could begin with detailed observation of the time budget, paying close attention to activities that conflict with incubation. One hopes that by analysis of observed variation or as a result of manipulation of such items as food supply, the constraints within which the time patterning of incubation duties must operate would gradually emerge. Logically, this parent-sided view should be combined with incubator experimentation with eggs to determine minimal requirements for successful incubation. The resistance to chilling and the need for ventilation come to mind. Whether the observed features of the rhythm of egg care are in keeping with the generalized rules formulated for circadian rhythms is another approach (Hoffmann, 1969) that deserves to be followed up. Changes in the pattern of egg care in the course of incubation have been correlated with endocrine changes, but far more work is needed to unravel the complex interaction between incubation behavior (onset, maintenance, modulation, and termination) and hormone levels. (Cheng, 1975).

Insight into the energetics of incubation, including an evaluation of the insulative capacity of the nest, can greatly aid the interpretation of the survival value of incubation rhythm from the viewpoint of the parent and is a further approach that can be considered the core problem for the ecology-minded field worker. The estimates obtained using the indirect method pioneered by Kendeigh (1963) show how revealing such data can be, and application of more direct measures of energy expenditure (recently collated by Gessaman, 1973) to the field situation should be pursued with energy. Study of the heat-exchange properties of the brood patch (or webs of the feet in some species) is a related topic for which adequate experimental techniques are available. Looking at energetics from the viewpoint of the embryo, the role of the yolk as an energy reserve during embryonic development and in the immediate posthatch period has barely been touched on so far (Grau, 1968). In short, the perspectives for further research on incubation are so promising that I fully expect a review in another ten years' time will require book length.

REFERENCES

Allan, R. G. (1962). The Madeiran Storm Petrel *Oceanodroma castro*. *Ibis* **103b**, 274–295.

Ar, A., Paganelli, C. V., Reeves, R. B., Greene, D. G., and Rahn, H. (1974). The avian egg: Water vapor conductance, shell thickness, and functional pore area. *Condor* **76**, 153–158.

Armstrong, E. A. (1964). Distraction display. *In* "A New Dictionary of Birds" (A. L. Thomson, ed.), p. 205. Nelson, London.

Aschoff, J. (1970). Temperatur Regulation *In* "Physiologie des Menschen" (O. H. Gauer, K. Kramer, and R. Jung, eds.), Vol. 2, pp. 43–116. Urban & Schwarzenberg, Berlin.

Aschoff, J., and Pohl, H. (1970). Der Ruheumsatz von Vögeln als Funktion der Tageszeit und der Körpergrösse. *J. Ornithol.* **11**, 38–47.

Axelsson, J. (1954). Influence of size of eggs on growth rate of embryos and chicks. *World's Poultry Congr., Proc., 10th, 1954* pp. 65–66.

Baerends, G. P. (1959). The ethological analysis of incubation behaviour. *Ibis* **11**, 357–368.

Baerends, G. P. (1970). A model of the functional organization of incubation behaviour. *Behaviour, Suppl.* **17**, 263–312.

Baerends, G. P. (1975). In preparation.

Baerends, G. P., Drent, R. H., Glas, P., and Groenewold, H. (1970). An ethological analysis of incubation behaviour in the Herring Gull. *Behaviour, Suppl.* **17**, 135–235.

Bailey, R. E. (1952). The incubation patch of passerine birds. *Condor* **54** 121–136.

Baker, J. R. (1968). Early embryology of the Adélie Penguin. *Antarctic J. U.S.* **3**, 133.

Balaban, M., and Hill, J. (1969). Perihatching behaviour patterns of chick embryos. (*Gallus domesticus*). *Anim. Behav.* **17**, 430–439.

Baldwin, S. J., and Kendeigh, S. C. (1932). Physiology of the temperature of birds. *Sci. Publ. Cleveland Mus. Natur. Hist.* **3**, 1–196.

Baltin, S. (1969). Zur Biologie und Ethologie des Telegalla-Huhns (*Alectura lathami* Gray) unter besonderer Berücksichtigung des Verhaltens während der Brutperiode. *Z. Tierpsychol.* **6**, 524–572.

Barott, H. G. (1937). Effects of temperature, humidity, and other factors on hatch of hens' eggs and on energy metabolism of chick embryos. *U.S., Dep. Agr., Tech. Bull.* **553**, 1–45.

Barry, T. W. (1962). Effect of late seasons on Atlantic Brant reproduction. *J. Wildl. Manage.* **26**, 19–26.

Bartels, H. (1970). "Prenatal Respiration." North-Holland Publ., Amsterdam.

Barth, E. K. (1949). Redetemperaturer og rugevaner. *Naturen (Bergen)* **73**, 81–95.

Barth, E. K. (1953). Calculation of egg volume based on loss of weight during incubation. *Auk* **70**, 151–159.

Barth, E. K. (1955). Egg-laying incubation and hatching of the Common Gull (*Larus canus*). *Ibis.* **97**, 222–239.

Beck, J. R., and Brown, D. W. (1971). The breeding biology of the Black-bellied Storm-petrel *Fregetta tropica*. *Ibis* **113**, 73–90.

Beck, J. R., and Brown, D. W. (1972). The biology of Wilson's Storm Petrel, *Oceanites oceanicus* (Kuhl), at Signy Island, South Orkney Islands. *Brit. Antarctic Surv. Sci. Rep.* **69**, 1–54.

Beer, C. G. (1964). "Incubation." *In* "A New Dictionary of Birds" (A. L. Thomson, ed.), pp. 396–398. Nelson, London.

Beer, C. G. (1965). Clutch size and incubation behavior in Black-billed Gulls (*Larus bulleri*). *Auk* **82**, 1–18.

Benedict, F. G. (1932). The physiology of large reptiles with special reference to the heat production of snakes, lizards and alligators. *Carnegie Inst. Wash. Publ.* **425**, 1–539.

Bergman, G. (1946). Der Steinwälzer, *Arenaria i. interpres* (L.), in seiner Beziehung zur Umwelt. *Acta Zool. Fenn.* **47**, 1–151.

Bjärvall, A. (1967). The critical period and the interval between hatching and exodus in Mallard ducklings. *Behaviour* **28**, 141–148.

Bjärvall, A. (1968). The hatching and nest-exodus behaviour of Mallard. *Wildfowl* **19**, 70–80.

Bligh, J., and Hartley, T. C. (1965). The deep body temperature of an unrestrained Ostrich *Struthio camelus* recorded continuously by a radio-telemetric technique. *Ibis* **107**, 104–105.

Bochenski, Z. (1961). Nesting biology of the Black-necked Grebe. *Bird Study* **8**, 6–15.

Bock, W. J., and Hikida, R. S. (1968). An analysis of twitch and tonus fibers in the hatching muscle. *Condor* **70**, 211–222.

Breckenridge, W. J. (1956). Nesting study of Wood Ducks. *J. Wildl. Manage.* **20**, 16–21.

Breitenbach, R. P., Nagra, C. L., and Meyer, R. K. (1965). Studies of incubation and broody behaviour in the Pheasant (*Phasianus colchicus*). *Anim. Behav.* **13**, 143–148.

Brenner, F. J. (1967). Seasonal correlations of reserve energy of the Red-winged Blackbird. *Bird-Banding* **38**, 195–211.

Brisbin, I. L. (1969). Bioenergetics of the breeding cycle of the Ring Dove. *Auk* **86**, 54–74.

Broekhuysen, G. J., and Frost, P. G. H. (1968a). Nesting behaviour of the Black-necked Grebe *Podiceps nigricollis* (Brehm) in southern Africa. 1. The reaction of disturbed incubating birds. *Bonn. Zool. Beitr.* **19**, 350–361.

Broekhuysen, G. J., and Frost, P. G. H. (1968b). Nesting behaviour of the Black-necked Grebe *Podiceps nigricollis* (Brehm) in southern-Africa. 2. Laying, clutch size, egg size, incubation and nesting success. *Ostrich* **39**, 242–252.

Brown, L. H., and Root, A. (1971). The breeding behaviour of the Lesser Flamingo *Phoeniconaias minor*. *Ibis* **113**, 147–272.

Brown, R. G. B. (1962). The aggressive and distraction behaviour of the Western Sandpiper, *Ereunetes mauri*. *Ibis* **104**, 1–12.

Buckley, P. A., and Buckley, F. G. (1972). Individual egg and chick recognition by adult Royal Terns (*Sterna maxima maxima*). *Anim. Behav.* **20**, 457–462.

Burke, E. (1925). A study of incubation. *Mont., Agr. Exp. Sta., Bull.* **178**, 1–44.

Burley, V. E. (1968). Incubation at low temperatures by the Zebra Finch. *Condor* **70**, 184.

Calder, W. A. (1964). Gaseous metabolism and water relations of the Zebra Finch, *Taeniopygia castanotis*. *Physiol. Zool.* **37**, 400–413.

Calder, W. A. (1971). Temperature relationships and nesting of the Calliope Hummingbird. *Condor* **73**, 314–321.

Calder, W. A. (1973). Microhabitat selection during nesting of hummingbirds in the Rocky Mountains. *Ecology* **54**, 127–134.

Calder, W. A., and J. Booser. (1973). Hypothermia of Broad-tailed Hummingbirds during incubation in nature with ecological correlations. *Science* **180**, 751–753.

Caldwell, P. J., and Cornwell, G. W. (1975). Incubation behavior and temperature of the Mallard Duck (*Anas platyrhynchos*). *Auk* **92**, (in press).

Cavé, A. J. (1968). The breeding of the Kestrel, *Falco tinnunculus* L., in the reclaimed area Oostelijk Flevoland. *Neth. J. Zool.* **18**, 313–407.

Chattock, A. P. (1924). On the physics of incubation. *Phil. Trans. Roy. Soc. London, Ser. B* **213**, 397–450.

Cheng, M. F. (1975). Induction of incubation behaviour in male Ring Doves (*Streptophelia risoria*): A behavioural analysis. *J. Reprod. Fert.* **42** (in press).

Clark, G. A., Jr. (1961). Occurrence and timing of egg teeth in birds. *Wilson Bull.* **73**, 268–278.

Collias, N. E., and Collias, E. C. (1964). Evolution of nest-building in the Weaverbirds (Ploceidae). *Univ. Calif., Berkeley, Publ. Zool.* **73**, 1–239.

Collias, N. E., and Collias, E. C. (1971). Ecology and behaviour of the Spotted-backed Weaverbird in the Kruger National Park. *Koedoe* **14**, 1–27.

Conrads, K. (1969). Beobachtungen am Ortolan (*Emberiza hortulana* L.) in der Brutzeit. *J. Ornithol.* **110**, 379–420.

Coulson, J. C. (1972). The significance of the pair-bond in the Kittiwake. *Proc. Int. Ornithol. Congr., 15th, 1970* pp. 424–433.

Crook, J. H. (1963). A comparative analysis of nest structure in the weaver birds (Ploceinae). *Ibis* **105**, 238–262.

Delius, J. D. (1963). Das Verhalten der Feldlerche. *Z. Tierpsychol.* **20**, 297–348.

de Reaumur, R. A. F. (1751). "Konst om Tamme-vogelen van Allerhandesoort in alle Jaartyden Uittebroeijen en Optebrengen zo door 't Middel van Mest als van't Gewone Vuur." s'Gravenhage (Pieter de Hondt). Vol. I, pp. 1–384.

Dorst, J. (1962a). A propos de la nidification hypogée de quelques oiseaux des Hautes Andes péruviennes. *Oiseau Rev. Fr. Ornithol.* **32**, 5–14.

Dorst, J. (1962b). Nouvelles recherches biologiques sur les trochilidés des Hautes Andes péruviennes (*Oreotrochilus estella*). *Oiseau Rev. Fr. Ornithol.* **32**, 95–126.

Drent, R. H. (1970). Functional aspects of incubation in the Herring Gull. *Behaviour, Suppl.* **17**, 1–132.

Drent, R. H. (1972). Adaptive aspects of the physiology of incubation. *Proc. Int. Ornithol. Congr., 15th, 1970* pp. 231–256.

Drent, R. H. (1973). The natural history of incubation. *In* "Breeding Biology of Birds" (D. S. Farner, ed.), pp. 262–311. Nat. Acad. Sci., Washington, D.C.

Drent, R. H., Postuma, K., and Joustra, T. (1970). The effect of egg temperature on incubation behaviour in the Herring Gull. *Behaviour, Suppl.* **17**, 237–261.

Eisner, E. (1969). The effect of hormone treatment upon the duration of incubation in the Bengalese Finch. *Behaviour* **33**, 262–276.

Eklund, C. R., and Charlton, F. E. (1959). Measuring the temperature of incubating penguin eggs. *Amer. Sci.* **42**, 80–86.

El-ibiary, H. M., Shaffner, C. S., and Godfrey, E. F. (1966). Pulmonary ventilation in a population of hatching chick embryos. *Brit. Poultry Sci.* **7**, 165–176.

El Wailly, A. (1966). Energy requirements for egg-laying and incubation in the Zebra Finch, *Taeniopygia castanotis*. *Condor* **68**, 582–594.

Evans, W. (1891). On the periods occupied by birds in the incubation of their eggs. *Ibis* **1891**, 52–93.

Eycleshymer, A. C. (1907). Some observations and experiments on the natural and artificial incubation of the Common Fowl. *Biol. Bull.* **12**, 360–375.

Fargeix, N. (1963). L'orientation dominante de l'embryon de la Caille domestique (*Coturnix coturnix japonica*) et la règle de von Baer. *C. R. Soc. Biol.* **157**, 1431–1434.

Farner, D. S. (1958). Incubation and body temperatures in the Yellow-eyed Penguin. *Auk* **75**, 249–262.

Faust, R. (1960). Brutbiologie des Nandus (*Rhea americana*) in Gefangenschaft. *Verh. Deut. Zool. Ges.* **54**, 398–401.

Fisher, H. I. (1958). The "hatching muscle" in the chick. *Auk* **75**, 391–399.

Fisher, H. I. (1961). The hatching muscle in North American grebes. *Condor* **63**, 227–233.

Fisher, H. I. (1962). The hatching muscle in Franklin's Gull. *Wilson Bull.* **74**, 166–172.

Fisher, H. I. (1966). Hatching and the hatching muscle in some North American ducks. *Trans. Ill. State Acad. Sci.* **59**, 305–325.

Fisher, H. I. (1967). Body weights in Laysan Albatrosses *Diomedea immutabilis*. *Ibis* **109**, 373–382.

Fisher, H. I. (1969). Eggs and egg-laying in the Laysan Albatross, *Diomedea immutabilis*. *Condor* **71**, 102–112.

Fisher, J. R. (1962). The hatching muscle in the American Coot. *Trans. Ill. State Acad. Sci.* **55**, 71–77.

Franks, E. C. (1967). The response of incubating Ringed Turtle Doves (*Streptopelia risoria*) to manipulated egg temperatures. *Condor* **69**, 268–276.

Frith, H. J. (1956a). Breeding habits in the family Megapodiidae. *Ibis* **98**, 620–640.

Frith, H. J. (1956b). Temperature regulation in the nesting mounds of the Mallee-Fowl, *Leipoa ocellata* Gould. CSIRO *Wildl. Res.* **1**, 79–95.

Frith, H. J. (1957). Experiments on the control of temperature in the mound of the Mallee-Fowl, *Leipoa ocellata* Gould. CSIRO *Wildl. Res.* **2**, 101–110.

Frith, H. J. (1959). Breeding of the Mallee Fowl *Leipoa ocellata* Gould. *CSIRO Wildl. Res.* **4**, 31–60.

Frith, H. J. (1962). "The Mallee-Fowl. The Bird that Builds an Incubator." Angus & Robertson, Sydney, Australia.

George, J. C., and Iype, P. T. (1963). The mechanism of hatching in the chick. *Pavo* **1**, 52–56.

Gessaman, J. A., ed. (1973). Ecological energetics of homeotherms. *Utah State Univ. Press, Mon. Ser.* **20**, 1–55.

Gibb, J. A. (1970). The turning down of marked eggs by Great Tits. *Bird-Banding* **41**, 40–41.

Goethe, F. (1953). Experimentelle Brutbeendigung und andere brutethologische Beobachtungen bei Silbermöwen (*Larus a. argentatus* Pontopp.). *J. Ornithol.* **94**, 160–174.

Goethe, F. (1955). Beobachtungen bei der Aufzucht junger Silbermöwen. *Z. Tierpsychol.* **12**, 402–433.

Gorman, M. L., and Milne, H. (1971). Seasonal changes in the adrenal steroid tissue of the Common Eider *Somateria mollissima* and its relation to organic metabolism in normal and oil-polluted birds. *Ibis* **113**, 218–228.

Gotzman, J. (1967). Remarks on ethology of the Red-backed Shrike, *Lanius collurio* L.—Nest defence and nest desertion. *Acta Ornithol.* (*Warszawa*) **10**, 83–96.

Gramza, A. F. (1967). Responses of brooding nighthawks to a disturbance stimulus. *Auk* **84**, 72–86.

Grassé, P.-P., ed. (1950). "Traité de Zoologie XV. Oiseaux." Masson, Paris.

Grau, C. R. (1968). Avian embryo nutrition. *Fed. Proc., Fed. Amer. Soc. Exp. Biol.* **27**, 185–192.

Groebbels, F. (1937). "Der Vogel, Bau, Funktion, Lebenserscheinung, Einpassung.

II. Geschlecht und Fortpflanzung." Borntraeger, Berlin.

Haftorn, S. (1966). Egglegging og ruging hos meiser baser på temperatur-målinger og direkte iakttagelser. *Sterna* 7, 49–102.

Haftorn, S. (1973). Lappmeisa *Parus cinctus* i hekketiden. *Sterna* 12, 91–155.

Hamburger, V., and Oppenheim, R. (1967). Prehatching motility and hatching behavior in the chick. *J. Exp. Zool.* 166, 171–204.

Hanson, H. L. (1959). The incubation patch of wild geese; its recognition and significance. *Arctic* 12, 139–150.

Harris, M. P. (1966). Breeding biology of the Manx Shearwater *Puffinus puffinus*. *Ibis* 108, 17–33.

Harris, M. P. (1969a). Food as a factor controlling the breeding of *Puffinus lherminieri*. *Ibis* 111, 139–156.

Harris, M. P. (1969b). The biology of storm petrels in the Galapagos Islands. *Proc. Calif. Acad. Sci.* [4] 39, 95–165.

Harvey, J. M. (1971). Factors affecting Blue Goose nesting success. *Can. J. Zool.* 49, 223–234.

Haverschmidt, F. (1957). Nachbarschaft von Vogelnesten und Wespennesten in Suriname. *J. Ornithol.* 98, 389–396.

Heinroth, O. (1922). Die Bezehiungen zwischen Vogelgewicht, Eigewicht, Gelegegewicht und Brutdauer. *J. Ornithol.* 70, 172–285.

Heinroth, O. (1938). "Aus dem Leben der Vögel." Springer-Verlag, Berlin and New York.

Heinroth, O., and Heinroth, M. (1926). "Die Vögel Mitteleuropas. I." H. Bermühler, Berlin.

Hess, E. H. (1972). The natural history of imprinting. *Ann. N.Y. Acad. Sci.* 193, 124–136.

Hinde, R. A. (1962). Temporal relations of brood patch development in domesticated canaries. *Ibis* 104, 90–97.

Hinde, R. A. (1967). Aspects of the control of avian reproductive development within the breeding season. *Proc. Int. Ornithol. Congr., 14th, 1966*, pp. 135–153.

Hinde, R. A., and Steele, E. A. (1964). Effect of exogenous hormones on the tactile sensitivity of the canary brood patch. *J. Endocrinol.* 30, 355–359.

Hinde, R. A., and Steele, E. A. (1966). Integration of the reproductive behaviour of female canaries. *Symp. Soc. Exp. Biol.* 20, 401–426.

Hinde, R. A., Bell, R. Q., and Steel, E. (1963). Changes in sensitivity of the Canary brood patch during the natural breeding season. *Anim. Behav.* 11, 553–560.

Hoesch, W. (1933). Ein Vogelnest mit verschliessbaren Eingang: Das Nest von *Anthoscopus caroli* (Sharpe). *Ornithol. Monatsber.* 41, 1–4.

Hoesch, W. (1940). Ueber den Einfluss der Zivilisation auf das Brutverhalten der Vögeln und über abweichende Brutgewohnheiten (Beobachtungen aus Süd-West Afrika). *J. Ornithol.* 88, 576–586.

Hoesch, W. (1956). Ueber das Nisten südwestafrikanischer Vögel unter dem Schutz der Wespe *Belonogaster rufipennis*. *J. Ornithol.* 97, 341–342.

Hoffmann, K. (1969). Zum Tagesrhythmus der Brutablösung beim Kaptäubchen (*Oena capensis* L.) und bei anderen Tauben. *J. Ornithol.* 110, 448–464.

Höhn, E. O. (1967). Observations on the breeding biology of Wilson's Phalarope (*Steganopus tricolor*) in central Alberta. *Auk* 84, 220–244.

Höhn, E. O. (1972). Prolactin lack in a brood parasite: A summary report and appeal for material. *Ibis* 114, 108.

Höhn, E. O., and Cheng, S. C. (1967). Gonadal hormones in Wilson's Phalarope (*Steganopus tricolor*) and other birds in relation to plumage and sex behaviour.

Gen. Comp. Endocrinol. **8,** 1–11.

Holcomb, L. C. (1969). Egg turning behavior of birds in response to color-marked eggs. *Bird-Banding* **40,** 105–113.

Holcomb, L. C. (1970). Prolonged incubation behaviour of Red-winged Blackbird incubating several egg sizes. *Behaviour* **36,** 74–83.

Holstein, V. (1942). "Duehøgen, *Astur gentilis dubius* (Sparrmann)." Hirschsprung, Copenhagen.

Holstein, V. (1944). "Hvepsevaagen, *Pernis apivorus apivorus* (L.)." Hirschsprung, Copenhagen.

Holstein, V. (1950). "Spurvehøgen, *Accipiter nisus nisus* (L.)." Hirschsprung, Copenhagen.

Horváth, O. (1964). Seasonal differences in Rufous Hummingbird nest height and their relation to nest climate. *Ecology* **45,** 235–241.

Howell, T. R. (1964). Notes on incubation and nestling temperatures and behavior of captive owls. *Wilson Bull.* **76,** 28–36.

Howell, T. R., and Bartholomew, G. A. (1961). Temperature regulation in nesting Bonin Island petrels, Wedge-tailed Shearwaters, and Christmas Island Shearwaters. *Auk* **78,** 343–354.

Howell, T. R., and Bartholomew, G. A. (1962a). Temperature regulation in the Red-tailed Tropic Bird and the Red-footed Booby. *Condor* **64,** 6–18.

Howell, T. R., and Bartholomew, G. A. (1962b). Temperature regulation in the Sooty Tern *Sterna fuscata. Ibis* **104,** 98–105.

Huggins, R. A. (1941). Egg temperatures of wild birds under natural conditions. *Ecology* **22,** 148–157.

Hutchison, R. E., Hinde, R. A., and Steele, E. (1967). The effects of oestrogen, progesterone and prolactin on brood patch formation in ovariectomized canaries. *J. Endocrinol.* **39,** 379–385.

Huxley, J. S. (1927). On the relation between egg-weight and body-weight in birds. *J. Linn. Soc. London, Zool.* **36,** 457–466.

Immelmann, K. (1965). Versuch einer ökologischen Verbreitungsanalyse beim australischen Zebrafinken, *Taeniopygia guttata castanotis* (Gould). *J. Ornithol.* **106,** 415–430.

Impekoven, M. (1973a). Response-contingent prenatal experience of maternal calls in the Peking Duck (*Anas platyrhynchos*). *Anim. Behav.* **21,** 164–168.

Impekoven, M. (1973b). The response of incubating Laughing Gulls (*Larus atricilla* L.) to calls of hatching chicks. *Behaviour* **46,** 94–113.

Irving, L. and Krog, J. (1956). Temperature during the development of birds in arctic nests. *Physiol. Zoöl.* **29,** 195–205.

Isenmann, P. (1970). Contribution à la biologie de reproduction du Petrel des neiges (*Pagodroma nivea* Forster). Le problème de la petite et de la grande forme. *Oiseau Rev. Fr. Ornithol.* **40,** 99–134.

Jander, R. (1968). Ueber die Ethometrie von Schlüsselreizen, die Theorie der telotaktischen Wahlhandlung und das Potensprinzip der terminalen Cumulation bei Arthropoden. *Z. Vergl. Physiol.* **59,** 319–356.

Jayakar, S. D., and Spurway, H. (1965a). The Yellow-wattled Lapwing, a tropical dry-season nester (*Vanellus malabaricus* (Boddaert), Charadriidae). I. The locality, and the incubatory adaptations. *Zool. Jahrb., Abt. Syst., Oekol. Geogr. Tiere* **92,** 53–72.

Jayakar, S. D., and Spurway, H. (1965b). The Yellow-wattled Lapwing (*Vanellus malabaricus* Boddaert) a tropical dry-season nester. II. Additional data on breeding biology. *J. Bombay Natur. Hist. Soc.* **62,** 1–14.

Jehl, J. R. (1968). The egg tooth of some charadriiform birds. *Wilson Bull.* **80,** 328–330.

Johns, J. E., and Pfeiffer, E. W. (1963). Testosterone-induced incubation patches of phalarope birds. *Science* **140**, 1225–1226.

Johnson, R. A. (1969). Hatching behavior of the Bobwhite. *Wilson Bull.* **81**, 79–86.

Jones, R. E. (1969a). Epidermal hyperplasia in the incubation patch of the California Quail, *Lophortyx californicus*, in relation to pituitary prolactin content. *Gen. Comp. Endocrinol.* **12**, 498–502.

Jones, R. E. (1969b). Hormonal control of incubation patch development in the California Quail *Lophortyx californicus*. *Gen. Comp. Endocrinol.* **13**, 1–13.

Jones, R. E. (1971). The incubation patch of birds. *Biol. Rev. Cambridge Phil. Soc.* **46**, 315–339.

Jones, R. E., Kreider, J. W., and Criley, B. B. (1970). Incubation patch of the chicken: Response to hormones *in situ* and transplanted to a dorsal site. *Gen. Comp. Endocrinol.* **15**, 398–403.

Kaltofen, R. S. (1961). "Het keren der eieren van de hen tijdens het broeden als mechanische factor van betekenis voor de broeduitkomsten," Inst. Veeteeltk. Ond. "Schoonoord" Publ. No. 133, Inst. Pluimveeteelt "Het Spelderholt" Meded. No. 88.

Kaltofen, R. S. (1969). The effect of air movements on hatchability and weight loss of chicken eggs during artificial incubation. *In* "The Fertility and Hatchability of the Hen's Egg" (T. C. Carter and B. M. Freeman, eds.), pp. 177–189. Oliver & Boyd, Edinburgh.

Kashkin, V. V. (1961). Heat exchange of bird eggs during incubation. *Biophysica* **6**, 57–63 (English version).

Kear, J. (1965). The internal food reserves of hatching Mallard ducklings. *J. Wildl. Manage.* **29**, 523–528.

Keil, W. (1964). Messung der Bruttemperatur bei einigen Singvogelarten. *Schriftenreihe Landesstelle Naturschutz Landschaftspflege in Nordrhein-Westfalen* No. 1, pp. 135–143.

Kemp, A. C. (1971). Some observations on the sealed-in nesting method of Hornbills (Family: Bucerotidae). *Ostrich, Suppl.* **8**, 149–155.

Kendeigh, S. C. (1940). Factors affecting length of incubation. *Auk* **57**, 499–513.

Kendeigh, S. C. (1952). Parental care and its evolution in birds. *Ill. Biol. Monogr.* **22**, 1–358.

Kendeigh, S. C. (1963). Thermodynamics of incubation in the House Wren, *Troglodytes aedon*. *Proc. Int. Ornithol. Congr., 13th, 1962* pp. 884–904.

Kendeigh, S. C. (1973). Discussion on incubation. *In* "Breeding Biology of Birds" (D. S. Farner, ed.), pp. 311–320. Nat. Acad. Sci., Washington, D.C.

Kendeigh, S. C., Kramer, J. C., and Hamerstrom, F. (1956). Variations in egg characteristics of the House Wren. *Auk* **73**, 42–65.

Kessler, F. W. (1960). Egg temperatures of the Ring-necked Pheasant obtained with a self-recording potentiometer. *Auk* **77**, 330–336.

King, J. R. (1973). Energetics of reproduction in birds. *In* "Breeding Biology of Birds" (D. S. Farner, ed.), pp. 78–107. Nat. Acad. Sci., Washington, D.C.

King, J. R., and Farner, D. S. (1961). Energy metabolism, thermoregulation, and body temperature. *In* "Biology and Comparative Physiology of Birds" (A. J. Marshall, ed.), Vol. 2, pp. 215–288. Academic Press, New York.

Kluijver, H. N. (1950). Daily routines of the Great Tit, *Parus m. major* L. *Ardea* **38**, 99–135.

Koch, A., and Steinke, L. (1944). Ueber Temperatur und Feuchtigkeit bei Natur- und Kunstbrut. *Arch. Kleintierzucht* **3**, 153–202.

Koepcke, M. (1972). Ueber die Resistenzformen der Vogelnester in einem begrenzten Gebiet des tropischen Regenwaldes in Peru. *J. Ornithol.* **113**, 138–160.

Komisaruk, B. R. (1967). Effects of local brain implants of progesterone on reproductive behavior in Ring Doves. *J. Comp. Physiol. Psychol.* **64,** 219–224.

Koskimies, J., and Routamo, E. (1953). Zur Fortpflanzungsbiologie der Samtente *Melanitta f. fusca* (L.). I. Allgemeine Nistökologie. *Pap. Game-Res.* **10,** 1–105.

Kossack, C. W. (1947). Incubation temperatures of Canada Geese. *J. Wildl. Manage.* **11,** 119–126.

Krüger, P. (1965). Ueber die Einwirkung der Temperatur auf das Brutgeschäft und das Eierlegen des Rebhuhnes. (*Perdix perdix* L.) *Acta Zool. Fenn.* **112,** 1–64.

Kruuk, H. (1964). Predators and anti-predator behaviour of the Black-headed Gull (*Larus ridibundus* L.) *Behaviour, Suppl.* **11,** 1–130.

Kutchai, H., and Steen, J. B. (1971). Permeability of the shell and shell membranes of hens' eggs during development. *Resp. Physiol.* **11,** 265–278.

Lack, D. (1947–48). The significance of clutch size. *Ibis* **89,** 302–352; *Ibis* **90,** 25–45.

Lack, D. (1967). Interrelationships in breeding adaptations as shown by marine birds. *Proc. Int. Ornithol. Congr., 14th, 1966* pp. 3–42.

Lack, D. (1968). "Ecological Adaptations for Breeding in Birds." Methuen, London.

Landauer, W. (1967). The hatchability of chicken eggs as influenced by environment and heredity. *Conn., Agr. Exp. Sta., Monogr.* **1,** 1–315 (revised).

Lange, B. (1928). Die Brutflecke der Vögel und die für sie wichtigen Hauteigentümlichkeiten. *Gegenbaurs Jahrb.* **59,** 601–712.

Lasiewski, R. C., and Dawson, W. R. (1967). A re-examination of the relation between standard metabolic rate and body weight in birds. *Condor* **69,** 13–23.

LeFebvre, E. A., Birkebak, R. C., and Dorman, F. D. (1967). A flight-time integrator for birds. *Auk* **84,** 124–128.

Lehrman, D. S. (1965). Interaction between internal and external environments in the regulation of the reproductive cycle of the Ring Dove. *In* "Sex and Behaviour" (F. A. Beach, ed.), pp. 355–380. Wiley, New York.

Lehrman, D. S., and Brody, P. N. (1964). Effect of prolactin on established incubation behavior in the Ringdove. *J. Comp. Physiol. Psychol.* **57,** 161–165.

Lehrman, D. S., and Wortis, R. P. (1967). Breeding experience and breeding efficiency in the Ring Dove. *Anim. Behav.* **15,** 223–228.

Lemmetyinen, R. (1971). Nest defence behaviour of Common and Arctic Terns and its effects on the success achieved by predators. *Ornis Fenn.* **48,** 13–24.

Ligon, J. D. (1968). The biology of the Elf Owl, *Micrathene whitneyi. Misc. Publ. Mus. Zool. Univ. Mich.* **136,** 1–70.

Lind, H. (1961). Studies on the behaviour of the Black-tailed Godwit (*Limosa limosa* (L)). *Medd. Naturfredningsradets Reservatudvalg* No. 66, 1–157.

Liversidge, R. (1961). Pre-incubation development of *Clamator jacobinus. Ibis* **103a,** 624.

Lloyd, J. A. (1965a). Seasonal development of the incubation patch in the Starling (*Sturnus vulgaris*). *Condor* **67,** 67–72.

Lloyd, J. A. (1965b). Effects of environmental stimuli on the development of the incubation patch in the European Starling (*Sturnus vulgaris*). *Physiol. Zool.* **38,** 121–128.

Löhrl, H. (1970). Unterschiedliche Bruthöhlenansprüche von Meisenarten und Kleibern als Beitrag zum Nischenproblem. *Verh. Deut. Zool. Ges.* **64,** 314–317.

Lokhorst, W., and Romijn, C. (1964). Some preliminary observations on barometric pressure and incubation. *In* "Energy Metabolism" (K. L. Blaxter, ed.), pp. 419–424. Academic Press, New York.

Lokhorst, W., and Romijn, C. (1969). Energy metabolism of eggs incubated in air with a low oxygen content. *In* "Energy Metabolism of Farm Animals" (K. L. Blaxter, J. Kielanowski, and G. Thorbeck, eds.), pp. 331–338. Oriel Press, London.

Lundy, H. (1969). A review of the effects of temperature, humidity, turning and gaseous environment in the incubator on the hatchability of the hen's egg. *In* "The Fertility and Hatchability of the Hen's Egg" (T. C. Carter and B. M. Freeman, eds.), pp. 143–176. Oliver & Boyd, Edinburgh.

McFarland, D. J. (1971). "Feedback Mechanisms in Animal Behaviour." Academic Press, New York.

McFarland, D. J., and Baker, E. (1968). Factors affecting feather posture in the Barbary Dove. *Anim. Behav.* **16,** 171–177.

MacLaren, P. I. R. (1950). Bird—ant nesting associations. *Ibis* **92,** 564–566.

Maclean, G. L. (1967). The breeding biology and behaviour of the Double-banded Courser *Rhinoptilus africanus* (Temminck). *Ibis* **109,** 556–569.

Maclean, G. L. (1970). The biology of the larks (Alaudidae) of the Kalahari sand-veld. *Zool. Afr.* **5,** 7–39.

MacRoberts, M. H., and MacRoberts, B. R. (1972). The relationship between laying date and incubation period in Herring and Lesser Black-backed Gulls. *Ibis* **114,** 93–97.

Manuwal, D. A. (1974). The incubation patches of Cassin's Auklet. *Condor* **76,** 481–484.

Marcström, V. (1960). Studies on the physiological and ecological background to the reproduction of the Capercaillie (*Tetrao urogallus* Linn.). *Viltrevy* **2,** 1–69.

Marcström, P. (1966). Mallard ducklings (*Anas platyrhychos* L.) during the first days after hatching. A physiological study with ecological considerations and a comparison with Capercaillie chicks (*Tetrao urogallus* L.). *Viltrevy* **4,** 343–370.

Maridon, B. and Holcomb, L. C. (1971). No evidence for incubation patch changes in Mourning Doves throughout reproduction. *Condor* **73,** 374–375.

Matthews, G. V. T. (1954). Some aspects of incubation in the Manx Shearwater *Procellaria puffinus,* with particular reference to chilling resistance in the embryo. *Ibis* **96,** 432–440.

Mertens, J. A. L. (1967). "Verslag betreffende het onderzoek over de temperatuur-regulatie en de energiebalans van de koolmees (*Parus major* L.) in 1965–1966," pp. 1–21. Oecologisch Instituut, Arnhem (unpublished report).

Mertens, J. A. L. (1970). Verslag van de werkzaamheden van het oecofysiologisch laboratorium in 1969. Oecologisch Instituut. Arnhem. Report, pp. 1–10.

Mertens, J. A. L. (1974). In preparation.

Murton, R. K. (1965). "The Wood Pigeon." Collins, London.

Needham, J. (1931). "Chemical Embryology," Vol. 1. Cambridge Univ. Press, London and New York.

Nelson, J. B. (1966). The breeding biology of the Gannet, *Sula bassana* on the Bass Rock, Scotland. *Ibis* **108,** 584–626.

Nelson, J. B. (1969). The breeding ecology of the Red-footed Booby in the Gala-pagos. *J. Anim. Ecol.* **38,** 181–198.

Nelson, J. B. (1970). The relationship between behaviour and ecology in the Sulidae with reference to other sea birds. *Oceanogr. Mar. Biol. Annu. Rev.* **8,** 501–574.

Nelson, J. B. (1971). The biology of Abbott's Booby *Sula abbotti. Ibis* **113,** 429–467.

New, D. A. T. (1957). A critical period for the turning of hens' eggs. *J. Embryol. Exp. Morphol.* **5,** 293–299.

Nice, M. M. (1954). Problems of incubation in North American birds. *Condor* **56,** 173–197.

Nice, M. M. (1962). Development of behavior in precocial birds. *Trans. Linn. Soc. N.Y.* **8,** 1–211.

Nicoll, C. S., Pfeiffer, E. W., and Fevold, H. R. (1967). Prolactin and nesting behavior in phalaropes. *Gen. Comp. Endocrinol.* **8,** 61–65.

Norton, D. W. (1970). Thermal regime of nests and bioenergetics of chick growth in the Dunlin (*Calidris alpina*) at Barrow, Alaska. M.Sc. Thesis, University of Alaska, pp. 1–78.

Norton, D. W. (1972). Incubation schedules of four species of calidridine sandpipers at Barrow Alaska. *Condor* 74, 164–176.

Oppenheim, R. W. (1970). Some aspects of embryonic behaviour in the duck (*Anas platyrhynchos*). *Anim. Behav.* 18, 335–352.

Oppenheim, R. W. (1972). Prehatching and hatching behaviour in birds: a comparative study of altricial and precocial species. *Anim. Behav.* 20, 644–655.

Oppenheim, R. W. (1973). Prehatching and hatching behavior: A comparative and physiological consideration. *In* "Studies on the Development of Behavior and the Nervous System" (G. Gottlieb, ed.), Vol. 2, pp. 163–244. Academic Press, New York.

Orr, Y. (1970). Temperature measurements at the nest of the Desert Lark (*Ammomanes deserti deserti*). *Condor* 72, 476–478.

Palmgren, M., and Palmgren, P. (1939). Ueber die Wärmeisolierungskapazität verschiedener Kleinvogel-nester. *Ornis Fenn.* 16, 1–6.

Parkes, K. C., and Clark, G. A., Jr. (1964). Additional records of avian egg teeth. *Wilson Bull.* 76, 147–154.

Parsons, J. (1972). Egg size, laying date and incubation period in the Herring Gull. *Ibis* 114, 536–541.

Payne, L. F. (1921). A study of multiple turning of incubated eggs. *J. Am. Instruct. Invest. Poult. Husb.* 7, 17–20.

Payne, R. B. (1966). Absence of brood patch in Cassin Auklets. *Condor* 68, 209–210.

Peterson, A. J. (1955). The breeding cycle in the Bank Swallow. *Wilson Bull.* 67, 235–286.

Ponomareva, T. S. (1972). Temperature conditions in nests of some bird species in arid regions. *Zool. Zh.* 51, 1846–1856. [In Russian.]

Portmann, A. (1955). Die postembryonale Entwicklung der Vögel als Evolutionsproblem. *Acta Int. Ornithol. Congr., 11th, 1954* pp. 138–151.

Prince, H. H., Siegel, P. B., and Cornwell, G. W. (1970). Inheritance of egg production and juvenile growth in Mallards. *Auk* 87, 342–352.

Proudfoot, F. G. (1969). The handling and storage of hatching eggs. *In* "The Fertility and Hatchability of the Hen's Egg" (T. C. Carter and B. M. Freeman, eds.), pp. 127–141. Oliver & Boyd, Edinburgh.

Ragosina, M. N. (1963). Die Entwicklung des Haushuhn-Embryos in seinen Beziehungen zum Dotter und zu den Eihäuten. *J. Ornithol.* 104, 82–84.

Rahn, H., and Ar, A. (1974). The avian egg: Incubation time and water loss. *Condor* 76 147–152.

Reid, B. E., and Bailey, C. (1966). The value of the yolk reserve in Adelie Penguin chicks. *Rec. Dominion Mus. (Wellington, N.Z.)* 5, 185–193.

Rensch, B. (1954). "Neuere Probleme der Abstammungslehre," pp. 170–172. Enke, Stuttgart.

Rice, D. W., and Kenyon, K. W. (1962). Breeding cycles and behavior of Laysan and Black-footed Albatrosses. *Auk* 79, 517–567.

Ricklefs, R. E. (1968a). Patterns of growth in birds. *Ibis* 110, 419–451.

Ricklefs, R. E. (1968b). On the limitation of brood size in passerine birds by the ability of adults to nourish their young. *Proc. Nat. Acad. Sci. U.S.* 61, 847–851.

Ricklefs, R. E. (1973). Patterns of growth in birds. II. Growth rate and mode of development. *Ibis* 115, 177–201.

Ricklefs, R. E., and Hainsworth, P. R. (1969). Temperature regulation in nestling Cactus Wrens: The nest environment. *Condor* 71, 32–37.

Ridpath, M. G. (1972). The Tasmanian Native Hen, *Tribonyx mortierii*. I. Patterns of behaviour. *CSIRO Wildl. Res.* 17, 1–51.

Riehm, H. (1970). Oekologie und Verhalten der Schwanzmeise (*Aegithalos caudatus* L.). *Zool. Jahrb. Abt. Syst. Oekol. Geogr. Tiere* 97, 338–400.

Rittinghaus, H. (1956). Untersuchungen am Seeregenpfeifer (*Charadrius alexandrinus* L.) auf der Insel Oldeoog. *J. Ornithol.* 97, 117–155.

Rittinghaus, H. (1961). "Der Seeregenpfeifer. (*Charadrius alexandrinus* L.)." Ziemsen, Wittenberg.

Robel, R. J. (1969). Nesting activities and brood movements of Black Grouse in Scotland. *Ibis* 111, 395–399.

Robertson, I. S. (1961a). The influence of turning on the hatchability of hens' eggs. I. The effect of rate of turning on hatchability. *J. Agr. Sci.* 57, 49–56.

Robertson, I. S. (1961b). The influence of turning on the hatchability of hens' eggs. II. The effect of turning frequency on the pattern of mortality, the incidence of malpositions, malformations and dead embryos with no somatic abnormality. *J. Agr. Sci.* 57, 57–69.

Robertson, I. S. (1961c). Studies of chick embryo orientation using X-rays. I. A preliminary investigation of presumed normal embryos. *Brit. Poultry Sci.* 2, 39–47.

Robertson, I. S. (1961d). Studies on the effect of humidity of the hatchability of hens' eggs. I. The determination of optimum humidity for incubation. *J. Agr. Sci.* 57, 185–194.

Robertson, I. S. (1961e). Studies on the effect of humidity on the hatchability of hens' eggs. II. A comparison of hatchability, weight loss and embryonic growth in eggs incubated at 40 and 70% R.H. *J. Agr. Sci.* 57, 185–198.

Romanoff, A. L. (1941). Development of homeothermy in birds. *Science* 94, 218–219.

Romanoff, A. L., and Romanoff, A. J. (1949). "The Avian Egg." Wiley, New York.

Romijn, C. (1950). Foetal respiration in the hen. Gas diffusion through the egg shell. *Poultry Sci.* 29, 42–51.

Romijn, C., and Lokhorst, W. (1955). Chemical heat regulation in the chick embryo. *Poultry Sci.* 34, 649–654.

Romijn, C., and Lokhorst, W. (1956). The caloric equilibrium of the chicken embryo. *Poultry Sci.* 35, 829–834.

Romijn, C., and Lokhorst, W. (1960). Foetal heat production in the Fowl. *J. Physiol. (London)* 145, 239–249.

Romijn, C., and Lokhorst, W. (1961). Dierenarts en kuikenbroederij. *Tijdschr. Diergeneesk.* 86, 995–1011.

Romijn, C., and Lokhorst, W. (1962). Humidity and incubation. *World's Poultry Congr., Proc., 12th, 1962* Sect. Pap., pp. 136–139.

Roseberry, J. L., and Klimstra, W. D. (1970). The nesting ecology and reproductive performance of the Eastern Meadowlark. *Wilson Bull.* 82, 241–352.

Royama, T. (1966). Factors governing feeding rate, food requirement and brood size of nestling Great Tits *Parus major*. *Ibis* 108, 313–347.

Russell, S. M. (1969). Regulation of egg temperatures by incubating White-winged Doves. *In* "Physiological Systems in Semiarid Environments" (C. C. Hoff and M. L. Riedesel, eds.), pp. 107–112. Univ. of New Mexico Press, Albuquerque.

Ryder, J. P. (1967). The breeding biology of Ross' goose in the Perry River region, Northwest Territories. *Canadian Wildl. Serv., Rep. Ser.* 3, 1–56.

Ryder, J. P. (1972). Biology of nesting Ross' Geese. *Ardea* 60, 185–215.

Saiz, F., and Hajek, E. R. (1968). Estudios ecológicos en Isla Robert (Shetland del Sur). 1. Observaciones de temperatura en nidos de Petrel gigante *Macronectes giganteus* (Gmelin). *Publ. Inst. Antarct. Chil.* 14, 1–15.

Schladweiler, P. (1968). Feeding behavior of incubating Ruffed Grouse females. *J. Wildl. Manage.* 32, 426–428.

Scott, D. A. (1970). The breeding biology of the Storm Petrel *Hydrobates pelagicus*. Ph.D. Thesis, Library Edward Grey Institute, Oxford.

Sealy, S. G. (1970). Egg teeth and hatching in some alcids. *Wilson Bull.* 82, 289–293.

Seel, D. C. (1968). Clutch-size, incubation and hatching success in the House Sparrow and Tree Sparrow *Passer* spp. at Oxford. *Ibis* 110, 270–282.

Selander, R. K., and Kuich, L. L. (1963). Hormonal control and development of the incubation patch in icterids, with notes on behavior of cowbirds. *Condor* 65, 73–90.

Selander, R. K., and Yang, S. Y. (1966). The incubation patch of the House Sparrow, *Passer domesticus* Linnaeus. *Gen. Comp. Endocrinol.* 6, 325–333.

Siegfried, W. R., and Frost, P. G. H. (1975a). Egg temperature and incubation behaviour of the Ostrich. (In press.)

Siegfried, W. R., and Frost, P. G. H. (1975b). Continuous breeding and incubation behaviour in the Moorhen. *Ibis* 117, 102–109.

Simmons, K. E. L. (1952). The nature of the predator-reactions of breeding birds. *Behaviour* 4, 161–171.

Simmons, K. E. L. (1955). The nature of the predator reactions of waders towards humans, with special reference to the role of the aggressive-, escape-, and brooding-drives. *Behaviour* 8, 130–173.

Simons, P. C. M. (1971). Ultrastructure of the hen eggshell and its physiological interpretation. *Agr. Res. Rep.* (*Wageningen*) 758, 1–136.

Skutch, A. F. (1957). The incubation patterns of birds. *Ibis* 99, 69–93.

Skutch, A. F. (1958). Roosting and nesting of Araçari Toucans. *Condor* 60, 201–219.

Skutch, A. F. (1962). The constancy of incubation. *Wilson Bull.* 74, 115–152.

Slater, P. J. B. (1967). External stimuli and readiness to incubate in the Bengalese Finch. *Anim. Behav.* 15, 520–526.

Snow, B. K. (1970). A field study of the Bearded Bellbird in Trinidad. *Ibis* 112, 299–329.

Southern, H. N. (1970). The natural control of a population of Tawny Owls (*Strix aluco*). *J. Zool.* 162, 197–285.

Spellerberg, I. F. (1969). Incubation temperatures and thermoregulation in the McCormick Skua. *Condor* 71, 59–67.

Spingarn, E. D. W. (1934). Some observations on the Semipalmated Plover (*Charadrius semipalmatus*) at St. Mary's Islands, Province of Quebec, Canada. *Auk* 51, 27–36.

Starck, D. (1965). "Embryologie, ein Lehrbuch auf allgemein biologischer Grundlage." Thieme, Stuttgart.

Steel, E. A., and Hinde, R. A. (1963). Hormonal control of brood patch and oviduct development in domesticated canaries. *J. Endocrinol.* 26, 11–24.

Steinmetz, H. (1930). Die Embryonalentwicklung des Blässhuhns (*Fulica atra*) unter besonderer Berücksichtigung der Allantois. *Gegenbaurs Jahrb.* 64, 275–338.

Stonehouse, B. (1956). The Brown Skua *Catharacta skua lönnbergi* (Mathews) of South Georgia. *Falkland Isl. Depend. Surv., Sci. Rep.* 14, 1–25.

Storteir, S., and Palmgren, A. (1971). Langtidsregistrering av ruvningsrytmik och matningsfrekvens med hjälp av radioaktiv märkning. *Ornis Fenn.* 48, 33–35.

Swanberg, P. O. (1950). On the concept of "incubation period." *Vår Fågelv.* 9, 63–80.

Swennen, C. (1968). Nest protection of eiderducks and shovelers by means of faeces. *Ardea* 56, 248–258.

Tangl, F., and von Mituch, A. (1908). Beiträge zur Energetik der Ontogenese. V. Mitteilung. Weitere Untersuchungen über die Entwicklungsarbeit und den

Stoffumsatz im bebrüteten Hühnerei. *Arch. Gesamte Physiol. Menschen Tiere* **121**, 437–458.

Thoresen, A. C. (1969). Observations on the breeding behavior of the Diving Petrel. *Pelecanoides u. urinatrix* (Gmelin). *Notornis* **16**, 241–260.

Tickell, W. L. N. (1962). The Dove Prion, *Pachyptila desolata* Gmelin. *Falkland Isl. Depend. Surv., Sci. Rep.* **33**, 1–55.

Tickell, W. L. N. (1968). The biology of the Great Albatrosses, *Diomedea exulans* and *Diomedea epomophora. Antarctic Res. Ser.* **12**, 1–55.

Tinbergen, N. (1948). Dierkundeles in het meeuwenduin. *Levende Natuur* **51**, 49–56.

Tinbergen, N. (1953). "The Herring Gull's World." Collins, London.

Tinbergen, N. (1967). Adaptive features of the Black-headed Gull *Larus ridibundus* L. *Proc. Int. Ornithol. Congr., 14th, 1966* pp. 43–59.

Tinbergen, N., Broekhuysen, G. J., Feekes, F., Houghton, J. C. W., Kruuk, H., and Szulc, E. (1962). Egg shell removal by the Black-headed Gull, *Larus ridibundus* L.; a behaviour component of camouflage. *Behaviour* **19**, 74–117.

Trost, C. H. (1972). Adaptations of Horned Larks (*Eremophila alpestris*) to hot environments. *Auk* **89**, 506–527.

Tschanz, B. (1968). Trottellumen. Die Entstehung der persönlichen Beziehungen zwischen Jungvogel und Eltern. *Z. Tierpsychol.,* Suppl. **4**, 1–103.

Tschanz, B., Ingold, P., and Lengacher, H. (1969). Eiform und Bruterfolg bei Trottellumen *Uria aalge aalge* Pnt. *Ornithol. Beob.* **66**, 25–42.

Tuck, L. M. (1961). "The Murres, Their Distribution, Population, and Biology, A Study of the Genus *Uria,*" Vol. 1, pp. 1–260. Canadian Wildlife Series. Queen's Printer, Ottawa, Canada.

Uspenski, S. M. (1958). "The Bird Bazaars of Novaya Zemlya" (transl. from the 1956 Russian ed..), Vol. 4, pp. 1–159. Can. Wildl. Serv. Transl., Ottawa.

van Kampen, M. and Romijn, C. (1970). Energy balance and heat regulation in the White Leghorn Fowl. *In* "Energy Metabolism of Farm Animals" (A. Schürch and C. Wenk, eds.), pp. 213–216. Juris, Zürich.

Verbeek, N. A. M. (1972). Daily and annual time budget of the Yellow-billed Magpie. *Auk* **89**, 567–582.

Vince, M. A. (1968). Retardation as a factor in the synchronization of hatching. *Anim. Behav.* **16**, 332–335.

Vince, M. A. (1969). Embryonic communication, respiration and the synchronization of hatching. *In* "Bird Vocalizations" (R. A. Hinde, ed.), pp. 233–260. Cambridge Univ. Press, London and New York.

Vince, M. A. (1970). Some aspects of hatching behaviour. *In* "Aspects of Poultry Behaviour" (B. M. Freeman and R. F. Gordon, eds.), pp. 33–62. Brit. Poultry Sci., Edinburgh.

Vince, M. A. (1973). Effects of external stimulation on the onset of lung ventilation and the time of hatching in the fowl, duck and goose. *Brit. Poultry Sci.* **14**, 389–401.

Visschedijk, A. H. J. (1962). Prenatale gaswisseling bij de kip. Ph.D. Thesis, Utrecht.

Visschedijk, A. H. J. (1968a). The air space and embryonic respiration. 2. The times of pipping and hatching as influenced by an artificially changed permeability of the shell over the air space. *Brit. Poultry Sci.* **9**, 185–196.

Visschedijk, A. H. J. (1968b). The air space and embryonic respiration. 3. The balance between oxygen and carbon dioxide in the air space of the incubating chicken egg and its role in stimulating pipping. *Brit. Poultry Sci.* **9**, 197–210.

von Haartman, L. (1956). Der Einfluss der Temperatur auf den Brutrhythmus experimentell nachgewiesen. *Ornis Fenn.* **33**, 100–107.

von Haartman, L. (1958). The incubation rhythm of the female Pied Flycatcher (*Ficedula hypoleuca*) in the presence and absence of the male. *Ornis Fenn.*

35, 71–76.

Wagner, H. O. (1955). Einfluss der Poikilothermie bei Kolibris auf ihre Brutbiologie. *J. Ornithol.* **96,** 361–368.

Walters, J. (1958). Ueber die Brutflecktemperaturen bei See- und Flussregenpfeifern, *Charadrius alexandrinus* und *dubius. Ardea* **46,** 124–138.

Wangensteen, O. D. (1972). Gas exchange by a bird's embryo. *Resp. Physiol.* **14,** 64–74.

Wangensteen, O. D., and Rahn, H. (1970). Respiratory gas exchange by the avian embryo. *Resp. Physiol.* **11,** 31–45.

Wangensteen, O. D., Wilson, D., and Rahn, H. (1970). Diffusion of gases across the shell of the hen's egg. *Resp. Physiol.* **11,** 16–30.

Wangensteen, O. D., Rahn, H., Burton, R. R., and Smith, A. H. (1974). Respiratory gas exchange of high altitude adapted chick embryos. *Respir. Physiol.* **21,** 61–70.

Ward, P. (1965). The breeding biology of the Black-faced Dioch *Quelea quelea* in Nigeria. *Ibis* **107,** 326–349.

Ward, P. (1969). The continuous recording of birds' nesting visits using radioactive tagging. *Ibis* **111,** 93–95.

Warham, J. (1971). Body temperatures of petrels. *Condor* **73,** 214–219.

Weeden, J. S. (1966). Diurnal rhythm of attentiveness of incubating female Tree Sparrows (*Spizella arborea*) at a northern latitude. *Auk* **83,** 368–388.

West, G. C. (1960). Seasonal variation in the energy balance of the Tree Sparrow in relation to migration. *Auk* **77,** 306–329.

Westerskov, K. (1956). Incubation temperatures of the Pheasant, *Phasianus colchicus. Emu* **56,** 405–420.

Wetherbee, D. K. (1959). Egg teeth and hatched shells of various bird species. *Bird-Banding* **30,** 119–121.

Wetherbee, D. K., and Bartlett, L. M. (1962). Egg teeth and shell rupture of the American Woodcock. *Auk* **79,** 117.

Wetmore, A. (1921). A study of the body temperature of birds. *Smithson. Misc. Collect.* **72,** (12), 1–51.

White, S. J., and Hinde, R. A. (1968). Temporal relations of brood patch development, nest-building and egg-laying in domesticated canaries. *J. Zool.* **155,** 145–155.

Willoughby, E. J., and Cade, T. J. (1964). Breeding behavior of the American Kestrel (Sparrow Hawk). *Living Bird* **3,** 75–96.

Wingstrand, K. G. (1943). Zur Diskussion über das Brüten des Hühnerhabichts, *Accipiter gentilis* (L.). *Kgl. Fysiogr. Saellsk. Lund, Foerh.* **13,** 220–228.

Winkel, W. (1970). Experimentelle Untersuchungen zur Brutbiologie von Kohl- und Blaumeise (*Parus major* und *P. caeruleus*). *J. Ornithol.* **111,** 154–174.

Witschi, E. (1949). Utilization of the egg albumen by the avian fetus, *In* "Ornithologie als biologische Wissenschaft" (E. Mayr and E. Schüz, eds.), pp. 111–122. Winter, Heidelberg.

Worth, C. B. (1940). Egg volumes and incubation periods. *Auk* **57,** 44–60.

Würdinger, I. (1970). Erzeugung, Ontogenie und Funktion der Lautäusserungen bei vier Gänsearten (*Anser indicus, A. caerulescens, A. albifrons* und *Branta canadensis*). *Z. Tierpsychol.* **27,** 257–302.

Würdinger, I. (1975). In preparation.

Chapter 7

ZOOGEOGRAPHY

François Vuilleumier

I. Introduction

In the last fifteen years, zoogeography has undergone what can rightly be called a revolution. Part of it is due to new thinking along theoretical lines and part to new approaches to old problems, especially the past connections between continents. Ten years after Darlington's landmark book (1957), the publication by MacArthur and Wilson of a treatise of the theory of island biogeography (1967) has directed the efforts of zoogeographers on a new route and has thus probably revitalized ecology (Hamilton, 1968). On the other hand, the recently formulated theory of plate tectonics has made it possible for biogeographers to hypothesize fruitfully about how continental drift has influenced the distribution of organisms.

Darlington (1957) studied the present distribution of vertebrates, including birds, and sought the most important patterns. On a global scale, these patterns suggested to him that the subdivision of the world into zoogeographic regions was useful conceptually and biologically valuable. Darlington's effort was perhaps spurred in part by a controversy that had its inception with Darwin (1859) and, much later, with Dunn (1922, 1931), who examined the herpetological fauna of the Americas, not in terms of the static concept of faunal regions, but rather in terms of dynamic faunal components within a fauna, each of which might have had different origins or histories.

Whether one uses the approach of faunal regions or that of faunal elements, one remains at the level of assemblages of species or other taxa, and one attempts to answer historical questions about these faunas. The merit of MacArthur and Wilson (1963, 1967) is that they have attempted to break away from the mold of this historicotaxonomic viewpoint. Instead of approaching zoogeography purely empirically, they have tried to identify the "fundamental processes, namely dispersal, invasion, competition, adaptation, and extinction" (MacArthur and Wilson, 1967, p. 4) by means of formal models, and so to develop by inductive and deductive methods the basis for a "theory of biogeography at the species level" (MacArthur and Wilson, 1967, p. 5). In other words, they were pursuing the ambitious goal of bringing together in a theoretical framework both the ecological and the genetic aspects of the distribution of species by relating these to factors of natural selection (see Fig. 1). MacArthur and Wilson (1967, pp. 5–6) were quick to point out how important it is to determine "whether biogeography has a solid enough empirical basis at the present time to make such an attempt." Thus, the break from more traditional zoogeography is only apparent, and empirical studies are needed, perhaps even more so than before. But by providing models that have their roots in the emergent field of population biology, where the ecological and genetic approaches are combined, MacArthur and Wilson (1963, 1967) have shown clearly how much a field as complex as biogeography can benefit from extensive reciprocal feedback between theoretical and empirical methods of enquiry. That this interaction is also beneficial to other fields of biology is suggested, for example, by the recent synthesis by Wilson (1971) of the social biology of insects.

The other facet of the revolution in zoogeography has had little influence so far on the thinking of avian biologists, although the views of a number of other biologists, including botanists, paleontologists, and

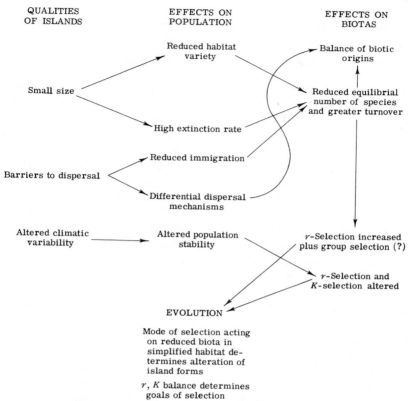

FIG. 1. Interrelationships of various aspects of species distribution on islands and the role of natural selection. [From "The Theory of Island Biogeography," by R. H. MacArthur and E. O. Wilson (copyright 1967 by Princeton University Press), Table 1, p. 6. Courtesy E. O. Wilson; reprinted by permission of Princeton University Press.]

entomologists, have already been permeated by the theory of plate tectonics. This model, recently developed by geologists and geophysicists (see Dickinson, 1971), offers biogeographers a base line from which to draw conclusions about the past history and evolution of distribution patterns of organisms. The immediate result of this theory is that the hypothetical reconstructions of land bridges between continents and archipelagos that were too often made to fit the distribution of individual taxa (see Schmidt, 1955, p. 772, for a review) are best forgotten once and for all. Jardine and McKenzie (1972) even go so far as to say that it is "no longer profitable for biologists to speculate about the past arrangements of land masses."

Thus, the earlier controversy between the proponents of faunal regions and the historical zoogeographers becomes obsolete. The distribution of living and extinct taxa and of faunas can now be examined against a background of plate tectonics theory, whereas, at a much finer level, the distribution of populations and of species can be viewed in the context of population biological theory. It seems, therefore, that conventional zoogeography is being replaced by a newer science that attempts integration at two major levels. At the present time, this integration is imperfect, because there is still a gap between the two levels and because there are discrepancies between empirical data and theory. This review will focus on the renewal that these developments are bringing to zoogeography, yet at the same time it will try to provide a balanced overview of the mosaic of which the field is now constituted.

The reader interested in learning about the distribution patterns of avian taxa should consult Darlington (1957) and Serventy (1960), who compiled detailed information about the geographical distribution of birds. More recent and up-to-date information on avian distribution on a global basis is not available, although Ashmole (1971) has discussed seabird distribution. A review of bird distribution comparable in scope and depth to some of the chapters on mammals in Keast *et al.* (1972) would be welcome. In view of the magnitude of the task, and because the reviews of Darlington (1957) and Serventy (1960) are still satisfactory as a source of data on bird distribution, I have not attempted to update these works for the present chapter. Instead, I have chosen to dwell upon the principles of zoogeography and to present the reader with a survey of the currently available analytical methods and theoretical constructs in zoogeography. I have selected this approach also because of the general scarcity of recent texts dealing with zoogeographical principles [the textbooks of Udvardy (1969), Bănărescu (1970), and MacArthur (1972), although offering much methodological and theoretical information, are either slightly outdated in some respects or else heavily biased toward some aspects of the field]. I would therefore recommend that this chapter should be read in connection with works in which emphasis is laid on the description of distribution patterns, for example, Serventy (1960) and Cracraft (1973a).

II. The Evolution of Zoogeography

The word zoogeography has quite different meanings to different biologists. To some, the emphasis should be on the geography (e.g., Crowley, 1967), and zoogeography means then the description of distribution patterns. To others, the description is not enough, and they seek to explain distribution in ecological or evolutionary terms or both (e.g., Mayr, 1965a; Brosset, 1969). To still others, causal explanations of empirical patterns do not constitute the complete scientific endeavor, and the patterns are used to construct predictive models (e.g., Chorley, 1964; MacArthur, 1972). These three fundamental approaches to biogeography have historical components as well as common empirical roots, so that the early biogeographers were also those who emphasized the geography, while the most recent stress theory, but the overlap in time as well as in content is considerable. So, although the division of zoogeography into three components may be somewhat arbitrary, it is a convenient classification for an exploration of the evolution of ideas and methods in zoogeography. Partial reviews of the history of biogeography can be found in Deevey (1949) and Schmidt (1955), and a very useful review of the literature in the period 1957 to 1964 was published by Niethammer and Kramer (1966). A good, modern review of the history of biogeographical ideas remains to be written.

A. DESCRIPTIVE ZOOGEOGRAPHY

The attention of early zoogeographers was directed to geographical zoology, that is to say, to a description of the geographical distribution of animals. Sclater (1858) and Huxley (1868) pioneered this approach using birds as examples to build a system of zoogeographical regions. In his classic work, Wallace (1876) discussed not only the faunas of the regions, but also the relevance of extinct mammalian faunas and the distribution of the various classes of vertebrates and of some insects and mollusks. In chapters on each of the zoogeographical regions (Palearctic, Ethiopian, Oriental, Australian, Neotropical, and Nearctic), Wallace reviewed the history of their respective faunas and gave extensive faunal lists. More recent summaries of the bird life of the zoogeographic regions can be found in Darlington (1957) and Serventy (1960).

From these early works to the present, a succession of authors have used the regional approach or a variant of it. A major difficulty of regional zoogeography has been (and is) what Mayr (1965a, p. 475) called "a purely static definition of faunas." Although Darwin (1859) had already long ago emphasized the heterogeneous nature of faunas (see Mayr, 1965a), zoogeographers have maintained until recently an active interest, almost an obsession, in the delimitation of faunal regions. As a result, vigorous controversies arose and are still being debated about such "problems" as the limit of the Nearctic and Neotropical regions, the limit between the Palearctic and the Ethiopian regions, or the boundary of the Oriental region.

Three points are worth mentioning about descriptive zoogeography; they are well illustrated by Rapoport (1968; see, also, 1971) in a review of the regional concept with special reference to the Neotropical region.

The first concerns the subdivisions of the region. The delimitation of boundaries between subregions and of boundaries within each of the subregions, is evidently an endless exercise. Compare, for instance, the subdivisions suggested by Sclater and Sclater (1899, for mammals), de Mello-Leitão (1931, 1937, for arachnids and scorpions), and Cabrera and Yepes (1940, for mammals), in Rapoport's Fig. 3 (1968, p. 69). Sclater and Sclater have four subdivisions, de Mello-Leitão's earlier scheme has five, his later one seven subdivisions, while Cabrera and Yepes have eleven. This would not be so bad if the subdivisions were concordant, but they are not; there is clearly utter chaos. The second point concerns the so-called "Sub-Tropical Line," or the delimitation between the Guyano–Brasilian and the Andean–Patagonian subregions. Rapoport's Fig. 4 (1968, p. 78) shows no fewer than four major and two minor modifications of this supposed boundary, which wanders in the north from northern Peru to northern Colombia, and in the southeast from south central Brazil and northeastern Argentina south of Buenos Aires. Clearly, one can make the boundary vary almost at will, depending upon the taxon selected for study and how one chooses to study it. The third point concerns the faunal relations among regions. Rapoport (1968, 1971) noted that a number of distribution patterns cannot be fitted to the classic regional scheme and proposed to designate two additional regions to accommodate distribution patterns found especially in some plant and invertebrate taxa.

It seems to me that the time is long overdue to recognize that the regional approach to zoogeography is artificial and that it should be dropped. Typology, in one form or another, appears to be the bane of several disciplines in evolutionary biology. Although typological thinking is largely dead in some fields (as in species-level systematics, see Selander, 1971), this cannot yet be said in biogeography. Fortunately for this field, other approaches do provide more satisfactory lines for research; these are described below.

B. ANALYTIC ZOOGEOGRAPHY

It is obvious that even descriptive zoogeographers use analytic methods. Yet the analysis of the origins and history of faunas from regions or other zoogeographic subdivisions is bound to be frustrated by the structure of these faunas (Mayr, 1965a, p. 474): "As a result of various historical forces, a fauna is composed of unequal elements and no fauna can be fully understood until it is segregated into its elements and until one has succeeded in explaining the separate history of each of these elements." Since Dunn (1922, 1931), ornithologists, largely lead by Mayr, have worked extensively on the historical zoogeography of avifaunas.

Recently, Mayr (1965a) discussed in detail the methods of inference in historical zoogeography, and presented a classification of faunal types (see Table I).

This sort of analysis seems perhaps more meaningful if the faunal types selected correspond to some kind of ecologic unit, such as the biome, vegetation formation, habitat, etc. One of the pioneers of this approach to faunal analysis was Stegmann (1932, 1936, 1938), who divided the Eurasian avifauna into seven types, each of which corresponds rather closely to one of the main vegetation zones. Radu (1962) applied Stegmann's scheme to an analysis of the Romanian avifauna, and similarly, Udvardy (1958, 1963) analyzed the bird faunas of North America, although he recognized additional faunal groups, seventeen in total. Voous (1960, 1963) used faunal types close to those of Stegmann as well as a number of others, based on a variety of criteria, in his well-known study of the European avifauna. Table II lists the faunal types of Voous. Although such studies are valuable, they are not entirely free from typological thinking, because it is necessary in this sort of work to assign species to one category among the ecological (or other) units adopted in the scheme. It seems to me that the larger the number of faunal types (or faunal groups, or faunal elements, depending on the authors' varied but sometimes confusing terminology), the greater the risk of arbitrary decisions about the classification of a given species. Furthermore, if the categories are ecological (in the sense of biome or habitat), one runs the additional risk of disregarding the reality of geographical variation in habitat preferences, which so many species exhibit. Thus the major weakness of this method comes from the apparently unavoidable subjectivity in the interpretation of the delimitation of "faunal types." The sort of practical question one must answer during the course of this type of work is, "Does this species inhabit rain forest or monsoon forest?" or "Should all forest types be grouped into one category, or should they be subdivided into several categories?" Each

TABLE I

FAUNAL TYPES RECOGNIZED BY MAYR (1965a) AND BASED ON INFERENCES ABOUT RELATIVE AGE OF THE TAXA, RELATIVE DISPERSAL CAPACITY OF THE ANIMALS, AND DISTRIBUTION OF RELATED TAXA[a]

1. Autochthonous adaptive radiation
2. Continued single origin colonization
3. Continued multiple origin colonization
4. Fusion of two faunas
5. Successive adaptation
6. Composite origin of various combinations of 1–5

[a] Courtesy of E. Mayr and of VEB Gustav Fischer Verlag.

TABLE II

FAUNAL TYPES DESIGNED BY VOOUS (1960) FOR AN ANALYSIS
OF THE EUROPEAN BIRD FAUNA[a]

Faunal type	Description
1. Arctic	Belonging to the fauna of the tundra climatic region and the birch zone of the boreal region of the Northern Hemisphere
2. Holarctic	Belonging to the fauna of the cold, temperate, and subtropical regions of the Northern Hemisphere
3. Siberian–Canadian	Belonging to the fauna of the boreal climatic zone in the holarctic region, notably in the coniferous forest belt
4. Siberian	Belonging to the fauna of the boreal climatic zone in the palaearctic region, notably of the taiga (coniferous forest belt)
5. Chinese–Manchurian	Belonging to the fauna of the mixed and coniferous forests of eastern Asia
6. Palaearctic	Belonging to the fauna of the cold, temperate, and subtropical regions of the northern half of the Old World
7. Nearctic	Belonging to the fauna of the cold, temperate, and subtropical regions of the northern half of the New World
8. North Atlantic	Belonging to the marine and coastal fauna of the northern Atlantic Ocean
9. European	Belonging to the fauna of the temperate and Mediterranean regions of Europe
10. European–Turkestanian	Belonging to the fauna of the temperate and Mediterranean regions of Europe and Southwest Asia
11. Turkestanian–Mediterranean	Belonging to the fauna of the summer-warm and summer-dry regions of southern Europe and southwest Asia, including the warm, low-lying steppes
12. Mediterranean	Belonging to the fauna of the Mediterranean region
13. Sarmatic	Belonging to the coastal fauna that in the late Tertiary and Pleistocene times inhabited the shallow, brackish, or salt Sarmatic inland sea. This sea, which was of variable extent, formed a continuation of the eastern Mediterranean Sea, stretching to the north of the Black Sea and possibly covering the present Hungarian Plain; eastward, it extended to the Caspian and Aral Seas, covering here the present Kara Kum and Kyzyl Kum deserts
14. Turkestanian	Belonging to the fauna of the low-lying steppe region of southwest Asia

TABLE II (*Continued*)

Faunal type	Description
15. Palaeoxeric	Belonging to the fauna of the steppes and deserts of the southern palaearctic region
16. Palaeoxeromontane	Belonging to the fauna of the arid slopes of the low mountains of the southern palaearctic region
17. Palaeomontane	Belonging to the fauna of the alpine or snow (nival) zones of the high mountains of the palaearctic region
18. Tibetan	Belonging to the fauna of the Tibetan Highlands in the tundra climatic zone
19. Mongolian–Tibetan	Belonging to the fauna of the middle asiatic cold, high steppes
20. Ethiopian	Belonging to the fauna of Africa south of the Sahara
21. Indian–African	Belonging to the fauna which is now largely discontinuous geographically but which in late Tertiary and Pleistocene times must have extended continuously from southern Asia to northern and central Africa
22. Old World	Belonging to the fauna of the great land masses of Eurasia and Africa combined
23. Antarctic	Belonging to the sea and shore fauna of the islands and coasts of the Antarctic continent
24. Cosmopolitan	Having such a wide distribution in all or all but one of the continents that the faunal origin can no longer be deduced from the present distribution range

[a] To these 24 faunal types, Voous (1960) has added another category, "Unknown," including five species "the present distribution ranges (of which) give no indication of their respective geographical origins or of the faunal group to which they belong." (Courtesy of K. H. Voous and of Thomas Nelson, Ltd.)

zoogeographer is likely to decide differently from the others, because his experience from museum and field will be a unique blend. It is thus important that each author state clearly his criteria of selection, so that even though others may disagree, they will be able to make useful comparisons between their work and the work of others. Two examples of such choices may be given here, not necessarily because they are the best ones, but chiefly because the examples will be discussed further in Section IV,A. The first example is Moreau's (1966, p. 80) selection of criteria to attribute bird species either to forest or to non-forest vegetation: "I have accepted as forest species those whose typical habitat is described in the standard works as primary or secondary forest (or both), whether or not they extend in some degree into 'gallery' or 'fringing' forest," and further: "But I have not counted as forest birds those species which typically inhabit this last vegetation type . . . and do not inhabit rain-forest (whether primary or secondary)." The second example is my own analysis of the páramo avifauna of the northern Andes, in which I considered as belonging to the páramo fauna the land birds "occurring in one of the following ecological categories: (a) grassland only, (b) grassland and open scrub, (c) grassland, scrub, and upper edge of the montane forests (ceja forests), and (d) edge only" (Vuilleumier, 1970a, p. 375). Surely some authors will take exception to Moreau's or my criteria, but one hopes the reasons for deciding how to assign species are stated sufficiently precisely so that others may use them in a comparative study of their own. For example, Zimmerman (1972, p. 325) has clearly stated why it is difficult to make direct comparisons between his data from the Kakamega forest and Moreau's data from other forests. (See also Keith *et al.*, 1969.)

In my view, one of the most fruitful attempts to analyze bird faunas in broad ecological terms is Moreau's (1966) study of the African avifaunas. I believe that his success is due to a combination of two things: (1) his superposition of a quasitrophic classification of bird species upon a very broad ecological classification, and (2) his extensive use of what is known or strongly inferred about the past distribution of vegetation. Thus, although Moreau's conclusions can be stated only in terms of probabilities that such or such event in the history of, say, lowland tropical forest birds, took place, his presentation of the arguments permits the reader to obtain a dynamic picture of the succession of taxa in diverse faunas on the African scene. His book is without doubt a landmark in analytic zoogeography, interweaving as it does hypotheses with facts and present evidence with past reconstructions. Further work on the African avifaunas can now be carried out, since an enormous amount of taxonomic and distributional material has been summarized by Hall and Moreau (1970) in their "Atlas of Speciation of African Birds." Comparable, but generally less extensive, studies have been made on the birds of Australia and Tasmania (Keast, 1961; Ridpath and Moreau, 1966), New Guinea (Diamond, 1972, 1973), and South America (Haffer, 1967a,b,c, 1969, 1970a,b; Vuilleumier, 1967, 1969b, 1972; see, also, Müller, 1971, 1972a,b, 1973; Müller and Schmithüsen, 1970).

It is important to note here that the methods used by Müller are quite different from those employed by the other authors cited above. The latter, especially Keast, Diamond, Haffer, and Vuilleumier, examined primarily patterns of speciation among the species making up the samples of the faunas they were studying. In other words, they were interested in determining the ecology and the geography of zones of "contact" between taxa, in particular those that are presently geographically isolated from each other and well marked in their morphological differentiation, and those that show evidence of either secondary hybridization or slight secondary overlaps in the contact area (see Mayr, 1963, for the background to this approach). From a comparison of the distribution patterns of these contact zones and the integration of data on the geological and botanical history, these authors then try to reconstruct what the history of the various elements of each fauna has been (element here means simply the sum of species showing consistent patterns of distribution centered around the contact zones). Müller, on the other hand, following the method of de Lattin (1957), attempted to analyze the distribution pattern of centers of dispersal (*Ausbreitungszentren*). To discover these centers, one proceeds primarily from the geographical technique of mapping the distribution of a large number of species, monotypic and polytypic separately, to the biological analysis of the ecology and geography of the centers, the latter being determined from the congruence of distribution patterns. These centers of dispersal are interpreted to represent areas in which the taxa (species, subspecies, or both) survived during past periods of "biotic constriction." Since, generally speaking, the geographical barriers that form the core of the methodology of the "speciationists" also correspond to unfavorable zones surrounding the normal habitat of species inhabiting more or less the same type of vegetation (broadly defined—e.g., rain forest, montane forest, savanna woodland, grassland), it is not so surprising that the two kinds of studies should yield parallel results. In both cases, areas that effectively (or supposedly) served as refuges during unfavorable periods of the past are uncovered, and in a number of instances these refuges (whether they are called centers of dispersal or merely refuges) are the same, regardless of the method used to detect them (see, for instance, Müller, 1972a, pp. 181–182). One should note that in the terminology of de Lattin, Müller, and others, faunal element refers to taxa that are distributed in accordance to given centers of dispersal.

In a way, the "speciationists" start from ecological and genetical considerations originating from the theory of speciation, whereas the "centrists" begin from geographical considerations stemming from accurate topographic mapping, without *a priori* assumptions about biological or evolutionary theories. The link between geography and evolutionary theory is provided during the second step of the analysis instead of in the first. Perhaps the comparative cartographic method is freer of bias about ecological preferences of species than the speciation method, but it is not so free of subjectivity when the decision to circumscribe centers of dispersal is made. In both methods, heavy reliance is made on the taxonomy, and taxonomic knowledge of species limits is an essential prerequisite to any mapping.

All of the analyses cited above are qualitative, even though some very elementary statistics may be used. There is thus no statistical theoretical basis by which one may be able to judge the "significance" of given null hypotheses, or at least, such a theory is still largely to be worked out. In order to eliminate some of the subjectivity in handling large amounts of zoogeographical data, several authors have begun to apply quantitative, or numerical, methods. The major difference between the approaches reviewed above and those to be cited now is that in the first the species are pooled to obtain distribution patterns, whereas in the second the species themselves are compared in different geographical areas. Thus, the faunal analysts making use of the theory of speciation

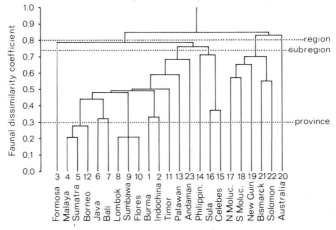

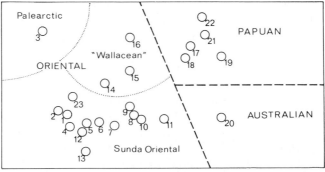

Fig. 2. (*Top*) Dendrogram obtained from single-link cluster analysis from the coefficients of faunal dissimilarity for the birds of the Indo-Australian area. (*Bottom*) Two-dimensional representation of primary areas obtained from the coefficients of faunal dissimilarity by nonmetric multidimensional scaling. The numerals correspond to the areas identified in the dendrogram. (From Holloway and Jardine, 1968. Courtesy J. D. Holloway and Academic Press, London.)

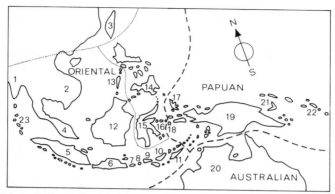

Fig. 3. Map of the Indo-Australian area showing regions and subregions derived from the dendrogram of Fig. 2. (From Holloway and Jardine, 1968. Courtesy J. D. Holloway and Academic Press, London.)

or of the hypothesis of centers of dispersal work with concepts of statistical averages and essentially ask questions like, "What fraction of species show discontinuous geographical distribution with taxonomic differentiation across a given eco-geographical barrier?" or "What fraction of species show congruent distribution patterns that can be interpreted to represent a center of dispersal?" By contrast, the student of faunal resemblance asks, "What fraction of taxa are shared among areas A, B, and C?" At first sight, the differences between the two approaches may not appear to be large, but in fact they are considerable. The speciationists and the centrists, each in their own methodological way, are trying to get away from the descriptive zoogeography of times past and to approach ultimately their subject with a blend of biology and geography. The numerical zoogeographers are trying, consciously or not, to refine the descriptive analysis by treating their data in repeatable, numerical ways. Their results show areas of greater or lesser sharing of species in common; and depending upon the methodology and the way of handling data, one may then, if one so wishes, arrive at an extremely split hierarchy of levels of faunal regions.

At first, these numerical analyses were restricted to studies of resemblance among faunas, and indices were devised to express numerically the degree of resemblance of faunas, treated as pairs (see, e.g., Ekman, 1940; Simpson, 1947, 1960; Burt, 1958; Stugren and Rădulescu, 1961, as cited in Bănărescu, 1970, p. 130; Peters 1968). More recently students of zoogeography have turned their attention to a number of multivariate techniques. Since much of this work has been reviewed by Peters (1971), we need only consider some of these techniques and some not mentioned by Peters (1971).

Phipps (1967, 1968a,b) established, for a variety of types of landscape in southern France, a series of biogeographical models that were subsequently analyzed by principal component and factorial analyses. Holloway and Jardine (1968) used cluster analysis and nonmetric multidimensional scaling in their study of the zoogeography of the Indo-Australian area, comparing birds, bats, and butterflies (for detailed information about these techniques, see Jardine and Sibson, 1971; Kruskal, 1964a,b). Figure 2 illustrates the dendrogram and the two-dimensional map of primary areas obtained from the coefficients of faunal dissimilarity, and Fig. 3 is the map of the Indo-Australian area showing the regional classification of birds there. Holloway further used cluster analysis in studies on the butterfly fauna of India (1969) and on the moth fauna of Mt. Kinabalu (1970). Kikkawa and Pearse (1969), in one of the very few numerical studies on birds, employed the technique of divisive information analysis to study avian distribution in Australia.

Although the number of such analyses is still small, especially in avian zoogeography, they appear to complement the classic analyses of regional zoogeography more than the analyses carried out on faunal elements, however these are defined. Holloway and Jardine (1968) have suggested that "both intuitive and numerical methods for grouping primary areas into zoogeographic regions are in themselves mainly of use for descriptive purposes. It is only when such methods are used in conjunction with more refined methods as multidimensional scaling of primary areas and grouping of taxa into faunal elements that detailed inferences about the history of particular taxonomic groups can be made." One may be tempted to challenge this view. If one examines the methods of analytic zoogeography, he cannot avoid the uneasy feeling that there is too much heterogeneity in methodology, a maze of empirical data, and a dearth of clear statements about the philosophy of the scientific endeavor. If, as I believe, and as I have explained earlier, descriptive zoogeography is not to be considered truly to be *bio*geography, then no matter how refined the technique, the goal will remain the same. Thus, I fail to see a great future in numerical techniques applied chiefly to the delimitation of zoogeographic regions. If, on the other hand, the goal of zoogeographers is to understand distribution patterns, then surely, as others have pointed out long ago, one must start from somewhere around the level of species, and so the map of the species range will remain as an important tool. From there on, much of what one does depends upon the theoretical framework one adheres to as a zoogeographer. Sadly, one is forced to remark that there is little cohesive theory behind much of what is analytical in zoogeography. If one accepts Mayr's (1965a, p. 474) statement quoted earlier about faunal elements and carries the analysis to its logical extreme, then he is forced to admit that the only "natural [in the sense of biological] faunal element" is the species. But at this level, unfortunately, everything becomes a particular case, and no generality is possible; hence, there may be little science. However, if the theoretical framework is built around the level of the species, but in terms of biological processes that may be common to all species, then meaning may again be given to biogeographic analysis. This is essentially the approach heralded by MacArthur and Wilson (1963, 1967) and reviewed below.

C. PREDICTIVE ZOOGEOGRAPHY

The bird faunas of the islands in the south Pacific, well studied from the taxonomic point of view, especially by Mayr (1940, 1942), and the fauna of the Galápagos, analyzed by Lack (1945, 1947), provided the impetus for a series of theoretical papers. In one of them, MacArthur and Wilson (1963) reconsidered the insular bird faunas of the south Pacific. They suggested that the apparent poverty in species numbers in the outlying islands, far away from the main sources of faunal flow, could be explained by an equilibrium theory, the antagonistic forces determining the balance being the two basic processes of immigration to an island and extinction on an island, the processes being the equivalent of birth and death of individuals in the theory of population dynamics. This model leads to a refinement of the notion that insular faunas are determined by factors that impinge upon these processes, especially the area of an island and its distance from a faunal source, because the genetic and demographic properties of the propagules themselves are included. The model was fully elaborated upon in their 1967 book.

Independently of MacArthur and Wilson, Hamilton and Rubinoff (1967) and Hamilton *et al.* (1964) studied the fauna and flora of the Galápagos Islands, as well as the fauna of a number of other archipelagos. Using analytic techniques of multiple regression, they suggested an equilibrium model somewhat at variance from that of MacArthur and Wilson, since it "assumes extinction of species to be relatively infrequent in nature, and thus a minor factor in controlling variation in insular number of species among islands" and "suggests that insular avifaunal size is regulated by equilibrium established between isolation and the ecologic diversity of an island" (see Hamilton, 1967, p. 46). Lack (1971) reached similar conclusions with a different reasoning.

These papers have opened a new era in zoogeography that is properly predictive. That is, through mathematical models having variables and parameters basic to the ecology and genetics of populations and of species, events such as colonization, immigration, success after colonization, and extinction can be quantified and their outcome predicted given certain conditions. This approach leads directly to the manipulation of biota for the testing of predictions derived from the models. Wilson and Simberloff (1969), Simberloff and Wilson (1969), and Simberloff (1969) have begun this kind of experimentation with island biota. Not surprisingly, their work was done with arthropods, and it appears unlikely that birds will be used for similar experimental purposes. However, the zeal of bird watchers is now rewarded, for comparisons of lists of species for insular faunas obtained at different times (Diamond, 1969, 1971), when analyzed in the light of the theory of MacArthur and Wilson, are yielding interesting empirical data that bear on the processes of immigration and extinction.

With this most recent development, biogeography has come full circle, from purely descriptive to experimental, and has clearly evolved. MacArthur's book (1972), spanning part of this evolution and interweaving empirical and theoretical aspects, illustrates the changes very dramatically.

III. The Significance of Taxonomy

All zoogeographic studies depend on empirical data, and much of the empirical evidence is derived ultimately from taxonomic work. Since it is often asserted that birds are the best-known animals from a taxonomic viewpoint, avian zoogeographers might appear to be privileged in comparison with other zoogeographers. Indeed, the number of new species of birds that are described each year is very small compared with the number of new species in most other taxa of similar rank. For example, of the fifty-one new species described from 1956 to 1965, only about thirty-five were retained as valid by Mayr (1971). This number represents only about 0.4% of the total number of avian species. By contrast, the taxonomy of birds above the species level, and especially above the genus level, is far less well advanced than that of many other groups of animals (see the remarks of Stresemann, 1959; Sibley, 1970; Sibley and Ahlquist, 1972; Cracraft, 1972b). There are several reasons for this state of affairs, the discussion of which would take us outside the scope of this chapter. But it is relevant to our theme to mention the view that the relative rarity of avian fossils cannot be invoked as the chief cause of the chaotic state of avian classification at higher categories. There are actually more bird fossils than is commonly believed (see the comments of Storer, 1960, pp. 50–51; Brodkorb, 1971, pp. 20–21; and others), and many of these remains are helpful (or, rather, could be more helpful if they were properly analyzed or restudied) in evaluating the taxonomic affinities of some taxa and/or of some faunal assemblages. Darlington (1957) was extremely cautious about the interpretation of fossil birds as biogeographic evidence and helped create a feeling that avian fossils were generally useless for zoogeographic studies. I believe that this thinking is often unwarranted, and that it is being dispelled by the work of some authors who are undertaking comprehensive and broadly comparative studies, either of some specific taxa (e.g., penguins, see Simpson, 1957, 1971a,b,c, 1972; Gruiformes, see Cracraft, 1968, 1969, 1971, 1972c, 1973c; cathartid vultures see Cracraft and Rich, 1972), or of some zoogeographically important faunas (e.g., the Tertiary faunas of Europe, see Ballmann, 1969a,b; Brunet, 1970; Cracraft, 1974), to cite just a few. I do not mean to imply by these comments, however, that fossils will solve the problems of classification of higher categories of birds, nor that they will provide us with answers to the major unsolved questions of avian distributional history. I do feel that careful studies of the rich extant fossil material will help us understand some issues in avian zoogeography, but only if these studies are carried out in the spirit of problem solving, the problems being both taxonomic and zoogeographic hypotheses in this case. Avian paleontologists must be bolder in their attempts at synthesis than they have been in the past. Useful discussions of the role of paleontology in systematics and in zoogeography can be found in Cracraft (1973a) and Schaeffer *et al.* (1972).

A. RECENT TAXONOMIC WORK

Avian zoogeographers have now at their disposal a much greater number of taxonomic revisions, monographs, and faunal lists (not to mention handbooks and field guides) than about twelve years ago, when Serventy (1960) wrote his review, or fifteen years ago, when Darlington (1957) summarized the distribution of birds. Perhaps more importantly, many of these recent taxonomic studies are thoroughly modern in approach, and far less typological than earlier ones. Thus the taxonomic foundation for avian biogeography is in a much better state today than it has ever been, although one must repeat the caveat that this is still true chiefly for taxonomy at the species level.

1. The Impact of Recent Techniques

Many taxonomic studies of the past ten years have been influenced by evolutionary thinking, especially Mayr's (1963). But the most recent techniques that have sprouted in taxonomic analysis have not yet had a profound impact in avian systematics, as noted by Selander (1971). Consequently, although recent work in avian taxonomy employs evolutionary theory as a framework, it is still largely carried out with methods similar to those used by earlier workers. As a result, few of the taxonomic revisions that are of interest to zoogeographers make extensive use of nonmorphological characters. Among the exceptions, one can cite Lanyon's (1967) attempt to reconstruct the distributional history of *Myiarchus* flycatchers in the West Indies, based extensively on data obtained from tape recordings of vocalizations complemented with playback experiments carried out in the field.

Among avian zoogeographers who have undertaken taxonomic studies on many species in a wide variety of higher taxa prior to zoogeographic analysis, the methods used have been chiefly to measure several morphological characters on study skins from museum collections, and to compare the means and ranges of variability among samples representative of the entire range of the species. In actual practice, these samples are rarely representative of the species distribution, and even if they are, the number of specimens per sample is often barely adequate, or inadequate, for statistical analysis of the numerical data. Often this museum work was supplemented by field work done by the zoogeographers themselves, either to collect additional museum material or to obtain data on the habitat preferences of the species studied, or both [see, for examples of methodology or for examples of zoogeographic analyses performed after such types of taxonomic surveys, the papers of Bond (1963), Diamond (1972), Dorst (1967), Grant (1966), Haffer (1969), Hall (1972), James (1970), Keast (1961), Mayr (1969), Moreau (1966), Olrog (1972), Salomonsen (1972), Serventy (1972), Vuilleumier (1972)]. Behavioral studies, biochemical assays of enzymes or other proteins, quantitative colorimetric techniques, or numerical taxonomic techniques (all described in Selander, 1971) have generally not been undertaken in such surveys. The reasons are simple enough; either the material was not extensive enough to warrant such sophistication (series of specimens of museums are often quite uneven in regard to sex, age, and stages of molt, and much too often have pitifully inadequate data on the labels about the state of the gonads, skull pneumatization, etc.), or an approach making full use of a gamut of techniques would make it necessary to spend so much time in the field, museum, laboratory, and computer center that it would preclude authors interested primarily in zoogeographic studies to do anything but exclusively taxonomic work. To give a specific example of these difficulties, James (1970) used four thousand specimens of twelve species in her analysis of trends of variation in birds from the eastern United States, but for only one species were there sufficiently large samples of skins to submit the data from measurements to detailed intralocality tests of variance.

2. Taxonomic Evidence in Zoogeography

As I said at the beginning of this section, taxonomy is truly at the base of all zoogeographic work. One of the kinds of zoogeographic questions in which taxonomic evidence is used is whether faunas A and B are more closely related to each other (i.e., have more taxa in common) than either is to fauna C. By fauna is meant here the sum of all taxa living in a given continental land mass, usually above the species level. In such instances, zoogeographers have concluded that if no taxa showed interfaunal relationships, then the two faunas must have been isolated from each other for a considerable period of time. An example of such reasoning is found in Mayr (1972b), who said the following about the avifaunas from Africa and South America: "The African avifauna is much more similar to those of Europe and Asia than to that of South America. Indeed, so far as I know, there is not a single African family or [sic] birds that is more closely related to a South American family than to an Asian family" (Mayr, 1972b, pp. 554–555). The first statement of the quotation is probably the reflection of a statistical truth, although to my knowledge the statistics are actually lacking (Mayr gives none). In other words I do not know of a taxonomist with an equally good knowledge of both the African and South American avifaunas who has made a quantitative estimate of the faunal resemblance. In fact, I suspect that careful, intercontinental studies of several taxa will reveal that more affinities do exist between the faunas of presently isolated land masses than orthodox zoogeography has it.

In the case of Africa versus South America, besides some casual papers (e.g., Winterbottom, 1965), very few thorough such transcontinental studies have been made (see Bock, on herons, 1956, and plovers, 1958; Voous 1964, 1966, on owls). More recently, after first-hand studies of both African and South American woodpeckers, Short (1970, p. 39) had this to say: "Ten of the 13 groups of African woodpeckers, considered to comprise three genera (*Campethera, Geocolaptes,* and *Dendropicos*), are rather closely interrelated. Except for *Picoides* (*Dendrocopos*), a widespread genus derived from *Dendropicos*, they have no apparent relatives in the Old World. Rather, their relationships, as indicated by their morphology, lie with New World colaptine (*Colaptes, Piculus, Veniliornis*) woodpeckers." Although it seems doubtful that a very large number of higher taxa are likely to have patterns similar to those uncovered by Short in the woodpeckers, it is not wise to guess at the number of such instances at present. Surely more patterns of this sort exist within the non-Passeriformes than the Passeriformes, but we will be able to assess the matter quantitatively only after many more thorough studies have been completed on the higher taxa of birds.

B. Taxonomy, Endemism, and Zoogeography[1]

Many faunal analyses are based on the assumption that the relative age of a taxon that is endemic to some geographical area is directly

---

[1] A paper by Horton (1973), pertinent to the theme of this section, was received after completion of the manuscript.

proportional to its rank in the taxonomic hierarchy. Thus an endemic subspecies is supposed to be, on the average, more recent than an endemic species, and the latter, in turn and again on average, more recent than an endemic genus, and so on (see, for example, Darlington, 1971). Assuming further a different average longevity for taxa of different ranks, increasing from subspecies to species and from species to genus, the fauna is then subdivided into "elements" showing its age stratification or structure. This structure is ordinarily expressed by the percentage of endemic subspecies or species in the area or in the fauna. Conclusions derived from analyses of this sort are varied, but often fall into two categories. The first is that if a fauna possesses a low percentage of endemic species and either many endemic subspecies or many nonendemic taxa, it is either of recent origin or has not been geographically isolated very strongly from other neighboring faunas with a similar taxonomic composition. In the latter case, there is much faunal exchange taking place between the two faunas. The second category of conclusions is that if a fauna has a high percentage of endemic species and either many endemic subspecies or few nonendemic taxa, it is either of older origin or has been strongly isolated geographically from other faunas, and faunal flow has been negligible.

Examples of papers based on the taxonomy–endemism concept include, among many others, those of Fleming (1962) on the history of the New Zealand avifauna, and of Mayr and Phelps (1967) on the origins of the Pantepui avifauna of southern Venezuela. Further details about this approach are given by Mayr (1965a, 1972b). In view of the broad use of this technique it is worthwhile to discuss it here from a taxonomic point of view.

There are at least three difficulties, taxonomically speaking, that must be overcome if the results obtained by the use of this method are to be accepted as statistically valid. The first is that subspecies are arbitrary taxonomic categories, as Selander (1971) has clearly shown. The second difficulty is that the "ages" of bird species found in different parts of the world and in different ecological conditions are extremely variable. The problem is perhaps most acute for species belonging to superspecies, the allospecies of Amadon (1966). The third is that judgments of taxonomists vary on what constitutes a species (especially the component members of superspecies) and a genus (Selander, 1971).

Because of the first difficulty it might be advisable to abandon in the future all comparisons involving endemic subspecies. But for the present time, when so few species of birds have been studied sufficiently thoroughly to enable one to quantify the levels of geographic variation in the entire range, there might be no other alternative but to include this category. As far as the second difficulty goes, since we cannot say

that under given conditions of geographic and ecological isolation speciation takes place at such or such a rate, we are forced to put all endemic
species in the same bag and to declare that it is probable that they
are older than the nonendemic species (Darlington, 1971). Finally, the
third difficulty might make one suppose that different zoogeographical
results could be obtained whether one, or more, taxonomists undertook
the revisions used as the taxonomic basis for deciding upon levels of
endemism. This would be especially true, of course, whether the taxonomist in question was a so-called "lumper" or a "splitter." In order to
test the possibility that differences in taxonomic treatment could result
in statistically significant differences in zoogeographical conclusions, I
compared three taxonomic arrangements of the birds living in *Nothofagus* (beech) forests of southern South America.

Table III gives the numbers (and percentages) of taxa that can be
grouped in one of three levels of endemism: endemic species (the species
occurs only in the *Nothofagus* fauna), nonendemic species with endemic
subspecies (the subspecies is restricted to the *Nothofagus* fauna), and
nonendemic species without endemic subspecies. Three classifications
are compared: those of Hellmayr (1932; who can be called a moderate
splitter), Philippi (1964; who can be called a moderate lumper), and
a third, not formally proposed in the literature, but which would correspond to that of a liberal lumper. It can be seen that the first two

TABLE III

Effects of Changes in Classification on the Interpretation of Levels
of Endemism in the *Nothofagus* Forest Fauna
of Southern South America

		Number (%) of	
Classification	Endemic species	Nonendemic species with endemic subspecies	Nonendemic species without endemic subspecies
1. Hellmayr (1932)[a]	26 (60.5)	9 (20.9)	8 (18.6)
2. Philippi (1964)[b]	26 (59.1)	7 (15.9)	11 (25.0)
3. Hypothetical (cf. F. Vuilleumier, unpublished)[c]	22 (51.2)	5 (11.6)	16 (37.2)

[a] Because Hellmayr (1932) did not recognize *Buteo ventralis* as a separate species,
there are 43 species in total. Hellmayr had 41 genera for these 43 species.

[b] Philippi (1964) recognized 44 species in 40 genera.

[c] This taxonomic treatment has not been advocated by any author, but corresponds
to a broad concept of the genus and the species. A total of 38 genera in 43 species are
recognized. There are 43 species because the *Pteroptochos tarnii-castaneus* superspecies
is treated as a single zoogeographical species.

classifications result in an 8–9% increase in the percentage of endemic species over the third classification. Similarly, the first two classifications allow for 4–9% more endemic subspecies. As a result of this comparison, one could conceivably conclude that the *Nothofagus* forest avifauna has a large percentage of endemic taxa, regardless of the classification employed as a basis of analysis, but that a broad genus and species concept (third classification) markedly reduces the absolute percentage of endemism. This may well be so, but it is interesting to note that none of the values obtained for heterogeneity chi-squares have a probability of a higher value larger than 0.10–0.50. Hence, the differences in taxonomic treatment appear not to be statistically significant.

If differences in taxonomy such as those affect zoogeographical interpretations only little, then it might not be unsound to compare levels of endemism in different faunas, regardless of their prior taxonomic treatments. If the differences in endemism observed between two or more faunas are significantly larger than those detected within one of the faunas, then surely the comparison is valid and the approach useful. I made a test of this supposition by comparing three isolated forest avifaunas, the Santa Marta montane forest avifauna, the Pantepui montane avifauna, and the *Nothofagus* avifauna. Table IV shows the numbers and percent of taxa grouped as in Table III: endemic species, endemic subspecies, and nonendemic species without endemic subspecies. The data for the *Nothofagus* fauna are not exactly alike in Tables III and IV. The reason is simply that in Table III the classification used was more extreme in the lumping, to oppose it more, for purposes of comparison, to those of Hellmayr and Phillippi, whereas in Table IV the classification falls more in line with published classifications in the other faunas.

Table IV shows clearly that the percentage of endemic species is higher in the Pantepui fauna than in the Santa Marta fauna, and highest in the *Nothofagus* fauna; furthermore, the percentages of endemic subspecies are much higher in the Santa Marta and Pantepui faunas than in the *Nothofagus* fauna; finally, the Santa Marta fauna has the largest proportion of nonendemic taxa and Pantepui the smallest. These differences are all statistically significant: all the values of the heterogeneity chi-squares have probabilities of a higher value smaller than 0.005. The differences in chi-square values between Table III, where within-fauna comparisons were made, and Table IV, where interfaunal endemism is compared, are striking. One may thus confidently conclude that the levels of stratification of the three faunas of Table IV are not the same and that zoogeographical reasons ought to be found to explain this observation. Although the absolute numbers of species assigned to each of the three categories and the statistics derived from them might not be

TABLE IV

LEVELS OF ENDEMISM IN THREE SOUTH AMERICAN AVIFAUNAS

Fauna	Endemic species	Number (%) of	
		Nonendemic species with endemic subspecies	Nonendemic species without endemic subspecies
1. Santa Marta, Colombia[a]	10 (9.9)	44 (43.6)	47 (46.5)
2. Pantepui, S. Venezuela[b]	29 (30.2)	55 (57.3)	12 (12.5)
3. *Nothofagus* covered region, Patagonia[c]	22 (51.2)	8 (18.6)	13 (30.2)

[a] From data in Todd and Carriker (1922), modified by F. Vuilleumier (unpublished). Total number of montane forest species considered is 101.

[b] From data in Mayr and Phelps (1967). Total number of montane species is 96.

[c] From unpublished studies of F. Vuilleumier. Total number of species is 43.

the same if workers applied slightly different taxonomic criteria to determine levels of endemism, it seems that the results would not be qualitatively different at least; thus, the methodology may be useful and provide the investigator with tools enabling him to construct hypotheses about the history of the faunas being studied (in the case of the faunas of Table IV, see preliminary assessments in Vuilleumier, 1967, 1972).

Considerable improvement might be brought to such studies of endemism if the taxonomic conclusions based on conventional taxonomic analysis could be supplemented with cytogenetic data. This approach is used in botany, for example, by Favarger (1969), who takes advantage of knowledge about levels of ploidy. Similar analyses have been carried out on mammals [see, for instance, Liapunova and Vorontsov (1970) on *Citellus,* or Jotterand (1972) on African *Mus*], lizards [see Gorman and Atkins (1969) on *Anolis*], and insects [see Carson (1970) on Hawaiian *Drosophila*]. In birds, unfortunately, information on caryotypes is so scanty that it is not possible to foresee when (or even if) cytogenetics will be an aid. Jovanović (1972) concluded that "today reliable karyological data exist for less than one percent of the living bird species."

C. PHYLOGENETIC SYSTEMATICS AND BIOGEOGRAPHY

The most original development in taxonomy that has influenced the thinking of zoogeographers is surely the theory of phylogenetic systematics of Hennig (1950, 1966). Hennig devised an extensive methodology to deal with distribution patterns of organisms, using these patterns as clues to the phylogeny of the groups studied. This method, or "biogeo-

graphic method (vicariance type doctrine)" is explained in detail by
Hennig (1966, p. 169ff). Using the distribution of taxonomic characters,
which can be either primitive (plesiomorph in Hennig's terminology)
or advanced (apomorph), among vicariant groups (that is, phyletically
related organisms that replace each other in different geographical regions), Hennig applies consistently the idea of dichotomic branching
sequences to the evolutionary history of organisms in order to reconstruct
their phylogenetic history. Thus, one must first search for the next sister
group of the group under study, and second, examine the distribution
of characters (whether apomorph or plesiomorph) in the two lineages.
This leads one to the identification of the geographic derivation of taxa.
Farris *et al.* (1970) have recently applied numerical techniques to the
method of phylogenetic systematics, thus refining the possible biogeographic analysis.

Hennig himself (1960) applied his theory and methodology to a study
of the Diptera of New Zealand, and other workers, following his lead,
studied the phylogeny and distributional history of several groups having
largely southern hemisphere distributions. Among some of the foremost
advocates of Hennig's method are Illies (1961a,b, 1965b), who worked
on Plecoptera, and Brundin (1965, 1966, 1967), who studied chironomid
midges (see also the work of Kiriakoff, 1964, 1967; Nelson, 1969, 1972).

Until recently, the taxonomists who pioneered or most extensively
used Hennig's theory were entomologists. However, other entomologists,
and notably Darlington, either made no mention of Hennig and his
school of thought (Darlington, 1965), or else vigorously attack his methods (Darlington, 1970; see the reply by Brundin, 1972). The dispute
over the validity of the methodology, which has raged in the pages
of *Systematic Zoology* since the late sixties, is of more than academic
interest, because what is at stake is not only a methodology of phylogenetic systematics [which has been reviewed and discussed by Bock
(1968), Nelson (1972), and Crowson (1970), among others], but also
the role played by Antarctica during earlier geologic epochs as a center
for evolution and migration of a number of animal and plant groups
having distribution patterns now centered around the southern tips of
South America, Africa, Australia, and on Tasmania, New Zealand, and
several subantarctic islands. The controversy is whether the southern
hemisphere distributions reflect the presence of a formerly important
southern hemisphere evolution center (as believed by Brundin, 1970),
or whether they do not (Darlington, 1965). In the first case, continental
drift was an integral part of the evolutionary history of the southern
hemisphere taxa, whereas in the second, drift is believed to have taken
place too early to have played any significant role, and dispersal is
supposed to have taken place largely over northern hemisphere routes.

Although some, perhaps many, taxonomists believe that biogeography and taxonomy are intimately tied together, very few of them have actually attempted to make the link within a common theoretical framework. Hennig (1966) and others after him (e.g., Kiriakoff, 1967; Brundin, 1966; Nelson, 1969; Cracraft, 1972b) did make this attempt, and have thus provided zoogeographers, who depend on taxonomy as we saw, with one methodology that appears more and more to be a useful tool. Most other taxonomists merely use zoogeographic data as a crutch to help them in taxonomic decisions about the possible relationships of taxa having discontinuous ranges. Similarly, many zoogeographers use taxonomic data as a crutch to zoogeographic decisions, and if the taxonomist and zoogeographer are the same person, the thinking becomes strangely circular.

Hennig's views have so far had very little impact on avian taxonomy (but see Meise, 1963). Thus, Selander (1971) does not discuss the possible significance of Hennig's approach, and Storer (1971a) does not cite Hennig's (1966) book in his chapter on avian classification. But things may be changing, for Cracraft (1972b) has written a thorough discussion of the applications of Hennig's methodology to avian systematics. More recently, J. Cracraft (1974), applying the method of phylogenetic systematics to the relationships of the Ratitae, has concluded, first, that these flightless birds are monophyletic, and second, that flightessness has evolved only once. On the basis of this taxonomic conclusion, Cracraft supposes that the zoogeography of the Ratitae is linked with the dispersal of their ancestors through southern land masses. Further studies of avian groups are in progress, for example on the Galliformes (J. Cracraft and L. L. Short, Jr., personal communication). Figure 4 illustrates

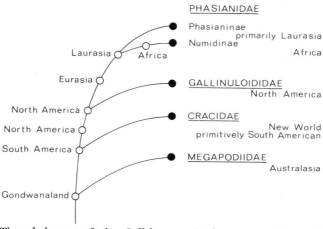

FIG. 4. The phylogeny of the Galliformes. Circles represent hypothetical (unknown) ancestors. The Phasianidae include the pheasants, guinea-fowl, grouse, and New World quail. [After J. Cracraft and L. L. Short, Jr., (unpublished) courtesy J. Cracraft.]

the phylogeny of the Galliformes and shows the graphic way of depicting the method of phylogenetic systematics.

IV.   Units of Analysis and Levels of Approach in Zoogeography

All types of zoogeographic analysis have been carried out on some unit. Table V lists six units in the left column and gives a possible hierarchy of three levels of approach to zoogeography. Generally speaking, these units and levels are those that have been used by the majority of zoogeographers working not only on birds, but also on other taxonomic groups. A cross section of the current zoogeographical literature would reveal the veracity of this assertion. The examples listed in Table V provide a sample of this literature, but only a very small one, of course, and the reader who would like to compile more extensive lists could do so by using the useful review of Niethammer and Kramer (1966). A division such as the one proposed in Table V is of course to some extent arbitrary, because overlaps between the different units and/or the levels exist, but for the purpose of this chapter it will permit the examination of some of the conventional and some of the newer aspects of zoogeography. Rather than to examine very superficially a rather large number of studies, I have chosen to select a few studies that appear to be representative of the three-level hierarchy of approach and of a broad geographical coverage, and to discuss them critically.

It will become apparent while proceeding through this section that there is a gap at present between the emphasis of historical zoogeography and that of theoretical zoogeography. These two aspects of biogeographic research are partially bridged by an approach combining historical methods still largely geared to a taxonomic framework and a more ecological one making use of trophic comparisons between taxa or faunas.

A.   HISTORICAL ZOOGEOGRAPHY

1. The Geographical Approach to Faunal Analysis

This kind of biogeographic work is the oldest and most traditionally employed among avian biogeographers. Earlier studies will not be reviewed here, because they are summarized in the reviews and texts of Darlington (1957), Serventy (1960), de Lattin (1967), Simpson (1967), Udvardy (1969), and Bănărescu (1970). Among the most recent studies, however, the approach is hardly different from that used by earlier authors. The major question posed is either, What is the origin of the fauna considered?, or, What are the elements composing the fauna and what are their respective origins? By "fauna" is to be

TABLE V
Units of Zoogeographical Analysis and Possible Levels
of Approach to Zoogeographic Problems[a]

Unit of analysis	Level of approach, types of questions asked, and examples discussed in detail in the text
1. *Fauna* of a zoogeographic region, a continental land mass, a large archipelago, etc. (e.g., Serventy, 1972, on the Australian avifauna; Bond, 1963, on the West Indian avifauna)	A. Historical zoogeography 1. Geographical approach (What are the geographical origins of the unit studied?) (e.g., avifaunas of the Americas and of the Provence)
2. *Fauna* of a subdivision of a zoogeographic region or of a political, or other, more or less arbitrary unit (e.g., Matvejev, 1961, on the fauna of Yugoslavia; Farkas, 1967, on the avifauna of Hungary; Mengel, 1965, on the avifauna of Kentucky)	2. Ecological approach (How did birds colonize a given vegetation type or habitat?) (e.g., avifaunas of the nonforest habitats of the high Andes, of the lowland forests of Africa)
3. *Fauna* of a vegetation zone or major habitat (e.g., Stegmann, 1958, on Eurasian steppe birds; Dorst, 1962, 1967, on high Andean nonforest birds)	B. Ecological zoogeography Trophic approach (Do units have ecological equivalence in different areas, and what are the ecological processes at work?) (e.g., ecological equivalents of Tyrannidae in Africa and Australia; trophic comparisons of zoogeographic regions; biogeographical energetics)
4. *Taxon* above the species level, e.g., genus, family (e.g., Brooke, 1971, on Apodidae; Larson, 1957, on arctic Charadrii; Voous and Payne, 1965, on Malagasy grebes; Ploeger, 1968, on Arctic Anatidae)	C. Theoretical zoogeography (What are the processes of biogeography and how do they interplay?) (e.g., mathematical modelling and theory testing by empirical studies)
5. *Taxon* at the species level, including superspecies and species group (e.g., Voous and van Marle, 1953, on *Sitta europaea;* Salomonsen, 1965, on *Fulmarus glacialis;* Paynter, 1972, on the *Atlapetes schistaceus* group)	
6. *Propagule* or *colonization unit* at the species level (e.g., Mayr, 1965b; Lack, 1969, 1971; MacArthur *et al.*, 1972; Ricklefs and Cox, 1972)	

[a] In the left column, a series of six units of analysis are listed, each lower unit representing a possible subset of the next higher one. In the right column, three levels of approach are considered and are discussed, with examples, in the text. It is obvious that in principle any unit of analysis could be chosen for any given level of approach, but traditionally, level A has used units 1–4, and especially 1–3, whereas the less traditional levels, B and C, use either units 5–6 (especially approach C), or else a combination of units (approach B).

understood an assemblage of species living within a geographical area, whether that area is defined by natural or artificial boundaries.

a. Fauna of the Americas. Mayr (1946) divided the North American fauna into elements from different geographical origins. Later (1964), he extended this study to the fauna of South America in an attempt to reconstruct the broad outlines of their history. His analysis was inspired by Simpson's (1950) work on mammals, showing that the South American continent had been colonized by successive waves of immigrants from North America during the Cenozoic Era.

The following major faunal elements were recognized by Mayr (1964): (A) primarily South American families, (B) Old World and primarily North American families, and (C) families with readily colonizing genera (see Table VI). After eliminating families "containing fewer than six living species, to guard against the danger of interpreting a relict distribution as indication of a center of origin" (p. 281), Mayr retained eight families of nonpasserines and fifteen of passerines: "Since these families were selected for their inability or unwillingness to cross water barriers, we are justified to infer that families in which the genera have a prevailingly South American distribution are of South American origin, and families in which the genera have a prevailingly North American distribution are North American in origin" (Mayr, 1964, p. 23). Table VI shows the classification of the twenty-three families into Mayr's faunal elements. Mayr's families and subfamilies differ somewhat from those of Storer (1971a), whose classification is generally used in "Avian Biology."

As can be seen from Table VI, some taxonomic groups fall clearly into one or the other of Mayr's elements. Thus, among the passerines of the suborder Tyranni, seven families (Dendrocolaptidae, Furnariidae, Formicariidae, Rhinocryptidae, Conopophagidae, Cotingidae, and Pipridae) have a distribution centered in South America, whereas only one (the Tyrannidae) is widely distributed in North America also. Mayr is probably right when he concludes from this pattern that "The South American origin of the Suboscines can hardly be questioned" (1964, p. 283).

Things appear less clear when one considers the so-called nine-primaried Oscines (comprising the Vireonidae, Thraupidae, Parulidae, Emberizinae, and Icteridae). Two of these groups (Vireonidae and Parulidae) have more genera in North than in South America (see Table VI), and so are classified as primarily North American. But elsewhere in his paper, Mayr also shows that the Parulidae have a large number of species in South America, and that the Vireonidae have actually more species in South than in North America (see Table 3 in Mayr, 1964).

TABLE VI

CLASSIFICATION OF TWENTY-THREE BIRD FAMILIES OF THE
AMERICAS INTO FAUNAL ELEMENTS[a]

Family	Essentially northern occurring		Essentially southern occurring	
	Not south of Panama	Also south of Panama	North beyond Nicaragua	Not north of Nicaragua
A. Primarily South American families				
Non-Passeres				
Tinamidae	0	0	2	7
Nyctibiidae	0	0	1	0
Bucconidae	0	0	1	9
Galbulidae	0	0	0	5
Ramphastidae	0	0	3	2
Passeres				
Dendrocolaptidae	0	0	7	6
Furnariidae	0	0	5	54
Formicariidae	0	0	10	43
Rhinocryptidae	0	0	0	10
Conopophagidae	0	0	0	2
Cotingidae	0	0	8	24
Pipridae	0	0	4	17
Total (12 families)	0	0	41	179
B. Old World (O.W.) and primarily North American families				
Trogonidae (O.W.)	1	1	1	0
Momotidae	3	1	2	0
Troglodytidae	6	3.5	2.5	3
Mimidae	5	0.5	0.5	1
Parulidae	15	4	2	0
Vireonidae	2	1	3	0
Total (6 families)	32	11	11	4
C. Families with readily colonizing genera				
Expanding South American				
Trochilidae	15	3.5	7.5	84
Tyrannidae	2	5	20	79
Secondarily South American				
Thraupidae	1	1.5	9.5	44
Pan-American				
Icteridae	5	5	6	17
Emberizidae	20	5	5	21
Total (5 families)	43	20	49	255

[a] The numbers represent the number of genera. A genus that cannot be classified easily into either northern or southern elements, is recorded as 0.5 under both. From Mayr (1964), courtesy of E. Mayr and The National Academy of Sciences U.S.A.

Further, the Thraupidae have far more genera (Table VI) and far more species (Table 3 in Mayr, 1964) in South than in North America. Yet the fact that they have one genus occurring in Middle America not south of Panama (see Table VI) qualifies the family as a member of the secondarily South American element, instead of the primarily South American element. The geographical distribution of the nine-primaried Oscines is evidently more complex than that of the Tyranni or Suboscines, and so it would appear difficult to draw inferences about their origins and history on the basis of these tabulations. Yet, Mayr (1964, p. 283) concludes that the nine-primaried Oscines as a whole have had a North American origin. This may well have been so, but the evidence he adduced in favor of this conclusion is not fully convincing. Much detail is blurred when considering faunal elements at the family level, and some imprecision is introduced when using the genus, which is a subjective category (Selander, 1971), and notoriously so in the systematics of South American birds. So it seems that it is still a good idea to heed Deevey's (1949) advice to consider the basic datum of zoogeography to be a map of the distribution of the species' range.

Such an approach to an analysis of the faunas of the Americas, employed unfortunately so far only for North America, has been pioneered by Simpson (1964), who worked on mammals, and was followed by Cook (1969) on birds. Instead of dividing the fauna into elements of different geographical origins, which are subjectively determined, Simpson (1964) and Cook (1969) worked on species and plotted, in the squares of a grid superposed upon a map of North America, the distribution of the numbers of species per square or quadrat. The isograms obtained in this fashion enable one to visualize on a map areas of high or low species diversity and permit one to generate hypotheses about the possible origins of the patterns in terms of vegetation and barriers of the present and past. A similar approach was adopted by Kaiser et al. (1972) for the birds and mammals of Canada. These authors carried the analysis a step further by calculating similarity coefficients that were then submitted to single link cluster analysis in order to obtain clusterings of like-similarity quadrats.

Clearly, the above type of work does not allow one to probe into events much older than the Pleistocene, or even the late Pleistocene, whereas the approach used by Mayr (1964) attempts to unravel the faunal history beyond the Quaternary. There is nevertheless need for integration of these two kinds of approaches, the historical one of Mayr, and the more ecological one of Simpson, Cook, and Kaiser et al. Whether this can be done in terms of faunal elements of different origins (sensu Mayr, 1946), or in terms of more precise taxonomic analysis on the historical events surrounding speciation (cf. Mengel, 1964, on Parulidae;

Short, 1965, on *Colaptes;* Mayr and Short, 1970, on superspecies among North American birds), remains to be seen. More use will have to be made of the existing fossil record, which is becoming better cataloged every day, in the link between the two approaches [cf. Mengel (1970) and Hubbard (1971) on the avifaunas of the Great Plains and the Appalachians, respectively].

b. Fauna of the Provence. Many avian biologists working on a relatively small territory attempt to answer the question of the origins of the fauna in their study area. The Provence of southern France, limited by relatively clear-cut natural features, is a privileged area for ecologists because it has a great variety of habitats within a small surface area. These habitats include types of landscape and vegetation as diverse as the pebbly steppes of the Crau, the deciduous forests along the banks of the Rhone River, the chaparral of the Alpilles, the lagoons of the Rhone delta, and the montane beech forests of the Ventoux (Blondel, 1970). Attracted by this diversity, Blondel attempted to analyze the origins of the Provence avifauna using the faunal types of Voous (1960) (see Table II). Eighteen of Voous' twenty-four types are represented in the Provence. Table VII lists the numbers of species for each of the faunal types and for different ecogeographical zones of the Provence.

From his analysis, Blondel (1970) suggested the following conclusions: (1) the birds of the montane and piedmont zones are members of cold or temperate faunal types from the northern part of Europe; (2) the birds of the plains (Basse Provence, between the Durance River and the Mediterranean) are chiefly members of faunal types from the warm and dry parts of Europe and southern Asia (some of them are members of the Ethiopian and Indo-African faunal types); (3) in the low alluvial plains, the mixture of faunal types is large, and elements from cold and temperate forests are found next to others from southern warm and dry types; (4) the Camargue (Rhone delta) is very recent in origin, and was higher up in Roman times than today, so that most marshy areas did not exist—its avifauna is of mixed origin, and there is every reason to believe that this brand new habitat has been colonized by all the faunal types that were present in the Provence at the time when it opened up for invasion; (5) the numerical dominance of Holarctic, Palearctic, European, and Old World faunal types in the Provence avifauna (94 species out of 174) shows the essentially European nature of the fauna.

One wonders whether more information might not have been gained by an examination of the Provence avifauna at the species level, in terms of ecological shifts, species diversity, superspecies distribution, and geographical barriers. Surely, the blending of criteria such as ecologi-

TABLE VII

CLASSIFICATION OF BIRD CATEGORIES FROM MOUNTAINS AND LOWLANDS OF THE PROVENCE AS A FUNCTION OF BIOGEOGRAPHIC ORIGINS OF SPECIES AS DETERMINED ACCORDING TO THE VOOUS LIST OF FAUNAL TYPES[a]

Faunal type[b]	Number of species							
	Montane	Piedmont	Mediter-ranean	Plains	Freshwater marshes	Brackish water habitats	Ubiquitous	Total
Siberian	1							1
Holarctic	1	4		1	3	1	4	14
Palaearctic	4	7	1	8	12	2	16	50
Old World				3	7		1	11
European	3	3		4	1		8	19
Palaeomontane	1			1				2
Palaeoxeromontane			3					3
Palaeoxeric			1	2				3
European Turkestanian		5	1	6	3		2	17
Turkestanian–Mediterranean			6	1	3	1	1	12
Sarmatic						4		4
Mediterranean			9	4			1	14
Ethiopian			1		1			2
Indian–African			1		2	1	1	5
Nearctic						1		1
Cosmopolitan			1		3	6	1	11
Total	10	19	24	30	35	16	35	169

[a] Sea birds and species of unknown origin are omitted. Two additional faunal types are represented in the Provence, the North Atlantic (with 1 species), and the "Unknown" (with 4 species). Thus the total number of species is 174. (After Blondel, 1970, courtesy of J. Blondel and the Société Ornithologique de France.)

[b] From Voous (1960).

cal preference, geographical distribution, supposed origins, and others that make up Voous' (1960) faunal types must blur the precision of an analysis done in a more ecological framework.

c. Discussion. The analyses reviewed above have shown the kind of approach used in historical geography, when the chief criteria are taxonomic affinities between taxa taken to represent the geographicohistorical relationships of elements within avifaunas. Two schemes, that of Mayr and that of Voous, have thus been inspected. There is no doubt whatsoever that faunas (when defined simply as the sum of species of birds breeding in a given geographical area) are stratified, and that they are composed of elements of different ages and origins, hence are heterogeneous in time and space. But the difficulty is that all too often the activity consisting in allocating a given species to this or that "faunal type" or "faunal element" is, first, pure guesswork, and second, almost pure typology. The successful analyses of historical zoogeography have been those on mammals (cf. Kurtén, 1968, 1972; Thenius, 1972; and the essays in the book edited by Keast *et al.*, 1972) because of the much larger fossil record available. Thus, neozoology and paleontology can be better integrated and there is far less guessing. I am sure that much more integration can be accomplished in avian zoogeography; the challenge must be picked up by an avian paleontologist who is also interested in contemporary faunas. Then the two approaches outlined above in the case of American birds may meet at the interface where the major biological factors are at work—dispersal, competition, extinction, and the like. In the meantime, we can investigate another way of looking into historical zoogeography, which is perhaps a bit more ecological than the one we have just seen.

2. The Ecological Approach to Faunal Analysis

Rather than consider the fauna of a region (zoogeographic or geographic), one can explore the history of an avifauna inhabiting a given ecological zone or area. Stegmann's (1938) study of Palearctic birds is an early example. I shall discuss here two analyses, one on the avifauna of the nonforest vegetation of the high Andes of South America, and the second on the avifauna of the lowland wet forests of Africa. These studies are less centered on the problem of the geographical origins of the birds within the fauna and do not dissect the latter into elements. Rather, the sort of question that is being asked is: How can one explain the observed distribution patterns of species in a given vegetation type in terms of known evolutionary phenomena? Nonforest and forest types of vegetation have been selected here, first because the problems inherent in the definition of what is or is not forest have already been dealt with earlier, and second, because in South America as in Africa, the respective avifaunas of forest and nonforest vegetation types are taxonomically quite unrelated.

   a. Fauna of the Nonforest Vegetation of the High Andes.   The history of the faunas living along the high Andes, now inhabiting more or less open types of vegetation, paramos in the north (Venezuela, Colombia, Ecuador), puna further south (Peru, Bolivia, Argentina, Chile), can be traced back to the late Pliocene–early Pleistocene (see references in Haffer, 1970b; B. S. Vuilleumier, 1971) when the Andean cordilleras began their last phases of uplift. Before that time, the high Andes were probably much less elevated above sea level than they are now, and large tracts of the summits were probably covered with forest vegetation in the north and perhaps forest or more open types of vegetation (wooded savanna?) further south. With the last uplift stages, the combined effects of altitude and climate determined physical conditions that were no longer suitable for a forest or forestlike flora, and the vegetation changed to open types, varying from wet grasslands to dry grasslands, open scrub, and desert, depending on the exposure of the slopes and plateaus to the dominant moisture-carrying winds and the presence of a dissected topography that would favor the formation of rain shadows.

   Thus, the open vegetation types that characterize the high Andean landscapes at present [see Troll (1959, 1968) for descriptions of the vegetation, and maps of their distribution; Koepcke (1954) and Dorst (1955, 1967) for descriptions of avian habitats] became available for colonization geologically recently, as Simpson (1967) has pointed out. Clearly, these high Andean surfaces did not all become available at the same time, and much of the change was progressive. Nevertheless these "new" habitats were invaded by a series of bird groups that were in many ways preadapted for life in open regions. That is, the birds that are inhabitants of forest or forestlike vegetation, either in the lowlands or in the highlands of South America, or, rather, their ancestors, were probably not the candidates for colonization of these nonforest habitats. This is strongly suggested by the great faunistic dissimilarity at the species and genus levels between the Andean forest and nonforest avifaunas, and between the Andean nonforest and lowland forest avifaunas. Rather, the high Andean bird fauna as a whole appears faunistically related, as judged by the percentage of taxa in common, to the bird fauna inhabiting the steppes and other nonforest vegetation types of Patagonia. Chapman (1917, 1926) was probably the first to perceive this clearly and to suggest that much of the high Andean nonforest bird fauna originated from elements that were southern in origin and taxonomically related to the present-day steppe fauna of Patagonia.

Since the early invasions of bird groups into the high Andean nonforest region, there has been diversification of parts of this fauna. A survey I conducted (Vuilleumier, 1969b) on a total of eighty-three species out of the approximately one hundred fifty-three in the avifauna (54%), permitted me to detect all sorts of intermediacies between groups that show no or little trace of evolutionary radiation in the high Andes (for example, the caracaras, where invasion has been followed by only moderate isolation and genetic differentiation; see Vuilleumier, 1970b) and others that show a complex adaptive radiation involving secondary sympatry and habitat overlap, including in some cases various mechanisms of habitat partitioning [for example, in the genus *Muscisaxicola* among the Tyrannidae; see Cody (1970) and F. Vuilleumier (1971) for a discussion of their habitat separation and of their modes of speciation, respectively]. The fundamental processes of biogeography—colonization, dispersion in the newly established habitats; differentiation, subsequent speciation with secondary overlaps followed by habitat co-occupancy; and extinction—are very clearly illustrated by high Andean birds (Cody, 1970; Dorst, 1967; Haffer, 1970b; F. Vuilleumier, 1969b, 1970a, 1972, and unpublished).

b. Fauna of the Lowland Forests of Africa.   The tropical rain forests of the world, and Africa is no exception, have widely been held in the past to be the oldest and most stable vegetation type (see, e.g., Chapin, 1932, p. 209). Moreau's analysis of the lowland forest avifaunas of Africa (1966, Chapter 9) indicates that the stability of African rain forests is a myth and that there have been wide fluctuations in the surface area and absolute distribution of this vegetation type in the Pleistocene. Figure 5 shows the extent of present-day lowland forests in Africa. Three major blocks can be recognized: the Upper Guinea Forest, the Nigerian Forest (separated from each other by the dry Dahomey gap), and the Congo Forest (the second and third blocks being separated by the Cameroon Highland area). There are, furthermore, isolated patches of lowland forest in the area east of Lake Albert, and along the east coast of the continent.

The Upper Guinea Forest has 182 species of birds. The northeasternmost part of the Congo Forest, at the other extremity of the distribution of the three major blocks, has 212 species. The total for the whole forest region is 266 species. Two points emerge from these figures: first, that the isolated Upper Guinea Forest is not especially depauperate, in spite of its geographical separation from the other forest blocks; second, that the region as a whole is relatively poor in species when it is compared with forest regions of the lowlands of the American tropics [Moreau cites Slud (1960) who recorded 269 species from a small area in Costa

456

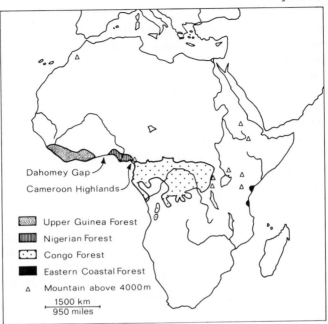

FIG. 5. Map of Africa showing the distribution of lowland forests. (After Moreau, 1966. Courtesy J. F. Monk and Academic Press, London.)

Rica; one can now add Koepcke's record (1972) of about 300 species "in an area of one square kilometer of natural rain forest" in eastern Peru, two-thirds of which appear to breed there].

Moreau contends that this relative poverty is due to comparatively recent disruptions, say during the last 100,000 years. A dry period during mid-Pleistocene times would have fragmented the forest region into more or less isolated segments. Furthermore, during the glaciations, the forest area would have been reduced to about one-half the size that it has at present, as a result of the encroachment of montane conditions at lower altitudes than at present, perhaps as low as 500 meters above sea level, and this probably occurred during each of the glacial episodes. Finally, an episode of drought about 11,000 years ago would have further reduced this area of the forest, especially along its southern margin.

Moreau's postulate that the "violent vicissitudes in the range of the biome might be expected to result in much extinction" seems contradicted by his statement that "the Upper Guinean forest bird fauna should be so comparable in number of species with that of the Congo block" (1966, p. 162). This contradiction appears even more evident when it is recalled that 157 of the 182 species (86%) found in the Upper Guinea forest block are shared with the Congo block. My own conclusion from these numbers would be either that the vicissitudes have not been as violent as that, and so extinction did not take place on a grand scale, or, if the vicissitudes did lead to large-scale extinction, then this phenomenon was balanced out afterwards by as much colonization (thus accounting for the relative richness in species of the Upper Guinea forest block when compared with the Congo block, and for the large percentage of the fauna shared by these two blocks).

Nevertheless, there are distributional gaps between the Upper Guinea and northeastern Congo forests, varying from about 960 to 2400 km, and involving 24 species. Since, according to Moreau, not all these gaps are due to lack of collecting records in the intervening area, there is a problem to be explained. This is even more acute since there are six endemics in the northeastern Congo forest area, which is not at all isolated at present from the remainder of the Congo block. Perhaps a northeastern refuge can be postulated to account for these distributions, in the Semliki Valley area, although only one of the six endemics occurs there today. Diamond (1972, 1973), working on montane forest birds in New Guinea, showed that distribution gaps most unlikely to be due to insufficient collecting efforts are actually of frequent occurrence. He postulated that local extinction, largely due to interspecies competition, could be the mechanism responsible for these gaps. Similarly, in high Andean birds living above the timber line and in birds living in montane forest vegetation along the Andean cordilleras, careful mapping reveals distribution gaps, which one cannot always easily dismiss as an artifact of collecting (Vuilleumier, 1968, 1969a). In these cases, as in those cited by Moreau in Africa, competitive situations are to be sought as possible factors. These considerations amply bear out Moreau's (1966, p. 369) remark that "the extent to which geographical and ecological problems are intermingled" is large.

c. Discussion. The work done on high Andean birds and African forest birds has been a blend of evolutionary and ecological studies throughout, based on a thorough background of taxonomy. In the early days of exploration (Chapman, 1917, and other works; Chapin, 1932, and other works) the emphasis was on the faunal survey and on the taxonomy, and the theory behind the work was somewhat old-fashioned by present-day standards. Yet the attack was fully along lines that integrate ecological data with historical ones. The more recent work has done away with the typological leftovers of Chapman's thinking (the life-zone approach) and of Chapin's faunal subdivisions, and deals with these faunas from the points of view of speciation, competition, and resource partitioning, without ceasing to be historical in its explanatory attempts. What is more, this background work is now sufficient to deal with at least part of these faunas from a more theoretical point of view, applying the predictions of MacArthur and Wilson (1963, 1967) to the more isolated habitat islands. Thus, the páramo avifauna has been analyzed in this regard (Vuilleumier, 1970a), and so has its African counterpart, the afroalpine avifauna of East African mountains (F. Vuilleumier and B. B. Simpson, unpublished). Actually, what these more recent studies have done is to show us that historical zoogeography is being amalgamated into theoretical zoogeography. But they are not the only way of effecting the transition, and the studies reviewed below will demonstrate another approach that has been developed during the last few years.

B.   ECOLOGICAL ZOOGEOGRAPHY—TROPHIC APPROACH

Inasmuch as ecological systems are composed of a series of trophic levels representing exchanges of energy or passage of energy from one level to another, one can wonder whether faunas, taxa, or other units of zoogeographical analysis found in different parts of the world are comparable in this regard. Especially, one may want to ask the questions whether evolutionary (historical) and ecological (environmental) processes have had parallel effects in different areas broadly resembling each other in geographical and physical characteristics, and conversely, to what degree one may invoke these same processes to be responsible for dissimilarities when these are observed. The first thing to do is to look for patterns, and the second is to try to explain the patterns (a) in ecological terms involving the trophic levels, as well as (b) in historical terms involving evolutionary processes (particularly speciation and adaptive radiation).

Lein (1972), for instance, started from the orthodox region concept and attempted to establish ecological equivalences between the passerines of the different regions. To this aim, he divided the birds into seven "broad trophic types," and computed indices of trophic comparisons in order to estimate quantitatively the amount of radiation and convergence when the faunal regions are compared among them. Although Lein contends that his approach is novel because it departs from the traditionally taxonomic way of looking at faunal regions, his methods remain, nevertheless, framed taxonomically. Indeed, his fundamental unit of analysis appears to be the family [or subfamily in the case of species-rich families, as the Muscicapidae, Thraupidae, Furnariidae, and Fringillidae (of Lein's usage)]. From this undoubtedly arises one of Lein's methodological difficulties, because the indices of trophic comparison (to establish "trophic roles") are "subjectively assigned" to each family "on the basis of qualitative observations on foods and feeding habits" taken from several general works.

Lein's two major conclusions are, first, that the avifaunas in the "tropical" regions are trophically more similar to each other than they are to the "temperate" regions, and second, that important trophic convergences exist between taxonomically dissimilar components of different avifaunal regions. The basis for the first conclusion is a difference in the percentage of trophic similarity, as indicated by the trophic indices. Thus the Neotropical region, for example, is 7–10% more similar trophically to the Australian, Oriental, and Ethiopian regions than to the Nearctic and Palearctic ones. This conclusion is perhaps quantitative, as Lein claims, but since there is no suggestion as to how these differences can be evaluated statistically, the value of the difference quoted is of unknown biological significance. The second conclusion is clearly qualitative. Perhaps the unit of analysis chosen by Lein does not enable him to get deeper and truly quantitatively into what is nevertheless a very interesting problem. Lein's methodology has been applied to the study of bats (Wilson, 1973).

Another method of approaching trophic comparisons of avifaunas is heralded by the work of Orians (1969), Karr (1968, 1971; personal communication, 1973), and Henry (1972). Clearly only the beginnings are at hand, but I think that the approach may lead to significant new ways of undertaking zoogeographical analysis. These authors are investigating the structure of avifaunas from temperate and tropical zone localities by examining the distribution of foraging types (Orians), or by studying the distribution of energy requirements among different trophic levels and foraging types (Karr), or else by analyzing the relationships between trophic isolation within components of the fauna and body size of the species involved (Henry). So far, the studies of Orians are restricted to the New World tropics (Costa Rica), and those of Henry to the Old World (Europe), while Karr has made comparisons of avifaunas from temperate and tropical areas, and from the Old and New Worlds (personal communication).

Karr indicates that the ratio of yearly existence energy requirements of tropical/temperate avifaunas (Panama/Illinois) is 1.48, and is "surprisingly similar to the tropical to temperate net primary productivity ratio of 1.40 suggested by data given by Golley" (J. R. Karr, personal communication, 1973). Karr thinks that the tropical surplus is due to species that are able to "switch resource utilization patterns with seasonal variation in resource abundance" and suggests that this situation may be comparable to that of the migrants leaving temperate forests. Studies on the relationships between migrants and resident species have been done in the Old World by Blondel (1969a,b) in Mediterranean types of vegetation of France, by Morel and Bourlière (1962), Morel (1968), and Morel and Morel (1972) in savanna vegetation (*sahel*) of Senegal, and by Brosset (1968) in the Gabon rainforest [see also the recent summary of Moreau (1972) on Palearctic migrants in Africa]; and in the New World by Leck (1972) and Willis (1966) in Panama, and by Miller (1963) in Colombia. Although to date this work is uneven in its thoroughness, geographically and temporally, it should be integrated with the approach of Orians and of Karr. Surely the coefficient proposed by Henry will be helpful in these comparisons. While it would be premature to attempt this review here, it appears that the trophic-energetic approach done by comparing species and disregarding the regional scheme will be more fruitful than that done by comparing families within zoogeographic regions. Yet some kinds of analyses involving families might be rewarding.

This is suggested by Keast's (1972a) study of the Tyrannidae of the New World and their ecological counterparts in Africa and Australia. Keast selected this family because it is one of the two dominant families of passerines in the New World tropics. By dominant here must be understood numerically dominant, (1) in terms of numbers of species, and (2) variety of habitats occupied. As Smith and I (1971) pointed out, the Tyrannidae comprise one out of ten species of South American birds, have successfully invaded every major vegetation formation of the continent, have had extensive adaptive radiations in most of them, and have ecological counterparts belonging to distinct avian families in other parts of the world.

In a sense, Keast's (1972a) goal was to quantify these statements, in the following ways: (1) by defining the adaptive zones (as the concept is used by Simpson, 1944, cited by Keast) of the tropical American flycatchers; (2) by documenting how flycatchers utilize space in regional faunas; (3) by finding out how the adaptive zones and space utilized by Tyrannidae in tropical America were filled, and by what other birds, in other continents, especially in tropical Africa and in Australia; (4) by defining what the nontyrannid competitors were, or had been, during the history of the group; and (5) by trying "in a final synthesis . . . to assess the significance of the associated avifauna in influencing, or limiting, the evolution and radiation of individual groups."

Keast (1972a) believes that the success of the Tyrannidae in tropical America, where they "occupy the adaptive zones, and ecological niches of some 8 to 10 families, and subfamilies" on the continents of Africa and Australia, can be understood by the respective histories of these land masses. During its long period of geographical isolation in the Tertiary, South America was the center of radiation of a number of taxa, one of them the Tyrannidae. Africa, on the other hand, has had repeated contacts with Eurasia during the same period, resulting in large-scale faunal exchanges between the two land masses. The competitive interactions following each new set of taxonomically varied colonists into the African scene would have provided a rather different background for adaptive radiation than the competitive interactions resulting from adaptive radiation within the Tyrannidae. The situation is not as clear for Australia, however. Although this continent has been isolated for much of its history, as has South America, its faunal diversity in avian types broadly comparable to Tyrannidae (in terms of adaptive zones) is more like that of Africa. This is illustrated in Fig. 6, showing a comparison of the three continental avifaunas, including Tyrannidae from tropical America, and several taxonomic groups from Africa and from Australia.

Although Keast's analysis is still very preliminary (partly because we know as yet so little about the breeding biology and feeding ecology of a large percentage of tyrannid species that trophic comparisons are made on a morphological basis only and do not include actual data on either food sizes or efficiency in food search), his results point out

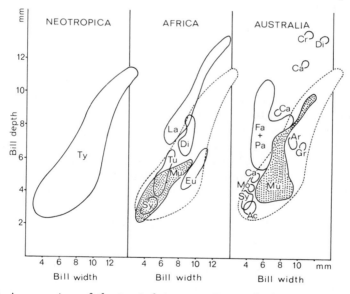

Fig. 6. A comparison of the tropical American Tyrannidae with African and Australian counterparts in other taxa in terms of bill depth and bill width. *Explanation of symbols:* in Neotropica, Ty = Tyrannidae; in Africa, the Tyrannidae (absent from this continent) are represented by the dotted line, which broadly overlaps with the other taxa as follows: La = Laniidae, Di = Dicruridae, Tu = Turdinae, Mu = Muscicapinae, Eu = Eurylaimidae, Sy = Sylviinae; in Australia the Tyrannidae (dotted line, absent from continent) overlap with the following taxa: Cr = Cracticidae, Di = Dicruridae, Ca = Campephagidae, Fa = Falcunculinae, Pa = Pachycephalinae, Ar = Artamidae, Gr = Grallinidae, Mo = Motacillidae, Ac = Acanthizinae, Mu = Muscicapinae, Sy = Sylviinae. (After Keast, 1972a. Courtesy A. Keast and Appleton-Century-Crofts/New Century.)

the intricate relations of the factors impinging upon the zoogeography of groups and faunas. Earlier zoogeographers were what I called geographical zoologists. There is little doubt that it is not possible, nor permissible, to ignore any longer the ecological aspects of biogeography (MacArthur, 1972). The phenomena of competition between species, niche subdivision, resource partitioning, adaptation to habitats, speciation and adaptive radiation, which are all germane to the ecology of communities [see MacArthur (1971) and the intercontinental comparisons of Cody (1968, 1974) for examples of analyses of community structure] are also germane to zoogeography.

We are thus directly led from empirical studies to theoretical ones. Having demonstrated that a series of factors recur again and again in all the kinds of zoogeographical analyses reviewed above, we must now pause and look into the processes themselves, from the point of view of theory, that is, by use of models that might illuminate how these processes work.

C. Theoretical Zoogeography

I indicated in the introduction to this chapter that the kind of thinking introduced by MacArthur and Wilson (1963, 1967) into biogeography represented an important conceptual advance. Until then, biogeographers had rarely attempted to build models in order to better grasp the essentials of their science. This seems to be true, not only of *bio*geographers, but also of other kinds of geographers, as Chorley (1964) pointed out.

It is chiefly at the level of the species, which Mayr (1963) considers the most important evolutionary unit, that the studies of MacArthur and Wilson (1963, 1967) and MacArthur (1972) have been done, although the approach to zoogeography that these authors developed is of course far more than just the use of species as units of analysis. Their approach is to build models expressed mathematically to describe the dynamics of processes in zoogeography, in a similar way as models have been made to describe the dynamics of population numbers in population ecology, or the dynamics of changes in allele frequencies in population genetics. Before reviewing their contribution, I should like to examine the state of mind of some theory-conscious biogeographers who were not trying to express the relationships of zoogeographical processes mathematically, yet attempted this verbally.

In a remarkable monograph on the evolution and ecology of animal populations of small Mediterranean islands, La Greca and Sacchi (1957, see especially Chapter 2, pp. 61–82) stress that the impoverishment in numbers of animal (and plant) species on small and medium-sized islands, as well as on some large islands, is due to the following basic mechanisms: (1) "difficulty of repopulation by invasion. . . , the more so when the island is far from the continent, small, with little water and vegetation" (p. 63); (2) "extinction of insular forms through predation" (p. 63) because of the principles of predator–prey relationships; (3) "insular biotic instability" (p. 65), or turnover, because of the importance of extinction which, unless it is counterbalanced by invasion, is total on islands, since there is usually no possibility of faunal exchange with adjacent biotopes in which the species would survive, as would be the case on the mainland (pp. 65–66). For La Greca and Sacchi (1957), the species–area curve thus permits a reasonable first approximation from which to derive the abundance relationships of animal species on Mediterranean islands. (4) As a consequence of the small area of islands and of the low diversity of insular habitats, "each species has a reduction in the density of population" (p. 74), reduction which depends on whether the species is mobile, of large size, predatory, etc.—this change in the structure of species populations has an important corollary, an increase in endogamy and a subsequent modification of the genetical structure of the island population of a given species.

It is evident that the verbal model of La Greca and Sacchi (1957) is a precursor of that of MacArthur and Wilson (1963). Similarly, Munroe (1963) notes that the differences and similarities in biotas were the result of a balance of three basic processes—evolution of populations (viewed both in the phyletic sense and in the sense of splitting of lineages through speciation), extinction (general or local), and immigration. "Immigration, like evolution, is a constructive factor, tending to increase the variety of the biota. These two factors are balanced against extinction, which tends to simplify the biota. On the balance that is struck depends not only the nature but also the number of species in the biota. Where evolution or immigration is favored, the biota will be rich; where more rigorous conditions give the advantage to extinction, the biota will be poor" (Munroe, 1963, pp. 304–305).

Munroe (1963) was perhaps inspired by Preston's (1962) study of the distribution of species among insular biota. Preston showed that the lognormal distribution gave a good fit to the observed distribution of numbers of species on islands, when the relative abundances of each species were taken into account. From this curve-fitting exercise, Preston derived a general and simple model,

$$S = CA^z$$

where S is the number of species, C is a constant, and A is the area of the island considered. The value of z predicted by Preston was about 0.27. Preston (1962) also suggested that the diversity of biotas could be the result of a balance between factors such as immigration and extinction.

A similar hypothesis was the basis of the model of MacArthur and Wilson (1963, 1967) for insular biogeography, illustrated in Fig. 7. "Both the immigration and extinction rates vary with the number of species present. The immigration rate (in new species per unit time) is a falling curve because as more species become established, fewer immigrants will belong to new species. We expect the curve to be concave, i.e., steeper at the left, because on the average the more rapidly dispersing species would become established first, causing a rapid initial drop in the initial immigration rate, while the later arrival of slow colonizers would drop the overall rate to an ever diminishing degree" (MacArthur and Wilson, 1967, p. 21). In their 1967 book, MacArthur and Wilson discuss in detail the species area curve, illustrating it with mathematical arguments derived from probability theory as well as with empirical data obtained from the case of Krakatau, which was partly destroyed (and whose flora and fauna were extirpated) by a volcanic explosion in 1883 and subsequently recolonized by plants and animals. This analysis led them to predict rapid extinction rates, defined as

$$X = 1.16\hat{S}/t_{0.90}$$

where $\hat{S}$ is the mean number of species at equilibrium, and $t_{0.90}$ is the time needed to reach 90% of the number of species at equilibrium.

Data recently published by Diamond (1969, 1971) seem to provide empirical verification of the prediction of high turnover rates in insular faunas by MacArthur and Wilson (1967). Diamond measured turnover

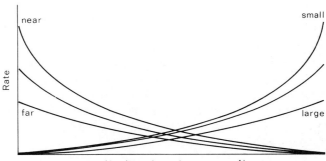

FIG. 7. MacArthur and Wilson's graphical model of equilibrium insular biogeography. The falling curves represent the rates of immigration (new species per unit time) to a set of islands, and the rising curves represent the rates of extinction on those islands. (From MacArthur and Wilson, 1963. Courtesy E. O. Wilson and *Evolution.*)

rates on islands from temperate and tropical regions, first (1969) by using censuses undertaken in 1917 on the Channel Islands of California by previous authors and in 1968 by himself, and second (1971), by comparing faunal lists from Karkar, an island 16 km offshore from New Guinea, gathered in 1914 by collectors for the Rothschild Museum and in 1969 by himself.

Diamond's studies show that turnover rates in the Channel Islands avifauna varied from 17% to 62% per 51 years, and that the rate on Karkar is about 17%. In order to compare more precisely a temperate island with a tropical one, Diamond chose to compare Santa Cruz with Karkar. The data for these two islands are summarized in Table VIII. Extinction was calculated as

$$e = 2E/(C + D)$$

where C and D are the numbers of species recorded in the earlier and later censuses, respectively, and E is the number of species that became extinct in the time interval t. As can be seen from Table VIII, the estimated extinction rates are 0.20% per year on Karkar and 0.32% per year on Santa Cruz. Immigration was calculated as the average fraction of species actually recorded on a given island over the number expected from the species pool of the nearest source area. For Karkar, Diamond took the lowlands of New Guinea as the source area, with a potential species pool of 228 species; for Santa Cruz, the lowlands of southern California were taken as the source, with a pool of 93 species. The estimated immigration rates are for Karkar, 0.10% of the potential species pool of the mainland per year, and for Santa Cruz, 0.20% of the potential pool per year.

TABLE VIII

EXTINCTION RATES IN THE AVIFAUNAS OF TWO ISLANDS OF COMPARABLE SIZE—TEMPERATE SANTA CRUZ IN THE CHANNEL ISLANDS OFF SOUTHERN CALIFORNIA AND TROPICAL KARKAR OFF NEW GUINEA[a]

Island (area in km²)	E	C	D	t (years)	e (%/year)
Karkar (227)	5	43	49	55	0.20
Santa Cruz (154)	6	36	37	51	0.32

[a] E = number of species that became extinct in the time interval between the two censuses; C = number of species recorded in the early census; D = number of species recorded in the later census; t = number of years between the two censuses; e = estimated extinction rate (for a discussion of how this rate is calculated, see text). (Data from Diamond, 1971.)

From the estimates for Karkar and Santa Cruz, Diamond concludes that they have similar extinction rates (the lower rate for Karkar being considered for various reasons as an underestimate), but different immigration rates (the lower estimate for Karkar being thought to be an overestimate "due to the lack of sight records for the earlier survey" and because Karkar is nearer the source area). Diamond (1971) seeks to explain the lower immigration rates for the tropical island as a function of its greater proportion of sedentary species compared with the temperate island, and as a function of the number of colonists that are birds of second-growth vegetation.

More evidence is clearly needed to document these differences and similarities between tropical and temperate islands, because the rapidity of turnover rates claimed by Diamond (1969) for the avifauna of the Channel Islands has been challenged by Lynch and Johnson (1974). These authors have reanalyzed Diamond's (1969) data and have found that the "small islands with few species have almost zero turnover and the large islands with many species have low turnover" (N. K. Johnson, personal communication). Such a divergent conclusion from Diamond's is due, in part, to the exclusion of extinctions and colonizations due to man. Thus, we are not in a good position to assess turnover rates in island avifaunas until a critical analysis of turnover parameters using several methods of measuring these parameters has been carried out. However, the rapidity of turnover rates has been demonstrated in invertebrate faunas. Wilson and Simberloff (1969) and Simberloff and Wilson (1969) obliterated the arthropod faunas in small mangrove islands, then monitored the colonization and disappearance of species, showing experimentally that some sort of equilibrium is reached very rapidly indeed, with turnover rates of the order of months. Recent data on turnover parameters (migration and extinction rates) in ant faunas from the Puerto Rican bank suggest high turnovers, as in the mangrove islands (Levins et al., 1973).

Another part of the theory of MacArthur and Wilson (1967) is concerned with the strategy of colonization. They attack the problem of colonization from the viewpoint of success; that is, they are interested in how long populations survive after having colonized a new island. (In time, obviously, all populations became extinct, and so, they argue, the approach that they favor is of greater interest.) They suggest a model based on the probabilities of birth and death, respectively λ and μ, and taking into account the growth of the colonizing population to reach the carrying capacity K. They predict that "the chance of a single pair reaching a population near size K and taking time T_k to go extinct is about $(\lambda - \mu)/\lambda$, while that of going extinct very rapidly is about μ/λ." So, when $(\lambda - \mu)/\lambda$ (which equals r/λ) is large, the success of colonization will be rapid. The properties of r, λ, and K, are important. It is advantageous for colonists to have a large r, or intrinsic rate of increase, but a small birth rate λ and a large K, for a small K would reduce the probability of longtime success or persistence on the island. MacArthur and Wilson (1967, Table 8), suggest that a large K is observed in birds thought to be good colonists, because of their ecological versatility and their seed-eating habits, and a large r/λ follows from flocking behavior (from data in Mayr, 1965b). Other evidence is adduced from ants and flowering plants, but so far there is still very little direct information bearing on the properties of r, λ, and K in insular colonists. One of the few papers in which the demographic properties of the model of MacArthur and Wilson are studied is Crowell's (1973) on mammals introduced onto small islands in Maine.

MacArthur and Wilson (1967) considered three further aspects of insular biogeography. The first is the number of species that an island can sustain. They studied this problem in terms of niche and competition theory. Following colonization, successful propagules should undergo a shift in niche, most often in the direction one would predict from the presence or absence of competing species on the island as compared with the larger pool on the mainland. Diamond (1970a,b, 1973) has documented these predictions with data on birds from New Guinea and islands surrounding it, and Keast (1971a) on birds from Tasmania.

The second aspect is the effect of various source areas in permitting faunal exchange between them. This problem is rendered difficult by the lack of quantitative data on dispersal, in spite of the wealth of qualitative information on the subject [Wallace (1876) devoted the second chapter of his book to dispersal, and in one of the most recent textbooks of zoogeography, that of Udvardy (1969), one chapter is devoted to dispersal]. Heatwole and Levins (1972) have begun to approach quantitatively the problem of dispersal of terrestrial animals of several taxa in the West Indies.

The third aspect of insular biogeography is the evolutionary consequences of colonization. MacArthur and Wilson (1967) considered the effects of natural selection in terms of r and/or K: "We have now replaced the classical population genetics of expanding populations, where fitness was r, as measured in an uncrowded environment, by an analogous population genetics of crowded populations where fitness is K" (p. 149). Since their book, the concept of r and K selection has been discussed by several authors, especially Pianka (1970, 1972), Roughgarden (1971), Gadgil and Solbrig (1972), and Schoener (1973). Evidence seems to build up to show that different genotypes in a population vary in regard to their r and K parameters, and that different environments will favor genotypes having a relatively greater r or K through selective mechanisms operating at the level of intra- or interspecific competition. Assuming then that these modes of selection (which should be viewed as complementary, rather than mutually exclusive) occur in populations, one can postulate a sequence of events following colonization. MacArthur and Wilson (1967) distinguished the following stages:

1. *Founder effect immediately after colonization by a small-sized propagule.* They assumed that some sort of sampling error might occur at this stage, but they were not ready to assess its importance. Nor are we truly in a better situation to do so now, several years later. Indeed, it may become very difficult to distinguish such random effects from selection effects, when the principal mode of selection is r selection, and K selection is not expected to have a large effect, as one would expect in a small colonizing population (see Clarke, 1972).

2. *Adaptation to the new environment.* The first trend appears to be a loss of dispersal power, and a resulting increase in differentiation of local populations. The second trend is character displacement, that is, "the shift in behavior or ecology which occurs as the species insinuates itself into its new set of competitors" (MacArthur and Wilson, 1967, p. 159), resulting in "greater phenotypic difference between pairs of related species where they occur together than where they occur apart" (p. 159). One of the consequences (or correlates?) of displacement follows from the observations that invading species tend to occur mostly in marginal habitats, then after colonization, and during the different stages of their "taxon cycle" [see Wilson's original analysis on ants (1961) and its application to birds of the Solomon Islands by Greenslade (1968) and to birds of the West Indies by Ricklefs (1970) and Ricklefs and Cox (1972)], they tend to occupy the more specialized habitats (such as montane rain forest) of the island, where divergence will take place; the latter process would occur mostly as a result of competitor pressure from newcomers. But character displacement need not be the only evolutionary development of island populations. Indeed, since islands generally have fewer species than mainlands, the absence of a greater number of competing species might result in "character release"; that is, the range of phenotypes in the island population is larger than that of its mainland counterpart, and the former may thus occupy a wider "niche" than the latter [an illustration was provided by Keast (1968) for some meliphagids].[2] This phenomenon

[2] A theoretical analysis by Roughgarden (1972) was received after this section had been written.

is actually not very different from the previously cited process of niche shift. And again, one might imagine cases where convergence, rather than displacement, would occur. MacArthur and Wilson (1967) argue that this could occur when the resources, such as food, available to two competing species are not only fine-grained but also rather similar. Divergence and specialization, the converse phenomenon, would take place when these fine-grained resources are more different. Some of these processes, taking place as they do within communities, have been discussed by MacArthur (1971), who, however, did not do full justice to the abundant literature on the subject, only some of which is on avian material. But since a recent critical and thorough review of the problem of character displacement, including references to most of the pertinent literature, has been published by Grant (1972) I cannot do better than to refer the interested reader to it.

3. *Speciation and adaptive radiation.* These processes produce the accumulation of species within the archipelago, a phenomenon distinct from continued immigration of propagules from some source area. MacArthur and Wilson (1967, p. 175) contend that "adaptive radiation will increase with distance from the major source region and, after corrections for area and climate, reach a maximum on archipelagos and large islands located in a circular zone close to the outermost dispersal range of the taxon." Perhaps this view is too much that of an "ideal" archipelago, for in most cases dispersal from a source area is a very complex affair, there being usually more than just one source area. One instance of a "continental archipelago" with a simplified dispersal pattern rather close to the ideal single source area is the páramo "islands" of the northern Andes, which I studied previously (Vuilleumier, 1970a). Although adaptive radiation has not yet proceeded very far in the birds of the páramo, the remark of MacArthur and Wilson appears to apply, and one can conclude that the páramo bird situation is quite close, on a continent, to that predicted by the MacArthur and Wilson equilibrium model of immigration and extinction. In the case of biota that are supposed to be in disequilibrium, two slightly different conditions may occur. In the first, the biota might actually be effectively cut off from all or most immigration because of stringent ecological isolation from any source region, and hence extinction alone is taking place at present (Brown, 1971). In the second case, immigration and extinction are both occurring, but since immigration takes place from multiple sources that are both intra- and interarchipelagic,

adaptive radiation is inhibited. Thus, the islandlike afroalpine moorlands of high mountains in East Africa support an avifauna in which extinction is, or has been, active, but in which adaptive radiation is prevented, not so much by repeated immigration within the montane archipelago, but by immigration from sources outside the moorland zone, especially it seems from suitable habitats occurring below the high-altitude belt of moorlands. However, the vascular plants of the same afroalpine zone do show adaptive radiation, and in their case, immigration has proceeded apparently only within the archipelago, so that the more or less cyclical alternations of glacial and interglacial episodes of the Pleistocene have coincided with periods of immigration versus isolation, respectively. As the phases of isolation have been sufficiently complete and long for divergence of some of the isolated mountaintop populations, speciation has been relatively important (see Hedberg, 1969, 1971; F. Vuilleumier and B. B. Simpson, unpublished; Moreau, 1966). These remarks on insular biota in continents were made to show that in fact there is every reason to suppose that the theory of MacArthur and Wilson can illuminate zoogeographical processes not only in truly insular situations but in continental ones as well. Thus, some earlier studies of speciation and adaptive radiation [e.g., those of Keast (1961) for Australian birds; Haffer (1969) for Amazonian forest birds; Vuilleumier (1969b) for high Andean birds; Hall and Moreau (1970) for African birds; and many others including fewer taxa, for instance those of Mengel (1964) on Parulidae; Hall (1963) on francolins; Selander (1964) on wrens], which all have as common denominator the theory of geographical speciation (Mayr, 1963), can now be reexamined in the additional light of the MacArthur and Wilson theory of insular zoogeography (1967). A review by Simberloff (1972) of theoretical models in biogeography appeared too late for inclusion in this section.

Very exciting beginnings have thus been made on island birds of the West Indies by Ricklefs and Cox (1972), and on continental birds of New Guinea by Diamond (1973). It is clear that a great deal more field work needs to be done to document, first, the results of competition among species, as this process is undoubtedly the key factor in natural selection, and second, the actual mechanisms of selection within communities. I would therefore conclude this section by saying that the theoretical level of approach appears relevant to all types of zoogeographic analysis that we have reviewed. This leaves us with the question of whether this could also be true of yet another "level," involving continental drift.

V. The Significance of Continental Drift

The present bird faunas of the world and the present distribution patterns of taxa are like snapshots taken in an instant of time when viewed against the very long evolutionary background. These faunas and these patterns have ultimately resulted from processes such as those that constitute the basic parameters manipulated by theorists of zoogeography. As the studies reviewed above seem to show, these processes, or at least some of them, take place in relatively short amounts of time, geologically speaking, and occur more or less simultaneously at any time, either within a local fauna, or among the faunas of a broad geographical area. In order to unroll the scroll that will permit us to have a glimpse of the long-term interactions of these zoogeographic processes for any fauna or any taxon, we must consider the methodological hurdles that stand in our way.

If we assume an earth with fixed climatic patterns and fixed continental margins throughout its history, or with smoothly changing parameters determining these patterns, it might be relatively easy to predict, for some taxa in some faunas, from a precise knowledge of factors such as immigration rate, extinction rate, speciation potential, and genetic potential in the face of competition, the equilibrium values at times $t - 1, t - 2, \ldots, t - n$. We would obtain a quantitative view of the vicissitudes of biota through estimates of variance in their equilibria for given periods of the geological past. These values could be checked directly against the values derived from observations based on fossil assemblages. But of course such a procedure is at present pure dream, because the qualitative aspects of the processes of extinction, immigration, speciation, and competition are such that they do not seem to make qualitative predictions possible (what were the taxa at equilibrium $\hat{Y}$ at time $t - n$? is an unanswerable question); also because the climatic and topographic patterns of the earth have neither remained fixed nor have changed only smoothly during the planet's history. Thus, for zoogeographers who are interested in qualitative questions, such as "what taxa occurred at such a place and at such a time and why," it is indispensable to know the answer to what are (a) basically historical questions about the past history of the earth's surface or (b) questions about the kinds of taxa present at past periods of the earth's history. Many of the solutions to problems that were alluded to in Sections III and IV are rooted in answers to these questions.

If this chapter were about mammals instead of birds, this section would probably be unnecessary, and the reader could choose from among many papers for information about mammalian distributions in relation to continental drift (e.g., those of Cox, 1970; Fooden, 1972; Hoffstetter, 1971, 1972; Jardine and McKenzie, 1972; Keast, 1971b, 1972b, 1973; McKenna, 1972). Thus, if we were mammalian biologists, we could proceed rapidly to an examination of past distribution patterns in terms of existing theory about the processes regulating distribution. This illustrates another consequence of the point made earlier (Section III) that birds are well known taxonomically only at the species level. This knowledge has permitted much of the theoretical advance of the last decade, since much of the empirical basis for the theorists was avian data. But the lack of adequate knowledge about the relationships of the higher categories of birds has hampered our understanding of the broader patterns of avian zoogeography, and ornithologists have to catch up to other specialists here.

A. THE DRIFT CONTROVERSY

Probably few topics in biogeography have been as controversial as the questions of whether continental drift occurred, and if it did, whether it influenced the distribution of present and/or past biota over the surface of the globe. A concise history of the controversy was written by Blackett in an introduction to a symposium on drift (Blackett *et al.*, 1965), and another was published at the same time by Illies (1965a).

Darlington (1957, p. 606) wrote as follows: "I have tried to keep my mind open on this subject and have made a new beginning by trying once more (as I have done before) to see if I can find any real signs of drift in the present distribution of animals. I can find none. So far as I can see, animal distribution now is fundamentally a product of movements of animals, not movement of land." Much later, Darlington (1965, p. 185) conceded that continental drift had taken place, but concluded: "I doubt if the distribution of plants and animals now has anything to do with continental drift. My point is not that existing distributions are incompatible with drift, but that they are not evidence of it."

On the basis of the same evidence used by Darlington (1964, 1965) to reject the view that continental drift has affected the distribution of existing organisms, other biogeographers have interpreted distributional patterns in terms of drift. They include, among others, zoologists (Ball and Fernando, 1969, working on triclads; Popham and Manly, 1969, working on Dermaptera; Illies, 1961a,b, 1965b, working on Plecoptera; Cracraft, 1973b, working on vertebrates; and others) and botanists (e.g., Good, 1966; Hawkes and Smith, 1965; Axelrod, 1970; Raven and Axelrod, 1972), as well as paleontologists.

Avian zoogeographers who examined distribution patterns and reviewed the fossil evidence have usually adhered to the widespread view among ornithologists that drift had virtually nothing to do with present bird distributions (see, for instance, Mayr, 1952). Wolfson (1955), however, tried to reinterpret the distribution of the North American bird fauna, previously analyzed by Mayr (1946), in terms of drift, and seems to have been one of the few avian biologists to have entertained, at that time, the view that drift constituted at least an alternative hypothesis to dispersal on fixed continents. Much more recently, in the wake of the surge of interest in the possibility that continental drift may have been important in the history of animal groups, Sauer and Rothe (1972) and Sauer (1972) have hypothesized that drift may be significant in understanding the dispersal of ratites in the Canary Islands, and Serventy (1972) has suggested that drift may permit one to account for some distribution patterns in Australian birds. Serventy's (1972) conclusions have been criticized by Mayr (1972a), who estimated that as far as he could ascertain "the long-standing thesis that Australia received nearly all of its bird life from southeastern Asia through island hopping is still fully valid. Only further research can elucidate whether (and if so, which) other routes were used in the Mesozoic and early Tertiary" (p. 28). But in a reply to both Serventy (1972) and Mayr (1972a), Cracraft (1972a) has pointed out, with justification, that neither of these authors have made an attempt "to incorporate the recent findings of plate tectonics" and that some of their conclusions about drift and avian distribution may have been based upon erroneous views.

B. THE THEORY OF PLATE TECTONICS

Recent studies by geologists, geophysicists, seismologists, and oceanographers have led to new concepts in explanations of geological history. The theory of plate tectonics, or global tectonics (for a summary, see Dickinson, 1971; also Hammond, 1971a,b), complementary to the theory of sea floor spreading (Dietz, 1961; Vine, 1966; Le Pichon, 1968; McKenzie 1972; and other authors), provides a mechanism which makes it possible to understand the motion of different parts of the earth's crust in relation to each other. Until recently, one of the major difficulties in explaining continental movements, sea-floor spreading, and anomalies in the position of the poles at former epochs has been the lack of a convincing model, but it seems that convection currents within the mantle of the earth offer such a model (see Runcorn, 1962, 1965b; Wilson, 1965b; for contrary views, however, see Worzel, 1965, cited by Martin, 1968, p. 48). Regardless of the validity of the model, there is a large body of evidence bearing out the suggestion that continents have drifted apart.

This evidence comes from diverse lines of investigation, especially geological (Westoll, 1965), tectonic (Wilson, 1965a), seismological (Isacks *et al.*, 1968), paleomagnetic (Irving, 1964; Creer, 1965a,b; Runcorn, 1965a), on the fit of continents (Bullard *et al.*, 1965; Smith and Hallam, 1970), and biostratigraphy (Reyment and Tait, 1972). Some of the most important new paleontological evidence has been the discovery of fossil vertebrates in Antarctica (Barrett *et al.*, 1968; Colbert, 1971, 1972; Elliot *et al.*, 1970; Kitching *et al.*, 1972). There is no doubt that continents have not remained fixed in their positions relative to each other. Let us see how, on present evidence, one can reconstruct the major episodes of these movements.

A convenient starting point is a quotation from Vine (1966) on the hypothesis of convection: "The hypothesis invokes slow convection within the upper mantle by creep processes, drift being initiated above an upwelling, and continental fragments riding passively away from such a rift on a conveyor belt of uppermantle material; movements of the order of a few centimeters per annum are required. Thus the oceanic crust is a surface expression of the upper mantle and is considered to be derived from it, in part by partial fusion, and in part by low-temperature modification" (p. 1405). As Dietz and Holden (1970) add, "plate tectonics may be envisioned as an extension of the sea-floor spreading concept in that it accepts the thesis of sea-floor spreading while adding that these 'conveyor belts' may themselves also be moving. It assumes that the crustal plates are largely interlocked on the earth's carapace so that most motions are accommodated globally" (p. 4940). Six plates were considered by Le Pichon (1968), whereas Dietz and Holden (1970) felt that subdivision of two plates was in order, thus bringing the total number of plates to nine. Regardless of the exact number of plates, Fig. 8 shows an idealized reconstruction of plate tectonics, based on Dietz and Holden (1970, which should be consulted for further details). Figure 9 illustrates Dietz and Holden's (1970) reconstructions of the continents at three different epochs. Some of the aspects of Dietz and Holden's (1970) reconstruction of the contacts between continents and the motions during drift have been criticized on various grounds. It would lead us too far, in the context of this chapter, to attempt a review of these points of view, especially since they have been discussed in other reviews (see, e.g., Keast, 1971b, 1972b, 1973; Cracraft, 1973a). We accept Dietz and Holden's (1970) reconstruction here as a convenient visualization of the most important episodes on a very large scale. Readers interested in a more thorough review are urged to consult Cracraft (1973a), who has published more detailed figures illustrating such reconstructions of the past positions of continents.

Figure 9A shows Pangaea, the supercontinent, at the end of the Per-

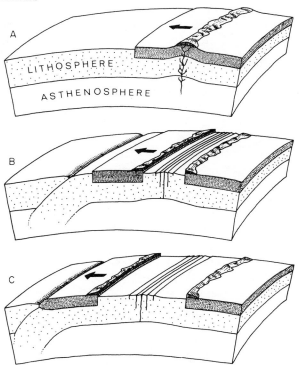

FIG. 8. A schematic representation of plate tectonics illustrating the processes of rifting and breaking up of a continent, the migration of a midocean rift, and drifting of continents. In A, a new rift appears beneath a single continent (heavy stippling) in a single crustal plate, and molten basalts spill out. In B, a trench appears (to the left) into which oceanic crust is absorbed; the new continent is rafted to the left (arrow), and a new ocean basin is created between the two continental masses (heavy stippling). In C, the ocean basin widens, and the midocean ridge migrates farther to the left. (After Dietz and Holden, 1970, *J. Geophys. Res.* **75**, Fig. 1, p. 4942, copyright by American Geophysical Union; and from Robert S. Dietz and John C. Holden, "The Breakup of Pangaea," *Scientific American*, Oct. 1970, illustration p. 32, copyright by Scientific American, Inc.; all rights reserved; courtesy of R. Dietz.)

mian, or about 225 million years ago. Figure 9B illustrates the position of the major continental land masses, comprising Gondwana and Laurasia, in the late Jurassic, or about 135 million years ago. And Fig. 9C shows the continents after having drifted apart further at about the end of the Cretaceous, or about 65 million years ago. If these reconstructions are broadly correct, the following four points are worth noting here for their relevance to avian biogeography: (1) All continents were fused together in a single land mass up to about the mid-Triassic (about 200 million years ago, or between the positions showed on Figs. 9A

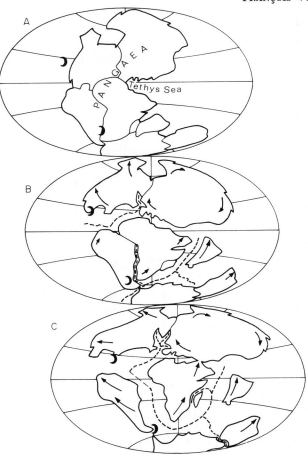

FIG. 9. Three episodes in the break up of Pangaea, shown on the Aitoff projection of the globe. The central meridian is 20°E, the parallels are the equator, the tropics, and the polar circles. The shoreline corresponds to the 1000 fathom isobath. Two geographic reference points are the Antilles and Scotia trenches (black arcs). New Guinea, New Zealand, and Southeast Asia are omitted. (A) Reconstruction of Pangaea at the end of the Permian, about 225 million years ago. (B) Reconstruction of the continents at the end of the Jurassic, about 135 million years ago. The arrows indicate the direction of drift during the Jurassic. (C) Reconstruction of the continents at the end of the Cretaceous, about 65 million years ago. (Simplified from Dietz and Holden, 1970, *J. Geophys. Res.* **75**, Fig. 2, p. 4943, Fig. 4, p. 4947, and Fig. 5, p. 4949, copyright by American Geophysical Union; courtesy of R. Dietz.)

and B) (see Dietz and Holden, 1970). (2) During the second half of the Triassic, two important events took place: (a) the rifting that opened up the North Atlantic Ocean, although North America and Europe remained fused, and (b) the rifting of the Pan-Antarctic system, with the initial separation of India from Africa and Antarctica. (3) In the Triassic, the drifting of continents was most marked for North America, Africa, and India, and least for Antarctica. Dietz and Holden (1970, p. 4947) consider that the Antarctic plate must have been surrounded in the Triassic (as it is today) by the Pan-Antarctic rift system. (4) Drifting continued in the Cretaceous, resulting in further opening of the Atlantic Ocean, separation of Madagascar from Africa (but see Cracraft, 1973a), and separation of Australia from Antarctica. The connection between Europe and North America was still effected, by way of Greenland.

C. Continental Drift and Avian Biogeography

The new evidence on continental drift makes it necessary to reconsider some of the formerly accepted ideas and to have a fresh look at bird distribution in past geological epochs. Avian zoogeographers have been largely influenced by Matthew (1915, 1939) and the other exponents of a school of thought (e.g., Mayr, Darlington, and Simpson) that is well known for a thesis that one might perhaps call the central dogma of biogeography. This thesis is tripartite, and the tryptic can be summarized as follows: First, the Old World tropics have been a (if not the) major center of evolution for vertebrates, which spread out from there (Darlington, 1957). Second, the Bering land bridge has played an extremely important role in the dispersal of organisms from the Old to the New World (Mayr, 1946; Simpson, 1947). Third, the Antarctic region has not been important either as an area for evolution or as a passage for migration of faunas between different parts of the southern hemisphere (Simpson, 1940). The advocates of this thesis have argued with many examples that if continental drift took place it played no or only an insignificant role in the evolutionary and biogeographical histories of modern vertebrates, because it occurred too early (see, e.g., Darlington, 1957, p. 607; Simpson, 1967, p. 213).

In spite of this near consensus, Wolfson (1955) tried to interpret some of Mayr's faunal elements (1946) in the North American avifauna in terms of drifting continents rather than in terms of fixed continents. He emphasized that the apparent dichotomy, permanence versus drift, was due only to our ignorance of what really happened, and that explanations based either on permanence alone or on drift alone were probably only partial explanations. [Note that Wolfson worked prior to the upsurge of interest in drift that resulted from studies on paleomagnetism, oceanic ridges, seismology, and fault systems, and that he used Du Toit's (1937) reconstruction.] Mayr (1946) used three elements, called, respectively, South American, North American, and Old World, for the taxa that had originated in one of these three land masses. Faunal elements that could not be classified into one of the above categories were placed in either the Panboreal, the Pan-American, the Pantropical, or the Unanalyzed elements.

As Mayr assumed relative permanence of continents, the dispersal of elements such as the Pantropical, Panboreal, and Old World was supposed to have taken place across the Bering land bridge at one time or another. But as Wolfson points out, this interpretation poses an ecological difficulty, for it means that the Bering land bridge had "a wide range of ecological conditions at any one period and/or in successive periods. That the Bering land bridge supported such diverse ecological conditions appears unlikely when one accepts the inclination of the earth's axis to the ecliptic as a stable feature in geological history" (Wolfson, 1955, p. 376). Another difficulty for Wolfson was the history and origin of "similar genera in the now widely separated land masses of the Southern Hemisphere" (pp. 376–377).

As an alternative to Mayr's views, Wolfson proposed to classify the North American avifauna into three elements, defined on the basis of the possibility that drift occurred, and these are described briefly below.

1. Wolfson's Original Older Element would include the groups having wide distributions in Mayr's Unanalyzed, Pantropical, and Old World elements. These groups are part of "an *original* avifauna that dispersed readily through Gondwana and/or Laurasia in the late Cretaceous and early Tertiary" (p. 377) (italics his). According to the reconstruction of Dietz and Holden (1970) the southern continents were already rather far apart at the end of the Cretaceous (see Fig. 9C), although other geologists do not agree with this view (J. Cracraft, personal communication). However, there was broad connection between Europe and North America across the North Atlantic, and some Pantropical and Old World elements could quite conceivably have crossed from Africa or Europe to North America via the North Atlantic land connection rather than via the Bering land connection. But it would not have been possible for elements to disperse from South America northward into North America other than by overseas movements.

2. Wolfson's second element, the Holarctic Element, is composed of taxa "which developed (or survived) in northern North America and/or Eurasia and dispersed freely across the North Atlantic connection (and its remaining remnant, the Bering Strait "bridge")" (p. 377). This kind of dispersal appears eminently possible (see Fig. 9C), for example for birds belonging to Mayr's

Panboreal Element, and the taxa of Mayr's Old World and North American Elements having a holarctic distribution.

3. The third element is the New World Element, which consists of taxa "which evolved in the New World, partly in the tropical southern half of North America and partly in South and Central America, with dispersal of families in both directions" (p. 377). Mayr's Pan-American Element, consolidated with several taxa from his North and South American Elements, would come close to Wolfson's New World Element.

From the point of view of continental drift, the most interesting element is Wolfson's Original Older Element. Although Wolfson's date for the supposed dispersal of this element (late Cretaceous or early Tertiary) appears to be too late, since the connections between South America and Africa probably began to be severed in the late Jurassic (Dietz and Holden, 1970), and the final link was broken in the early Upper Cretaceous (Lower Turonian, according to Reyment and Tait, 1972), his suggestion is probably applicable to a number of taxa that in fact appear to have histories going farther back in geological time than is commonly believed by ornithologists. The time is now ripe for a reexamination of the distribution of birds, especially those belonging to old families, against the background offered by the theory of plate tectonics.

Cracraft (1973a) has recently begun just such an analysis. After a thorough review of the pertinent literature on plate tectonics, Cracraft reviewed in detail the dispersal of birds such as the penguins, ratites, galliformes, parrots, pigeons, cuckoos, and suboscine passerines, and attempted to document the history of the various continental avifaunas.

Cracraft starts from the premise that only the method of phylogenetic systematics, or cladistics, as briefly reviewed in Section III,C, can offer the means of obtaining data pertinent to an understanding of biogeographic patterns. It is therefore imperative to document the presence of primitive versus derived character states in the taxa one is studying, and to establish on this basis "the distribution patterns of primitive and derived taxa and thus arrive at a more parsimonious hypothesis regarding centres of origin." Cracraft (1973a) states further, "Once we accept the importance of determining cladistic relationships and then, from this, deriving information about centres of origin and dispersal, we may ask how such information relates to continental history."

So far, only two groups of birds appear to have been analyzed from a cladistic viewpoint, the Ratitae (Cracraft, 1973, 1974), and the Galliformes (J. Cracraft and L. L. Short, Jr., personal communication, 1973). In spite of this drawback, Cracraft (1973a) attempts to reconstruct the possible or probable histories of a number of groups which could have been affected in their dispersal by moving continents. Whether the phylogenetic method is the only cure to the ills of historical biogeography, as Cracraft (1973a) claims, can be and undoubtedly will be debated by some avian taxonomists. Regardless of the method(s) used, it is evident, as I have pointed out elsewhere in this chapter, and as Cracraft (1973a) emphasises in his review, that the relationships of birds above the species level are much in need of further work. Since the numerous "traditional" taxonomic analyses of family- or order-level relationships, based on anatomy, plumage, or biochemistry, have so far yielded more alternative hypotheses (see the reviews in Sibley, 1970; Sibley and Ahlquist, 1972) than one parsimonious probable hypothesis about the origin and dispersal of taxon after taxon, avian taxonomists have clearly nothing to lose by approaching higher-level systematics through the phylogenetic or cladistic method.

Although avian zoogeographers are not yet in a position to describe with reasonable confidence the history of a taxon, or the history of faunal successions in a continental land mass, it is clear that many earlier hypotheses should be revised, because there is now evidence, not only that continental drift has occurred, but also that it has taken place in a not too remote past, at periods when many birds were actively evolving. To quote Cracraft (1973a); ". . . many families of nonpasserines had their origins in the Cretaceous. Consequently, these families were in existence at a time when most of the southern continents were still interconnected and when it would have been possible to live at high latitudes." On the basis of fossil evidence and what is presently known of relationships, Cracraft (1973a) concludes that birds such as diatrymids, idiornithids, cathartids, and some Galliformes dispersed between North America and Eurasia across the North Atlantic landbridge rather than across Beringia, and that a southern hemisphere dispersal route was most probable for birds such as the penguins, ratites, some Galliformes and the suboscines. Many other birds might have used a southern hemisphere dispersal route, but so far, evidence in favor of this view is inconclusive. The taxa involved are pigeons, parrots, and cuckoo-like birds.

Thus, on balance, and taking into account the still precarious state of affairs, it appears that much avian dispersal and history in the Tertiary and before occurred according to the reconstructions of biogeographers like Mayr and Darlington, but that these authors have minimized the actual importance, (a) of early history of many groups (i.e., in the Cretaceous), and (b) of dispersal routes over the North Atlantic and the southern hemisphere. Finally, it appears that the distribution patterns of taxa at the species or genus levels, especially the diversity gradients observed at present, are only remotely related to Tertiary dispersal, but are instead the result of Pleistocene events (Cracraft, 1973a). I agree with Cracraft on this point. Although Pleistocene climatic changes and their effects on avian speciation were considered by Selander (1965, 1971) and need not be repeated here, it is perhaps necessary to emphasize again the conclusion of numerous authors that much Pleistocene speciation has taken place in the tropics (see, e.g., Haffer, 1969; Hall, 1972; Keast, 1961; Mayr, 1969; Moreau, 1966; Vanzolini, 1970; B. S. Vuilleumier, 1971; F. Vuilleumier, 1972). Thus, we have alternative hypotheses to the ones offered by earlier biographers, especially Darlington's view (1957) that the Old World tropics have been a major evolutionary center (see Cracraft, 1973c, for arguments about alternative hypotheses about centers of origin of vertebrates).

D. Continental Drift and Biogeographic Theory

Having examined the evidence about continental drift and discussed some of the possible implications to the zoogeography of birds, we must now enquire about the possible link between biogeographic theory (see Section IV,C) and patterns of dispersion of taxa related to drift. So far as I am aware, such a link has been sought chiefly for marine invertebrates, and so the approach is not directly applicable to birds, but I believe that avian biologists might benefit by gaining familiarity with it.

The approach can be summarized by the following quotation from Valentine (1971, p. 253): ". . . the two most fundamental physical aspects of biogeography, climate and land–sea relations, must have undergone significant changes throughout much of geologic time, and these changes can be related to plate tectonic processes. The geographic consequences of plate movements and associated processes, must have led to the present biogeographic patterns." In other words, one should not be content merely to describe qualitatively the historical events of origin and dispersal of taxa and faunal assemblages through time and space in terms of past geological history, but one should relate these events to the quantitative patterns that one might expect given the interaction of biological processes like immigration, extinction, speciation, and adaptation viewed against the broader environmental background. This is obviously an ambitious undertaking, but it appears certainly worthwhile to try. Valentine (1969, 1971) and Valentine and Moores (1970, 1972) have used shelf-benthic marine invertebrates as organisms for their study.

Valentine and Moores (1972) argue that highly fluctuating environments have populations with broad food tolerances, broad habitat tolerances, high reproductive potential, and individual robustness. In this case, one might expect the major selective mode to be r selection rather than K selection, although Valentine and Moores do not mention the possibility in these terms. The species in fluctuating environments are characteristically generalists. Thus, few trophic levels can be supported, and one should expect relatively low species diversity. By contrast, in a stable environment, populations have narrower food preferences, narrower habitat preferences, and narrow tolerances for environmental changes. Species are more specialized and diversity will be comparatively high.

Valentine and Moores (1972) argue further that speciation is enhanced when dispersal barriers created by plate tectonic activity parcel the range of species that were formerly spread across the barrier. "When barriers appear that endure for long periods of geological time, they permit species diversity to increase by an amount potentially equal to the number of species that are cut into separate populations" (p. 177). The kinds of barriers that Valentine and Moores have in mind operate of course for marine shelf-benthos invertebrates and not for birds, and their analysis cannot be applied directly to land organisms. But in a general way, for both sea invertebrates and land vertebrates one can distinguish topographic and climatic kinds of barriers.

Then Valentine and Moores (1972) attempt to trace the greater trends of adaptive radiation in terms of feeding adaptations. This is where their analysis is least applicable to birds, but surely a similar trend analysis could be undertaken for birds, perhaps by reviewing the kinds of adaptive types in the class Aves as Storer (1971b) did in Chapter 4, Volume 1 of this treatise, but in relation to the time factor.

Valentine and Moores (1972) finally analyse the variation in diversity of families (in the absence of sufficient data about the diversity of species) as provided in the fossil record in terms of the expected factors that would favor low versus high diversity, low versus high speciation, and that would permit adaptation of organisms into diverse feeding types. Although it might be premature to apply this approach to birds because the relationships of the higher categories are still uncertain in so many cases, it might permit, in at least some instances, a view of the influence of plate tectonics and continental drift in another way besides the more conventional one of historical zoogeography.

ACKNOWLEDGMENTS

This chapter is dedicated to P. Géroudet, who introduced me to ornithology, and to P. J. Darlington, whose book stimulated my interest in zoogeography.

I am very grateful to D. S. Farner (Seattle) and J. Bergerard (Roscoff) for the facilities they put at my disposal while writing this chapter. For help in locating references, in supplying me with copies of published or unpublished material, or in suggesting references, I am especially indebted to J. Cracraft, R. Furrer, J. R. Karr, S. G. Kiriakoff, A. Keast, L. Malyshew, and P. Müller. Other correspondents and friends, too numerous to list here, generously sent me reprints of papers that were unavailable to me.

The responsibility for the choice of subjects and for the selection of material to be included in this chapter is entirely mine, and I apologize to the numerous authors of important works that I reluctantly had to delete from the bibliography for space or other reasons.

M. Cusin helped with some of the early bibliographic work, and J. Bourdon typed a draft of the manuscript. I thank J. Cracraft for critical comments on the manuscript, and Y. Michelhaciski for preparing the figures.

REFERENCES

Amadon, D. (1966). The superspecies concept. *Syst. Zool.* **15**, 245–249.

Ashmole, N. P. (1971). Sea bird ecology and the marine environment. *In* "Avian Biology" (D. S. Farner and J. R. King, eds.), Vol. 1, Chapter 6, pp. 223–286. Academic Press, New York.

Axelrod, D. I. (1970). Mesozoic paleogeography and early Angiosperm history. *Bot. Rev.* **36**, 277–316.

Ball, I. R., and Fernando, C. H. (1969). Freshwater triclads (Platyhelminthes, Turbellaria) and continental drift. *Nature* (*London*) **221**, 1143–1144.

Ballmann, P. (1969a). Die Vögel aus der altburdigalen Spaltenfüelung von Wintershof (West) bei Eichstätt in Bayern. *Zitteliana* (*Munich*) **1**, 5–60.

Ballmann, P. (1969b). Les oiseaux miocènes de La Grive-Saint-Alban (Isère). *Geobios* (*Lyon*) **2**, 157–204.

Bănărescu, P. (1970). "Principii şi Probleme de Zoogeografie." Acad. Repub. Soc. Romania, Bucharest.

Barrett, P. J., Baillie, R. J., and Colbert, E. H. (1968). Triassic amphibian from Antarctica. *Science* **161**, 460–462.

Blackett, P. M. S., Bullard, E., and Runcorn, S. K., eds. (1965). A symposium on continental drift. *Phil. Trans. Roy. Soc. London, Ser. A* **258**, 1–324.

Blondel, J. (1969a). "Synécologie des Passereaux Résidents et Migrateurs dans le Midi Méditerranéen Français." Centre Régional de Documentation Pédagogique, Marseille.

Blondel, J. (1969b). Les migrations transcontinentales d'oiseaux vues sous l'angle écologique. *Bull. Soc. Zool. Fr.* **94**, 577–598.

Blondel, J. (1970). Biogéographie des oiseaux nicheurs en Provence occidentale, du Mont Ventoux à la Mer Méditerranée. *Oiseau Rev. Fr. Ornithol.* **40**, 1–47.

Bock, W. J. (1956). A generic review of the family Ardeidae (Aves). *Amer. Mus. Nov.* **1779**, 1–49.

Bock, W. J. (1958). A generic review of the plovers (Charadriinae, Aves). *Bull. Mus. Comp. Zool., Harvard Univ.* **118**, 27–97.

Bock, W. J. (1968). Phylogenetic systematics, cladistics and evolution (review of W. Hennig, Phylogenetic Systematics). *Evolution* **22**, 646–648.

Bond, J. (1963). Derivation of the Antillean avifauna. *Proc. Acad. Natur. Sci. Philadelphia* **115**, 79–98.

Brodkorb, P. (1971). Origin and evolution of birds. *In* "Avian Biology" (D. S. Farner and J. R. King, eds.), Vol. 1, Chapter 2, pp. 19–55. Academic Press, New York.

Brooke, R. K. (1971). Zoogeography of the swifts. *Ostrich. Suppl.* **8**, 47–54.

Brosset, A. (1968). Localisation écologique des oiseaux migrateurs dans la forêt équatoriale du Gabon. *Biol. Gabonica* **4**, 211–216.

Brosset, A. (1969). La biogéographie vue par un écologiste. *C. R. Soc. Biogéogr.* **45**, 46–52.

Brown, J. H. (1971). Mammals on mountaintops: Nonequilibrium insular biogeography. *Amer. Natur.* **105**, 467–478.

Brundin, L. (1965). On the real nature of transantarctic relationships. *Evolution* 19, 496–505.

Brundin, L. (1966). Transantarctic relationships and their significance, as evidenced by chironomid midges with a monograph of the subfamilies Podonomidae and Aphroteniidae and the austral Heptagyiae. *Kgl. Sv. Ventenskapsakad., Handl.* [4] 11, No. 1.

Brundin, L. (1967). Insects and the problem of austral disjunctive distribution. *Annu. Rev. Entomol.* 12, 149–168.

Brundin, L. (1970). Antarctic land faunas and their history. *In* "Antarctic Ecology" (M. W. Holdgate, ed.), Vol. 1, pp. 41–53. Academic Press, New York.

Brundin, L. (1972). Phylogenetics and biogeography. *Syst. Zool.* 21, 69–79.

Brunet, J. (1970). Oiseaux de l'éocène supérieur du bassin de Paris. *Ann. Paléontol.* 56, 1–57.

Bullard, E., Everett, J. E., and Smith, A. G. (1965). The fit of the continents around the Atlantic. *Phil. Trans. Roy. Soc. London, Ser. A* 258, 41–51.

Burt, W. H. (1958). The history and affinities of the recent land mammals of western North America. *In* "Zoogeography" (C. L. Hubbs, ed.), Publ. No. 51, pp. 131–154. Amer. Ass. Advan. Sci., Washington, D.C.

Cabrera, A., and Yepes, J. (1940). Cited by Rapoport (1968).

Carson, H. L. (1970). Chromosome tracers of the origin of species. *Science* 168, 1414–1418.

Chapin, J. P. (1932). The birds of the Belgian Congo. Part 1. *Bull. Amer. Mus. Natur. Hist.* 65, 1–756.

Chapman, F. M. (1917). The distribution of bird-life in Colombia: A contribution to a biological survey of South America. *Bull. Amer. Mus. Natur. Hist.* 36, 1–729.

Chapman, F. M. (1926). The distribution of bird-life in Ecuador. *Bull. Amer. Mus. Natur. Hist.* 55, 1–784.

Chorley, R. J. (1964). Geography and analogue theory. *Ann. Ass. Amer. Geogr.* 54, 127–137.

Clarke, B. (1972). Density-dependent selection. *Amer. Natur.* 106, 1–13.

Cody, M. L. (1968). On the methods of resource division in grassland bird communities. *Amer. Natur.* 102, 107–147.

Cody, M. L. (1970). Chilean bird distribution. *Ecology* 51, 455–464.

Cody, M. L. (1974). "Competition and the Structure of Bird Communities." Princeton Univ. Press, Princeton, New Jersey.

Colbert, E. H. (1971). Tetrapods and continents. *Quart. Rev. Biol.* 46, 250–269.

Colbert, E. H. (1972). Lystrosaurus and Gondwanaland. *Evol. Biol.* 6, 157–177.

Cook, R. E. (1969). Variation in species density of North American birds. *Syst. Zool.* 18, 63–84.

Cox, C. B. (1970). Migrating marsupials and drifting continents. *Nature (London)* 226, 767–770.

Cracraft, J. (1968). A review of the Bathornithidae (Aves, Gruiformes), with remarks on the relationships of the suborder Cariamae. *Amer. Mus. Nov.* 2326, 1–46.

Cracraft, J. (1969). Systematics and evolution of the Gruiformes (Class Aves). 1. The Eocene family Geranoididae and the early history of the Gruiformes. *Amer. Mus. Nov.* 2388, 1–41.

Cracraft, J. (1971). Systematics and evolution of the Gruiformes (Class Aves). 2. Additional comments on the Bathornithidae, with descriptions of new species. *Amer. Mus. Nov.* 2449, 1–14.

Cracraft, J. (1972a). Continental drift and Australian avian biogeography. *Emu* 72, 171–174.

Cracraft, J. (1972b). The relationships of the higher taxa of birds: Problems in phylogenetic reasoning. *Condor* **74**, 379–392.

Cracraft, J. (1972c). Phylogenetic relationships within the Gruiformes. *Proc. Int. Ornithol. Congr., 15th, 1970* p. 639.

Cracraft, J. (1973a). Continental drift, paleoclimatology, and the evolution and biogeography of birds. *J. Zool.* **169**, 455–545.

Cracraft, J. (1973b). Vertebrate evolution and biogeography in the Old World tropics: Implications of continental drift and palaeoclimatology. *In* "Implications of Continental Drift to the Earth Sciences" (D. H. Tarling and S. K. Runcorn, eds.), Vol. 1, pp. 373–393. Academic Press, New York.

Cracraft, J. (1973c). Systematics and evolution of the Gruiformes (Class Aves). 3. Phylogeny of the suborder Grues. *Bull. Amer. Mus. Natur. Hist.* **151**, 1–127.

Cracraft, J. (1974). Phylogeny and evolution of the ratite birds. *Ibis* **116**, 494–521.

Cracraft, J., and Rich, P. V. (1972). The systematics and evolution of the Cathartidae in the Old World Tertiary. *Condor* **74**, 272–283.

Creer, K. M. (1965a). Palaeomagnetism and the time of the onset of continental drift. *Nature (London)* **207**, 51.

Creer, K. M. (1965b). Palaeomagnetic data from the Gondwanic continents. *Phil. Trans. Roy. Soc. London, Ser. A* **258**, 27–40.

Crowell, K. L. (1973). Experimental zoogeography: Introductions of mice to small islands. *Amer. Natur.* **107**, 535–558.

Crowley, J. M. (1967). La biogéographie vue par un géographe. *C. R. Soc. Biogéogr.* **44**, 20–28.

Crowson, R. A. (1970). "Classification and Biology." Atherton Press, New York.

Darlington, P. J., Jr. (1957). "Zoogeography: The Geographical Distribution of Animals." Wiley, New York.

Darlington, P. J., Jr. (1964). Drifting continents and late Paleozoic geography. *Proc. Nat. Acad. Sci. U.S.* **52**, 1084–1091.

Darlington, P. J., Jr. (1965). "Biogeography of the Southern End of the World." Harvard Univ. Press, Cambridge, Massachusetts.

Darlington, P. J., Jr. (1970). A practical criticism of Hennig-Brundin "Phylogenetic Systematics" and Antarctic biogeography. *Syst. Zool.* **19**, 1–18.

Darlington, P. J., Jr. (1971). Interconnected patterns of biogeography and evolution. *Proc. Nat. Acad. Sci. U.S.* **68**, 1254–1258.

Darwin, C. (1859). "On the Origin of Species." Murray, London.

Deevey, E. S., Jr. (1949). Biogeography of the Pleistocene. Part 1. Europe and North America. *Bull. Geol. Soc. Amer.* **60**, 1315–1416.

Diamond, J. M. (1969). Avifaunal equilibria and species turnover rates on the Channel Islands of California. *Proc. Nat. Acad. Sci. U.S.* **64**, 57–63.

Diamond, J. M. (1970a). Ecological consequences of island colonization by Southwest Pacific birds. I. Types of niche shifts. *Proc. Nat. Acad. Sci. U.S.* **67**, 529–536.

Diamond, J. M. (1970b). Ecological consequences of island colonization by Southwest Pacific birds. II. The effect of species diversity on total population density. *Proc. Nat. Acad. Sci. U.S.* **67**, 1715–1721.

Diamond, J. M. (1971). Comparison of faunal equilibrium turnover rates on a tropical island and a temperate island. *Proc. Nat. Acad. Sci. U.S.* **68**, 2742–2745.

Diamond, J. M. (1972). Avifauna of the eastern highlands of New Guinea. *Publ. Nuttall Ornithol. Club Cambridge, Mass.* No. 12, pp. 1–438.

Diamond, J. M. (1973). Distributional ecology of New Guinea birds. *Science* **179**, 759–769.

Dickinson, W. R. (1971). Plate tectonics in geologic history. *Science* **174**, 107–113.

Dietz, R. S. (1961). Continent and ocean basin evolution by spreading of the sea floor. *Nature (London)* **190**, 854–857.

Dietz, R. S., and Holden, J. C. (1970). Reconstruction of Pangaea: Breakup and dispersion of continents, Permian to present. *J. Geophys. Res.* **75**, 4939–4956.

Dorst, J. (1955). Recherches écologiques sur les oiseaux des hauts plateaux péruviens. *Trav. Inst. Français Etud. Andines Lima* **5**, 83–140.

Dorst, J. (1962). Considérations sur le peuplement avien des hautes Andes péruviennes. *C. R. Soc. Biogéogr.* **38**, 23–28.

Dorst, J. (1967). Considérations zoogéographiques et écologiques sur les oiseaux des hautes Andes. *In* "Biologie de l'Amérique Australe" (C. Delamare–Deboutteville and E. Rapoport, eds.), Vol. III, pp. 471–524. CNRS, Paris.

Dunn, E. R. (1922). A suggestion to zoogeographers. *Science* **56**, 336–338.

Dunn, E. R. (1931). The herpetological fauna of the Americas. *Copeia* pp. 106–119.

Du Toit, A. L. (1937). "Our Wandering Continents." Oliver & Boyd, Edinburgh.

Ekman, S. (1940). Begründung einer statistischen Methode in der regionalen Tiergeographie. Nebst einer Analyse der paläarktischen Steppen– und Wüstenfauna. *Nova Acta Regiae Soc. Sci. Uppsala.* [4] **12**, 1–117.

Elliot, D. H., Colbert, E. H., Breed, W. J., Jensen, J. A., and Powell, J. S. (1970). Triassic tetrapods from Antarctica: Evidence for continental drift. *Science* **169**, 1197–1201.

Farkas, T. (1967). "Ornithogeographie Ungarns." Duncke & Humboldt, Berlin.

Farris, J. S., Kluge, A. G., and Eckardt, M. J. (1970). A numerical approach to phylogenetic systematics. *Syst. Zool.* **19**, 172–189.

Favarger, C. (1969). L'endémisme en géographie botanique. *Scientia (Milan)* [7] **104**, 1–16.

Fleming, C. A. (1962). History of the New Zealand land bird fauna. *Notornis* **9**, 270–274.

Fooden, J. (1972). Breakup of Pangaea and isolation of relict mammals in Australia, South America, and Madagascar. *Science* **175**, 894–898.

Gadgil, M., and Solbrig, O. T. (1972). The concept of *r*- and *K*-selection: Evidence from wild flowers and some theoretical considerations. *Amer. Natur.* **106**, 14–31.

Good, R. (1966). The botanical aspects of continental drift. *Sci. Progr. (London)* **54**, 315–324.

Gorman, G. C., and Atkins, L. (1969). The zoogeography of Lesser Antillean *Anolis* lizards: An analysis based upon chromosomes and lactic dehydrogenases. *Bull. Mus. Comp. Zool., Harvard Univ.* **138**, 53–80.

Grant, P. R. (1966). Ecological compatibility of bird species on islands. *Amer. Natur.* **100**, 451–462.

Grant, P. R. (1972). Convergent and divergent character displacement. *Biol. J. Linn. Soc.* **4**, 39–68.

Greenslade, P. J. M. (1968). Island patterns in the Solomon Islands bird fauna. *Evolution* **22**, 751–761.

Haffer, J. (1967a). Speciation in Colombian forest birds west of the Andes. *Amer. Mus. Nov.* **2294**, 1–57.

Haffer, J. (1967b). Some allopatric species pairs of birds in northwestern Colombia. *Auk* **84**, 343–365.

Haffer, J. (1967c). Zoogeographical notes on the 'non-forest' lowland bird faunas of northwestern South America. *Hornero* **10**, 315–333.

Haffer, J. (1969). Speciation in Amazonian forest birds. *Science* **165**, 131–137.

Haffer, J. (1970a). Art-Entstehung bei einigen Waldvögeln Amazoniens. *J. Ornithol.* **111**, 285–331.

Haffer, J. (1970b). Entstehung und Ausbreitung nord-Andiner Bergvögel. *Zool. Jahrb., Abt. Syst., Oekol. Geogr. Tiere* **97**, 301–337.

Hall, B. P. (1963). The Francolins, a study in speciation. *Bull. Brit. Mus. (Natur. Hist.) Zool.* **10**, 108–204.

Hall, B. P. (1972). Causal ornithogeography of Africa. *Proc. Int. Ornithol. Congr., 15th, 1970* pp. 585–593.

Hall, B. P., and Moreau, R. E. (1970). "Atlas of Speciation of African Birds." British Museum (Natur. Hist.), London.

Hamilton, T. H. (1967). "Process and Pattern in Evolution." Macmillan, New York.

Hamilton, T. H. (1968). Biogeography and ecology in a new setting (review of R. H. MacArthur and E. O. Wilson, "The Theory of Island Biogeography"). *Science* **159**, 71–72.

Hamilton, T. H., and Rubinoff, I. (1967). On predicting insular variation in endemism and sympatry for the Darwin Finches in the Galapagos Archipelago. *Amer. Natur.* **101**, 161–172.

Hamilton, T. H., Barth, R. H., Jr., and Rubinoff, I. (1964). The environmental control of insular variation in bird species abundance. *Proc. Nat. Acad. Sci. U.S.* **52**, 132–140.

Hammond, A. L. (1971a). Plate tectonics: The geophysics of the Earth's surface. *Science* **173**, 40–41.

Hammond, A. L. (1971b). Plate tectonics (II): Mountain building and continental geology. *Science* **173**, 133–134.

Hawkes, J. G., and Smith, P. (1965). Continental drift and the age of angiosperm genera. *Nature (London)* **207**, 48–50.

Heatwole, H., and Levins, R. (1972). Biogeography of the Puerto Rican bank: Flotsam transport of terrestrial animals. *Ecology* **53**, 112–117.

Hedberg, O. (1969). Evolution and speciation in a tropical high mountain flora. *Biol. J. Linn. Soc.* **1**, 135–148.

Hedberg, O. (1971). Evolution of the Afroalpine flora. *In* "Adaptive Aspects of Insular Evolution" (W. L. Stern, ed.), pp. 16–23. Washington State Univ. Press, Pullman.

Hellmayr, C. E. (1932). The birds of Chile. *Field Mus. Natur. Hist., Publ.* **308**, Zool. Ser. 19, 1–472.

Hennig, W. (1950). "Grundzüge einer Theorie der phylogenetischen Systematik." Deutscher Zentralverlag, Berlin.

Hennig, W. (1960). Die Dipteren-Fauna von Neuseeland als systematisches und tiergeographisches Problem. *Beitr. Entomol.* **10**, 221–329.

Hennig, W. (1966). "Phylogenetic Systematics." Univ. of Illinois Press, Urbana.

Henry, C. (1972). Le poids corporel et la longueur de l'aile chez les Passereaux d'Europe: Application à l'étude de l'isolement trophique. *Terre Vie* **26**, 571–582.

Hoffstetter, R. (1971). Le peuplement mammalien de l'Amérique du Sud. Rôle des continents austraux comme centres d'origine, de diversification et de dispersion pour certains groupes mammaliens. *An. Acad. Brasil. Cienc.* **43**, 125–144.

Hoffstetter, R. (1972). Données et hypothèses concernant l'origine et l'histoire biogéographique des marsupiaux. *C. R. Acad. Sci.* **274**, 2635–2638.

Holloway, J. D. (1969). A numerical investigation of the biogeography of the butterfly fauna of India, and its relation to continental drift. *Biol. J. Linn. Soc.* **1**, 373–385.

Holloway, J. D. (1970). The biogeographical analysis of a transect sample of the moth fauna of Mt. Kinabalu, Sabah, using numerical methods. *Biol. J. Linn. Soc.* **2**, 259–286.

Holloway, J. D., and Jardine, N. (1968). Two approaches to zoogeography: A study based on the distributions of butterflies, birds and bats in the Indo-Australian area. *Proc. Linn. Soc. London* **179**, 153–188.

Horton, D. R. (1973). Endemism and zoogeography. *Syst. Zool.* **22**, 84–86.

Hubbard, J. P. (1971). The avifauna of the Southern Appalachians: Past and present. *In* "The Distributional History of the Biota of the Southern Appalachians" (P. C. Holt, ed.), Res. Div. Monogr. No. 4, pp. 197–232. Virginia Polytechnic Institute and State University, Blacksburg.

Huxley, T. H. (1868). On the classification and distribution of the Alectoromorphae and Heteromorphae. *Proc. Zool. Soc. London* pp. 294–319.

Illies, J. (1961a). Phylogenie und Verbreitungsgeschichte der Ordnung Plecoptera. *Zool. Anz., Suppl.* **25**, 384–394.

Illies, J. (1961b). Verbreitungsgeschichte der Plecopteren auf der Südhemisphäre. *Proc. Int. Congr. Entomol., 11th, 1960* pp. 476–480.

Illies, J. (1965a). Die Wegenersche Kontinentalverschiebungstheorie im Lichte der modernen Biogeographie. *Naturwissenschaften* **18**, 505–511.

Illies, J. (1965b). Phylogeny and zoogeography of the Plecoptera. *Annu. Rev. Entomol.* **10**, 117–140.

Irving, E. (1964). "Paleomagnetism." Wiley, New York.

Isacks, B., Olliver, J., and Sykes, L. (1968). Seismology and the new global tectonics. *J. Geophys. Res.* **73**, 5855–5899.

James, F. C. (1970). Geographic size variation in birds and its relationship to climate. *Ecology* **51**, 365–390.

Jardine, N., and McKenzie, O. (1972). Continental drift and the dispersal and evolution of organisms. *Nature (London)* **235**, 20–24.

Jardine, N., and Sibson, R. (1971). "Mathematical Taxonomy." Wiley, New York.

Jotterand, M. (1972). Le polymorphisme chromosomique des *Mus* (Leggadas) africains. Cytogénétique, zoogéographie, évolution. *Rev. Suisse Zool.* **79**, 287–359.

Jovanović, V. (1972). Chromosome cytology in the class Aves. *Proc. Int. Ornithol. Congr., 15th, 1970* p. 658.

Kaiser, G. W., Lefkovitch, L. P., and Howden, H. F. (1972). Faunal provinces in Canada as exemplified by mammals and birds: A mathematical consideration. *Can. J. Zool.* **50**, 1087–1104.

Karr, J. R. (1968). Habitat and avian diversity on strip-mined land in east-central Illinois. *Condor* **70**, 348–357.

Karr, J. R. (1971). Structure of avian communities in selected Panama and Illinois habitats. *Ecol. Monogr.* **41**, 207–233.

Keast, A. (1961). Bird speciation on the Australian continent. *Bull. Mus. Comp. Zool., Harvard Univ.* **123**, 305–495.

Keast, A. (1968). Competitive interactions and the evolution of ecological niches as illustrated by the Australian Honeyeater genus *Melithreptus* (Meliphagidae). *Evolution* **22**, 762–784.

Keast, A. (1971a). Adaptive evolution and shifts in niche occupation in island birds. *In* "Adaptive Aspects of Insular Evolution" (W. L. Stern, ed.), pp. 39–53. Washington State Univ. Press, Pullman.

Keast, A. (1971b). Continental drift and the evolution of the biota on southern continents. *Quart. Rev. Biol.* **46**, 335–378.

Keast, A. (1972a). Ecological opportunities and dominant families, as illustrated by the neotropical Tyrannidae (Aves). *Evol. Biol.* **5**, 229–277.

Keast, A. (1972b). Continental drift and the evolution of the biota on southern continents. *In* "Evolution, Mammals, and Southern Continents" (A. Keast, F. C. Erk, and B. Glass, eds.), pp. 23–87. State Univ. of New York Press, Albany, New York.

Keast, A. (1973). Contemporary biotas and the separation sequence of southern continents. In "Implications of Continental Drift to the Earth Sciences" (D. H. Tarling and S. K. Runcorn, eds.), Vol. 1, pp. 305–338. Academic Press, New York.

Keast, A., Erk, F. C., and Glass, B., eds. (1972). "Evolution, Mammals, and Southern Continents." State Univ. of New York Press, Albany, New York.

Keith, S., Twomey, A., Friedmann, H., and Williams, J. (1969). The avifauna of the Impenetrable Forest, Uganda. Amer. Mus. Nov. 2389, 1–41.

Kikkawa, J., and Pearse, K. (1969). Geographical distribution of land birds in Australia—a numerical analysis. Aust. J. Zool. 17, 821–840.

Kiriakoff, S. G. (1964). La vicariance géographique et la taxonomie. C. R. Soc. Biogéogr. 41, 103–115.

Kiriakoff, S. G. (1967). Biogeography and taxonomy. Bull. Nat. Inst. Sci. India 34, 219–224.

Kitching, J. W., Collinson, J. W., Elliot, D. H., and Colbert, E. H. (1972). Lystrosaurus zone (Triassic) fauna from Antarctica. Science 175, 524–527.

Koepcke, M. (1954). Corte ecológico transversal en los Andes del Perú central con especial consideración de las Aves. Parte I. Costa, vertientes occidentales y región altoandina. Mem. Mus. Hist. Nat. Javier Prado, Lima, 3, 3–119.

Koepcke, M. (1972). On the types of bird's nests found side by side in a Peruvian rain forest. Proc. Int. Ornithol. Congr., 15th, 1970 pp. 662–663.

Kruskal, J. B. (1964a). Multidimensional scaling by optimising goodness-of-fit to a nonmetric hypothesis. Psychometrika 129, 1–27.

Kruskal, J. B. (1964b). Nonmetric multidimensional scaling: A numerical method. Psychometrika 129, 115–129.

Kurtén, B. (1968). "Pleistocene Mammals of Europe." Aldine, Chicago, Illinois.

Kurtén, B. (1972). "The Age of Mammals." Columbia Univ. Press, New York.

Lack, D. (1945). The Galapagos Finches (Geospizinae) a study in variation. Occas. Pap. Calif. Acad. Sci. 21, 1–151.

Lack, D. (1947). "Darwin's Finches." Cambridge Univ. Press, London and New York.

Lack, D. (1969). Population changes in the land birds of a small island. J. Anim. Ecol. 38, 211–218.

Lack, D. (1971). Island birds. In "Adaptive Aspects of Insular Evolution" (W. L. Stern, ed.), pp. 29–31. Washington State Univ. Press, Pullman.

La Greca, M., and Sacchi, C. F. (1957). Problemi del poplamento animale nelle piccole isole mediterranee. Ann. Ist. Mus. Zool. Univ. Napoli 9, 1–189.

Lanyon, W. E. (1967). Revision and probable evolution of the Myiarchus flycatchers of the West Indies. Bull. Amer. Mus. Natur. Hist. 136, 329–370.

Larson, S. (1957). The suborder Charadrii in Arctic and Boreal areas during Tertiary and Pleistocene. A zoogeographic study. Acta Vertebr. 1, 1–84.

Lattin, G. de (1957). Die Ausbreitungszentren der holarktischen Landtierwelt. Verh. Deut. Zool. Ges. pp. 380–410.

Lattin, G. de (1967). "Grundriss der Zoogeographie." Fischer, Stuttgart.

Leck, C. F. (1972). The impact of some North American migrants at fruiting trees in Panama. Auk 89, 842–850.

Lein, M. R. (1972). A trophic comparison of avifaunas. Syst. Zool. 21, 135–150.

Le Pichon, X. (1968). Sea-floor spreading and continental drift. J. Geophys. Res. 73, 3661–3697.

Levins, R., Pressick, M. L., and Heatwole, H. (1973). Coexistence patterns in insular ants. Amer. Sci. 61, 463–472.

Liapunova, E. A., and Vorontsov, N. N. (1970). Chromosomes and some issues of the evolution of the Ground Squirrel genus *Citellus* (Rodentia: Sciuridae). *Experientia* **26**, 1033–1038.

Lynch, J. F., and Johnson, N. K. (1974). Turnover and equilibria in insular avifaunas, with special reference to the California Channel Islands. *Condor* **76**, 370–384.

MacArthur, R. H. (1971). Patterns of terrestrial bird communites. *In* "Avian Biology" (D. S. Farner and J. R. King, eds.), Vol. 1, Chapter 5, pp. 189–221. Academic Press, New York.

MacArthur, R. H. (1972). "Geographical Ecology. Patterns in the Distribution of Species." Harper, New York.

MacArthur, R. H., and Wilson, E. O. (1963). An equilibrium theory of insular zoogeography. *Evolution* **17**, 373–387.

MacArthur, R. H., and Wilson, E. O. (1967). "The Theory of Island Biogeography." Princeton Univ. Press, Princeton, New Jersey.

MacArthur, R. H., Diamond, J. M., and Karr, J. R. (1972). Density compensation in island faunas. *Ecology* **53**, 330–342.

McKenna, M. C. (1972). Was Europe connected directly to North America prior to the Middle Eocene? *Evol. Biol.* **6**, 179–189.

McKenzie, D. P. (1972). Plate tectonics and sea-floor spreading. *Amer. Sci.* **60**, 425–435.

Martin, H. (1968). A critical review of the evidence for a former direct connection of South America with Africa. *In* "Biogeography and Ecology in South America" (E. J. Fittkau *et al.*, eds.), Vol. I, pp. 25–53. Junk, The Hague.

Matthew, W. D. (1915). Climate and evolution. *Ann. N.Y. Acad. Sci.* **24**, 171–318.

Matthew, W. D. (1939). Climate and evolution. *Spec. Publ., N.Y. Acad. Sci.* **1**, 1–148 (reprint).

Matvejev, S. D. (1961). Biogeography of Yugoslavia. *Biol. Inst. N. R. Serb., Belgrade, Monogr.* **9**, 1–232.

Mayr, E. (1940). The origin and the history of the bird fauna of Polynesia. *Proc. Pac. Sci. Congr., 6th, 1939* Vol. 4, pp. 197–216.

Mayr, E. (1942). "Systematics and the Origin of Species." Columbia Univ. Press, New York.

Mayr, E. (1946). History of the North American bird fauna. *Wilson Bull.* **58**, 1–41.

Mayr, E., ed. (1952). The problem of land connections across the South Atlantic, with special reference to the Mesozoic. *Bull. Amer. Mus. Natur. Hist.* **99**, 81–258.

Mayr, E. (1963). "Animal Species and Evolution." Belknap Press, Cambridge, Massachusetts.

Mayr, E. (1964). Inferences concerning the Tertiary American bird faunas. *Proc. Nat. Acad. Sci. U.S.* **51**, 280–288.

Mayr, E. (1965a). What is a fauna? *Zool. Jahrb., Abt. Syst., Oekol. Geogr. Tiere* **92**, 473–486.

Mayr, E. (1965b). The nature of colonizations in birds. *In* "The Genetics of Colonizing Species" (H. G. Baker and G. L. Stebbins, eds.), pp. 29–47. Academic Press, New York.

Mayr, E. (1969). Bird speciation in the tropics. *Biol. J. Linn. Soc.* **1**, 1–17.

Mayr, E. (1971). New species of birds described from 1956 to 1965. *J. Ornithol.* **112**, 302–316.

Mayr, E. (1972a). Continental drift and the history of the Australian bird fauna. *Emu* **72**, 26–28.

Mayr, E. (1972b). Geography and ecology as faunal determinants. *Proc. Int. Ornithol. Congr., 15th, 1970* pp. 551–561.

Mayr, E., and Phelps, W. H. (1967). The origin of the bird fauna of the South Venezuelan highlands. *Bull. Amer. Mus. Natur. Hist.* **136**, 269–335.

Mayr, E., and Short, L. L. (1970). Species taxa of North American birds: A contribution to comparative systematics. *Publ. Nuttall Ornithol. Club Cambridge, Mass.* No. 9, pp. 1–127.

Meise, W. (1963). Verhalten der straussartigen Vögel und Monophylie der Ratitae. *Proc. Int. Ornithol. Congr., 13th, 1962* pp. 115–125.

Mello-Leitão, C. de (1931). Cited by Rapoport (1968).

Mello-Leitão, C. de (1937). Cited by Rapoport (1968).

Mengel, R. M. (1964). The probable history of species formation in some northern Wood Warblers (Parulidae). *Living Bird* **3**, 9–43.

Mengel, R. M. (1965). The birds of Kentucky. *Ornithol. Monogr.* No. 3, pp. 1–581.

Mengel, R. M. (1970). The North American central plains as an isolating agent in bird speciation. *In* "Pleistocene and Recent Environments of the Central Great Plains," Spec. Publ. No. 3, pp. 279–340. Dept. Geol., University of Kansas, Lawrence.

Miller, A. H. (1963). Seasonal activity and ecology of the avifauna of an American equatorial cloud forest. *Univ. Calif., Berkeley, Publ. Zool.* **66**, 1–78.

Moreau, R. E. (1966). "The Bird Faunas of Africa and its Islands." Academic Press, New York.

Moreau, R. E. (1972). "The Palaearctic-African Bird Migration System." Academic Press, New York.

Morel, G. (1968). Contribution à la synécologie des oiseaux du Sahel sénégalais. *Mém. ORSTOM.* **29**, 1–179.

Morel, G., and Bourlière, F. (1962). Relations écologiques des avifaunes sédentaire et migratrice dans une savane sahélienne du Bas Sénégal. *Terre Vie* **16**, 371–393.

Morel, G., and Morel, M.-Y. (1972). Recherches écologiques sur une savane sahélienne du Ferlo septentrional, Sénégal: L'avifaune et son cycle annuel. *Terre Vie* **26**, 410–439.

Müller, P. (1971). Ausbreitungszentren und Evolution in der Neotropis. *Mitt. Biogeogr. Abt. Geogr. Inst. Univ. Saarlandes* **1**, 1–20.

Müller, P. (1972a). Centres of dispersal and evolution in the Neotropical Region. *Stud. Neotrop. Fauna* **7**, 173–185.

Müller, P. (1972b). Der neotropische Artenreichtum als biogeographisches Problem. *Zool. Mededel.* **47**, 88–110.

Müller, P. (1973). The dispersal centers of terrestrial vertebrates in the Neotropical realm. Junk, The Hague.

Müller, P., and Schmithüsen, J. (1970). Probleme der Genese südamerikanischer Biota. *In* "Deutsche Geographische Forschung in der Welt von Heute" (E. Gentz Festschrift), pp. 109–122. Verlag Ferdinand Hirt, Kiel.

Munroe, E. (1963). Perspectives in biogeography. *Can. Entomol.* **95**, 299–308.

Nelson, G. J. (1969). The problem of historical biogeography. *Syst. Zool.* **18**, 243–246.

Nelson, G. J. (1972). Comments on Hennig's "Phylogenetic systematics" and its influence on ichthyology. *Syst. Zool.* **21**, 364–374.

Niethammer, G., and Kramer, H. (1966). Tiergeographie (Bericht über die Jahre 1957–1964). *Fortschr. Zool.* **18**, 1–138.

Olrog, C. C. (1972). Causal ornithogeography of South America. *Proc. Int. Ornithol. Congr., 15th, 1970* pp. 562–573.

Orians, G. (1969). The number of bird species in some tropical forests. *Ecology* **50**, 783–801.

Paynter, R. A., Jr. (1972). Biology and evolution of the *Atlapetes schistaceus* species-group (Aves: Emberizinae). *Bull. Mus. Comp. Zool., Harvard Univ.* **143,** 297–320.

Peters, J. A. (1968). A computer program for calculating degree of biogeographic resemblance between areas. *Syst. Zool.* **17,** 64–69.

Peters, J. A. (1971). A new approach in the analysis of biogeographic data. *Smithson. Contrib. Zool.* **107,** 1–28.

Philippi, R. A. (1964). Catálogo de las aves chilenas con su distributión geográfica. *Invest. Zool. Chil.* **11,** 1–179.

Phipps, M. (1967). Introduction au concept de modèle biogéographique. *Actes Symp. Int. Photointerprét., 2nd, 1966* Publ. IV, 41–49. *Editions Technip, Paris.*

Phipps, M. (1968a). Recherche de la structure d'un paysage local par les méthodes d'analyse multivariable. *C. R. Acad. Sci.* **226,** 224–227.

Phipps, M. (1968b). Analyse d'une structure régionale de modèles bio-géographiques. *Vie Milieu, Sér. C* **19,** 303–330.

Pianka, E. R. (1970). On *r*- and *K*-selection. *Amer. Natur.* **104,** 592–597.

Pianka, E. R. (1972). *r* and *K* selection or *b* and *d* selection? *Amer. Natur.* **106,** 581–588.

Ploeger, P. L. (1968). Geographical differentiation in arctic Anatidae as a result of isolation during the last glacial. *Ardea* **56,** 1–159.

Popham, E. J., and Manly, B. F. J. (1969). Geographical distribution of the Dermaptera and the continental drift hypothesis. *Nature (London)* **222,** 981–982.

Preston, F. W. (1962). The canonical distribution of commonness and rarity. Part I and Part II. *Ecology* **43,** 185–215 and 410–432.

Radu, D. (1962). Originea geograficaă şi dinamica fenologică a păsărilor din R. P. R. *In* "Probleme de Biologie" (A. Săvulescu, ed.), pp. 513–574. Acad. Repub. Pop. Romania, Bucharest.

Rapoport, E. H. (1968). Algunos problemas biogeográficos del Nuevo Mundo con especial referencia a la región neotropical. *In* "Biologie de l'Amérique Australe" (C. Delamare–Deboutteville and E. H. Rapoport, eds.), pp. 53–110. CNRS, Paris.

Rapoport, E. H. (1971). The geographical distribution of Neotropical and Antarctic Collembola. *Pac. Insect Monogr.* **25,** 99–118.

Raven, P. H., and Axelrod, D. I. (1972). Plate tectonics and Australasian paleobiogeography. *Science* **176,** 1379–1386.

Reyment, R. A., and Tait, E. A. (1972). Biostratigraphical dating of the early history of the South Atlantic Ocean. *Phil. Trans. Roy. Soc. London, Ser. B* **264,** 55–95.

Ricklefs, R. E. (1970). Stage of taxon cycle and distribution of birds on Jamaica, Greater Antilles. *Evolution* **24,** 475–477.

Ricklefs, R. E., and Cox, G. W. (1972). Taxon cycle in the West Indian avifauna. *Amer. Natur.* **106,** 195–219.

Ridpath, M. G., and Moreau, R. E. (1966). The birds of Tasmania: Ecology and evolution. *Ibis* **108,** 348–393.

Roughgarden, J. (1971). Density-dependent natural selection. *Ecology* **52,** 453–469.

Roughgarden, J. (1972). Evolution of niche width. *Amer. Natur.* **106,** 683–718.

Runcorn, S. K. (1962). Towards a theory of continental drift. *Nature (London)* **193,** 311–314.

Runcorn, S. K. (1965a). Palaeomagnetic comparisons between Europe and North America. *Phil. Trans. Roy. Soc. London, Ser. A* **258,** 1–11.

Runcorn, S. K. (1965b). Changes in the convection patterns in the Earth's mantle and continental drift: Evidence for a cold origin of the Earth. *Phil. Trans. Roy. Soc. London, Ser. A* **258,** 228–251.

Salomonsen, F. (1965). The geographical variation of the Fulmar (*Fulmarus gla-*

cialis) and the zones of marine environment in the North Atlantic. *Auk* **82,** 327–355.

Salomonsen, F. (1972). Zoogeographical and ecological problems in Arctic birds. *Proc. Int. Ornithol. Congr., 15th, 1970* pp. 25–74.

Sauer, E. G. F. (1972). Ratite eggshells and phylogenetic questions. *Bonn. Zool. Beitr.* **23,** 3–48.

Sauer, E. G. F., and Rothe, P. (1972). Ratite eggshells from Lanzarote, Canary Islands. *Science* **176,** 43–45.

Schaeffer, B., Hecht, M. K., and Eldredge, N. (1972). Phylogeny and paleontology. *Evol. Biol.* **6,** 31–46.

Schmidt, K. P. (1955). Animal geography. *In* "A Century of Progress in the Natural Sciences, 1853–1953," pp. 767–794. Calif. Acad. Sci., San Francisco.

Schoener, T. W. (1973). Population growth regulated by intraspecific competition for energy or time: Some simple presentations. *Theor. Popul. Biol.* **4,** 56–84.

Sclater, P. L. (1858). On the general geographical distribution of the members of the class Aves. *J. Proc. Linn. Soc. Zool.* **2,** 130–145.

Sclater, W. L., and Sclater, P. L. (1899). Cited by Rapoport (1968).

Selander, R. K. (1964). Speciation in wrens of the genus *Campylorhynchus*. *Univ. Calif., Berkeley, Publ. Zool.* **74,** 1–305.

Selander, R. K. (1965). Avian speciation in the Quaternary. *In* "The Quaternary of the United States" (H. E. Wright and D. G. Frey, eds.), pp. 527–542. Princeton Univ. Press, Princeton, New Jersey.

Selander, R. K. (1971). Systematics and speciation in birds. *In* "Avian Biology" (D. S. Farner and J. R. King, eds.), Vol. 1, Chapter 3, pp. 57–147. Academic Press, New York.

Serventy, D. L. (1960). Geographical distribution of living birds. *In* "Biology and Comparative Physiology of Birds" (A. J. Marshall, ed.), Vol. 1, pp. 95–126. Academic Press, New York.

Serventy, D. L. (1972). Causal zoogeography in Australia. *Proc. Int. Ornithol. Congr., 15th, 1970* pp. 574–584.

Short, L. L., Jr. (1965). Hybridization in the flickers (*Colaptes*) of North America. *Bull. Amer. Mus. Natur. Hist.* **129,** 307–428.

Short, L. L. (1970). The affinity of African with Neotropical woodpeckers. *Ostrich, Suppl.* **8,** 35–40.

Sibley, C. G. (1970). A comparative study of the egg-white proteins of passerine birds. *Bull. Peabody Mus. Natur. Hist., Yale Univ.* **32,** 1–131.

Sibley, C. G., and Ahlquist, J. E. (1972). A comparative study of the egg-white proteins of non-Passerine birds. *Bull. Peabody Mus. Natur. Hist., Yale Univ.* **39,** 1–276.

Simberloff, D. S. (1969). Experimental zoogeography of islands. A model for insular colonization. *Ecology* **50,** 296–314.

Simberloff, D. S. (1972). Models in biogeography. *In* "Models in Paleobiology" (T. J. M. Schopf, ed.), Chap. 9, pp. 160–191. Freeman, San Francisco California.

Simberloff, D. S., and Wilson, E. O. (1969). Experimental zoogeography of islands. The colonization of empty islands. *Ecology* **50,** 278–296.

Simpson, G. G. (1940). Antarctica as a faunal migration route. *Proc. Pac. Sci. Congr., 6th, 1939* **2,** 755–768.

Simpson, G. G. (1944). Cited by Keast (1972a).

Simpson, G. G. (1947). Holarctic mammalian faunas and continental relationships during the Cenozoic. *Bull. Geol. Soc. Amer.* **58,** 613–688.

Simpson, G. G. (1950). History of the fauna of Latin America. *Amer. Sci.* **38,** 361–389.

Simpson, G. G. (1957). Australian fossil penguins, with remarks on penguin evolution and distribution. *Rec. S. Aust. Mus.* **13**, 51–70.

Simpson, G. G. (1960). Notes on the measurement of faunal resemblance. *Amer. J. Sci., Bradley Vol.* **258A**, 300–311.

Simpson, G. G. (1964). Species density of North American recent mammals. *Syst. Zool.* **13**, 57–73.

Simpson, G. G. (1967). 'The Geography of Evolution." Capricorn, New York.

Simpson, G. G. (1971a). A review of the pre-Pliocene penguins of New Zealand. *Bull. Amer. Mus. Natur. Hist.* **144**, 319–378.

Simpson, G. G. (1971b). Review of fossil penguins from Seymour Island. *Proc. Roy. Soc., London Ser. B* **178**, 357–387.

Simpson, G. G. (1971c). Fossil penguin from the late Cenozoic of South Africa. *Science* **171**, 1144–1145.

Simpson, G. G. (1972). Conspectus of Patagonian fossil penguins. *Amer. Mus. Nov.* **2488**, 1–37.

Slud, P. (1960). The birds of Finca "La Selva" Costa Rica: A tropical wet forest locality. *Bull. Amer. Mus. Natur. Hist.* **121**, 49–148.

Smith, A. G., and Hallam, A. (1970). The fit of the southern continents. *Nature (London)* **225**, 139–144.

Smith, W. J., and Vuilleumier, F. (1971). Evolutionary relationships of some South American ground tyrants. *Bull. Mus. Comp. Zool., Harvard Univ.* **141**, 179–268.

Stegmann, B. (1932). Die Herkunft der paläarktischen Taiga-Vögel. *Arch. Naturgesch.* [N.F.] **1**, 355–398.

Stegmann, B. (1936). Ueber das Prinzip der zoogeographischen Einteilung des paläarktischen Gebietes unter Zugrundelegung ornithologischen Faunentypen. *Bull. Acad. Sci. USSR* pp. 523–563 (German summary).

Stegmann, B. (1938). Grundzüge der ornithogeographischen Gliederung der paläarktischen Gebietes. *Faune URSS Oiseaux* **1**, 1–156 (German summary).

Stegmann, B. (1958). Die Herkunft der eurasiatischen Steppenvögel. *Bonn. Zool. Beitr.* **9**, 208–230.

Storer, R. W. (1960). The classification of birds. *In* "Biology and Comparative Physiology of Birds" (A. J. Marshall, ed.), Vol. 1, pp. 57–93. Academic Press, New York.

Storer, R. W. (1971a). Classification of birds. *In* "Avian Biology" (D. S. Farner and J. R. King, eds.), Vol. 1, Chapter 1, pp. 1–18. Academic Press, New York.

Storer, R. W. (1971b). Adaptive radiation of birds. *In* "Avian Biology" (D. S. Farner and J. R. King, eds.), Vol. 1, Chapter 4, pp. 149–188. Academic Press, New York.

Stresemann, E. (1959). The status of avian systematics and its unsolved problems. *Auk* **76**, 269–280.

Stugren, B., and Rădulescu, M. (1961). Metode matematice în zoogeografia regională. *Stud. Cercet. Biol., Cluj* **12**, 8–24 (cited by Bănărescu, 1970).

Thenius, E. (1972). "Grundzüge der Verbreitungsgeschichte der Säugetiere. Eine historische Tiergeographie." Fischer, Stuttgart.

Todd, W. E. C., and Carriker, M. A., Jr. (1922). The birds of the Santa Marta region of Colombia: A study in altitudinal distribution. *Ann. Carnegie Mus.* **14**, 1–611.

Troll, C. (1959). Die tropischen Gebirge. Ihre dreidimensionale, klimatische und pflanzengeographische Zonierung. *Bonn. Geogr. Abh.* **25**, 1–93.

Troll, C., ed. (1968). Geo-ecology of the mountainous regions of the tropical Americas. *Colloq. Geogr.* **9**, 1–222.

Udvardy, M. D. F. (1958). Ecological and distributional analysis of North American birds. *Condor* 60, 50–66.

Udvardy, M. D. F. (1963). Bird faunas of North America. *Proc. Int. Ornithol. Congr., 13th, 1962* pp. 1147–1167.

Udvardy, M. D. F. (1969). "Dynamic Zoogeography with Special Reference to Land Animals." Van Nostrand-Reinhold, Princeton, New Jersey.

Valentine, J. W. (1969). Patterns of taxonomic and ecological structure of the shelf benthos during Phanerozoic time. *Palaeontology* 12, 684–709.

Valentine, J. W. (1971). Plate tectonics and shallow marine diversity and endemism, an actualistic model. *Syst. Zool.* 20, 253–264.

Valentine, J. W., and Moores, E. M. (1970). Plate-tectonic regulation of faunal diversity and sea-level: A model. *Nature (London)* 228, 657–659.

Valentine, J. W., and Moores, E. M. (1972). Global tectonics and the fossil record. *J. Geol.* 80, 167–184.

Vanzolini, P. E. (1970). "Zoologia, Sistemática, Geografia e a Origem das Espécies." Univ. São Paulo, Brazil.

Vanzolini, P. E., and Williams, E. E. (1970). South American anoles: the geographic differentiation and evolution of the *Anolis chrysolepis* species group (Sauria: Iguanidae). *Arq. Zool. São Paulo* 19, 1–124.

Vine, F. J. (1966). Spreading of the ocean floor: new evidence. *Science* 154, 1405–1415.

Voous, K. H. (1960). "Atlas of European Birds." Nelson, London.

Voous, K. H. (1963). The concept of faunal elements or faunal types. *Proc. Int. Ornithol. Congr., 13th, 1962* pp. 1104–1108.

Voous, K. H. (1964). Wood owls of the genera *Strix* and *Ciccaba*. *Zool. Mededel.* 39, 471–478.

Voous, K. H. (1966). The distribution of owls in Africa in relation to general zoogeographical problems. *Ostrich, Suppl.* 6, 499–506.

Voous, K. H., and Payne, H. A. W. (1965). The grebes of Madagascar. *Ardea* 53, 9–30.

Voous, K. H., and van Marle, J. G. (1953). The distributional history of the Nuthatch, *Sitta europaea. Ardea* 41, 1–68.

Vuilleumier, B. S. (1971). Pleistocene changes in the fauna and flora of South America. *Science* 173, 771–780.

Vuilleumier, F. (1967). Phyletic evolution in modern birds of the Patagonian forests. *Nature (London)* 215, 247–248.

Vuilleumier, F. (1968). Population structure of the *Asthenes flammulata* superspecies (Aves: Furnariidae). *Breviora* 297, 1–21.

Vuilleumier, F. (1969a). Systematics and evolution in *Diglossa* (Aves, Coerebidae). *Amer. Mus. Nov.* 2381, 1–44.

Vuilleumier, F. (1969b). Pleistocene speciation in birds living in the high Andes. *Nature (London)* 223, 1179–1180.

Vuilleumier, F. (1970a). Insular biogeography in continental regions. I. The northern Andes of South America. *Amer. Natur.* 104, 373–388.

Vuilleumier, F. (1970b). Generic relations and speciation patterns in the Caracaras (Aves: Falconidae). *Breviora* 335, 1–29.

Vuilleumier, F. (1971). Generic relationships and speciation patterns in *Ochthoeca, Myiotheretes, Xolmis, Neoxolmis, Agriornis*, and *Muscisaxicola. Bull. Mus. Comp. Zool., Harvard Univ.* 141, 181–232.

Vuilleumier, F. (1972). Speciation in South American birds: A progress report. *Acta Congr. Latinoamer. Zool., 4th, 1968* Vol. I, pp. 239–255.

Wallace, A. R. (1876). "The Geographical Distribution of Animals," 2 vols. Harper, New York.

Westoll, T. S. (1965). Geological evidence bearing upon continental drift. *Phil. Trans. Roy. Soc. London, Ser. A* **258**, 12–26.

Willis, E. O. (1966). The role of migrant birds at swarms of army ants. *Living Bird* **5**, 187–231.

Wilson, D. E. (1973). Bat faunas: A trophic comparison. *Syst. Zool.* **22**, 14–29.

Wilson, E. O. (1961). The nature of the taxon cycle in the Melanesian ant fauna. *Amer. Natur.* **95**, 169–193.

Wilson, E. O. (1971). "The Insect Societies." Harvard Univ. Press, Cambridge, Massachusetts.

Wilson, E. O., and Simberloff, D. S. (1969). Experimental zoogeography of islands. Defaunation and monitoring techniques. *Ecology* **50**, 267–278.

Wilson, J. T. (1965a). A new class of faults and their bearing on continental drift. *Nature (London)* **207**, 343–347.

Wilson, J. T. (1965b). Evidence from ocean islands suggesting movements in the Earth. *Phil. Trans. Roy. Soc. London, Ser. A* **258**, 145–167.

Winterbottom, J. M. (1965). Avifaunal relationships between the Neotropical and Ethiopian regions. *Hornero* **10**, 209–214.

Wolfson, A. (1955). Origin of the North American bird fauna: Critique and reinterpretation from the standpoint of continental drift. *Amer. Midl. Natur.* **53**, 353–380.

Worzel, D. D. (1965). Cited by Martin (1968).

Zimmerman, D. A. (1972). The avifauna of the Kakamega Forest, western Kenya, including a bird population study. *Bull. Amer. Mus. Natur. Hist.* **149**, 255–340.

ADDITIONAL REFERENCES[3]

Carlquist, S. (1974). "Island biology." Columbia Univ. Press, New York. [A detailed review of trends of insular biogeography, with analyses of numerous examples (Section II, B.]

Croizat, L., Nelson, G., and Rosen, D. E. (1974). Centers of origin and related concepts. *Syst. Zool.* **23**, 265–287. [This article and the one by Nelson (1973) are excellent summaries of Croizat's views of biogeographic analysis, and of the place of these views in general biogeographic theory (Section II).]

Haffer, J. (1974). Avian speciation in tropical South America. *Publ. Nuttall Ornithol. Club*, No. 14, pp. 1–390. [This book summarizes Haffer's views on the zoogeography of South American forest avifaunas against a background of Pleistocene vicissitudes (Section II, B).]

Horton, D. (1973). The concept of zoogeographic subregions. *Syst. Zool.* **22**, 191–195. [Another discussion of the relative validity of different subregions (Sections II,A and II,B)].

Nelson, G. J. (1973). Comments on Leon Croizat's biogeography. *Syst. Zool.* **22**, 312–320. [Section II.]

Rotramel, G. L. (1973). The development and application of the area concept in biogeography. *Syst. Zool.* **22**, 227–232. [A general paper on the relationship of biogeography to geography (section II).]

Short, L. L. (1975). A zoogeographical analysis of the South American chaco avifauna. *Bull. Amer. Mus. Nat. Hist.* **154**, 163–352. [An important addition to the literature mentioned in Section IV,A,2.]

[3] Among the many books and articles that have come to my attention since this chapter was completed, these are of special interest and must be cited to bring this review up to date.

AUTHOR INDEX

Numbers in italics refer to the pages on which the complete references are listed.

INDEX TO BIRD NAMES

SUBJECT INDEX

Errata to Volume III

In Chapter 4, Neuroendocrinology in Birds, by Hideshi Kobayashi and Masaru Wada, the following corrections should be noted:

On p. 296, line 7 *should read:* and the capillaries in the neural lobe. These. . .

On p. 300, lines 30–31 *should read:* in the reticular layer, there are axons containing the small, intermediate, or larger granules. These three kinds of axons and processes of the. . .

On p. 344, line 48 read acetylcholinesterase for asetylcholinesterase.

In Chapter 5, Avian Vision, by Arnold J. Stillman the following corrections should be noted:

On p. 352, lines 30–31 *should read:* teleost fishes, for example, the lens is moved forward toward the cornea as an object moves closer. In mammals, on the other hand, the. . .

On page 360, Figure 7 should be rotated 90°, counterclockwise.

In the Author Index on p. 550, Owen, W. C. *should read* Owen, W. G.